Mariano Giaquinta
Giuseppe Modica

Mathematical Analysis

Linear and Metric Structures
and Continuity

Birkhäuser
Boston • Basel • Berlin

Mariano Giaquinta
Scuola Normale Superiore
Dipartimento di Matematica
I-56100 Pisa
Italy

Giuseppe Modica
Università degli Studi di Firenze
Dipartimento di Matematica Applicata
I-50139 Firenze
Italy

Cover design by Alex Gerasev.

Mathematics Subject Classification (2000): 00A35, 15-01, 32K99, 46L99, 32C18, 46E15, 46E20

Library of Congress Control Number: 2006927565

ISBN-10: 0-8176-4375-3
ISBN-13: 978-0-8176-4375-1   ISBN 978-0-8176-4514-4 (eBook)

Printed on acid-free paper.

9 8 7 6 5 4 3 2 1

www.birkhauser.com

(MP)

Preface

One of the fundamental ideas of mathematical analysis is the notion of a *function*; we use it to describe and study relationships among variable quantities in a system and transformations of a system. We have already discussed real functions of one real variable and a few examples of functions of several variables[1], but there are many more examples of functions that the real world, physics, natural and social sciences, and mathematics have to offer:

(a) not only do we associate numbers and points to points, but we associate numbers or vectors to vectors,

(b) in the *calculus of variations* and in *mechanics* one associates an *energy* or *action* to each curve $y(t)$ connecting two points $(a, y(a))$ and $(b, y(b))$:

$$\Phi(y) := \int_a^b F(t, y(t), y'(t))\, dt$$

in terms of the so-called Lagrangian $F(t, y, p)$,

(c) in the theory of *integral equations* one maps a function into a new function

$$x(s) \;\rightarrow\; \int_a^b K(s, \tau) x(\tau)\, d\tau$$

by means of a *kernel* $K(s, \tau)$,

(d) in the theory of *differential equations* one considers transformations of a function $x(t)$ into the new function

$$t \;\rightarrow\; \int_a^t f(s, x(s))\, ds,$$

where $f(s, y)$ is given.

---

[1] in M. Giaquinta, G. Modica, *Mathematical Analysis. Functions of One Variable*, Birkhäuser, Boston, 2003, which we shall refer to as [GM1] and in M. Giaquinta, G. Modica, *Mathematical Analysis. Approximation and Discrete Processes*, Birkhäuser, Boston, 2004, which we shall refer to as [GM2].

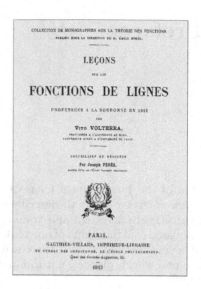

Figure 0.1. Vito Volterra (1860–1940) and the frontispiece of his *Leçons sur les fonctions de lignes.*

Of course all the previous examples are covered by the abstract setting of functions or mappings from a set X (of numbers, points, functions, ...) with values in a set Y (of numbers, points, functions, ...). But in this general context we cannot grasp the richness and the specificity of the different situations, that is, the *essential* ingredients from the point of view of the question we want to study. In order to continue to treat these specificities in an abstract context in mathematics, but also use them in other fields, we proceed by identifying specific *structures* and studying the properties that only depend on these structures. In other words, we need to identify the relevant relationships among the elements of X and how these relationships reflect on the functions defined on X.

Of course we may define many intermediate structures. In this volume we restrict ourselves to illustrating some particularly important structures: that of a *linear* or *vector space* (the setting in which we may consider linear combinations), that of a *metric space* (in which we axiomate the notions of limit and continuity by means of a *distance*), that of a *normed vector space* (that combines linear and metric structures), that of a *Banach space* (where we may operate linearly and pass to the limit), and finally, that of a *Hilbert space* (that allows us to operate not only with the length of vectors, but also with the angles that they form).

The study of *spaces of functions* and, in particular, of *spaces of continuous functions* originating in Italy in the years 1870–1880 in the works of among others Vito Volterra (1860–1940), Giulio Ascoli (1843–1896), Cesare Arzelà (1847–1912) and Ulisse Dini (1845–1918), is especially relevant in the previous context.

A descriptive diagram is the following:

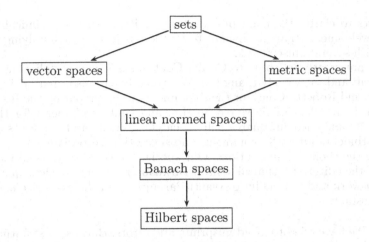

Accordingly, this book is divided into three parts. In the first part we study the linear structure. In the first three chapters we discuss basic ideas and results, including Jordan's canonical form of matrices, and in the fourth chapter we present the spectral theorem for self-adjoint and normal operators in finite dimensions.

In the second part, we discuss the fundamental notions of general topology in the metric context in Chapters 5 and 6, continuous curves in Chapter 7, and finally, in Chapter 8 we illustrate the notions of homotopy and degree, and Brouwer's and Borsuk's theorems with a few applications to the topology of $\mathbb{R}^n$.

In the third part, after some basic preliminaries, we discuss in Chapter 9 the Banach space of continuous functions presenting some of the classical fixed point theorems that play a relevant role in the solvability of functional equations and, in particular, of differential equations. In Chapter 10 we deal with the theory of Hilbert spaces and the spectral theory of compact operators. Finally, in Chapter 9 we survey some of the important applications of the ideas and techniques that we previously developed to the study of geodesics, nonlinear ordinary differential and integral equations and trigonometric series.

In conclusion, this volume[2] aims at studying continuity and its implications both in finite- and infinite-dimensional spaces. It may be regarded as a companion to [GM1] and [GM2], and as a reference book for multi-dimensional calculus, since it presents the abstract context in which concrete problems posed by multi-dimensional calculus find their natural setting.

Though this volume discusses more advanced material than [GM1,2], we have tried to keep the same spirit, always providing examples and

---

[2] This book is a translation and revised edition of M. Giaquinta, G. Modica, *Analisi Matematica, III. Strutture lineari e metriche, continuità*, Pitagora Ed., Bologna, 2000.

exercises to clarify the main presentation, omitting several technicalities and developments that we thought to be too advanced and supplying the text with several illustrations.

We are greatly indebted to Cecilia Conti for her help in polishing our first draft and we warmly thank her. We would like to also thank Fabrizio Broglia and Roberto Conti for their comments when preparing the Italian edition; Laura Poggiolini, Marco Spadini and Umberto Tiberio for their comments and their invaluable help in catching errors and misprints and Stefan Hildebrandt for his comments and suggestions, especially those concerning the choice of illustrations. Our special thanks also go to all members of the editorial technical staff of Birkhäuser for the excellent quality of their work and especially to Avanti Paranjpye and the executive editor Ann Kostant.

Note: We have tried to avoid misprints and errors. But, as most authors, we are imperfect authors. We will be very grateful to anybody who wants to inform us about errors or just misprints or wants to express criticism or other comments. Our e-mail addresses are

> `giaquinta@sns.it`   `modica@dma.unifi.it`

We shall try to keep up an errata corrige at the following webpages:

> `http://www.sns.it/~giaquinta`
>
> `http://www.dma.unifi.it/~modica`

Mariano Giaquinta
Giuseppe Modica
Pisa and Firenze
October 2006

Contents

Part I

Linear Algebra

William R. Hamilton (1805–1865), James Joseph Sylvester (1814–1897) and Arthur Cayley (1821–1895).

1. Vectors, Matrices and Linear Systems

The early developments of *linear algebra*, and related to it those of *vectorial analysis*, are strongly tied, on the one hand, to the geometrical representation of complex numbers and the need for more abstraction and formalization in geometry and, on the other hand, to the newly developed theory of electromagnetism. The names of William R. Hamilton (1805–1865), August Möbius (1790–1868), Giusto Bellavitis (1803–1880), Adhémar de Saint Venant (1797–1886) and Hermann Grassmann (1808–1877) are connected with the beginning of linear algebra, while J. Willard Gibbs (1839–1903) and Oliver Heaviside (1850–1925) established the basis of modern vector analysis motivated by the then recent *Treatise in Electricity and Magnetism* by James Clerk Maxwell (1831–1879). The subsequent formalization is more recent and relates to the developments of functional analysis and quantum mechanics.

Today, linear algebra appears as a language and a collection of results that are particularly useful in mathematics and in applications. In fact, most modeling, which is done via linear programming of ordinary or partial differential equations or control theory, can be treated numerically by computers only after it has been transformed into a linear system; in the end most of the modeling on computers deals with linear systems.

Our aim here is not to present an extensive account; for instance, we shall ignore the computational aspects (error estimations, conditioning, etc.), despite their relevance, but rather we shall focus on illustrating the language and collecting a number of useful results in a wider sense.

There is a strict link between linear algebra and linear systems. For this reason in this chapter we shall begin by discussing linear systems in the context of vectors in $\mathbb{R}^n$ or $\mathbb{C}^n$.

1.1 The Linear Spaces $\mathbb{R}^n$ and $\mathbb{C}^n$

a. Linear combinations

Let $\mathbb{K}$ be the field of real numbers or complex numbers. We denote by $\mathbb{K}^n$ the space of *ordered n-tuples* of elements of $\mathbb{K}$,

$$\mathbb{K}^n := \Big\{ \mathbf{x} \,|\, \mathbf{x} := (x^1, x^2, \ldots, x^n), \ x^i \in \mathbb{K}, \ i = 1, \ldots, n \Big\}.$$

The elements of $\mathbb{K}^n$ are often called *points* or *vectors* of $\mathbb{K}^n$; in the latter case we think of a point in $\mathbb{K}^n$ as the end-point of a vector applied at the origin. In this context the real or complex numbers are called *scalars* as they allow us to regard a vector at different *scales*.

We can sum points of $\mathbb{K}^n$, or multiply them by a scalar by summing their coordinates or multiplying the coordinates by λ:

$$\mathbf{x} + \mathbf{y} := (x^1 + y^1, x^2 + y^2, \ldots, x^n + y^n), \qquad \lambda\mathbf{x} := (\lambda x^1, \lambda x^2, \ldots, \lambda x^n),$$

if $\mathbf{x} = (x^1, x^2, \ldots, x^n)$, $\mathbf{y} = (y^1, y^2, \ldots, y^n)$, $\lambda \in \mathbb{K}$.

Of course $\forall \mathbf{x}, \mathbf{y}, \mathbf{z} \in \mathbb{K}^n$ and $\forall \lambda, \mu \in \mathbb{K}$, we have

- $(\mathbf{x} + \mathbf{y}) + \mathbf{z} = \mathbf{x} + (\mathbf{y} + \mathbf{z})$, $\mathbf{x} + \mathbf{y} = \mathbf{y} + \mathbf{x}$,
- $\lambda(\mathbf{x} + \mathbf{y}) = \lambda\mathbf{x} + \lambda\mathbf{y}$, $(\lambda + \mu)\mathbf{x} = \lambda\mathbf{x} + \mu\mathbf{x}$, $(\lambda\mu)\mathbf{x} = \lambda(\mu\mathbf{x})$,
- if $\mathbf{0} := (0, \ldots, 0)$, then $\mathbf{x} + \mathbf{0} = \mathbf{0} + \mathbf{x} = \mathbf{x}$,
- $1 \cdot \mathbf{x} = \mathbf{x}$ and, if $-\mathbf{x} := (-1)\mathbf{x}$, then $\mathbf{x} + (-\mathbf{x}) = \mathbf{0}$.

We write $\mathbf{x} - \mathbf{y}$ for $\mathbf{x} + (-\mathbf{y})$ and, from now on, the vector $\mathbf{0}$ will be simply denoted by 0.

1.1 Example. If we identify $\mathbb{R}^2$ with the plane of geometry via a Cartesian system, see [GM1], the sum of vectors in $\mathbb{R}^2$ corresponds to the sum of vectors acccording to the *parallelogram law*, and the multiplication of $\mathbf{x}$ by a scalar λ to a dilatation by a factor $|\lambda|$ in the same sense of $\mathbf{x}$ if $\lambda > 0$ or in the opposite sense if $\lambda < 0$.

1.2 About the notation. A list of vectors in $\mathbb{K}^n$ will be denoted by a lower index, $\mathbf{v}_1, \mathbf{v}_2, \ldots, \mathbf{v}_k$, and a list of scalars with an upper index $\lambda^1, \lambda^2, \ldots, \lambda^k$. The components of a vector $\mathbf{x}$ will be denoted by upper indices. In connection with the *product row by columns*, see below, it is useful to display the components as a *column*

$$\mathbf{x} = \begin{pmatrix} x^1 \\ x^2 \\ \vdots \\ x^n \end{pmatrix} \in \mathbb{K}^n.$$

However, since this is not very convenient typographically, if not strictly necessary, we shall write instead $\mathbf{x} = (x^1, x^2, \ldots, x^n)$.

Given k scalars $\lambda^1, \lambda^2, \ldots, \lambda^k$ and k vectors $\mathbf{v}_1, \mathbf{v}_2, \ldots, \mathbf{v}_k$ of $\mathbb{R}^n$, we may form the *linear combination* vector of $\mathbf{v}_1, \mathbf{v}_2, \ldots, \mathbf{v}_k$ with coefficients $\lambda^1, \lambda^2, \ldots, \lambda^k$ given by

$$\sum_{j=1}^{k} \lambda^j \mathbf{v}_j \in \mathbb{K}^n.$$

1.3 Definition. (i) *We say that $W \subset \mathbb{K}^n$ is a* linear subspace, *or simply a* subspace of $\mathbb{K}^n$, *if all finite linear combinations of vectors in W belong to W.*

Figure 1.1. Giusto Bellavitis (1803–1880) and a page from his *Nuovo metodo di geometria analitica*.

(ii) *Given a subset $S \subset \mathbb{K}^n$ we call the* span *of S the subset of $\mathbb{K}^n$, denoted by* Span S, *of all finite combinations of vectors in S. If $W = $ Span S, we say that the elements of S span W, or that S is a set of* generators *for W.*

(iii) *We say that k vectors $\mathbf{v}_1, \mathbf{v}_2, \ldots, \mathbf{v}_k \in \mathbb{K}^n$ are* linearly dependent *if there exist scalars $\lambda^1, \lambda^2, \ldots, \lambda^k$, not all zero, such that $\sum_{j=1}^{k} \lambda^j \mathbf{v}_j = 0$.*

(iv) *If k vectors $\mathbf{v}_1, \mathbf{v}_2, \ldots, \mathbf{v}_k$ are* not *linearly dependent, that is*

$$\lambda^1 \mathbf{v}_1 + \cdots + \lambda^k \mathbf{v}_k = 0 \qquad implies \qquad \lambda^1 = \lambda^2 = \cdots = \lambda^k = 0,$$

then $\mathbf{v}_1, \mathbf{v}_2, \ldots, \mathbf{v}_k$ are called linearly independent.

(v) *A subset $S \subset \mathbb{K}^n$ is* a set of linearly independent vectors *if any finite choice of vectors $\mathbf{v}_1, \mathbf{v}_2, \ldots, \mathbf{v}_k \in S$ is made of linearly independent vectors.*

(vi) *Let W be a linear subspace of $\mathbb{K}^n$. A subset $S \subset W$ of linearly independent vectors that form a set of generators for W is called a* basis *of W.*

Observe the following.

(i) W is a linear subspace of $\mathbb{K}^n$ if and only if for all $\mathbf{x}, \mathbf{y} \in W$ and all $\lambda, \mu \in \mathbb{K}$, we have $\lambda \mathbf{x} + \mu \mathbf{y} \in W$.

(ii) If W is a linear subspace of $\mathbb{K}^n$, then $0 \in W$ and moreover $\lambda \mathbf{v} \in W$ for all $\lambda \in \mathbb{K}$ if $\mathbf{v} \in W$.

(iii) $\mathbf{v}_1, \mathbf{v}_2, \ldots, \mathbf{v}_k \in \mathbb{K}^n$ are linearly dependent if and only if one of the vectors $\mathbf{v}_1, \mathbf{v}_2, \ldots, \mathbf{v}_k$ is a linear combination of the others.

 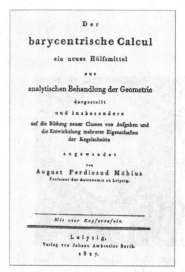

Figure 1.2. A page from *Mémoire sur les sommes et les différences géométriques* by Adhémar de Saint Venant (1797–1886) and the frontispiece of the *Barycentrische Calcul* by August Möbius (1790–1868).

(iv) If k vectors $\mathbf{v}_1, \mathbf{v}_2, \ldots, \mathbf{v}_k$ are linearly independent, then necessarily $\mathbf{v}_1, \mathbf{v}_2, \ldots, \mathbf{v}_k$ are distinct and not zero. Moreover, any choice of h, $1 \le h \le k$, yields a linearly independent set.

(v) Let $S \subset \mathbb{K}^n$. Then $W := \operatorname{Span} S$ is a linear subspace of $\mathbb{K}^n$. More explicitly $W = \operatorname{Span} S$ if and only if for every $w \in W$ there exist $k \in \mathbb{N}$, scalars $\lambda^1, \lambda^2, \ldots, \lambda^k \in \mathbb{K}$ and vectors $\mathbf{v}_1, \mathbf{v}_2, \ldots, \mathbf{v}_k \in S$ such that $w = \sum_{j=1}^k \lambda^j \mathbf{v}_j$.

b. Basis

We shall now discuss the crucial notion of basis.

1.4 Definition. *Let $\mathcal{C} \subset \mathbb{K}^n$. A subset $\mathcal{B} \subset \mathcal{C}$ is a* maximal set of linearly independent vectors *of $\mathcal{C}$ if $\mathcal{B}$ is a set of linearly independent vectors and, for every $\mathbf{w} \in \mathcal{C} \setminus \mathcal{B}$, the set $\mathcal{B} \cup \{\mathbf{w}\}$ is not a set of linearly independent vectors.*

1.5 Proposition. *Let W be a linear subspace of $\mathbb{K}^n$. $\mathcal{B}$ is a basis of W if and only if $\mathcal{B}$ is a maximal set of linearly independent vectors of W.*

Proof. Let $\mathcal{B}$ be a basis of W. $\mathcal{B}$ is a set of linearly independent vectors. Moreover, since for every $\mathbf{w} \in W$ there are k vectors $\mathbf{v}_1, \mathbf{v}_2, \ldots, \mathbf{v}_k \in \mathcal{B}$ and k scalars $\mu^1, \mu^2, \ldots, \mu^k$ such that $\mathbf{w} = \sum_{j=1}^k \mu^j \mathbf{v}_j$; the vectors $\mathbf{v}_1, \mathbf{v}_2, \ldots, \mathbf{v}_k, \mathbf{w}$ are not linearly independent, i.e., $\mathcal{B} \cup \{\mathbf{w}\}$ is not a set of linearly independent vectors.

Conversely, suppose that $\mathcal{B}$ is a maximal set of linearly independent vectors. To prove that $\mathcal{B}$ is a basis, it suffices to prove that $\operatorname{Span} \mathcal{B} = W$ and, actually, that $W \subset$

Span $\mathcal{B}$. If $\mathbf{w} \in W$, by assumption $\mathcal{B} \cup \{\mathbf{w}\}$ is not a set of linearly independent vectors. Then there exist $\mathbf{v}_1, \mathbf{v}_2, \ldots, \mathbf{v}_k \in \mathcal{B}$ and scalars $\alpha, \lambda^1, \lambda^2, \ldots, \lambda^k$ such that

$$\alpha w + \sum_{i=1}^{k} \lambda^i v_i = 0.$$

On the other hand $\alpha \neq 0$, since otherwise $\mathbf{v}_1, \mathbf{v}_2, \ldots, \mathbf{v}_k$ would be linearly dependent, hence

$$w = -\frac{1}{\alpha} \sum_{i=1}^{k} \lambda^i v_i,$$

and $w \in \text{Span}\,\mathcal{B}$. Therefore $W \subset \text{Span}\,\mathcal{B}$. $\qquad\square$

Using Zorn's lemma, see e.g., [GM2], one can show that every subspace of a linear space, including of course the linear space itself, has a *basis*. In the present situation, Proposition 1.5 allows us to select a *basis* of a subspace W by selecting a maximal set of linearly independent vectors in W. For instance, if $W = \text{Span}\{\mathbf{v}_1, \mathbf{v}_2, \ldots, \mathbf{v}_k\}$, we select the first nonzero element, say $\mathbf{w}_1 := \mathbf{v}_1$, then successively choose $\mathbf{w}_2, \mathbf{w}_3, \ldots, \mathbf{w}_h$ in the list $(\mathbf{v}_1, \mathbf{v}_2, \ldots, \mathbf{v}_k)$ so that $\mathbf{w}_2$ is not a multiple of $\mathbf{v}_1$, and by induction, $\mathbf{w}_j$ is not a linear combination of $\mathbf{w}_1, \mathbf{w}_2, \ldots, \mathbf{w}_{j-1}$. This is not a very efficient method, but it works. For more efficient methods, see Exercises 1.46 and 3.28.

1.6 ¶. Define the notion of *minimal set of generators* and show that it is equivalent to the notion of *basis*.

c. Dimension

We shall now show that all bases of a subspace W of $\mathbb{K}^n$ have the same number of elements, called the *dimension of W*, denoted by $\dim W$, and that $\dim W \leq n$.

1.7 Theorem. *We have the following.*

(i) *Let $\{\mathbf{w}_1, \mathbf{w}_2, \ldots, \mathbf{w}_k\}$ be a basis of a subspace $W \subset \mathbb{K}^n$ and let $\mathbf{v}_1, \mathbf{v}_2, \ldots, \mathbf{v}_p$ be p vectors of W that are linearly independent with $p < n$. Then one can complete the list $\mathbf{v}_1, \mathbf{v}_2, \ldots, \mathbf{v}_p$ with $n - p$ vectors among $\mathbf{w}_1, \mathbf{w}_2, \ldots, \mathbf{w}_k$ to form a new basis of W.*

(ii) *k vectors of $\mathbb{K}^n$ are always linearly dependent if $k > n$.*

(iii) *All bases of a subspace $W \subset \mathbb{K}^n$ have the same number of elements k and $k \leq n$.*

Proof. (i) We proceed by induction on $p = 1, 2, \ldots, n$.

Let $p = 1$ and $\mathbf{v}_1 \neq 0$. Since $\mathbf{w}_1, \mathbf{w}_2, \ldots, \mathbf{w}_k$ is a basis of W, we have

$$\mathbf{v}_1 = x^1 \mathbf{w}_1 + \cdots + x^k \mathbf{w}_k \qquad (1.1)$$

with at least one of the x^i not zero. Without loss of generality, we can assume that $x^1 \neq 0$, hence

$$\mathbf{w}_1 := \frac{1}{x^1} \mathbf{v}_1 - \sum_{j=2}^{k} \frac{x^j}{x^1} \mathbf{w}_j,$$

and the vectors $\mathbf{v}_1, \mathbf{w}_2, \mathbf{w}_3, \ldots, \mathbf{w}_k$ span W. The vectors $\mathbf{v}_1, \mathbf{w}_2, \mathbf{w}_3, \ldots, \mathbf{w}_k$ are also independent. In fact, if $\lambda^1, \lambda^2, \ldots, \lambda^n$ are such that

$$\lambda^1 \mathbf{v}_1 + \sum_{j=2}^{k} \lambda^j \mathbf{w}_j = 0,$$

then (1.1) yields

$$\lambda^1 x^1 \mathbf{w}_1 + \sum_{i=2}^{k} (x^i \lambda^1 + \lambda^i) \mathbf{w}_i = 0,$$

and this implies $\lambda^1 = 0$ since $x^1 \neq 0$; consequently $\sum_{j=2}^{k} \lambda^j \mathbf{w}_j = 0$, hence $\lambda^2 = \cdots = \lambda^k = 0$.

Assume now the inductive hypothesis, that is, that we can choose $n-p$ vectors of the basis $\mathbf{w}_1 = (1, 0, \ldots, 0)$, $\mathbf{w}_2 = (0, 1, \ldots, 0)$, ..., $\mathbf{w}_n = (0, 0, \ldots, 1)$, say $\mathbf{w}_{p+1}, \ldots, \mathbf{w}_n$, such that $\{\mathbf{v}_1, \ldots, \mathbf{v}_p, \mathbf{w}_{p+1}, \ldots, \mathbf{w}_n\}$ is a basis of W. Let us prove the claim for $p+1$ vectors. Since $\mathbf{v}_{p+1}$ is independent of $\mathbf{v}_1, \mathbf{v}_2, \ldots, \mathbf{v}_p$, we infer by the induction hypothesis that

$$\mathbf{v}_{p+1} = \sum_{i=1}^{p} x^i \mathbf{v}_i + \sum_{j=p+1}^{k} y^j \mathbf{w}_j \tag{1.2}$$

where at least one of the y^i is not zero. Assuming, without loss of generality, that $y^{p+1} \neq 0$, we have

$$\mathbf{w}_{p+1} := \frac{1}{y^{p+1}} \mathbf{v}_{p+1} - \sum_{j=p+2}^{k} \frac{y^j}{y^{p+1}} \mathbf{w}_j - \sum_{i=1}^{p} \frac{x^i}{y^{p+1}} \mathbf{v}_i,$$

and the vectors $\mathbf{v}_1, \ldots, \mathbf{v}_{p+1}, \mathbf{w}_{p+2}, \ldots, \mathbf{w}_k$ span W. Let us finally show that these vectors are also independent. If

$$\sum_{i=1}^{p} \lambda^i \mathbf{v}_i + \lambda^{p+1} \mathbf{v}_{p+1} + \sum_{j=p+2}^{k} \lambda^j \mathbf{w}_j = 0,$$

(1.2) yields

$$\sum_{i=1}^{p} (\lambda^i + \lambda^{p+1} x^i) \mathbf{v}_i + \lambda^{p+1} y^{p+1} \mathbf{w}_{p+1} + \sum_{j=p+2}^{k} (\lambda^j + \lambda^{p+1} y^j) \mathbf{w}_j = 0.$$

Since $\{\mathbf{v}_1, \ldots, \mathbf{v}_p, \mathbf{w}_{p+1}, \ldots, \mathbf{w}_n\}$ is a basis of W, because of the induction assumption, and $y^{p+1} \neq 0$ by construction, we conclude that $\lambda^{p+1} = 0$, and consequently $\lambda^i = 0$ for all indices i.

(ii) Assume that the $\mathbf{v}_1, \mathbf{v}_2, \ldots, \mathbf{v}_k \in \mathbb{K}^n$ are independent and $k > n$. By (i) we can complete the basis $\{\mathbf{w}_1, \mathbf{w}_2, \ldots, \mathbf{w}_n\}$ of $\mathbb{K}^n$ to form a basis of Span $\{\mathbf{v}_1, \mathbf{v}_2, \ldots, \mathbf{v}_k\}$ with k elements; this is a contradiction since $\{\mathbf{e}_1, \mathbf{e}_2, \ldots, \mathbf{e}_n\}$ is already a basis of $\mathbb{K}^n$, hence a maximal system of linearly independent vectors of $\mathbb{K}^n$.

(iii) follows as (ii). Let us prove that two bases of W have the same number of elements. Suppose that $\{\mathbf{v}_1, \mathbf{v}_2, \ldots, \mathbf{v}_p\}$ and $\{\mathbf{e}_1, \mathbf{e}_2, \ldots, \mathbf{e}_k\}$ are two bases of W with $p < k$. By (i) we may complete $\mathbf{v}_1, \mathbf{v}_2, \ldots, \mathbf{v}_p$ with $k-p$ vectors chosen among $\mathbf{e}_1, \mathbf{e}_2, \ldots, \mathbf{e}_k$ to form a new basis $\{\mathbf{v}_1, \mathbf{v}_2, \ldots, \mathbf{v}_p, \mathbf{e}_{p+1}, \ldots, \mathbf{e}_k\}$ of W; but this is a contradiction since $\{\mathbf{e}_1, \mathbf{e}_2, \ldots, \mathbf{e}_p\}$ is already a basis of W, hence a maximal system of linearly independent vectors of W, see Proposition 1.5. Similarly, and we leave it to the reader, one can prove that $k \leq n$. □

1.8 Definition. *The number of the elements of a (all) basis of a linear subspace W of $\mathbb{K}^n$ is called the* dimension *of W and denoted by* dim W.

1.9 Corollary. *The linear space $\mathbb{K}^n$ has dimension n and, if W is a linear subspace of $\mathbb{K}^n$, then $\dim W \leq n$. Moreover,*

(i) *there are k linearly independent vectors $\mathbf{v}_1, \mathbf{v}_2, \ldots, \mathbf{v}_k \in W$,*

(ii) *a set of k linearly independent vectors $\mathbf{v}_1, \mathbf{v}_2, \ldots, \mathbf{v}_k \in W$ is always a basis of W,*

(iii) *any p vectors $\mathbf{v}_1, \mathbf{v}_2, \ldots, \mathbf{v}_p \in W$ with $p > k$ are always linearly dependent,*

(iv) *if $\mathbf{v}_1, \mathbf{v}_2, \ldots, \mathbf{v}_p$ are p linearly independent vectors of W, then $p \leq k$,*

(v) *for every subspace $V \subset \mathbb{K}^n$ such that $V \subset W$ we have $\dim V \leq k$,*

(vi) *let V, W be two subspaces of $\mathbb{K}^n$; then $V = W$ if and only if $V \subset W$ and $\dim V = \dim W$.*

1.10 ¶. Prove Corollary 1.9.

d. Ordered basis

Until now, a basis S of a linear subspace W of $\mathbb{K}^n$ is just a finite set of linearly independent generators of W; every $x \in W$ is a *unique* linear combination of the basis elements. Here, uniqueness means uniqueness of the value of each coefficient in front of each basis element. To be precise, one would write

$$\mathbf{x} = \sum_{\mathbf{v} \in S} \lambda^{(\mathbf{v})} \mathbf{v}.$$

It is customary to index the elements of S with natural numbers, i.e., to consider S as a list instead of as a set. We call any list made with the elements of a basis S an *ordered basis*. The order just introduced is then used to link the coefficients to the corresponding vectors by correspondingly indexing them. This leads to the simpler notation

$$\mathbf{x} = \sum_{i=1}^{k} \lambda^i \mathbf{v}_i$$

we have already tacitly used. Moreover,

1.11 Proposition. *Let W be a linear subspace of $\mathbb{K}^n$ of dimension k and let $(\mathbf{v}_1, \mathbf{v}_2, \ldots, \mathbf{v}_k)$ be an ordered basis of W. Then for every $\mathbf{x} \in W$ there is a unique vector $\lambda \in \mathbb{K}^k$, $\lambda := (\lambda^1, \lambda^2, \ldots, \lambda^k)$ such that $x = \sum_{i=1}^{n} \lambda^i \mathbf{v}_i$.*

1.12 Example. The list $(\mathbf{e}_1, \mathbf{e}_2, \ldots, \mathbf{e}_n)$ of vectors of $\mathbb{K}^n$ given by $\mathbf{e}_1 := (1, 0, \ldots, 0)$, $\mathbf{e}_2 := (0, 1, \ldots, 0), \ldots, \mathbf{e}_n = (0, 0, \ldots, 1)$ is an ordered basis of $\mathbb{K}^n$. In fact $\mathbf{e}_1, \mathbf{e}_2, \ldots, \mathbf{e}_n$ are trivially linearly independent and span $\mathbb{K}^n$ since

$$\mathbf{x} = \begin{pmatrix} x_1 \\ x_2 \\ \vdots \\ x_n \end{pmatrix} = x^1 \begin{pmatrix} 1 \\ 0 \\ \vdots \\ 0 \end{pmatrix} + x^2 \begin{pmatrix} 0 \\ 1 \\ \vdots \\ 0 \end{pmatrix} + \cdots + x^n \begin{pmatrix} 0 \\ 0 \\ \vdots \\ 1 \end{pmatrix} = \sum_{i=1}^{n} x^i \mathbf{e}_i$$

for all $\mathbf{x} \in \mathbb{K}^n$. $(\mathbf{e}_1, \mathbf{e}_2, \ldots, \mathbf{e}_n)$ is called the *canonical* or *standard* basis of $\mathbb{K}^n$. We shall always think of the canonical basis as an ordered basis.

1.2 Matrices and Linear Operators

Following Arthur Cayley (1821–1895) we now introduce the calculus of matrices. An $m \times n$ *matrix* $\mathbf{A}$ with entries in X is an ordered table of elements of X arranged in m *rows* and n *columns*. It is customary to index the rows from top to bottom from 1 to m and the columns from left to right from 1 to n. If $\{a_j^i\}$ denotes the element or entry in the ith row and the jth column, we write

$$\mathbf{A} = \begin{pmatrix} a_1^1 & a_2^1 & \cdots & a_n^1 \\ a_1^2 & a_2^2 & \cdots & a_n^2 \\ \vdots & \vdots & \ddots & \vdots \\ a_1^m & a_2^m & \cdots & a_n^m \end{pmatrix} \qquad \text{or } \mathbf{A} = [a_j^i], \ i = 1, \ldots, m, \ j = 1, \ldots, n.$$

Given a matrix $\mathbf{A}$ we write $\mathbf{A}_j^i$ for the entry (i, j) of $\mathbf{A}$, and denote the set of matrices with m rows and n columns with $M_{m,n}(X)$. Usually X will be the field of real or complex numbers $\mathbb{K}$, but a priori one allows other entries.

1.13 Remark (Notation). The common agreement about the indices of the elements of a matrix is that the row is determined by the upper index and the column by the lower index. Later, we shall also consider matrices with two lower indices or two upper indices, $\mathbf{A} = [a_{ij}]$ or $\mathbf{A} = [a^{ij}]$. In both cases the agreement is that the first index identifies the row. These agreements turn out to be particularly useful to keep computation under control. The three different types of notation correspond to different mathematical objects represented by those matrices. But for the moment we shall not worry about this.

If $\mathbf{A} = [a_j^i] \in M_{p,n}$ and $\mathbf{B} = [b_j^i] \in M_{p,m}$ are two matrices with the same number of rows p, we denote by $[\mathbf{A} \,|\, \mathbf{B}]$, or by $\left(\boxed{\mathbf{A}} \ \boxed{\mathbf{B}} \right)$, the matrix with p rows and $(n + m)$ columns defined by

$$\left[\mathbf{A} \,\middle|\, \mathbf{B} \right] := \left(\boxed{\mathbf{A}} \ \boxed{\mathbf{B}} \right) = \begin{pmatrix} a_1^1 & a_2^1 & \cdots & a_n^1 & b_1^1 & b_2^1 & \cdots & b_m^1 \\ a_1^2 & a_2^2 & \cdots & a_n^2 & b_1^2 & b_2^2 & \cdots & b_m^2 \\ \vdots & \vdots & & \vdots & \vdots & \vdots & \ddots & \vdots \\ a_1^p & a_2^p & \cdots & a_n^p & b_1^p & b_2^p & \cdots & b_m^p \end{pmatrix}$$

or shortly by

$$\left[\mathbf{A} \,\middle|\, \mathbf{B} \right]_j^i := \begin{cases} a_j^i & \text{if } 1 \le j \le n \\ b_{j-n}^i & \text{if } n + 1 \le j \le n + m \end{cases}, \qquad i = 1, \ldots, p.$$

Similarly, given $\mathbf{A} = [a_j^i] \in M_{p,n}$ and $\mathbf{B} = [b_j^i] \in M_{q,n}$, we denote by

$$\begin{pmatrix} \boxed{\mathbf{A}} \\ \boxed{\mathbf{B}} \end{pmatrix}$$

the $(p + q) \times n$ matrix $\mathbf{C} = [c_j^i]$ defined by

$$c_j^i = \begin{cases} a_j^i & \text{if } 1 \le i \le p, \\ b_j^{i-p} & \text{if } p+1 \le i \le p+q. \end{cases}$$

a. The algebra of matrices

Two matrices $\mathbf{A} := [a_j^i]$, $\mathbf{B} = [b_j^i]$ in $M_{m,n}(\mathbb{K})$ can be *summed* by setting

$$\mathbf{A} + \mathbf{B} = [c_j^i] \qquad \text{where } c_j^i := a_j^i + b_j^i, \qquad i = 1, \dots, m, \ j = 1, \dots, n.$$

Moreover, one can multiply a matrix $\mathbf{A} \in M_{m,n}(\mathbb{K})$ by a scalar $\lambda \in \mathbb{K}$ by setting

$$\lambda \mathbf{A} := \left[\lambda\, a_j^i \right],$$

that is, each entry of $\lambda \mathbf{A}$ is the corresponding entry of $\mathbf{A}$ multiplied by λ.

Notice that the sum of matrices is meaningful if and only if both matrices have the same number of rows and columns.

Putting the rows one after the other, we can identify $M_{m,n}(\mathbb{K})$ with $\mathbb{K}^{nm}$ as a set, and the operations on the matrices defined above correspond to the sum and the multiplication by a scalar in $\mathbb{K}^{nm}$. Thus $M_{m,n}(\mathbb{K})$, endowed with the two operations $(\mathbf{A}, \mathbf{B}) \to \mathbf{A} + \mathbf{B}$ and $(\lambda, \mathbf{A}) \to \lambda \mathbf{A}$, is essentially $\mathbb{K}^{nm}$. A basis for $M_{m,n}(\mathbb{K})$ is the set of $m \times n$ matrices $\{\mathbf{I}_j^i\}$ where $\mathbf{I}_j^i$ has entries 1 at the (i,j) position and zero otherwise.

1.14 Definition (Product of matrices). *If the number of rows of* $\mathbf{A}$ *is the same as the number of the columns of* $\mathbf{B}$, $A = [a_j^i] \in M_{p,n}(\mathbb{K})$, $\mathbf{B} = [b_j^i] \in M_{n,q}(\mathbb{K})$, *we define the* product matrix $\mathbf{AB} \in M_{p,q}$ *by setting*

$$\mathbf{AB} = [c_j^i] \qquad \text{where } c_j^i = \sum_{k=1}^{n} a_k^i b_j^k.$$

Notice that if $(a_1^i, a_2^i, \dots, a_n^i)$ is the ith row of $\mathbf{A}$ and $(b_j^1, b_j^2, \dots, b_j^n)$ is the jth column of $\mathbf{B}$, then

$$\mathbf{AB} := \begin{pmatrix} a_1^1 & a_2^1 & \cdots & a_n^1 \\ & & \vdots & \\ \boxed{a_1^i \quad a_2^i \quad \cdots \quad a_n^i} \\ & & \vdots & \\ a_1^p & a_2^p & \cdots & a_n^p \end{pmatrix} \begin{pmatrix} b_1^1 & \boxed{b_j^1} & b_q^1 \\ b_1^2 & \boxed{b_j^2} & b_q^2 \\ \vdots & \cdots \ \boxed{\vdots} \ \cdots & \vdots \\ b_1^n & \boxed{b_j^n} & b_q^n \end{pmatrix},$$

where
$$(\mathbf{AB})_j^i := c_j^i = a_1^i b_j^1 + a_2^i b_j^2 + \cdots + a_n^i b_j^n.$$

For this reason the product of matrices is called the *product rows by columns*.

It is easily seen that the product of matrices is *associative* and *distributive* i.e.,

$$(\mathbf{AB})\mathbf{C} = \mathbf{A}(\mathbf{BC}) =: \mathbf{ABC}, \qquad \mathbf{A}(\mathbf{B} + \mathbf{C}) = \mathbf{AB} + \mathbf{AC}$$

but, in general, it is not commutative,

$$\mathbf{AB} \neq \mathbf{BA},$$

as simple examples show. Indeed, we may not be able to form $\mathbf{BA}$ even if $\mathbf{AB}$ is meaningful. Moreover, $\mathbf{AB}$ may equal 0, 0 being the matrix with zero in all entries, although $\mathbf{A} \neq 0$ and $\mathbf{B} \neq 0$.

b. A few special matrices

For future purposes, it is convenient to single out some special matrices.

A square $n \times n$ matrix $\mathbf{A}$ with nonzero entries only on the *principal diagonal* is called a *diagonal matrix* and denoted by

$$\mathbf{A} = \operatorname{diag}(\lambda_1, \ldots, \lambda_n) := \begin{pmatrix} \lambda_1 & 0 & \cdots & 0 \\ 0 & \lambda_2 & \cdots & 0 \\ \vdots & \vdots & \ddots & \vdots \\ 0 & 0 & \cdots & \lambda_n \end{pmatrix};$$

in short, $\mathbf{A}$ is diagonal iff $A = [a_j^i], a_j^i := \lambda^j \delta_j^i$ where δ_j^i is the Kronecker symbol,

$$\delta_j^i := \begin{cases} 1 & \text{if } i = j \\ 0 & \text{if } i \neq j. \end{cases}$$

The $n \times n$ matrix

$$\operatorname{Id}_n := \operatorname{diag}(1, \ldots, 1) = \left[\delta_j^i \right]$$

is called the *identity matrix* since for every $\mathbf{A} \in M_{p,n}(\mathbb{K})$ and $\mathbf{B} \in M_{n,q}$ we have

$$\mathbf{A}\operatorname{Id}_n = \mathbf{A}, \qquad \operatorname{Id}_n \mathbf{B} = \mathbf{B}.$$

We say that $\mathbf{A} = [a_j^i]$ is *upper triangular* if $a_j^i = 0$ for all (i,j) with $i > j$ and *lower triangular* if $a_j^i = 0$ for all (i,j) with $j < i$.

1.15 Definition. *We say that a $n \times n$ square matrix $\mathbf{A} \in M_{n,n}(\mathbb{K})$ is invertible if there exists $\mathbf{B} \in M_{n,n}(\mathbb{K})$ such that $\mathbf{AB} = \operatorname{Id}_n$ and $\mathbf{BA} = \operatorname{Id}_n$. Since the inverse is trivially unique, we call $\mathbf{B}$ the inverse of $\mathbf{A}$ and we denote it by $\mathbf{A}^{-1}$.*

1.16 ¶. Show that an upper (lower) triangular matrix $\mathbf{A} = [a^i_j] \in M_{n,n}(\mathbb{K})$ is invertible if and only if $a^i_i \neq 0$ $\forall i = 1, \ldots, n$. Show that, if $\mathbf{A}$ is invertible, $\mathbf{A}^{-1}$ is upper (lower) triangular.

1.17 Definition. *Let* $\mathbf{A} = [a^i_j] \in M_{m,n}(\mathbb{K})$. *The* transpose $\mathbf{A}^T$ *of* $\mathbf{A}$ *is the matrix* $\mathbf{A}^T := [b^i_j] \in M_{n,m}(\mathbb{K})$ *where* $b^i_j = a^j_i$ $\forall i = 1, \ldots, n$, $\forall j = 1, \ldots, m$.

We obtain $\mathbf{A}^T$ from $\mathbf{A}$ by *exchanging rows with columns*, that is, writing the successive columns of $\mathbf{A}$ from left to right as successive rows from top to bottom.

It is easily seen that

(i) $(\mathbf{A}^T)^T = \mathbf{A}$,
(ii) $(\lambda \mathbf{A} + \mu \mathbf{B})^T = \lambda \mathbf{A}^T + \mu \mathbf{B}^T$ $\forall \lambda, \mu \in \mathbb{K}$,
(iii) $(\mathbf{A}\mathbf{B})^T = \mathbf{B}^T \mathbf{A}^T$ $\forall \mathbf{A}, \mathbf{B}$,
(iv) $\mathbf{A}$ is invertible if and only if $\mathbf{A}^T$ is invertible and $(\mathbf{A}^{-1})^T = (\mathbf{A}^T)^{-1}$.

In particular, in the context of matrices with one upper and one lower index, the transposition operation exchanges upper and lower indices; thus in the case of row- and column-vectors we have

$$(a_1, a_2, \ldots, a_n)^T = \begin{pmatrix} a^1 \\ a^2 \\ \vdots \\ a^n \end{pmatrix}.$$

c. Matrices and linear operators

A map $\Lambda : \mathbb{K}^n \to \mathbb{K}^m$ is said to be *linear* if

$$A(\lambda \mathbf{x} + \mu \mathbf{y}) = \lambda\, A(\mathbf{x}) + \mu\, A(\mathbf{y}) \qquad \forall \mathbf{x}, \mathbf{y} \in \mathbb{K}^n,\ \forall \lambda, \mu \in \mathbb{K}.$$

In particular $A(0) = 0$. By induction it is easily seen that A is linear if and only if

$$A\Big(\sum_{j=1}^{k} \lambda^j \mathbf{v}_j \Big) = \sum_{j=1}^{k} \lambda^j A(\mathbf{v}_j)$$

for any $k = 1, 2, 3, \ldots$, for any $\mathbf{v}_1, \mathbf{v}_2, \ldots, \mathbf{v}_k \in \mathbb{K}^n$ and scalars $\lambda^1, \ldots, \lambda^k$.

Linear maps from $\mathbb{K}^n$ into $\mathbb{K}^m$ and $m \times n$ matrices are in one-to-one correspondence. In fact, we have the following.

1.18 Proposition. *Let* $A : \mathbb{K}^n \to \mathbb{K}^m$ *be a linear map. Then the matrix*

$$\mathbf{A} := [A(\mathbf{e}_1)|A(\mathbf{e}_2)|\ldots|A(\mathbf{e}_n)], \tag{1.3}$$

where $(\mathbf{e}_1, \mathbf{e}_2, \ldots, \mathbf{e}_n)$ *is the canonical basis of* $\mathbb{K}^n$, *is the unique matrix such that*

$$A(\mathbf{x}) = \mathbf{A}\mathbf{x}, \qquad \mathbf{x} = (x^1, x^2, \ldots, x^n). \tag{1.4}$$

Conversely, if $\mathbf{A} \in M_{n,m}(\mathbb{K})$, *then the linear map in* (1.4) *is a linear map from* $\mathbb{K}^n$ *into* $\mathbb{K}^m$.

Proof. Assuming $\mathbf{A}$ is defined by (1.3), we have for all $\mathbf{x} = (x^1, x^2, \ldots, x^n) \in \mathbb{K}^n$

$$A(\mathbf{x}) = A\Big(\sum_{i=1}^{n} x^i \mathbf{e}_i\Big) = \sum_{i=1}^{n} x^i A(\mathbf{e}_i) = \mathbf{Ax}.$$

Actually, $\mathbf{A}$ is characterized by (1.4), since if $A(\mathbf{x}) = \mathbf{Bx}\ \forall x$, then $A(\mathbf{e}_i) = \mathbf{Be}_i\ \forall i = 1, \ldots, n$, hence $\mathbf{A}$ and $\mathbf{B}$ have the same columns.

Conversely, given $\mathbf{A} \in M_{m,n}(\mathbb{K})$, it is trivial to check that the map $\mathbf{x} \to \mathbf{Ax}$ is a linear map from $\mathbb{K}^n$ into $\mathbb{K}^m$.   □

1.19 Remark. The map $A \to \mathbf{A}$ that relates linear operators and matrices is tied to the (*ordered*) canonical basis of $\mathbb{K}^n$ and $\mathbb{K}^m$.

If A and $\mathbf{A}$ are related by (1.4), we refer to A and $\mathbf{A}$ respectively as the linear map associated to the matrix $\mathbf{A}$ and the matrix associated to the linear map A.

If we denote by $\mathbf{a}_1, \mathbf{a}_2, \ldots, \mathbf{a}_n$ the columns of $\mathbf{A}$ indexed from the left so that $\mathbf{A} = [\mathbf{a}_1 \,|\, \mathbf{a}_2 \,|\, \ldots \,|\, \mathbf{a}_n]$, then

$$A(\mathbf{x}) = \mathbf{Ax} = \sum_{i=1}^{n} x^i \mathbf{a}_i, \qquad \mathbf{x} = (x^1, x^2, \ldots, x^n), \tag{1.5}$$

that is, for every $\mathbf{x} = (x^1, x^2, \ldots, x^n)$, $A(\mathbf{x})$ is the linear combination of vectors $\mathbf{a}_1, \mathbf{a}_2, \ldots, \mathbf{a}_n$ of $\mathbb{K}^m$ with scalars $x^1, x^2, \ldots, x^n$ as coefficients. Observe that $A(\mathbf{e}_1) = \mathbf{a}_1, \ldots, A(\mathbf{e}_n) = \mathbf{a}_n$, where $(\mathbf{e}_1, \mathbf{e}_2, \ldots, \mathbf{e}_n)$ is the canonical basis of $\mathbb{K}^n$.

1.20 Proposition. *Under the correspondence* (1.4) *between matrices and linear maps, the sum of two matrices corresponds to the sum of the associated operators, and the product of two matrices corresponds to the composition product of the associated operators.*

Proof. (i) Let $\mathbf{A}, \mathbf{B} \in M_{m,n}(\mathbb{K})$ and let $A(\mathbf{x}) := \mathbf{Ax}$ and $B(\mathbf{x}) := \mathbf{Bx}$. Then we have $(A + B)(\mathbf{x}) := A(\mathbf{x}) + B(\mathbf{x}) = \mathbf{Ax} + \mathbf{Bx} = (\mathbf{A} + \mathbf{B})\mathbf{x}$.

(ii) Let $\mathbf{A} \in M_{m,n}(\mathbf{K})$, $\mathbf{B} \in M_{p,m}(\mathbb{K})$, $A(\mathbf{x}) := \mathbf{Ax}$ and $B(\mathbf{y}) := \mathbf{By}\ \forall \mathbf{x} \in \mathbb{K}^n$, $\forall \mathbf{y} \in \mathbb{K}^m$. Then

$$(B \circ A)(\mathbf{x}) = B(A(\mathbf{x})) = \mathbf{B}(A(\mathbf{x})) = \mathbf{B}(\mathbf{Ax}) = (\mathbf{BA})\mathbf{x}.$$

□

1.21 ¶. Give a few examples of 2×3 and 3×2 matrices, their sums and their products (whenever this is possible). Show examples for which $\mathbf{AB} \neq \mathbf{BA}$ and $\mathbf{AB} = 0$ although $\mathbf{A} \neq 0$ and $\mathbf{B} \neq 0$. Finally, show that $\mathbf{Ax} = 0\ \forall \mathbf{x} \in \mathbb{K}^n$ implies $\mathbf{A} = 0$. [*Hint:* Compare Exercises 1.76, 1.79 and 1.81.]

1.22 ¶. Show that $\varphi : \mathbb{R}^2 \to \mathbb{R}$ is linear if and only if there exist $a, b \in \mathbb{R}$ such that $\varphi((x, y)) = ax + by\ \forall x, y \in \mathbb{R}$.

1.23 ¶. Show that $\varphi : \mathbb{R}^2 \to \mathbb{R}$ is linear if and only if
   (i) $\varphi((\lambda x, \lambda y)) = \lambda \varphi(x, y)\ \forall (x, y) \in \mathbb{R}^2$ and $\forall \lambda \in \mathbb{R}_+$,
   (ii) there exist A and $\tau \in \mathbb{R}$ such that $\varphi((cos\theta, \sin\theta)) = A\cos(\theta + \tau)\ \forall \theta \in \mathbb{R}$.

1.24 ¶. The reader is invited to find the form of the associated matrices corresponding to the linear operators sketched in Figure 1.3.

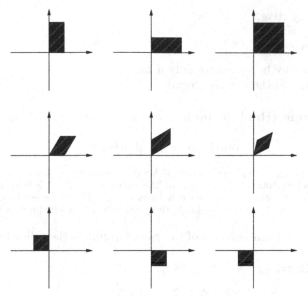

Figure 1.3. Some linear transformations of the plane. In the figure the possible images of the square $[0, 1] \times [0, 1]$ are in shadow.

d. Image and kernel

Let $\mathbf{A} \in M_{m,n}(\mathbb{K})$ and let $A(\mathbf{x}) := \mathbf{A}\mathbf{x}$, $\mathbf{x} \in \mathbb{K}^n$ be the linear associated operator. The *kernel* and the *image* of A (or $\mathbf{A}$) are respectively defined by

$$\ker A = \ker \mathbf{A} := \Big\{ \mathbf{x} \in \mathbb{K}^n \ \Big| \ A(\mathbf{x}) = 0 \Big\},$$
$$\operatorname{Im} A = \operatorname{Im} \mathbf{A} := \Big\{ \mathbf{y} \in \mathbb{K}^m \ \Big| \ \exists\, \mathbf{x} \in \mathbb{K}^n \text{ such that } A(\mathbf{x}) = \mathbf{y} \Big\}.$$

Trivially, $\ker A$ is a linear subspace of the source space $\mathbb{K}^n$, and it easy to see that the following three claims are equivalent:

(i) A is injective,
(ii) $\ker \mathbf{A} = \{0\}$,
(iii) $\mathbf{a}_1, \mathbf{a}_2, \ldots, \mathbf{a}_n$ are linearly independent in $\mathbb{K}^m$.

If one of the previous claims holds, we say that A is *nonsingular*, although in the current literature nonsingular usually refers to square matrices. Also observe that A may be nonsingular only if $m \geq n$.

Also $\operatorname{Im} A = \operatorname{Im} \mathbf{A}$ is a linear subspace of the target space $\mathbb{K}^m$, and by definition $\operatorname{Im} A = \operatorname{Span} \{\mathbf{a}_1, \mathbf{a}_2, \ldots, \mathbf{a}_n\}$. The dimension of $\operatorname{Im} A = \operatorname{Im} \mathbf{A}$ is called the *rank* of A (or of $\mathbf{A}$) and is denoted by $\operatorname{Rank} A$ (or $\operatorname{Rank} \mathbf{A}$). By definition $\operatorname{Rank} \mathbf{A}$ is the *maximal number of linearly independent columns of* $\mathbf{A}$, in particular $\operatorname{Rank} \mathbf{A} \leq \min(n, m)$. Moreover, it is easy to see that the following claims are equivalent

(i) A is surjective,
(ii) $\operatorname{Im} \mathbf{A} = \mathbb{K}^m$,
(iii) $\operatorname{Rank} \mathbf{A} = m$.

Therefore A may be surjective only if $m \leq n$.
   The following theorem is crucial.

1.25 Theorem (Rank formula). *For every matrix* $\mathbf{A} \in M_{m,n}(\mathbb{K})$ *we have*
$$\dim \operatorname{Im} \mathbf{A} = n - \dim \ker \mathbf{A}.$$

Proof. Let $(\mathbf{v}_1, \mathbf{v}_2, \ldots, \mathbf{v}_k)$ be a basis of $\ker \mathbf{A}$. According to Theorem 1.7 we can choose $(n - k)$ vectors $\mathbf{e}_{k+1}, \ldots, \mathbf{e}_n$ of the standard basis of $\mathbb{K}^n$ in such a way that $\mathbf{v}_1, \mathbf{v}_2, \ldots, \mathbf{v}_k, \mathbf{e}_{k+1}, \ldots, \mathbf{e}_n$ form a basis of $\mathbb{K}^n$. Then one easily checks that $(\mathbf{A}(\mathbf{e}_{k+1}), \ldots, \mathbf{A}(\mathbf{e}_n))$ is a basis of $\operatorname{Im} \mathbf{A}$, thus concluding that $\dim \operatorname{Im} \mathbf{A} = n - k$. $\square$

A first trivial consequence of the rank formula is the following.

1.26 Corollary. *Let* $\mathbf{A} \in M_{m,n}(\mathbb{K})$.

(i) *If* $m < n$, *then* $\dim \ker \mathbf{A} > 0$.
(ii) *If* $m \geq n$, *then* $\mathbf{A}$ *is nonsingular, i.e.,* $\ker \mathbf{A} = \{0\}$, *if and only if* $\operatorname{Rank} \mathbf{A}$ *is maximal,* $\operatorname{Rank} \mathbf{A} = n$.
(iii) *If* $m = n$, *i.e.,* $\mathbf{A}$ *is a square matrix, then the following two equivalent claims hold:*
   a) Let $A(\mathbf{x}) := \mathbf{A}\mathbf{x}$ *be the associated linear map. Then* A *is surjective if and only if* A *is injective.*
   b) $\mathbf{A}\mathbf{x} = \mathbf{b}$ *is solvable for any choice of* $\mathbf{b} \in \mathbb{K}^m$ *if and only if* $A(\mathbf{x}) = 0$ *has zero as a unique solution.*

Proof. (i) From the rank formula we have $\dim \ker \mathbf{A} = n - \dim \operatorname{Im} \mathbf{A} \geq n - m > 0$.

(ii) Again from the rank formula, $\dim \operatorname{Im} A = n - \dim \ker A = n = \min(n, m)$.

(iii) (a) Observe that A is injective if and only if $\ker A = \{0\}$, equivalently if and only if $\dim \ker A = 0$, and that A is surjective if and only if $\operatorname{Im} A = \mathbb{K}^m$, i.e., $\dim \operatorname{Im} \mathbf{A} = m = n$. The conclusion follows from the rank formula.

(iii) (b) The equivalence between (iii) (a) and (iii) (b) is trivial. $\square$

Notice that (i) and (ii) imply that $A : \mathbb{K}^n \to \mathbb{K}^m$ may be injective and surjective only if $n = m$.

1.27 ¶. Show the following.

Proposition. *Let* $\mathbf{A} \in M_{n,n}(\mathbb{K})$ *and* $A(\mathbf{x}) := \mathbf{A}\mathbf{x}$. *The following claims are equivalent:*
   (i) A *is injective and surjective,*
   (ii) $\mathbf{A}$ *is nonsingular, i.e.,* $\ker A = \{0\}$,
   (iii) A *is surjective,*
   (iv) *there exists* $\mathbf{B} \in M_{n,n}(\mathbb{K})$ *such that* $\mathbf{B}\mathbf{A} = \operatorname{Id}_n$,
   (v) *there exists* $\mathbf{B} \in M_{n,n}(\mathbb{K})$ *such that* $\mathbf{A}\mathbf{B} = \operatorname{Id}_n$,
   (vi) $\mathbf{A}$ *is invertible, i.e., there exists a matrix* $\mathbf{B} \in M_{n,n}(\mathbb{K})$ *such that* $\mathbf{B}\mathbf{A} = \mathbf{A}\mathbf{B} = \operatorname{Id}_n$.

An important and less trivial consequence of the rank formula is the following.

1.28 Theorem (Rank of the transpose). *Let* $\mathbf{A} \in M_{m,n}$. *Then we have*

(i) *the maximum number of linearly independent columns and the maximum number of linearly independent rows are equal, i.e.,*

$$\operatorname{Rank} \mathbf{A} = \operatorname{Rank} \mathbf{A}^T,$$

(ii) *let* $p := \operatorname{Rank} \mathbf{A}$. *Then there exists a nonsingular* $p \times p$ *square submatrix of* $\mathbf{A}$.

Proof. (i) Let $\mathbf{A} = [a^i_j]$, let $\mathbf{a}_1, \mathbf{a}_2, \ldots, \mathbf{a}_n$ be the columns of $\mathbf{A}$ and let $p := \operatorname{Rank} \mathbf{A}$. We assume without loss of generality that the first p columns of $\mathbf{A}$ are linearly independent and we define $\mathbf{B}$ as the $m \times p$ submatrix formed by these columns, $\mathbf{B} := [\mathbf{a}_1 \,|\, \mathbf{a}_2 \,|\, \ldots \,|\, \mathbf{a}_p]$. Since the remaining columns of $\mathbf{A}$ depend linearly on the columns of $\mathbf{B}$, we have

$$a^k_j = \sum_{i=1}^{p} r^i_j a^k_i \qquad \forall k = 1, \ldots, m, \ \forall j = p+1, \ldots, n$$

for some $\mathbf{R} = [r^i_j] \in M_{p,n-p}(\mathbb{K})$. In terms of matrices,

$$\left[\mathbf{a}_{p+1} \,\middle|\, \mathbf{a}_{p+2} \,\middle|\, \ldots \,\middle|\, \mathbf{a}_n\right] = \left[\mathbf{a}_1 \,\middle|\, \ldots \,\middle|\, \mathbf{a}_p\right] \mathbf{R} = \mathbf{B}\mathbf{R},$$

hence

$$\mathbf{A} = \left[\mathbf{B} \,\middle|\, \mathbf{B}\mathbf{R}\right] = \mathbf{B}\left[\operatorname{Id}_p \,\middle|\, \mathbf{R}\right].$$

Taking the transposes, we have $\mathbf{A}^T \in M_{n,m}(\mathbb{K})$, $\mathbf{B}^T \in M_{p,m}(\mathbb{K})$ and

$$\mathbf{A}^T = \left(\frac{\operatorname{Id}_p}{\mathbf{R}^T}\right) \mathbf{B}^T. \tag{1.6}$$

Since $[\operatorname{Id}_p \,|\, \mathbf{R}]^T$ is trivially injective, we infer that $\ker \mathbf{A}^T = \ker \mathbf{B}^T$, hence by the rank formula

$$\operatorname{Rank} \mathbf{A}^T = m - \dim \ker \mathbf{A}^T = m - \dim \ker \mathbf{B}^T = \operatorname{Rank} \mathbf{B}^T,$$

and we conclude that

$$\operatorname{Rank} \mathbf{A}^T = \operatorname{Rank} \mathbf{B}^T \leq \min(m, p) = p = \operatorname{Rank} \mathbf{A}.$$

Finally, by applying the above to the matrix $\mathbf{A}^T$, we get the opposite inequality $\operatorname{Rank} \mathbf{A} = \operatorname{Rank}(\mathbf{A}^T)^T \leq \operatorname{Rank} \mathbf{A}^T$, hence the conclusion.

(ii) With the previous notation, we have $\operatorname{Rank} \mathbf{B}^T = \operatorname{Rank} \mathbf{B} = p$. Thus $\mathbf{B}$ has a set of p independent rows. The submatrix $\mathbf{S}$ of $\mathbf{B}$ made by these rows is a square $p \times p$ matrix with $\operatorname{Rank} \mathbf{S} = \operatorname{Rank} \mathbf{S}^T = p$, hence nonsingular. $\quad\square$

1.29 ¶. Let $\mathbf{A} \in M_{m,n}(\mathbb{K})$, let $A(\mathbf{x}) := \mathbf{A}\mathbf{x}$ and let $(\mathbf{v}_1, \mathbf{v}_2, \ldots, \mathbf{v}_n)$ be a basis of $\mathbb{K}^n$. Show the following:

(i) A is injective if and only if the vectors $A(\mathbf{v}_1), A(\mathbf{v}_2), \ldots, A(\mathbf{v}_n)$ of $\mathbb{K}^m$ are linearly independent,

(ii) A is surjective if and only if $\{A(\mathbf{v}_1), A(\mathbf{v}_2), \ldots, A(\mathbf{v}_n)\}$ spans $\mathbb{K}^m$,

(iii) A is bijective iff $\{A(\mathbf{v}_1), A(\mathbf{v}_2), \ldots, A(\mathbf{v}_n)\}$ is a basis of $\mathbb{K}^m$.

e. Grassmann's formula

Let U and V be two linear subspaces of $\mathbb{K}^n$. Clearly, both $U \cap V$ and

$$U + V := \left\{ \mathbf{x} \in \mathbb{K}^n \,\middle|\, \mathbf{x} = \mathbf{u} + \mathbf{v} \text{ for some } \mathbf{u} \in U \text{ and } \mathbf{v} \in V \right\}$$

are linear subspaces of $\mathbb{K}^n$. When $U \cap V = \{0\}$, we say that $U + V$ is the *direct sum* of U and V and we write $U \oplus V$ for $U + V$. If moreover $U \oplus V = \mathbb{K}^n$, we say that U and V are *supplementary subspaces*. The following formula is very useful.

1.30 Proposition (Grassmann's formula). *Let U and V be linear subspaces of $\mathbb{K}^n$. Then*

$$\dim(U + V) + \dim(U \cap V) = \dim U + \dim V.$$

Proof. Let $(\mathbf{u}_1, \mathbf{u}_2, \ldots, \mathbf{u}_h)$ and $(\mathbf{v}_1, \mathbf{v}_2, \ldots, \mathbf{v}_k)$ be two bases of U and V respectively. The vectors $\mathbf{u}_1, \mathbf{u}_2, \ldots, \mathbf{u}_h, \mathbf{v}_1, \mathbf{v}_2, \ldots, \mathbf{v}_k$ span $U + V$, and a subset of them form a basis of $U + V$. In particular, $\dim(U + V) = \operatorname{Rank} \mathbf{L}$ where $\mathbf{L}$ is the $n \times (h + k)$ matrix defined by

$$\mathbf{L} := \left[\mathbf{u}_1 \,\middle|\, \ldots \,\middle|\, \mathbf{u}_h \,\middle|\, -\mathbf{v}_1 \,\middle|\, \ldots \,\middle|\, -\mathbf{v}_k \right].$$

Moreover, a vector $\mathbf{x} = \sum_{i=1}^{n} x^i \mathbf{u}_i \in \mathbb{K}^n$ is in $U \cap V$ if and only if there exist unique $y^1, y^2, \ldots, y^k$ such that

$$\mathbf{x} = x^1 \mathbf{u}_1 + \ldots x^h \mathbf{u}_h = y^1 \mathbf{v}_1 + \cdots + y^k \mathbf{v}_k,$$

thus, if and only if the vector $\mathbf{w} := (-x^1, -x^2, \ldots, -x^h, y^1, y^2, \ldots, y^k) \in \mathbb{K}^{h+k}$ belongs to $\ker \mathbf{L}$. Consequently, the linear map $\phi : \mathbb{K}^{h+k} \to \mathbb{K}^n$,

$$\phi(\mathbf{x}, \mathbf{y}) := \sum_{i=1}^{h} x^i \mathbf{u}_i$$

is injective and surjective from $\ker \mathbf{L}$ onto $U \cap V$. It follows that $\dim(U \cap V) = \dim \ker \mathbf{L}$ and, by the rank formula,

$$\dim(U \cap V) + \dim(U + V) = \dim \ker \mathbf{L} + \operatorname{Rank} \mathbf{L} = h + k = \dim U + \dim V.$$

$\square$

1.31 ¶. Notice that the proof of Grassmann's formula is in fact a procedure to compute two bases of $U + V$ and $U \cap V$ starting from two bases of U and V. The reader is invited to choose two subspaces U and V of $\mathbb{K}^n$ and to compute the basis of $U + V$ and of $U \cap V$.

f. Parametric and implicit equations of a subspace

1.32 Parametric equation of a straight line in $\mathbb{K}^n$. Let $\mathbf{a} \neq 0$ and let $\mathbf{q}$ be two vectors in $\mathbb{K}^n$. The *parametric equation of a straight line through $\mathbf{q}$ and direction $\mathbf{a}$* is the map $\mathbf{r} : \mathbb{K} \to \mathbb{K}^n$ given by $\mathbf{r}(\lambda) := \lambda \mathbf{a} + \mathbf{q}$, $\lambda \in \mathbb{K}$. The image of $\mathbf{r}$

$$\left\{ \mathbf{x} \in \mathbb{R}^n \,\middle|\, \exists \lambda \text{ such that } \mathbf{x} = \lambda \mathbf{a} + \mathbf{q} \right\}$$

is the *straight line through $\mathbf{q}$ and direction $\mathbf{a}$*.

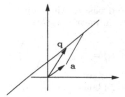

Figure 1.4. Straight line through **q** and direction **a**.

We have $\mathbf{r}(0) = \mathbf{q}$ and $\mathbf{r}(1) = \mathbf{a} + \mathbf{q}$. In other words, $\mathbf{r}(t)$ passes through $\mathbf{q}$ and $\mathbf{a} + \mathbf{q}$. Moreover, $\mathbf{x}$ is on the straight line passing through $\mathbf{q}$ and $\mathbf{a} + \mathbf{q}$ if and only if there exists $t \in \mathbb{K}$ such that $\mathbf{x} = t\mathbf{a} + \mathbf{b}$, or, more explicitly

$$\begin{cases} x^1 = t\,a^1 + q^1, \\ x^2 = t\,a^2 + q^2, \\ \quad\vdots \\ x^n = t\,a^n + q^n. \end{cases} \tag{1.7}$$

In kinematics, $\mathbb{K} = \mathbb{R}$ and the map $t \to \mathbf{r}(t) := t\mathbf{a} + \mathbf{q}$ gives the position at time t of a point moving with constant vector velocity $\mathbf{a}$ starting at $\mathbf{q}$ at time $t = 0$ on the straight line through $\mathbf{q}$ and $\mathbf{a} + \mathbf{q}$.

1.33 Implicit equation of a straight line in $\mathbb{K}^n$. We want to find a representation of the straight line (1.7) which makes no use of the free parameter t.

Since $\mathbf{a} \neq 0$, one of its components is nonzero. Assume for instance $a^1 \neq 0$, we can solve the first equation in (1.7) to get $t = (q^1 - x^1)/a^1$ and, substituting the result into the last $(n - 1)$ equations, we find a system of $(n - 1)$ *constraints* on the variable $\mathbf{x} = (x^1, x^2, \ldots, x^n) \in \mathbb{K}^n$,

$$\begin{cases} x^2 = \frac{(q^1 - x^1)}{a^1} a^2 + q^2, \\ x^3 = \frac{(q^1 - x^1)}{a^1} a^3 + q^3, \\ \quad\vdots \\ x^n = \frac{(q^1 - x^1)}{a^1} a^n + q^n. \end{cases}$$

The previous linear system can be written as $\mathbf{A}(\mathbf{x} - \mathbf{q}) = 0$ where $\mathbf{A} \in M_{n-1,n}(\mathbb{K})$ is the matrix defined by

$$\mathbf{A} = \begin{pmatrix} -a^2/a^1 & -1 & 0 & 0 & \ldots & 0 \\ -a^3/a^1 & 0 & -1 & 0 & \ldots & 0 \\ -a^4/a^1 & 0 & 0 & -1 & \ldots & 0 \\ \vdots & \vdots & \vdots & \vdots & \ddots & \vdots \\ -a^n/a^1 & 0 & 0 & 0 & \ldots & -1 \end{pmatrix}.$$

1.34 ¶. Show that there are several parametric equations of a given straight line. A parametric equation of the straight line through $\mathbf{a}$ and $\mathbf{b} \in \mathbb{R}^n$ is given by $t \to \mathbf{r}(t) := \mathbf{a} + t(\mathbf{b} - \mathbf{a})$, $t \in \mathbb{R}$.

1.35 Parametric and implicit equations of a 2-plane in $\mathbb{K}^3$. Given two linearly independent vectors $\mathbf{v}_1, \mathbf{v}_2$ in $\mathbb{R}^3$ and a point $\mathbf{q} \in \mathbb{R}^3$, we call the *parametric equation* of the plane directed by $\mathbf{v}_1, \mathbf{v}_2$ and passing through $\mathbf{q}$, the map $\varphi : \mathbb{K}^2 \to \mathbb{K}^3$ defined by $\varphi((\alpha, \beta)) := \alpha \mathbf{v}_1 + \beta \mathbf{v}_2 + \mathbf{q}$, or in matrix notation

$$\varphi((\alpha, \beta)) = \left[\mathbf{v}_1 \,\middle|\, \mathbf{v}_2 \right] \begin{pmatrix} \alpha \\ \beta \end{pmatrix} + \mathbf{q}.$$

Of course φ is linear iff $\mathbf{q} = 0$. The 2-plane determined by this parametrization is defined by

$$\Pi := \operatorname{Im} \varphi = \left\{ \mathbf{x} \in \mathbb{R}^3 \,\middle|\, \mathbf{x} - \mathbf{q} \in \operatorname{Im} \mathbf{A} \right\}.$$

Suppose $\mathbf{v}_1 = (a, b, c)$ and $\mathbf{v}_2 = (d, e, f)$ so that

$$\mathbf{A} = \left[\mathbf{v}_1 \,\middle|\, \mathbf{v}_2 \right] = \begin{pmatrix} a & d \\ b & e \\ c & f \end{pmatrix}.$$

Because of Theorem 1.28, there is a nonsingular 2×2 submatrix $\mathbf{B}$ of $\mathbf{A}$ and, without loss of generality, we can suppose that $\mathbf{B} = \begin{pmatrix} a & d \\ b & e \end{pmatrix}$. We can then solve the system

$$\begin{cases} x^1 - q^1 = a\alpha + d\beta, \\ x^2 - q^2 = b\alpha + e\beta \end{cases}$$

in the unknown (α, β), thus finding α and β as linear functions of $x^1 - q^1$ and $x^2 - q^2$. Then, substituting into the third equation, we can eliminate (α, β) from the last equation, obtaining an *implicit equation*, or *constraint*, on the independent variables, of the form

$$r \left(x^1 - q^1 \right) + s \left(x^2 - q^2 \right) + t \left(x^3 - q^3 \right) = 0,$$

that describes the 2-plane without any further reference to the free parameters (α, β).

More generally, let W be a linear subspace of dimension k in $\mathbb{K}^n$, also called a *k-plane* (through the origin) of $\mathbb{K}^n$. If $\mathbf{v}_1, \mathbf{v}_2, \ldots, \mathbf{v}_k$ is a basis of W, we can write $W = \operatorname{Im} \mathbf{L}$ where

$$\mathbf{L} := \left[\mathbf{v}_1 \,\middle|\, \mathbf{v}_2 \,\middle|\, \ldots \,\middle|\, \mathbf{v}_k \right].$$

We call $\mathbf{x} \to L(\mathbf{x}) := \mathbf{Lx}$ the *parametric equation* of W generated by $(\mathbf{v}_1, \mathbf{v}_2, \ldots, \mathbf{v}_k)$. Of course a different basis of W yields a different parametrization.

We can also write any subspace W of dimension k as $W = \ker \mathbf{A}$ where $\mathbf{A} \in M_{n-k,n}(\mathbb{K})$. We call it an *implicit representation* of W. Notice that since $\ker \mathbf{A} = W$, we have $\operatorname{Rank} \mathbf{A}^T = \operatorname{Rank} \mathbf{A} = n - k$ by Theorem 1.28 and the rank formula. Hence the rows of $\mathbf{A}$ are $n - k$ linearly independent vectors of $\mathbb{K}^n$.

1.36 Remark. A k-dimensional subspace of $\mathbb{K}^n$ is represented by means of k *free parameters*, i.e., the image of $\mathbb{K}^k$ through a nondegenerate parametric equation, or by a set of independent $(n - k)$ constraints given by linearly independent scalar equations in the ambient variables.

1.37 Parametric and implicit representations. One can go back and forth from the parametric to the implicit representation in several ways. For instance, start with $W = \operatorname{Im} \mathbf{L}$ where $\mathbf{L} \in M_{n,k}(\mathbb{K})$ has maximal rank, $\operatorname{Rank} \mathbf{L} = k$. By Theorem 1.28 there is a $k \times k$ nonsingular matrix submatrix $\mathbf{M}$ of $\mathbf{L}$. Assume that $\mathbf{M}$ is made by the first few rows of $\mathbf{L}$ so that

$$\mathbf{L} = \begin{pmatrix} \mathbf{M} \\ \mathbf{N} \end{pmatrix}$$

where $\mathbf{N} \in M_{n-k,k}(\mathbb{K})$. Writing $\mathbf{x}$ as $\mathbf{x} = (\mathbf{x}', \mathbf{x}'')$ with $\mathbf{x}' \in \mathbb{K}^k$ and $\mathbf{x}'' \in \mathbb{K}^{n-k}$, the parametric equation $\mathbf{x} = \mathbf{Lt}$, $\mathbf{t} \in \mathbb{K}^k$, writes as

$$\begin{cases} \mathbf{x}' = \mathbf{Mt}, \\ \mathbf{x}'' = \mathbf{Nt}. \end{cases} \tag{1.8}$$

As $\mathbf{M}$ is invertible,

$$\begin{cases} t = \mathbf{M}^{-1}\mathbf{x}', \\ \mathbf{NM}^{-1}\mathbf{x}' = \mathbf{x}''. \end{cases}$$

We then conclude that $\mathbf{x} \in \operatorname{Im} \mathbf{L}$ if and only if $\mathbf{NM}^{-1}\mathbf{x}' = \mathbf{x}''$. The latter is an implicit equation for W, that we may write as $\mathbf{Ax} = 0$ if we define $\mathbf{A} \in M_{n-k,n}(\mathbb{K})$ by

$$\mathbf{A} = \begin{pmatrix} \mathbf{NM}^{-1} & -\operatorname{Id}_k \end{pmatrix}.$$

Conversely, let $W = \ker \mathbf{A}$ where $\mathbf{A} \in M_{n,k}(\mathbb{K})$ has $\operatorname{Rank} \mathbf{A} = n - k$. Select $n - k$ independent columns, say the first $n - k$ on the left, call $\mathbf{B} \in M_{n-k,n-k}(\mathbb{K})$ the square matrix made by these columns, and split $\mathbf{x}$ as $\mathbf{x} = (\mathbf{x}', \mathbf{x}'')$ where $\mathbf{x}' \in \mathbb{K}^{n-k}$ and $\mathbf{x}'' \in \mathbb{K}^k$. Thus $\mathbf{Ax} = 0$ rewrites as

$$\begin{pmatrix} \mathbf{B} & \mathbf{C} \end{pmatrix} \begin{pmatrix} \mathbf{x}' \\ \mathbf{x}'' \end{pmatrix} = 0, \quad \text{or} \quad \mathbf{Bx}' + \mathbf{Cx}'' = 0.$$

As $\mathbf{B}$ is invertible, the last equation rewrites as $\mathbf{x}' = -\mathbf{B}^{-1}\mathbf{Cx}''$, Therefore $\mathbf{x} \in \ker \mathbf{A}$ if and only if

$$\mathbf{x} = \begin{pmatrix} -\mathbf{B}^{-1}\mathbf{C} \\ \operatorname{Id}_k \end{pmatrix} \mathbf{x}'' := \mathbf{Lx}'', \quad \mathbf{x}'' \in \mathbb{K}^k$$

i.e., $W = \operatorname{Im} \mathbf{L}$.

1.3 Matrices and Linear Systems

a. Linear systems and the language of linear algebra

Matrices and linear operators are strongly tied to linear systems. A linear system of m equations and n unknowns has the form

$$\begin{cases} a_1^1 x^1 + a_2^1 x^2 + \cdots + a_n^1 x^n = b^1, \\ a_1^2 x^1 + a_2^2 x^2 + \cdots + a_n^2 x^n = b^2, \\ \vdots \\ a_1^m x^1 + a_2^m x^2 + \cdots + a_n^m x^n = b^m. \end{cases} \tag{1.9}$$

The m-tuple $(b^1, \ldots, b^m)$ is the given right-hand side, the n-tuple $(x^1, \ldots, x^n)$ is the unknown and the numbers $\{a_j^i\}$, $i = 1, \ldots, m$, $j = 1, \ldots, n$ are given and called the coefficients of the system. If we think of the coefficients as the entries of a matrix $\mathbf{A}$,

$$\mathbf{A} := [a_j^i] = \begin{pmatrix} a_1^1 & a_2^1 & \cdots & a_n^1 \\ a_1^2 & a_2^2 & \cdots & a_n^2 \\ \vdots & \vdots & \ddots & \vdots \\ a_1^m & a_2^m & \cdots & a_n^m \end{pmatrix}, \tag{1.10}$$

and we set $\mathbf{b} := (b^1, b^2, \ldots, b^m) \in \mathbb{K}^m$, $\mathbf{x} := (x^1, x^2, \ldots, x^n) \in \mathbb{K}^n$, then the system can be written in a compact way as

$$\mathbf{A}\mathbf{x} = \mathbf{b}. \tag{1.11}$$

Introducing the linear map $A(\mathbf{x}) := \mathbf{A}\mathbf{x}$, (1.9) can be seen as a *functional equation*

$$A(\mathbf{x}) = \mathbf{b} \tag{1.12}$$

or, denoting by $\mathbf{a}_1, \mathbf{a}_2, \ldots, \mathbf{a}_n$ the n-columns of $\mathbf{A}$ indexed from left to right, as

$$x^1 \mathbf{a}_1 + x^2 \mathbf{a}_2 + \cdots + x^n \mathbf{a}_n = \mathbf{b}. \tag{1.13}$$

Thus, the discussion of linear systems, linear independence, matrices and linear maps are essentially the same, in different languages. The next proposition collects these equivalences.

1.38 Proposition. *With the previous notation we have:*

(i) $\mathbf{A}\mathbf{x}$ *is a linear combination of the columns of* $\mathbf{A}$.
(ii) *The following three claims are equivalent:*
    a) the system (1.11) *or* (1.9), *is solvable, i.e., there exists* $\mathbf{x} \in \mathbb{K}^n$ *such that* $\mathbf{A}\mathbf{x} = \mathbf{b}$;
    b) $\mathbf{b}$ *is a linear combination of* $\mathbf{a}_1, \mathbf{a}_2, \ldots, \mathbf{a}_n$;
    c) $\mathbf{b} \in \operatorname{Im} A$.
(iii) *The following four claims are equivalent:*

a) $\mathbf{A}\mathbf{x} = \mathbf{b}$ *has at most one solution,*
b) $\mathbf{A}\mathbf{x} = 0$ *implies* $\mathbf{x} = 0$,
c) $A(\mathbf{x}) = 0$ *has a unique solution,*
d) $\ker \mathbf{A} = \{0\}$,
e) $\mathbf{a}_1, \mathbf{a}_2, \ldots, \mathbf{a}_n$ *are linearly independent.*

(iv) $\ker \mathbf{A}$ *is the set of all solutions of the system* $\mathbf{A}\mathbf{x} = 0$.

(v) $\operatorname{Im} \mathbf{A}$ *is the set of all* $\mathbf{b}$'s *such that the system* $A\mathbf{x} = \mathbf{b}$ *has at least one solution.*

(vi) *Let* $x_0 \in \mathbb{K}^n$ *be a solution of* $\mathbf{A}\mathbf{x}_0 = \mathbf{b}$. *Then the set of all solutions of* $\mathbf{A}\mathbf{x} = \mathbf{b}$ *is the set*

$$\{\mathbf{x}_0\} + \ker \mathbf{A} := \{\mathbf{x} \in \mathbb{K}^n \mid \mathbf{x} - \mathbf{x}_0 \in \ker \mathbf{A}\}.$$

With the previous notation, we see that $\mathbf{b}$ is linearly dependent of $\mathbf{a}_1, \mathbf{a}_2, \ldots, \mathbf{a}_n$ if and only if

$$\operatorname{Rank} \left[\mathbf{a}_1 \mid \ldots \mid \mathbf{a}_n\right] = \operatorname{Rank} \left[\mathbf{a}_1 \mid \ldots \mid \mathbf{a}_n \mid \mathbf{b}\right].$$

Thus from Proposition 1.38 (ii) we infer the following.

1.39 Proposition (Rouché–Capelli). *With the previous notation, the system* (1.9) *or* (1.11) *is solvable if and only if*

$$\operatorname{Rank} \left[\mathbf{a}_1 \mid \ldots \mid \mathbf{a}_n\right] = \operatorname{Rank} \left[\mathbf{a}_1 \mid \ldots \mid \mathbf{a}_n \mid \mathbf{b}\right].$$

The $m \times (n+1)$ matrix

$$\left[\mathbf{a}_1 \mid \ldots \mid \mathbf{a}_n \mid \mathbf{b}\right] := \begin{pmatrix} a_1^1 & a_2^1 & \ldots & a_n^1 & b^1 \\ a_1^2 & a_2^2 & \ldots & a_n^2 & b^2 \\ \vdots & \vdots & \ddots & \vdots & \vdots \\ a_1^m & a_2^m & \ldots & a_n^m & b^m \end{pmatrix}$$

is often called the *complete matrix of the system* (1.9).

1.40 ¶. Prove all claims in this section.

1.41 Solving linear systems. Let us return to the problem of solving

$$\mathbf{A}\mathbf{x} = \mathbf{b}, \qquad \text{where} \qquad \mathbf{A} \in M_{m,n}(\mathbb{K}), \ \mathbf{b} \in \mathbb{K}^m.$$

If $n = m$ and $\mathbf{A}$ is nonsingular, then the unique solution is $\mathbf{x}_0 := \mathbf{A}^{-1}b$. In the general case, according to Proposition 1.39, the system is solvable if and only if $\operatorname{Rank} \mathbf{A} = \operatorname{Rank} [\mathbf{A} \mid \mathbf{b}]$, and if $x_0 \in \mathbb{K}^n$ is a solution, the set of all solutions is given by $\{x_0\} + \ker \mathbf{A}$.

Let $r := \operatorname{Rank} \mathbf{A}$. Since $\operatorname{Rank} \mathbf{A}^T = r$, we may assume without loss of generality that the first r rows of $\mathbf{A}$ are linearly independent and the other rows depend linearly on the first r rows. Therefore, if we solve the system of r equations

$$
\begin{pmatrix} a_1^1 & a_2^1 & \cdots & a_n^1 \\ a_1^2 & a_2^2 & \cdots & a_n^2 \\ \vdots & \vdots & \ddots & \vdots \\ a_1^r & a_2^r & \cdots & a_n^r \end{pmatrix} \begin{pmatrix} x^1 \\ x^2 \\ \vdots \\ x^n \end{pmatrix} = \begin{pmatrix} b^1 \\ b^2 \\ \vdots \\ b^r \end{pmatrix}, \tag{1.14}
$$

the remaining equations are automatically fulfilled. So it is enough to solve $\mathbf{Ax} = \mathbf{b}$ in the case where $\mathbf{A} \in M_{r,n}(\mathbb{K})$ and Rank $\mathbf{A}^T = $ Rank $\mathbf{A} = r$.

We have two cases. If $r = n$, then $\mathbf{A} \in M_{r,r}$ is nonsingular, consequently $\mathbf{Ax} = \mathbf{b}$ has a unique solution $\mathbf{x} = \mathbf{A}^{-1}\mathbf{b}$. If $r < n$, then $\mathbf{A}$ has r linearly independent columns, say the first r. Denote by $\mathbf{R}$ the $r \times r$ nonsingular matrix made by these columns, and decompose $\mathbf{x} = (\mathbf{x}', \mathbf{x}'')$ with $\mathbf{x}' \in \mathbb{K}^r$ and $\mathbf{x}'' \in \mathbb{K}^{n-r}$. Then $\mathbf{Ax} = \mathbf{b}$ writes as

$$
\left(\boxed{\ \ \mathbf{R}\ \ } \ \ \boxed{\ \ \mathbf{S}\ \ } \right) \begin{pmatrix} \mathbf{x}' \\ \mathbf{x}'' \end{pmatrix} = \mathbf{b},
$$

i.e., $\mathbf{Rx}' + \mathbf{Sx}'' = \mathbf{b}$, or $\mathbf{x}' = \mathbf{R}^{-1}(\mathbf{b} - \mathbf{Rx}'')$. Therefore,

$$
\mathbf{x} = \left(\begin{matrix} \boxed{-\mathbf{R}^{-1}\mathbf{S}} \\ \boxed{\mathrm{Id}_{n-r}} \end{matrix} \right) \mathbf{x}' + \begin{pmatrix} -\mathbf{R}^{-1}\mathbf{b} \\ 0 \end{pmatrix} =: \mathbf{Lx}' + \mathbf{x}_0,
$$

concluding that the set of all solutions of the system $\mathbf{Ax} = \mathbf{b}$ is

$$
\{\mathbf{x} \,|\, \mathbf{x} - \mathbf{x}_0 \in \ker \mathbf{A}\} = \left\{ \mathbf{x} \,\Big|\, \mathbf{x} - \mathbf{x}_0 \in \mathrm{Im}\,\mathbf{L} \right\}.
$$

b. The Gauss elimination method

As we have seen, linear algebra yields a proper language to discuss linear systems, and conversely, most of the constructions in linear algebra reduce to solving systems. Moreover, the proofs we have presented are constructive and become useful from a numerical point of view if one is able to efficiently solve the following two questions:

(i) find the solution of a nonsingular square system $\mathbf{Ax} = \mathbf{b}$,
(ii) given a set of vectors $T \subset \mathbb{K}^n$, find a subset $S \subset T$ such that $\mathrm{Span}\,S = \mathrm{Span}\,T$.

In this section we illustrate the classical *Gauss elimination method* which efficiently solves both questions.

1.42 Example. Let us begin with an example of how to solve a linear system. Consider the linear system

$$
\begin{cases} 6x + 18y + 6z & = b_1, \\ 3x + 8y + 6z & = b_2, \\ 2x + y + z & = b_3, \end{cases} \quad \text{or} \quad \mathbf{Ax} = \mathbf{b}
$$

where $\mathbf{x} := (x, y, x)$, $\mathbf{b} := (b_1, b_2, b_3)$ and

$$A = \begin{pmatrix} 6 & 18 & 6 \\ 3 & 8 & 6 \\ 2 & 1 & 1 \end{pmatrix}.$$

We subtract from the second and third equations the first one multiplied by 1/2 and 1/3 respectively to get the new equivalent system:

$$\begin{cases} 6x + 18y + 6z & = b_1, \\ 3x + 8y + 6z - \frac{1}{2}(6x + 18y + 6z) & = -\frac{1}{2}b_1 + b_2, \\ 2x + y + z - \frac{1}{3}(6x + 18y + 6z) & = -\frac{1}{3}b_1 + b_3, \end{cases}$$

i.e.,

$$\begin{cases} 6x + 18y + 6z & = b_1, \\ -y + 3z & = -\frac{1}{2}b_1 + b_2, \\ -5y - z & = -\frac{1}{3}b_1 + b_3. \end{cases} \tag{1.15}$$

This essentially requires us to solve the system of the last two equations

$$\begin{cases} -y + 3z & = -\frac{1}{2}b_1 + b_2, \\ -5y - z & = -\frac{1}{3}b_1 + b_3. \end{cases}$$

We now apply the same argument to this last system, i.e., we subtract from the last equation the first one multiplied by 5 to get

$$\begin{cases} 6x + 18y + 6z & = b_1, \\ -y + 3z & = -\frac{1}{2}b_1 + b_2, \\ -5y - z - 5(-y + 3z) & = -\frac{1}{2}b_1 + b_3 - 5(-\frac{1}{2}b_1 + b_2), \end{cases}$$

i.e.,

$$\begin{cases} 6x + 18y + 6z & = b_1, \\ -y + 3z & = -\frac{1}{2}b_1 + b_2, \\ -16z & = 2b_1 - 5b_2 + b_3. \end{cases}$$

This system has exactly the same solution as the original one and, moreover, it is easily solvable starting from the last equation. Finally, we notice that the previous method produced two matrices

$$U := \begin{pmatrix} 6 & 18 & 6 \\ 0 & -1 & 3 \\ 0 & 0 & -16 \end{pmatrix}, \qquad L := \begin{pmatrix} 1 & 0 & 0 \\ -\frac{1}{2} & 1 & 0 \\ 2 & -5 & 1 \end{pmatrix}.$$

U is *upper triangular* and L is *lower triangular* with 1 in the principal diagonal, so the original system $Ax = b$ rewrites as

$$Ux = Lb.$$

Since $L = [l_j^i]$ is invertible ($l_i^i = 1 \; \forall i$) and x is arbitrary, we can rewrite the last formula as a decomposition formula for A,

$$A = L^{-1}U.$$

The algorithm we have just described in Example 1.42, that transforms the proposed 3×3 square system into a triangular system, extends to systems with an arbitrary number of unknowns and equations, and it is called the *Gauss elimination method*. Moreover, it is particularly efficient, but does have some drawbacks from a numerical point of view.

Let

$$\mathbf{A}\mathbf{x} = 0 \tag{1.16}$$

be a linear homogeneous system with m equations, n unknowns and a coefficient matrix given by

$$\mathbf{A} = \begin{pmatrix} a_1^1 & a_2^1 & \ldots & a_n^1 \\ a_1^2 & a_2^2 & \ldots & a_n^2 \\ \vdots & \vdots & \ddots & \vdots \\ a_1^m & a_2^m & \ldots & a_n^m \end{pmatrix}.$$

Starting from the left, we denote by j_1 the index of the *first* column of $\mathbf{A}$ that has at least one nonzero element. Then we reorder the rows into a new matrix $\mathbf{B}$ of the same dimensions, in such a way that the element $b_{j_1}^1$ is nonzero and all columns with index less than j_1 are zero,

$$\mathbf{B} = [b_j^i] = \begin{pmatrix} 0 & \ldots & 0 & b_{j_1}^1 & * & \ldots & * \\ 0 & \ldots & 0 & * & * & \ldots & * \\ \vdots & \vdots & \vdots & \vdots & \vdots & \ddots & \vdots \\ 0 & \ldots & 0 & * & * & \ldots & * \end{pmatrix},$$

where $*$ denotes the unspecified entries. We then set $p_1 := b_{j_1}^1$, and for $i = 2, \ldots, m$ we subtract from the ith row of $\mathbf{B}$ the first row of $\mathbf{B}$ multiplied by $-b_{j_1}^i/p_1$. The resulting matrix therefore has the following form

$$\mathbf{A}_1 := \begin{pmatrix} 0 & \ldots & 0 & p_1 & * & \ldots & * \\ 0 & \ldots & 0 & 0 & * & \ldots & * \\ \vdots & \vdots & \vdots & \vdots & \vdots & \ddots & \vdots \\ 0 & \ldots & 0 & 0 & * & \ldots & * \end{pmatrix}$$

where $p_1 \neq 0$, below p_1 all entries are zero and $*$ denotes the unspecified entries.

We then transform $\mathbf{A}_1$ into $\mathbf{A}_2$, $\mathbf{A}_2$ into $\mathbf{A}_3$, and so on, operating as previously, but on the submatrix of $\mathbf{A}_1$ of rows of index respectively larger than $2, 3, \ldots$. The algorithm of course stops when there are no more rows and/or columns.

The resulting matrix produced this way is not uniquely determined as there is freedom in exchanging the rows. However, a resulting matrix that we call a *Gauss reduced matrix*, is clearly upper triangular if $\mathbf{A}$ is a square matrix, $m = n$, and in general has the following *stair-shaped form*

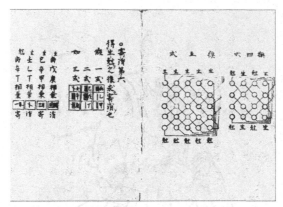

Figure 1.5. Two pages of the Japanese mathematician Takakazu Seki (1642–1708) who apparently dealt with determinants before Gauss.

$$
\mathbf{G_A} := \begin{pmatrix}
0 & \cdots & p_1 & * & \cdots & * & * & \cdots & * & \cdots & * & \cdots & * \\
0 & \cdots & 0 & 0 & \cdots & p_2 & * & \cdots & * & \cdots & * & \cdots & * \\
0 & \cdots & 0 & 0 & \cdots & 0 & 0 & \cdots & p_3 & \cdots & * & \cdots & * \\
0 & \cdots & 0 & 0 & \cdots & 0 & 0 & \cdots & 0 & \cdots & 0 & \cdots & p_r \\
\vdots & \vdots & \vdots & \vdots & \vdots & \vdots & \vdots & \vdots & \vdots & \vdots & \vdots & \ddots & \vdots \\
0 & \cdots & 0 & 0 & \cdots & 0 & 0 & \cdots & 0 & \cdots & 0 & \cdots & 0
\end{pmatrix}
$$

$$(1.17)$$

where $*$ denotes the unspecified entries; the nonzero numbers $p_1, p_2, \ldots, p_r$ are called the *pivots* of the stair-shaped matrix $\mathbf{G_A}$.

Finally, since

○ multiplying one of the equations of the system $\mathbf{Ax} = 0$ by a nonzero scalar,
○ exchanging the order of the equations,
○ summing a multiple of one equation and another equation,

produces a linear system with the same solution as the initial system, and observing that the Gauss elimination procedure operates with transformations of this type, we conclude that $\mathbf{G_A x} = 0$ has the same solution as the initial system.

We now discuss the solvability of the system $\mathbf{Lx} = \mathbf{b}$, if $\mathbf{L}$ is stair-shaped.

1.43 Proposition. *Let $\mathbf{L}$ be a stair-shaped $m \times n$ matrix. Suppose that $\mathbf{L}$ has r pivots, $r \le \min(n, m)$. Then a basis of $\operatorname{Im} \mathbf{L}$ is given by the r columns containing the pivots, and the system $\mathbf{Lx} = \mathbf{b}$, $\mathbf{b} = (b^1, b^2, \ldots, b^m)^T$, has a solution if and only if $b^{r+1} = \cdots = b^m = 0$.*

Proof. Since there are r pivots and at most one pivot per row, the last rows of $\mathbf{L}$ are identically zero, hence $\operatorname{Im} \mathbf{L} \subset \{\mathbf{b} \in \mathbb{K}^m \mid \mathbf{b} = (b_1, b_2, \ldots, b_r, 0, \ldots, 0)\}$. Consequently,

Figure 1.6. Takakazu Seki (1642–1708).

$\dim \operatorname{Im} \mathbf{L} \leq r$. On the other hand the r columns that contain the pivots are in $\operatorname{Im} \mathbf{L}$ and are linearly independent, hence $\operatorname{Rank} \mathbf{L} = r$, and $\operatorname{Im} \mathbf{L} = \{(b^1, b^2, \ldots, b^r, 0, \ldots, 0) \mid b^i \in \mathbb{K} \ \forall i, \ i = 1, \ldots, r\}$. □

The Gauss elimination procedure preserves several properties of the original matrix $\mathbf{A}$.

1.44 Theorem. *Let* $\mathbf{A} \in M_{m,n}(\mathbb{K})$ *and let* $\mathbf{G_A}$ *be one of the matrices resulting from the Gauss elimination procedure. Then*

(i) $\ker \mathbf{A} = \ker \mathbf{G_A}$,
(ii) $\operatorname{Rank} \mathbf{A} = \operatorname{Rank} \mathbf{G_A} = $ *number of pivots of* $\mathbf{G_A}$,
(iii) *let* $j_1, j_2, \ldots, j_r$ *be the indices of the columns of the pivots of* $\mathbf{G_A}$, *then the columns of* $\mathbf{A}$ *with the same indices are linearly independent,*

Proof. (i) is a rewriting of the equivalence of $\mathbf{Ax} = 0$ with $\mathbf{G_A x} = 0$.

(ii) Because of (i), the rank formula yields $\operatorname{Rank} \mathbf{A} = \operatorname{Rank} \mathbf{G_A}$, and $\operatorname{Rank} \mathbf{G_A}$ equals the number of pivots by Proposition 1.43.

(iii) Let $\mathbf{A} = [\mathbf{a}_1 \mid \mathbf{a}_2 \mid \ldots \mid \mathbf{a}_n]$ and let

$$\mathbf{B} := \left[\mathbf{a}_{j_1} \mid \mathbf{a}_{j_2} \mid \ldots \mid \mathbf{a}_{j_k} \right].$$

Following the Gauss elimination procedure we used on $\mathbf{A}$, we easily see that the columns of $\mathbf{B}$ transform into the columns of the pivots which are linearly independent. By (i) $\ker \mathbf{B} = \{0\}$. □

1.45 ¶. Let $\mathbf{A} \in M_{m,n}(\mathbb{K})$. Show a procedure to find a basis for $\operatorname{Rank} \mathbf{A}$ and $\operatorname{Rank} \mathbf{A}^T$.

1.46 ¶. Let $W = \operatorname{Span} \{\mathbf{v}_1, \mathbf{v}_2, \ldots, \mathbf{v}_k\}$ be a subspace of $\mathbb{K}^n$. Show a procedure to find a basis of W among the vectors $\mathbf{v}_1, \mathbf{v}_2, \ldots, \mathbf{v}_k$.

1.47 ¶. Let $\mathbf{A} \in M_{m,n}(\mathbb{K})$. Show a procedure to find a basis of $\ker \mathbf{A}$.

1.48 ¶. Let $\mathbf{v}_1, \mathbf{v}_2, \ldots, \mathbf{v}_k \in \mathbb{K}^m$ be k linearly independent vectors. Show a procedure to complete them with $n - k$ vectors of the canonical basis of $\mathbb{R}^n$ in order to form a new basis of $\mathbb{R}^n$. [*Hint:* Apply the Gauss elimination procedure on the matrix

$$
\mathbf{A} := \left(
\begin{array}{cccc|cccc}
 & & & & 1 & 0 & \ldots & 0 \\
 & & & & 0 & 1 & \ldots & 0 \\
\mathbf{v}_1 & \mathbf{v}_2 & \ldots & \mathbf{v}_k & \vdots & \vdots & \ddots & \vdots \\
 & & & & 0 & 0 & \ldots & 1
\end{array}
\right) .]
$$

1.49 ¶. Show that $\mathbf{A} \in M_{n,n}(\mathbb{K})$ is invertible if and only if a Gauss reduced matrix of $\mathbf{A}$ has n pivots.

c. The Gauss elimination procedure for nonhomogeneous linear systems

Now consider the problem of solving $\mathbf{A}(\mathbf{x}) = \mathbf{b}$, where $\mathbf{A} \in M_{m,n}(\mathbb{K})$, $\mathbf{x} \in \mathbb{K}^n$ and $\mathbf{b} \in \mathbb{K}^m$. We can equivalently write it as

$$
\begin{pmatrix}
a_1^1 & a_2^1 & a_3^1 & \ldots & a_n^1 & 1 & 0 & \ldots & 0 \\
a_1^2 & a_2^2 & a_3^2 & \ldots & a_n^2 & 0 & 1 & \ldots & 0 \\
\vdots & \vdots & \ddots & \vdots & \vdots & \vdots & \ddots & \vdots \\
a_1^m & a_2^m & a_3^m & \ldots & a_n^m & 0 & 0 & \ldots & 1
\end{pmatrix}
\begin{pmatrix}
x^1 \\ x^2 \\ \vdots \\ x^n \\ -b^1 \\ -b^2 \\ \vdots \\ -b^m
\end{pmatrix}
= 0.
$$

If one computes a Gauss reduced form of the $m \times (n + m)$ matrix

$$
\mathbf{B} := \Big[\mathbf{A} \,\Big|\, \mathrm{Id}_n \Big] =
\begin{pmatrix}
a_1^1 & a_2^1 & \ldots & a_n^1 & 1 & 0 & \ldots & 0 \\
a_1^2 & a_2^2 & \ldots & a_n^2 & 0 & 1 & \ldots & 0 \\
\vdots & \vdots & \ddots & \vdots & \vdots & \vdots & \ddots & \vdots \\
a_1^m & a_m^2 & \ldots & a_n^m & 0 & 0 & \ldots & 1
\end{pmatrix}
$$

we find, on account of Theorem 1.44, that

$$
\mathbf{G_B} := \left(\begin{array}{c|c} \mathbf{G_A} & \mathbf{S} \end{array} \right)
$$

where $\mathbf{G_A} \in M_{m,n}(\mathbb{K})$ is a Gauss reduced matrix of $\mathbf{A}$ and $\mathbf{S} \in M_{m,m}(\mathbb{K})$. Moreover, *if the elimination procedure has been carried out without any permutation of the rows*, then $\mathbf{S}$ is a lower triangular matrix with 1 as entries in the principal diagonal, hence it is invertible. Since for every $\mathbf{b}$ the system $\mathbf{Ax} = \mathbf{b}$ is equivalent to $\mathbf{G_A x} = \mathbf{Sb}$, we then have

$$
\mathbf{G_A x} = \mathbf{Sb} = \mathbf{SAx} \qquad \forall \mathbf{x} \in \mathbb{K}^n,
$$

thus concluding that $\mathbf{A} = \mathbf{S}^{-1} \mathbf{G_A}$. In particular,

1.50 Proposition (*LR decomposition*). *Let* $\mathbf{A} \in M_{n,n}(\mathbb{K})$ *be a square matrix. If the elimination procedure proceeds without any permutation of the rows, we can decompose* $\mathbf{A}$ *as* $\mathbf{A} = \mathbf{LR}$, *where* $\mathbf{R} = \mathbf{G_A}$ *is the resulting Gauss reduced matrix and* $\mathbf{L}$ *is a suitable lower triangular matrix with* 1 *as entries in the principal diagonal of* $\mathbf{L}$.

In general, howewer, the permutation of the rows must be taken into account. For this purpose, let us fix some notation. Recall that a *permutation* of $\{1,\dots,m\}$ is a one-to-one map $\sigma : \{1,\dots,m\} \to \{1,\dots,m\}$. The set of all permutations of m elements is denoted by $\mathcal{P}_m$. For every permutation σ of m elements, define the associated *permutation matrix* $\mathbf{R}_\sigma \in M_{m,m}(\mathbb{K})$ by

$$\mathbf{R}_\sigma := \left[\mathbf{e}_{\sigma_1} \,\middle|\, \mathbf{e}_{\sigma_2} \,\middle|\, \dots \,\middle|\, \mathbf{e}_{\sigma_m}\right],$$

where $(\mathbf{e}_1, \mathbf{e}_2, \dots, \mathbf{e}_m)$ is the canonical basis of $\mathbb{K}^m$.

Let $\mathbf{A} \in M_{m,n}(\mathbb{K})$. If σ permutes the indices of the rows of $\mathbf{A}$, then the resulting matrix is $\mathbf{R}_\sigma \mathbf{A}$. Now denote by $\mathcal{G}(\mathbf{A})$ the Gauss reduced matrix, if it exists, obtained by the Gauss elimination procedure starting from the top row and proceeding without any permutation of the rows.

Let $\mathbf{G_A}$ be a Gauss reduced form of $\mathbf{A}$. Then $\mathbf{G_A} = \mathcal{G}(\mathbf{R}_\sigma \mathbf{A})$ for some permutation σ of m elements.

Now fix a Gauss reduced form $\mathbf{G_A}$ of $\mathbf{A}$, and let σ be such that $\mathbf{G_A} = \mathcal{G}(\mathbf{R}_\sigma \mathbf{A})$. Write $\mathbf{Ax} = \mathbf{y}$ as $(\mathbf{R}_\sigma \mathbf{A})\mathbf{x} = \mathbf{R}_\sigma \mathbf{y} = \mathrm{Id}_m(\mathbf{R}_\sigma \mathbf{y})$ and let

$$\mathbf{B} := \left[\mathbf{R}_\sigma \mathbf{A} \,\middle|\, \mathrm{Id}_m\right].$$

Then $\mathbf{B}$ and $\mathbf{R}_\sigma \mathbf{A}$ may be reduced without any permutation of the rows, hence by the above

$$\mathcal{G}(\mathbf{B}) = \left[\mathcal{G}(\mathbf{R}_\sigma \mathbf{A}) \,\middle|\, \mathbf{S}\right] = \left[\mathbf{G_A} \,\middle|\, \mathbf{S}\right]$$

where $\mathbf{S}$ is lower triangular with all entries in the principal diagonal equal to 1. Therefore $\mathbf{G_A x} = \mathbf{S R}_\sigma \mathbf{y} = \mathbf{S R}_\sigma \mathbf{A x} \,\forall \mathbf{x}$, that is,

$$\mathbf{G_A} = \mathbf{S R}_\sigma \mathbf{A}. \tag{1.18}$$

When $\mathbf{A} \in M_{n,n}(\mathbb{K})$ is a square matrix, (1.18) shows that $\mathbf{A}$ is invertible if and only if a Gauss reduced form $\mathbf{G_A}$ of $\mathbf{A}$ is invertible and

$$\mathbf{A}^{-1} = \mathbf{G_A}^{-1} \mathbf{S R}_\sigma.$$

In practice, let $(\mathbf{e}_1, \mathbf{e}_2, \dots, \mathbf{e}_n)$ be the canonical basis of $\mathbb{K}^n$ and let $\mathbf{A}^{-1} =: [\mathbf{v}_1 \,|\, \mathbf{v}_2 \,|\, \dots \,|\, \mathbf{v}_n]$. Let $i = 1, \dots n$. To compute $\mathbf{v}_i$, we observe that $\mathbf{v}_i = \mathbf{A}^{-1} \mathbf{e}_i$, i.e., $\mathbf{A v}_i = \mathbf{e}_i$. Thus, using the Gauss elimination procedure, from (1.18) $\mathbf{v}_i$ is a solution of $\mathbf{G_A v}_i = \mathbf{S R}_\sigma \mathbf{e}_i$. Now, since $\mathbf{G_A}$ is upper triangular, this last system is easily solved by inductively computing the components of $\mathbf{v}_i$ starting from the last, upward.

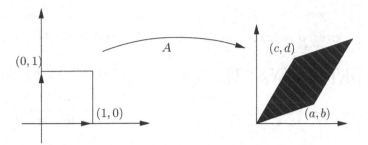

Figure 1.7. The area transformation.

1.4 Determinants

The notion of *determinant* originated with the observation of Gabriel Cramer (1704–1752) and Carl Friedrich Gauss (1777–1855) that the process of elimination of unknowns when solving a linear system of equations amounts to operating with the indices of the associated matrix. Developments of this observation due to Pierre-Simon Laplace (1749–1827), Alexandre Vandermonde (1735–1796), Joseph-Louis Lagrange (1736–1813) and Carl Friedrich Gauss (1777–1855), who introduced the word *determinant*, were then accomplished with a refined study of the properties of the determinant by Augustin-Louis Cauchy (1789–1857) and Jacques Binet (1786–1856). Here we illustrate the main properties of the determinant.

1.51 Determinant and area in $\mathbb{R}^2$. Let

$$\mathbf{A} = \begin{pmatrix} a & b \\ c & d \end{pmatrix} \tag{1.19}$$

be a 2×2 matrix. It is easily seen that $\mathbf{A}$ is not singular, i.e., the linear homogeneous system

$$\begin{cases} ax + by = 0, \\ cx + dy = 0 \end{cases}$$

has zero, $(x, y) = (0, 0)$, as a unique solution if and only if $ad - bc \neq 0$. The number $ad - bc$ is the *determinant* of the matrix $\mathbf{A}$,

$$\det \mathbf{A} = \det \begin{pmatrix} a & b \\ c & d \end{pmatrix} := ad - bc.$$

One immediately notices the combinatorial characteristic of this definition: if $\mathbf{A} = [a_j^i]$, then $\det A = a_1^1 a_2^2 - a_2^1 a_1^2$.

Let $\mathbf{a} := (a, c)$ and $\mathbf{b} := (b, d)$ be the two columns of the matrix $\mathbf{A}$ in (1.19). The elementary area of the parallelogram spanned by $\mathbf{a}$ and $\mathbf{b}$ with vertices $(0, 0)$, (a, c), (b, d) and $(a + b, c + d)$, is given by

$$\text{Area}\,(T) = |\mathbf{a}|\,|\mathbf{b}|\,|\sin \theta|$$

where θ is the angle $\widehat{\mathbf{a}\mathbf{b}}$, irrespective of the orientation, see Figure 1.7. On the other hand, by Carnot's formula $|\mathbf{a}|\,|\mathbf{b}|\cos \theta = ab + cd$, hence

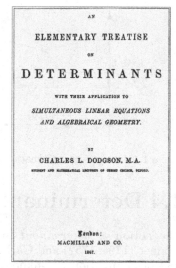

Figure 1.8. Frontispieces of two books on determinants respectively by Ernesto Pascal and Charles L. Dodgson, better known as Lewis Caroll.

$$\text{Area}\,(T)^2 = (a^2 + c^2)(b^2 + d^2)(1 - \cos^2\theta)$$
$$= a^2b^2 + a^2d^2 + b^2c^2 + c^2d^2 - (ab + cd)^2$$
$$= a^2d^2 + b^2c^2 - 2abcd$$
$$= (ad - bc)^2 = \det A^2,$$

i.e.,

$$\text{Area}\,(T) = |\det \mathbf{A}|.$$

We may think of $\det \mathbf{A}$ as of the *area of T with sign*. In fact, the sign of $\det \mathbf{A}$ may be used to define the sign of the angle formed by the vectors $\mathbf{a}$ and $\mathbf{b}$. The angle $\widehat{\mathbf{ab}}$ is *positively* (*negatively*) oriented if $\det[\mathbf{a}\,|\,\mathbf{b}] > 0$ ($\det[\mathbf{a}\,|\,\mathbf{b}] < 0$).

Angles with sign in geometry are also modelled by complex multiplication, identifying $\mathbb{R}^2$ with $\mathbb{C}$. Using the previous notation, setting $z := a + ib$, $w = c + id$ we have

$$z\overline{w} = (a + ib)(c - id) = (ac + bd) + i(bc - ad) = (\mathbf{a} \bullet \mathbf{b})_{\mathbb{R}^2} + i\det \mathbf{A}.$$

Let $\mathbf{v}_1, \mathbf{v}_2 \in \mathbb{R}^2$. As we have seen, the determinant of the matrix $[\mathbf{v}_1\,|\,\mathbf{v}_2]$ is not zero iff $\mathbf{v}_1$ and $\mathbf{v}_2$ are linearly independent. Actually, for any $n \geq 1$ there is a real function defined on $n \times n$ matrices that tells us whether the n columns of the matrix are linearly independent: the *determinant*. One of the simplest ways to define it is as follows.

We recall that a *permutation* of $\{1, \ldots, n\}$ is a one-to-one map σ : $\{1, \ldots, n\} \to \{1, \ldots, n\}$. The set of permutations of n objects, denoted by $\mathcal{P}_n$ is a group with respect to the operation of composition. A permutation that exchanges two adjacent indices and leaves the other indices unchanged is called a *transposition*. Transpositions are elementary permutations in the sense that each permutation σ can be obtained by composing subsequent transpositions. Of course, there are several ways to decompose a given permutation into elementary transpositions, but the *parity* or *oddity* of the number of transpositions needed to realize a given permutation σ depends only on the permutation σ. We define the *signature*, or *sign*, of the permutation σ the number

$$(-1)^\sigma := \begin{cases} +1 & \text{if } \sigma \text{ decomposes in an even number of transpositions,} \\ -1 & \text{if } \sigma \text{ decomposes in an odd number of transpositions.} \end{cases}$$

1.52 Definition. *Let* $\mathbf{A} = [a_j^i] \in M_{n,n}(\mathbb{K})$, $n \geq 1$. *The* determinant *of* $\mathbf{A}$ *is then defined by*

$$\det \mathbf{A} := \sum_{\sigma \in \mathcal{P}_n} (-1)^\sigma a^1_{\sigma(1)} a^2_{\sigma(2)} \dots a^n_{\sigma(n)}. \tag{1.20}$$

Notice that $\det \mathbf{A}$ is a sum of products and each product contains just one element from each row and each column, and the sum, apart from the sign, is extended to all possible choices.

1.53 Example. Of course for matrix $\mathbf{A}$ in (1.19) we again get $\det \mathbf{A} = ad - bc$. Going back to the area, one shows that given 3 vectors $\mathbf{v}_1, \mathbf{v}_2, \mathbf{v}_3 \in \mathbb{R}^3$ and denoting by T the polyhedra generated by these vectors, we still have

$$\text{Vol}_3(T) = |\det[\mathbf{v}_1 \,|\, \mathbf{v}_2 \,|\, \mathbf{v}_3]|.$$

For n vectors $\mathbf{v}_1, \mathbf{v}_2, \dots, \mathbf{v}_n \in \mathbb{R}^n$, let

$$\mathbf{L} := \left[\mathbf{v}_1 \,\Big|\, \mathbf{v}_2 \,\Big|\, \dots \,\Big|\, \mathbf{v}_n \right] \in M_{n,n}(\mathbb{R})$$

and let $L(x) := \mathbf{L}\mathbf{x}$. If Q is the unit cube of $\mathbb{R}^n$,

$$Q := \left\{ \mathbf{x} = (x^1, x^2, \dots, x^n) \,\Big|\, 0 \leq x^i \leq 1 \;\forall i \right\},$$

we define the n-dimensional volume of $T := L(Q)$ by

$$\text{Vol}_n(T) := |\det \mathbf{L}|.$$

It is useful to think of the determinant as a function of the columns of the matrix. In fact, we have the following.

1.54 Theorem. *The determinant on* $n \times n$ *matrices is the unique function* $\det : M_{n,n}(\mathbb{K}) \to \mathbb{K}$ *such that, when seen as a function of columns, it is*

(i) (LINEAR ON EACH FACTOR)*: for all* $\mathbf{a}'_i, \mathbf{a}''_i \in \mathbb{K}^n$, $i = 1, \dots, n$, *and* $\lambda \in \mathbb{K}$

$$\det \left[\dots \,\Big|\, \mathbf{a}'_i + \mathbf{a}''_i \,\Big|\, \dots \right] = \det \left[\dots \,\Big|\, \mathbf{a}'_i \,\Big|\, \dots \right] + \det \left[\dots \,\Big|\, \mathbf{a}''_i \,\Big|\, \dots \right],$$

$$\det \left[\dots \,\Big|\, \lambda \mathbf{a}_i \,\Big|\, \dots \right] = \lambda \det \left[\dots \,\Big|\, \mathbf{a}_i \,\Big|\, \dots \right],$$

(ii) (ALTERNATING): *by exchanging two adjacent columns the determinant changes sign,*

$$\det \left[\ldots \,\middle|\, \mathbf{a}_i \,\middle|\, \mathbf{a}_{i+1} \,\middle|\, \ldots \right] = -\det \left[\ldots \,\middle|\, \mathbf{a}_{i+1} \,\middle|\, \mathbf{a}_i \,\middle|\, \ldots \right],$$

(iii) (NORMALIZED): $\det \mathrm{Id}_n = 1$.

Notice that because of (i) the alternating property can be equivalently formulated by saying that $\det \mathbf{A} = 0$ if $\mathbf{A}$ has two equal columns.

Proof. Clearly the right-hand side of (1.20) fulfills the conditions (i), (ii), (iii). To prove uniqueness, suppose that $D : M_{n,n}(\mathbb{K}) \to \mathbb{K}$ fulfills (i), (ii), (iii) of Theorem 1.54. Write $\mathbf{A} = [a_j^i] \in M_{n,n}(\mathbb{K})$ as $\mathbf{A} = [\mathbf{a}_1 \,|\, \mathbf{a}_2 \,|\, \ldots \,|\, \mathbf{a}_n]$ where

$$\mathbf{a}_i = \sum_{j=1}^n a_i^j \mathbf{e}_j,$$

$(\mathbf{e}_1, \mathbf{e}_2, \ldots, \mathbf{e}_n)$ being the canonical basis of $\mathbb{K}^n$. Then by (i)

$$D(\mathbf{A}) = \sum_{\sigma(1),\ldots,\sigma(n)} a_{\sigma(1)}^1 a_{\sigma(2)}^2 \cdots a_{\sigma(n)}^n D([\mathbf{e}_1 \,|\, \ldots \,|\, \mathbf{e}_n])$$

where $\sigma(1), \sigma(2), \ldots, \sigma(n)$ vary in $\{1, \ldots, n\}$. Since by (ii) $D(\mathbf{A}) = 0$ if $\mathbf{A}$ has two equal columns, we infer that $\sigma(i) \neq \sigma(j)$ if $i \neq j$, i.e., that σ is a permutation of $(1, 2, \ldots, n)$. Since $D([\mathbf{e}_{\sigma(1)} \,|\, \ldots \,|\, \mathbf{e}_{\sigma(n)}]) = (-1)^\sigma D([\mathbf{e}_1 \,|\, \ldots \,|\, \mathbf{e}_n])$ and $D([\mathbf{e}_1 \,|\, \ldots \,|\, \mathbf{e}_n]) = 1$, we conclude that $D(\mathbf{A})$ agrees with the right-hand side of (1.20), hence $D(\mathbf{A}) = \det \mathbf{A}$. $\square$

The determinant can also be computed by means of an inductive formula.

1.55 Definition. *Let* $\mathbf{A} = [a_j^i] \in M_{n,n}(\mathbb{K})$, $n \geq 1$. *A* r-*minor of* $\mathbf{A}$ *is a* $r \times r$ *submatrix of* A, *that is a matrix obtained by choosing the common entries of a choice of* r *rows and* r *columns of* $\mathbf{A}$ *and relabeling the indices from 1 to* r. *For* $i, j = 1, \ldots, n$ *we define the* complementing (i, j)-*minor of the matrix* $\mathbf{A}$, *denoted by* $\mathbf{M}_i^j(\mathbf{A})$, *as the* $(n - 1) \times (n - 1)$-*minor obtained by removing the* ith *row and the* jth *column from* $\mathbf{A}$.

1.56 Theorem (Laplace). *Let* $\mathbf{A} \in M_{n,n}(\mathbb{K})$, $n \geq 1$. *Then*

$$\det \mathbf{A} := \begin{cases} \mathbf{A} & \text{if } n = 1, \\ \sum_{j=1}^n (-1)^{j+1} a_j^1 \det \mathbf{M}_j^1(\mathbf{A}) & \text{if } n > 1. \end{cases} \tag{1.21}$$

Proof. Denote by $D(\mathbf{A})$ the right-hand side of (1.21). Let us prove that $D(\mathbf{A})$ fulfills the conditions (i), (ii) and (iii) of Theorem 1.54, thus $D(A) = \det \mathbf{A}$. The conditions (i) and (ii) of Theorem 1.54 are trivially fulfilled by $D(\mathbf{A})$. Let us also show that (iii) holds, i.e., if $\mathbf{a}_j = \mathbf{a}_{j+1}$ for some j, then $D(\mathbf{A}) = 0$. We proceed by induction on j. By the induction step, $\det \mathbf{M}_h^1(\mathbf{A}) = 0$ for $h \neq j, j+1$, hence $D(\mathbf{A}) = (-1)^{j+1} a_j^1 \det \mathbf{M}_j^1(\mathbf{A}) + (-1)^j a_{j+1}^1 \det \mathbf{M}_{j+1}^1(\mathbf{A})$. Since $a_j^1 = a_{j+1}^1$, and, consequently, $\mathbf{M}_j^1(\mathbf{A}) = \mathbf{M}_{j+1}^1(\mathbf{A})$, we conclude that $D(\mathbf{A}) = 0$. $\square$

From (1.20) we immediately infer the following.

1.57 Theorem (Determinant of the transpose). *We have*

$$\det \mathbf{A}^T = \det \mathbf{A} \qquad \text{for all } \mathbf{A} \in M_{n,n}(\mathbb{K}).$$

One then shows the following important theorem.

1.58 Theorem (Binet's formula). *Let* $\mathbf{A}$ *and* $\mathbf{B}$ *be two* $n \times n$ *matrices. Then*

$$\det(\mathbf{BA}) = \det \mathbf{B} \det \mathbf{A}.$$

Proof. Let $\mathbf{A} = [a_j^i] = [\mathbf{a}_1 \,|\, \dots \,|\, \mathbf{a}_n]$, $\mathbf{B} = [b_j^i] = [\mathbf{b}_1 \,|\, \dots \,|\, \mathbf{b}_n]$ and let $(\mathbf{e}_1, \dots, \mathbf{e}_n)$ be the canonical basis of $\mathbb{K}^n$. Since

$$\sum_{j=1}^n (\mathbf{BA})_i^j \mathbf{e}_j = \sum_{j,r=1}^n b_r^j a_i^r \mathbf{e}_j = \sum_{r=1}^n a_i^r \mathbf{b}_r,$$

we have

$$\det(\mathbf{BA}) = \det\left(\left[\sum_{r=1}^n a_1^r \mathbf{b}_r \,\middle|\, \dots \,\middle|\, \sum_{r=1}^n a_n^r \mathbf{b}_r\right]\right)$$

$$= \sum_{\sigma \in \mathcal{P}_n} a_{\sigma(1)}^1 a_{\sigma(2)}^2 \cdots a_{\sigma(n)}^n \det[\mathbf{b}_{\sigma(1)} \,|\, \dots \,|\, \mathbf{b}_{\sigma(n)}]$$

$$= \sum_{\sigma \in \mathcal{P}_n} (-1)^\sigma a_{\sigma(1)}^1 a_{\sigma(2)}^2 \cdots a_{\sigma(n)}^n \det \mathbf{B} = \det \mathbf{A} \det \mathbf{B}.$$

$\square$

As stated in the beginning, the determinant gives us a criterion to decide whether a matrix is nonsingular or, equivalently, whether n vectors are linearly independent.

1.59 Theorem. *A* $n \times n$ *matrix* $\mathbf{A}$ *is nonsingular if and only if* $\det \mathbf{A} \neq 0$.

Proof. If $\mathbf{A}$ is nonsingular, there is a $\mathbf{B} \in M_{n,n}(\mathbb{K})$ such that $\mathbf{AB} = \mathrm{Id}_n$, see Exercise 1.27; by Binet's formula $\det \mathbf{A} \det \mathbf{B} = 1$. In particular $\det \mathbf{A} \neq 0$.

Conversely, if the columns of $\mathbf{A}$ are linearly dependent, then it is not difficult to see that $\det \mathbf{A} = 0$ by using Theorem 1.54. $\square$

Let $\mathbf{A} = [a_j^i]$ be an $m \times n$ matrix. We say that the *characteristic* of $\mathbf{A}$ is r if all p-minors with $p > r$ have zero determinant and there exists a r-minor with nonzero determinant.

1.60 Theorem (Kronecker). *The rank and the characteristic of a matrix are the same.*

Proof. Let $\mathbf{A} \in M_{m,n}(\mathbb{K})$ and let $r := \mathrm{Rank}\,\mathbf{A}$. For any minor $\mathbf{B}$, trivially $\mathrm{Rank}\,\mathbf{B} \leq \mathrm{Rank}\,\mathbf{A} = r$, hence every p-minor is singular, i.e., has zero determinant, if $p > n$. On the other hand, Theorem 1.28 implies that there exists a nonsingular r-minor $\mathbf{B}$ of $\mathbf{A}$, hence with $\det \mathbf{B} \neq 0$. $\square$

The defining inductive formula (1.21) requires us to compute the determinant of the complementing minors of the elements of the first row; on account of the alternance, we can use any row, and on account of Theorem 1.57, we can use any column. More precisely,

1.61 Theorem (Laplace's formulas). *Let* $\mathbf{A}$ *be an* $n \times n$ *matrix. We have for all* $h, k = 1, \ldots, n$

$$\delta_{kh} \det \mathbf{A} = \sum_{j=1}^{n} (-1)^{h+j} a_j^h \det \mathbf{M}_j^k(\mathbf{A}),$$

$$\delta_{kh} \det \mathbf{A} = \sum_{i=1}^{n} (-1)^{i+h} a_k^i \det \mathbf{M}_h^i(\mathbf{A}),$$

where δ_{hk} *is Kronecker's symbol.*

1.62 ¶. To compute the determinant of a square $n \times n$ matrix $\mathbf{A}$ we can use a Gauss reduced matrix $\mathbf{G_A}$ of $\mathbf{A}$. Show that $\det \mathbf{A} = (-1)^{\sigma} \prod_{i=1}^{n} (\mathbf{G_A})_i^i$ where σ is the permutation of rows needed to compute $\mathbf{G_A}$, and the product is the product of the pivots.

It is useful to rewrite Laplace's formulas using matrix multiplication. Denote by $\mathbf{cof}(\mathbf{A}) = [c_j^i]$ the square $n \times n$ matrix, called the *matrix of cofactors* of $\mathbf{A}$, defined by

$$c_j^i := (-1)^{i+j} \det \mathbf{M}_i^j(\mathbf{A}).$$

Notice the exchange between the row and column indices: the (i, j) entry of $\mathbf{cof}(\mathbf{A})$ is $(-1)^{i+j}$ times the determinant of the complementing (j, i)-minor. Using the cofactor matrix, Laplace's formulas in Theorem 1.61 rewrite in matrix form as

1.63 Theorem (Laplace's formulas). *Let* $\mathbf{A}$ *be an* $n \times n$ *matrix. Then we have*

$$\mathbf{cof}(\mathbf{A}) \, \mathbf{A} = \mathbf{A} \, \mathbf{cof}(\mathbf{A}) = \det \mathbf{A} \, \mathrm{Id}_n. \tag{1.22}$$

We immediately infer the following.

1.64 Proposition. *Let* $\mathbf{A} = [\mathbf{a}_1 \,|\, \mathbf{a}_2 \,|\, \ldots \,|\, \mathbf{a}_n] \in M_{n,n}(\mathbb{K})$ *be nonsingular.*

(i) *We have*

$$\mathbf{A}^{-1} = \frac{1}{\det \mathbf{A}} \mathbf{cof}(\mathbf{A}).$$

(ii) (CRAMER'S RULE) *The system* $\mathbf{A}\mathbf{x} = \mathbf{b}$, $\mathbf{b} \in \mathbb{K}^n$, *has a unique solution given by*

$$\mathbf{x} = (x^1, x^2, \ldots, x^n), \qquad x^i = \frac{\det \mathbf{B}_i}{\det \mathbf{A}},$$

where

$$\mathbf{B}_i := \Big[\mathbf{a}_1 \,\Big|\, \ldots \,\Big|\, \mathbf{a}_{i-1} \,\Big|\, \mathbf{b} \,\Big|\, \mathbf{a}_{i+1} \,\Big|\, \ldots \,\Big|\, \mathbf{a}_n \Big].$$

Proof. (i) follows immediately from (1.22). (ii) follows from (i), but it is better shown using linearity and the alternating property of the determinant. In fact, solving $\mathbf{A}\mathbf{x} = \mathbf{b}$ is equivalent to finding $\mathbf{x} = (x^1, x^2, \ldots, x^n)$ such that $\mathbf{b} = \sum_{i=1}^{n} x^i \mathbf{a}_i$. Now, linearity and the alternating property of the determninant yield

$$\det \mathbf{B}_i = \det \left[\ldots \left| \mathbf{a}_{i-1} \right| \sum_{j=1}^{n} x^j \mathbf{a}_j \left| \mathbf{a}_{i+1} \right| \ldots \right] = \sum_{j=1}^{n} x^j \det \left[\ldots \left| \mathbf{a}_{i-1} \right| \mathbf{a}_j \left| \mathbf{a}_{i+1} \right| \ldots \right].$$

Since the only nonzero addend on the right-hand side is the one with $j = i$, we infer

$$\det \mathbf{B}_i = x^i \det \left[\mathbf{a}_1 \left| \ldots \right| \mathbf{a}_{i-1} \left| \mathbf{a}_i \right| \mathbf{a}_{i+1} \left| \ldots \right| \mathbf{a}_n \right] = x^i \det \mathbf{A}.$$

$\square$

1.65 ¶. Show that $\det \mathbf{cof}(\mathbf{A}) = (\det \mathbf{A})^{n-1}$.

1.5 Exercises

1.66 ¶. Find the values of $x, y \in \mathbb{R}$ for which the three vectors $(1, 1, 1)$, $(1, x, x^2)$, $(1, y, y^2)$ form a basis of $\mathbb{R}^3$.

1.67 ¶. Let $\alpha_1, \alpha_2 \in \mathbb{C}$ be distinct and nonzero. Show that $e^{\alpha_1 t}$, $e^{\alpha_2 t}$, $t \in \mathbb{R}$, are linearly independent on $\mathbb{C}$. [*Hint:* See [GM2] Corollary 5.54.]

1.68 ¶. Write the parametric equation of a straight line
  (i) through $\mathbf{b} = (1, 1, 1)$ and with direction $\mathbf{a} = (1, 0, 0)$,
  (ii) through $\mathbf{a} = (1, 1, 1)$ and $\mathbf{b} = (1, 0, 0)$.

1.69 ¶. Describe in a parametric or implicit way in $\mathbb{R}^3$,
  ○ a straight line through two points,
  ○ the intersection of a plane and a straight line,
  ○ a straight line that is parallel to a given plane,
  ○ a straight line on a plane,
  ○ a plane through three points,
  ○ a plane through a point containing a given straight line,
  ○ a plane perpendicular to a straight line.

1.70 ¶ Affine transformations. An *affine transformation* $\varphi : \mathbb{K}^n \to \mathbb{K}^m$ is a map of the type $\varphi(\mathbf{x}) := L(\mathbf{x}) + \mathbf{q}_0$ where $L : \mathbb{K}^n \to \mathbb{K}^m$ is linear and $\mathbf{q}_0 \in \mathbb{K}^m$. Show that φ is an affine transformation if and only if φ maps straight lines onto straight lines.

1.71 ¶. Let P_1 and P_2 be two $(n-1)$-planes in $\mathbb{R}^n$. Show that either $P_1 = P_2$ or $P_1 \cap P_2 = \emptyset$ or $P_1 \cap P_2$ has dimension $n - 2$.

1.72 ¶. In $\mathbb{R}^4$ find
  (i) two 2-planes through the origin that meet only at the origin,
  (ii) two 2-planes through the origin that meet along a straight line.

1.73 ¶. In $\mathbb{R}^2$ write the 2×2 matrix associated with the counterclockwise rotations of angles $\pi/2$, π, $3\pi/2$, and, in general, $\theta \in \mathbb{R}$.

1.74 ¶. Write the matrix associated with the axial symmetry in $\mathbb{R}^3$ and to plane symmetries.

1.75 ¶. Write down explicit linear systems of 3, 4, 5 equations with 4 or 5 unknowns, and use the Gauss elimination procedure to solve them.

1.76 ¶. Let $\mathbf{A} \in M_{n,n}(\mathbb{K})$. Show that if $\mathbf{AB} = 0\ \forall \mathbf{B} \in M_{n,n}(\mathbb{K})$, then $\mathbf{A} = 0$.

1.77 ¶. Let $\mathbf{A} = \begin{pmatrix} 1 & -1 & 2 \\ 0 & -1 & 3 \end{pmatrix}$ and $\mathbf{B} = \begin{pmatrix} 3 & -1 & 2 \\ 5 & 2 & 3 \end{pmatrix}$. Compute $\mathbf{A} + \mathbf{B}, \sqrt{2}\mathbf{A} + \mathbf{B}$.

1.78 ¶. Let

$$\mathbf{A} = \begin{pmatrix} 3 & 3 & 2 & 3 \\ 2 & 2 & 0 & 2 \\ -1 & 0 & 1 & 1 \end{pmatrix} \qquad \mathbf{B} = \begin{pmatrix} 2 & -1 \\ 3 & 2 \\ 1 & -1 \\ 2 & 5 \end{pmatrix}.$$

Compute $\mathbf{AB}, \mathbf{BB}^T, \mathbf{B}^T\mathbf{B}$.

1.79 ¶. Let

$$\mathbf{A} = \begin{pmatrix} 0 & 0 & 0 \\ 0 & 1 & 0 \\ 0 & 0 & 0 \end{pmatrix} \qquad \mathbf{B} = \begin{pmatrix} 1 & 0 & 0 \\ 0 & 0 & 0 \\ 0 & 0 & 1 \end{pmatrix}.$$

Show that $\mathbf{AB} = 0$.

1.80 ¶. Let $\mathbf{A}, \mathbf{B} \in M_{n,n}(\mathbb{K})$. Show that if $\mathbf{AB} = 0$ and $\mathbf{A}$ is invertible, then $\mathbf{B} = 0$.

1.81 ¶. Let

$$\mathbf{A} := \begin{pmatrix} 0 & 1 \\ -1 & 0 \end{pmatrix}, \qquad \mathbf{B} := \begin{pmatrix} 0 & i \\ i & 0 \end{pmatrix}, \qquad \mathbf{C} := \begin{pmatrix} i & 0 \\ 0 & -i \end{pmatrix}.$$

Show that
- $\mathbf{A}^2 = \mathbf{B}^2 = \mathbf{C}^2 = -\operatorname{Id}$,
- $\mathbf{AB} = -\mathbf{BA} = \mathbf{C}$,
- $\mathbf{BC} = -\mathbf{CB} = \operatorname{Id}$,
- $\mathbf{CA} = -\mathbf{AC} = \mathbf{B}$.

1.82 ¶. Let $\mathbf{A}, \mathbf{B} \in M_{n,n}$. We say that $\mathbf{A}$ is *symmetric* if $\mathbf{A} = \mathbf{A}^T$. Show that, if $\mathbf{A}$ is symmetric, then $\mathbf{AB}$ is symmetric if and only if $\mathbf{A}$ and $\mathbf{B}$ *commute*, i.e., $\mathbf{AB} = \mathbf{BA}$.

1.83 ¶. Let $\mathbf{M} \in M_{n,n}(\mathbb{K})$ be an upper triangular matrix with all entries in the principal diagonal equal to 1. Suppose that for some k we have $\mathbf{M}^k = \mathbf{M}\,\mathbf{M} \cdots \mathbf{M} = \operatorname{Id}_n$. Show that $\mathbf{M} = \operatorname{Id}_n$.

1.84 ¶. Let $\mathbf{A}, \mathbf{B} \in M_{n,n}(\mathbb{K})$. In general $\mathbf{AB} \neq \mathbf{BA}$. The $n \times n$ matrix $[\mathbf{A}, \mathbf{B}] := \mathbf{AB} - \mathbf{BA}$ is called the *commutator* or the *Lie bracket* of $\mathbf{A}$ and $\mathbf{B}$. Show that
  (i) $[\mathbf{A}, \mathbf{B}] = -[\mathbf{B}, \mathbf{A}]$,
  (ii) (JACOBI'S IDENTITY) $[[\mathbf{A}, \mathbf{B}], \mathbf{C}] + [[\mathbf{B}, \mathbf{C}], \mathbf{A}] + [[\mathbf{C}, \mathbf{A}], \mathbf{B}] = 0$,
  (iii) the trace of $[\mathbf{A}, \mathbf{B}]$ is zero. The *trace* of a $n \times n$ matrix $\mathbf{A} = [a_j^i]$ is defined as $\operatorname{tr} \mathbf{A} := \sum_i a_i^i = 0$.

1.85 ¶. Let $\mathbf{A} \in M_{n,n}$ be diagonal. Show that $\mathbf{B}$ is diagonal if and only if $[\mathbf{A}, \mathbf{B}] = 0$.

1.86 ¶ Block matrices. Write a $n \times n$ matrix as

$$\mathbf{A} = \begin{pmatrix} \mathbf{A}_1^1 & \mathbf{A}_2^1 \\ \mathbf{A}_1^2 & \mathbf{A}_2^2 \end{pmatrix}$$

where $\mathbf{A}_1^1$ is the submatrix of the first k rows and h columns, $\mathbf{A}_2^1$ is the submatrix of the first k rows and $n - h$ columns, etc. Show that

$$\begin{pmatrix} \mathbf{A}_1^1 & \mathbf{A}_2^1 \\ \mathbf{A}_1^2 & \mathbf{A}_2^2 \end{pmatrix} \begin{pmatrix} \mathbf{B}_1^1 & \mathbf{B}_2^1 \\ \mathbf{B}_1^2 & \mathbf{B}_2^2 \end{pmatrix} = \begin{pmatrix} \mathbf{A}_1^1\mathbf{B}_1^1 + \mathbf{A}_2^1\mathbf{B}_1^2 & \mathbf{A}_1^1\mathbf{B}_2^1 + \mathbf{A}_2^1\mathbf{B}_2^2 \\ \mathbf{A}_1^2\mathbf{B}_1^1 + \mathbf{A}_2^2\mathbf{B}_1^2 & \mathbf{A}_1^2\mathbf{B}_2^1 + \mathbf{A}_2^2\mathbf{B}_2^2 \end{pmatrix}.$$

1.87 ¶. Let $\mathbf{A} \in M_{k,k}(\mathbb{K})$, $\mathbf{B} \in M_{n,n}(\mathbb{K})$ and

$$\mathbf{C} = \begin{pmatrix} \mathbf{A} & 0 \\ 0 & \mathbf{B} \end{pmatrix}.$$

Compute $\det \mathbf{C}$.

1.88 ¶. Let $\mathbf{A} \in M_{k,k}(\mathbb{K})$, $\mathbf{B} \in M_{n,n}(\mathbb{K})$, $\mathbf{C} \in M_{k,n}(\mathbb{K})$ and

$$\mathbf{M} = \begin{pmatrix} \mathbf{A} & \mathbf{C} \\ 0 & \mathbf{B} \end{pmatrix}.$$

Compute $\det \mathbf{M}$.

1.89 ¶ Vandermonde determinant. Let $\lambda_1, \lambda_2, \ldots, \lambda_n \in \mathbb{K}$ and

$$\mathbf{A} := \begin{pmatrix} 1 & 1 & 1 & \cdots & 1 \\ \lambda_1 & \lambda_2 & \lambda_3 & \cdots & \lambda_n \\ \lambda_1^2 & \lambda_2^2 & \lambda_3^2 & \cdots & \lambda_n^2 \\ \lambda_1^3 & \lambda_2^3 & \lambda_3^3 & \cdots & \lambda_n^3 \\ \vdots & \vdots & \vdots & \ddots & \vdots \\ \lambda_1^n & \lambda_2^n & \lambda_3^n & \cdots & \lambda_n^n \end{pmatrix}.$$

Prove that $\det \mathbf{A} = \prod_{i<j}(\lambda_i - \lambda_j)$. [*Hint:* Proceed by induction on n. Notice that $\det \mathbf{A}$ is a polynomial in λ_n and use the principle of identity for polynomials.]

1.90 ¶. Compute the rank of the following matrices

$$\begin{pmatrix} 2 & 1 & 3 & 1 \\ 2 & 1 & -3 & 1 \\ 3 & 3 & 1 & -1 \\ 5 & 4 & -2 & 0 \end{pmatrix} \qquad \begin{pmatrix} 2 & 3 & 1 & 3 \\ 3 & 1 & -1 & 2 \\ -1 & 2 & 2 & 1 \\ 1 & 5 & 3 & 4 \end{pmatrix} \qquad \begin{pmatrix} 3 & 3 & 3 & 1 \\ 1 & 3 & 3 & 3 \\ 3 & 3 & 3 & 3 \\ 1 & 1 & 1 & 1 \end{pmatrix}.$$

1.91 ¶. Solve the following linear systems

$$\begin{cases} 3x - y + 2z + t = 1, \\ x + 2y - z + 2t = 2, \\ x - 5y + 4z - 3t = 1, \end{cases} \qquad \begin{cases} 2x + 4y + 3z - 2t = 3, \\ 2x + 2y - 3z + 3t = 3, \\ x + 2y - z + 3t = 2, \\ x - 3y + 2z + 2t = -4, \\ 4x + y - 2z + 8t = 1. \end{cases}$$

2. Vector Spaces and Linear Maps

The linear structure of $\mathbb{K}^n$ is shared by several mathematical objects. We have already noticed that the set of $m \times n$ matrices satisfies the laws of sum and multiplication by scalars. The aim of this chapter is to introduce abstract language and illustrate some facts related to *linear structure*. In particular, we shall see that in every finite-dimensional vector space we can introduce the coordinates related to a basis and explain how the coordinates description of intrinsic objects changes when we change the coordinates, i.e., the basis.

2.1 Vector Spaces and Linear Maps

a. Definition

Let $\mathbb{K}$ be a commutative field, here it will be either $\mathbb{R}$ or $\mathbb{C}$.

2.1 Definition. *A vector space over the field $\mathbb{K}$ is a set X endowed with*

(i) *an operation $+ : X \times X \to X$, called the* sum, *that makes X a commutative group, i.e.,*
   a) *$(x + y) + z = x + (y + z)$, $x + y = y + x$, $\forall x, y \in X$,*
   b) *there exists an element $0 \in X$ called the zero element, such that $x + 0 = 0 + x = x$ $\forall x \in X$,*
   c) *for every $x \in X$ there exists $-x \in X$ such that $x + (-x) = 0$,*
(ii) *an operation of* multiplication by a scalar $\cdot : \mathbb{K} \times X \to X$ *that associates to every $\lambda \in \mathbb{K}$ and $x \in X$ an element of X denoted by λx such that*
   a) *$\lambda(x + y) = \lambda x + \lambda y$, $(\lambda + \mu)x = \lambda x + \mu x$,*
   b) *$\lambda(\mu x) = (\lambda \mu)x$, $1 \cdot x = x$.*

In particular, $(-1)x = -x$ $\forall x \in X$; we therefore write $x - y$ instead of $x + (-y)$.

The elements of a vector space over $\mathbb{K}$ are called *vectors*, and the elements of $\mathbb{K}$ are called *scalars*. The product of a vector by scalars allows us to regard a vector at all scales.

2.2 Example. As we have seen, $\mathbb{K}^n$ for $n \geq 1$, and all the linear subspaces of $\mathbb{K}^n$ are vector spaces over $\mathbb{K}$. Also, the space of $m \times n$ matrices with entries in $\mathbb{K}$, $M_{m,n}(\mathbb{K})$, is a vector space over $\mathbb{K}$, with the two operations of sum of matrices and multiplication of a matrix by a scalar, see Section 1.2.

2.3 Example. Let X be any set. Then the class $\mathcal{F}(X, \mathbb{K})$ of all functions $\varphi : X \to \mathbb{K}$ is a vector space with the two operations of sum and multiplication by scalars defined by

$$(\varphi + \psi)(x) := \varphi(x) + \psi(x), \qquad (\lambda\varphi)(x) := \lambda\varphi(x) \qquad \forall x \in X.$$

Several subclasses of functions are vector spaces, actually linear subspaces of $\mathcal{F}(X, \mathbb{K})$. For instance,

o the set $C^0([0,1], \mathbb{R})$ of all continuous functions $\varphi : [0,1] \to \mathbb{R}$, the set of k-differentiable functions from $[0,1]$ into $\mathbb{R}$, the set $C^k([0,1], \mathbb{R})$ of all functions with continuous derivatives up to the order k, the set $C^\infty([0,1], \mathbb{R})$ of infinitely differentiable functions,

o the set of polynomials of degree less than k, the set of all polynomials,

o the set of all complex trigonometric polynomials,

o the set of Riemann summable functions in $]0,1[$,

o the set of all sequences with values in $\mathbb{K}$.

We now begin the study of properties that depend only on the linear structure of a vector space, independently of specific examples.

b. Subspaces, linear combinations and bases

2.4 Definition. *A subset W of a vector space X is called a* linear subspace, *or shortly a* subspace *of X, if*

   (i) $0 \in W$,
   (ii) $\forall \ x, y \in W$ *we have* $x + y \in W$,
   (iii) $\forall \ x \in W$ *and* $\forall \ \lambda \in \mathbb{K}$ *we have* $\lambda x \in W$.

Obviously the element 0 is the zero element of X and the operations of sum and multiplication by scalars are as those in X.

In a vector space we may consider the *finite linear combinations* of elements of X with coefficients in $\mathbb{K}$, i.e.,

$$\sum_{i=1}^{n} \lambda^i v_i \in X$$

where $\lambda^1, \lambda^2, \ldots, \lambda^n \in \mathbb{K}$ and $v_1, v_2, \ldots, v_n \in X$. Notice that we have indexed both the vectors and the relative coefficients, and we use the standard notation on the indices: a list of vectors has lower indices and a list of coefficients has upper indices.

It is readily seen that a subset $W \subset X$ is a subspace of X if and only if all finite linear combinations of elements of X with coefficients in $\mathbb{K}$ belong to W. Moreover, given a set $S \subset X$, the family of all finite linear combinations of elements of S is a subspace of X called the *span* of S and denoted by $\mathrm{Span}\,S$.

We say that a finite number of vectors are *linearly dependent* if there are scalars, not all zero, such that

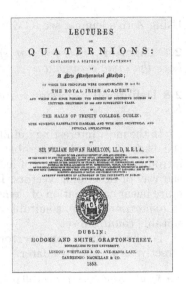

Figure 2.1. Arthur Cayley (1821–1895) and the *Lectures on Quaternions* by William R. Hamilton (1805–1865).

$$\lambda^1 v_1 + \cdots + \lambda^n v_n = 0,$$

or, in other words, if one vector is a linear combination of the others. If n vectors are not linearly dependent, we say that they are *linearly independent*. More generally, we say that *a set S of vectors is a set of linearly independent vectors* whenever any finite list of elements of S is made by linearly independent vectors. Of course linearly independent vectors are distinct and nonzero.

2.5 Definition. *Let X be a vector space. A set S of linearly independent vectors such that* $\operatorname{Span} S = X$ *is called a* basis *of X.*

A set $\mathcal{A} \subset X$ is a maximal independent set *of X if $\mathcal{A}$ is a set of linearly independent vectors and, whenever we add to it a vector $w \in X \backslash \mathcal{A}$, $\mathcal{A} \cup \{w\}$ is not a set of linearly independent vectors.*

Thus a basis of X is a subset $S \subset X$ such that

(i) every $x \in X$ is a finite linear combination of some elements of S. Equivalently, for every $x \in X$ there is a map $\lambda : S \to \mathbb{K}$ such that $x = \sum_{v \in S} \lambda(v)v$ and $\lambda(v) = 0$ except for a finite number of elements (depending on x) of S,

(ii) each finite subset of S is a set of linearly independent vectors.

It is easy to prove that for every $x \in X$ the representation $x = \sum_{v \in S} \lambda(v)v$ is unique if S is a basis of X.

Using the same proof as in Proposition 1.5 we then infer

2.6 Proposition. *Let X be a vector space over $\mathbb{K}$. Then $S \subset X$ is a basis of X if and only if S is a maximal independent set.*

Using Zorn's lemma, see [GM2], one can also show the following.

2.7 Theorem. *Every vector space X has a basis. Moreover, two bases have the same cardinality.*

2.8 Definition. *A vector space X is* finite dimensional *if X has a finite basis.*

In the most interesting infinite-dimensional vector spaces, one can show the basis has nondenumerable cardinality. Later, we shall see that the introduction of the notion of limit, i.e., of a new structure on X, improves the way of describing vectors. Instead of trying to see every $x \in X$ as a finite linear combination of elements of a nondenumerable basis, it is better to *approximate* it by a suitable sequence of *finite* linear combinations of a suitable *countable* set.

For finite-dimensional vector spaces, Theorem 2.7 can be proved more directly, as we shall see later.

2.9 ¶. Show that the space of all polynomials and $C^0([0,1],\mathbb{R})$ are infinite-dimensional vector spaces.

c. Linear maps

2.10 Definition. *Let X and Y be two vector spaces over $\mathbb{K}$. A map $\varphi : X \to Y$ is called $\mathbb{K}$-linear, or* linear *for short, if*

$$\varphi(x + y) = \varphi(x) + \varphi(y) \qquad and \qquad \varphi(\lambda x) = \lambda\varphi(x)$$

for any $x, y \in X$ and $\lambda \in \mathbb{K}$.

A linear map that is injective and surjective is called a (linear) isomorphism.

Of course, if $\varphi : X \to Y$ is linear, we have $\varphi(0) = 0$ and, by induction,

2.11 Proposition. *Let $\varphi : X \to Y$ be linear. Then*

$$\varphi\left(\sum_{i=1}^{k} \lambda^i e_i\right) = \sum_{i=1}^{k} \lambda^i \varphi(e_i)$$

for any $\lambda^1, \lambda^2, \ldots, \lambda^n \in \mathbb{K}$ and $e_1, e_2, \ldots, e_n \in X$. In particular, a linear map is fixed by the values it takes on a basis.

The space of linear maps $\varphi : X \to Y$ between two vector spaces X and Y, denoted by $\mathcal{L}(X, Y)$, is a vector space over $\mathbb{K}$ with the operations of sum and multiplication by scalars defined in terms of the operations on Y by

$$(\varphi + \psi)(x) := \varphi(x) + \psi(y), \qquad (\lambda\varphi)(x) = \lambda\varphi(x)$$

for all $\varphi, \psi \in \mathcal{L}(X, Y)$ and $\lambda \in \mathbb{R}$. Notice also that the composition of linear maps is again a linear map, and that, if $\varphi : X \to Y$ is an isomorphism, then the inverse map $\varphi^{-1} : Y \to X$ is also an isomorphism.

It is easy to check the following.

2.12 Proposition. *Let $\varphi : X \to Y$ be a linear map.*

(i) *If $S \subset X$ spans $W \subset X$, then $\varphi(S)$ spans $\varphi(W)$.*

(ii) *If $e_1, e_2, \ldots, e_n$ are linearly dependent in X, then $\varphi(e_1), \ldots, \varphi(e_n)$ are linearly dependent in Y.*

(iii) *φ is injective if and only if any list $(e_1, e_2, \ldots, e_n)$ of linearly independent vectors in X is mapped into a list $(\varphi(e_1), \varphi(e_2), \ldots, \varphi(e_n))$ of linearly independent vectors in X.*

(iv) *The following claims are equivalent*
   a) φ is an isomorphism,
   b) $S \subset X$ is a basis of X if and only if $\varphi(S)$ is a basis of Y.

2.13 ¶. Show that the following maps are linear

(i) the derivation map $D : C^1([0,1]) \to C^0([0,1])$ that maps a C^1-function into its derivative, $f \to f'$.

(ii) the map that associates to every function of class $C^0([0,1])$ its integral over $[0,1]$,

$$f \to \int_0^1 f(t)\,dt,$$

(iii) the primitive map $C^0([0,1]) \to C^1([0,1])$ that associates to every continuous function the primitive function

$$f(x) \to F(x) := \int_0^x f(t)\,dt.$$

2.14 Definition. *Let $\varphi : X \to Y$ be a linear map. The* kernel of φ *and the* image of φ *are respectively*

$$\ker \varphi :- \Big\{ x \in X \,\Big|\, \varphi(x) - 0 \Big\},$$

$$\operatorname{Im} \varphi := \Big\{ y \in Y \,\Big|\, \exists\, x \in X : \varphi(x) = y \Big\}.$$

It is easily seen that $\ker \varphi$ is a linear subspace of the source space X and that $\ker \varphi = \{0\}$ if and only if φ is injective. Also, $\operatorname{Im} \varphi$ is a linear subspace of the target space Y, and $\operatorname{Im} \varphi = Y$ if and only if φ is surjective. If $\operatorname{Im} \varphi$ has finite dimension, its dimension is called the *rank* of φ and denoted by $\operatorname{Rank} \varphi$. Of course φ is surjective if and only if $\dim Y = \operatorname{Rank} \varphi$, provided $\dim \operatorname{Im} \varphi < +\infty$.

d. Coordinates in a finite-dimensional vector space

Let X be a finite-dimensional vector space over $\mathbb{K}$ and let $(e_1, e_2, \ldots, e_n)$ be an *ordered* basis on X. Then every vector $x \in X$ writes uniquely as $x = \sum_{i=1}^n x^i e_i$, where $x_1, x_2, \ldots, x_n \in \mathbb{K}$. Then $(e_1, e_2, \ldots, e_n)$ defines a map $\mathcal{E} : X \to \mathbb{K}^n$ characterized by

$$\mathcal{E}(x) = \mathbf{x} = (x^1, x^2, \ldots, x^n) \quad \text{if and only if} \quad x = \sum_{i=1}^n x^i e_i.$$

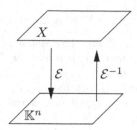

Figure 2.2. Coordinate system in a finite-dimensional vector space.

It is trivial to verify that $\mathcal{E}$ is linear, injective and surjective, hence an isomorphism, together with its inverse

$$\mathcal{E}^{-1} : \mathbb{K}^n \to X, \qquad \mathcal{E}^{-1}(\mathbf{x}) = \sum_{i=1}^{n} x^i e_i, \quad \mathbf{x} = (x^1, x^2, \dots, x^n).$$

We call $\mathcal{E}$ the *coordinate system* related to the ordered basis $(e_1, e_2, \dots, e_n)$ and refer to $\mathcal{E}(x)$ as to the *coordinate vector* of x with respect to the basis $(e_1, e_2, \dots, e_n)$. Notice that $\mathcal{E}$ maps e_i to the ith vector $\mathbf{e}_i$ of the canonical basis of $\mathbb{K}^n$.

Also notice that ordered bases and isomorphims $\mathcal{E} : X \to \mathbb{K}^n$ are in one-to-one correspondence. In particular, any isomorphism $\mathcal{E} : X \to \mathbb{K}^n$ is a coordinate system related to a suitable basis. In fact, the vectors $e_1, e_2, \dots, e_n$ of X defined for $i = 1, \dots, n$ by $e_i := \mathcal{E}^{-1}(\mathbf{e}_i)$ form a basis of X by Proposition 2.12 (iii) and it is easy to check that

$$\mathcal{E}(x) = \mathbf{x} = (x^1, x^2, \dots, x^n) \qquad \text{if and only if} \qquad x = \sum_{i=1}^{n} x^i e_i.$$

The use of a basis, or, equivalently, of a coordinate system, allows us to transfer definitions and results in $\mathbb{K}^n$ to similar definitions and claims in X. We have the following.

2.15 Proposition. *Let X be a finite-dimensional vector space.*

(i) *Let $(e_1, e_2, \dots, e_n)$ be an ordered basis of X and let $v_1, v_2, \dots, v_p \in X$, $p \le n$, be p linearly independent vectors. Then we can choose $n - p$ elements among $e_1, e_2, \dots, e_n$, say $e_1, e_2, \dots, e_{n-p}$, such that $(v_1, v_2, \dots, v_p, e_1, e_2, \dots, e_{n-p})$ is a basis of X.*

(ii) *Assume that $v_1, v_2, \dots, v_k$ spans X. Possibly eliminating some of the $v_1, v_2, \dots, v_k$, we get a basis of X.*

(iii) *Any two bases of X have the same number of elements.*

The number of elements of a basis of a finite-dimensional space X is called the *dimension* of X and denoted by $\dim X$.

The following corollaries follow from Proposition 2.15.

2.16 Corollary. *Let X be a vector space of dimension n, let $\mathcal{E} : X \to \mathbb{K}^n$ be a coordinate system on X and let W be a subspace of X. Then $E(W)$ is a subspace of $\mathbb{K}^n$ and $\dim W = \dim \mathcal{E}(W)$.*

2.17 Corollary. *Let X be a vector space of dimension n. Then*

  (i) *n linearly independent vectors of X form a basis of X,*
  (ii) *if $k > n$, then k vectors of X are always linearly dependent,*
  (iii) *for every subspace W of X we have $\dim W \le n$,*
  (iv) *let V, W be two subspaces of X. Then $V = W$ if and only if $V \subset W$ and $\dim V = \dim W$.*

Let U and V be two subspaces of a vector space X. Then both $U \cap V$ and

$$U + V := \left\{ x \in X \,\middle|\, x = u + v, \ u \in U, \ v \in V \right\}$$

are linear subspaces of X. When $U \cap V = \{0\}$, we say that $U + V$ is the *direct sum* of U and V and we write $U \oplus V$ instead of $U + V$. Moreover if $X = U \oplus V$, we say that U and V are supplementary. Thus $X = U \oplus V$ means that every $x \in X$ decomposes *uniquely* as $x = u + w$, $u \in U$, $v \in V$.

2.18 Corollary (Grassmann's formula). *Let U and V be two finite-dimensional subspaces of a vector space X. Then*

$$\dim U + \dim V = \dim(U \cap V) + \dim(U + V).$$

2.19 ¶. Show that every n-dimensional vector space is the direct sum of n subspaces of dimension 1.

2.20 ¶. Let $e_1, e_2, \ldots, e_n$ be distinct vectors of X and let $1 < p < n$. Then, trivially, $\operatorname{Span}\{e_1, e_2, \ldots, e_p\} + \operatorname{Span}\{e_{p+1}, \ldots, e_n\} = \operatorname{Span}\{e_1, e_2, \ldots, e_n\}$. Show that, if the e_i's are linearly independent, then

$$\operatorname{Span}\{e_1, e_2, \ldots, e_p\} \oplus \operatorname{Span}\{e_{p+1}, \ldots, e_n\} = \operatorname{Span}\{e_1, e_2, \ldots, e_n\}.$$

2.21 ¶. Let V_1, V_2 be two subspaces of a vector space V of finite dimension and assume that $V = V_1 \oplus V_2$. Then every vector $v \in V$ decomposes uniquely as $v = v_1 + v_2$ with $v_i \in V$. Show that the coordinate maps $\pi_i : V \to V_i$, $i = 1, 2$, $\pi_i(v) = v_i$, are linear.

2.22 ¶. Let $\varphi : X \to Y$ be an isomorphism from X onto Y. Show that $\dim X = \dim Y$.

e. Matrices associated to a linear map

Let X, Y be two vector spaces of dimension n and m respectively. We shall now show that every choice of an oriented basis, equivalently of a coordinate system, in X and Y yields an identification between linear maps and matrices.

Let $(e_1, e_2, \ldots, e_n)$ be an oriented basis in X, $(f_1, f_2, \ldots, f_m)$ be an oriented basis in Y and let $\mathcal{E} : X \to \mathbb{K}^n$, $\mathcal{F} : Y \to \mathbb{K}^m$ be the corresponding

Figure 2.3. Hermann Grassmann (1808–1877) and his *Ausdenungslehre*.

coordinate systems. To every linear map $\ell : X \to Y$ one associates the linear map $L : \mathbb{K}^n \to \mathbb{K}^m$ defined by

$$L := \mathcal{F} \circ \ell \circ \mathcal{E}^{-1}, \tag{2.1}$$

see Figure 2.4, that maps the coordinates of a vector $x \in X$, relative to the basis $(e_1, e_2, \ldots, e_n)$, into the coordinates of $\ell(x) \in Y$, relative to the basis $(f_1, f_2, \ldots, f_m)$, and then, see Proposition 1.18, an $m \times n$ matrix $\mathbf{L}$ such that $L(\mathbf{x}) = \mathbf{L}\mathbf{x}$. We call L and $\mathbf{L}$ respectively, the *map* and the *matrix associated to ℓ using the coordinate systems $\mathcal{E}$ and $\mathcal{F}$*, or, equivalently, *using $(e_1, e_2, \ldots, e_n)$ and $(f_1, f_2, \ldots, f_m)$ as a basis in X and in Y, respectively*.

Since $\mathcal{E}^{-1}$ maps the ith vector $\mathbf{e}_i$ of the canonical basis of $\mathbb{K}^n$ to e_i, $L(\mathbf{e}_i)$ is the coordinate vector of $\ell(e_i)$ in the basis $(f_1, f_2, \ldots, f_m)$, hence $\mathbf{L} = [L_j^i]$ where

$$\ell(e_j) = \sum_{i=1}^{n} L_j^i f_i. \tag{2.2}$$

Equivalently, see Proposition 1.18,

$$\mathbf{L} = \left[L(\mathbf{e}_1) \,\middle|\, L(\mathbf{e}_2) \,\middle|\, \ldots \,\middle|\, L(\mathbf{e}_n) \right].$$

Since $\mathcal{E} : X \to \mathbb{K}^n$ and $\mathcal{F} : Y \to \mathbb{K}^m$ are isomorphisms, we trivially have

$$\mathcal{E}(\ker \ell) = \ker \mathbf{L}, \qquad \mathcal{F}(\operatorname{Im} \ell) = \operatorname{Im} \mathbf{L}.$$

Hence, recalling Theorem 1.25, we have the following.

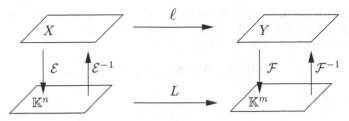

Figure 2.4. The matrix associated to a linear map.

2.23 Theorem (Rank formula). *Let $\ell : X \to Y$ be a linear map between linear spaces. If X is finite dimensional, then*

$$\operatorname{Rank} \ell = \dim \operatorname{Im} \ell = \dim X - \dim \ker \ell.$$

Proof. Let $(e_1, e_2, \ldots, e_n)$ be a basis of X. Then $\operatorname{Im} \ell = \operatorname{Span} \{\ell(e_1), \ldots, \ell(e_n)\}$, hence $\dim \operatorname{Im} \ell < +\infty$. Now choose a basis $(f_1, f_2, \ldots, f_m)$ of $\operatorname{Im} \ell$ and consider the linear associated map $L : \mathbb{K}^n \to \mathbb{K}^m$ using the two bases $(e_1, e_2, \ldots, e_n)$ on X and $(f_1, f_2, \ldots, f_m)$ on $\operatorname{Im} \ell$. Then Theorem 1.25 yields

$$\dim \operatorname{Im} \ell = \dim \operatorname{Im} L = n - \dim \ker L = n - \dim \ker \ell.$$

$\square$

f. The space $\mathcal{L}(X, Y)$

Let X and Y be vector spaces of dimension n and m, and let $(e_1, e_2, \ldots, e_n)$ and $(f_1, f_2, \ldots, f_m)$ be two bases in X and Y. Then (2.2) defines a map $\mathcal{M} : \mathcal{L}(X, Y) \to M_{m,n}(\mathbb{K})$ which is trivially injective and surjective. Since $\mathcal{M}$ is also linear, we deduce that $\mathcal{L}(X, Y)$ and $M_{m,n}(\mathbb{K})$ are isomorphic. In particular, the vector space $\mathcal{L}(X, Y)$ has dimension mn. A basis of $\mathcal{L}(X, Y)$ is given by the mn maps $\{\varphi_j^i\}$, $j = 1, \ldots, n$, $i = 1, \ldots, m$, defined in terms of the bases as

$$\varphi_j^i(e_k) = \delta_k^i f_j = \begin{cases} 0 & \text{if } k \neq i, \\ f_j & \text{if } k = i, \end{cases} \qquad k = 1, \ldots, n.$$

The matrix associated to φ_j^i is the $m \times n$ matrix with all entries 0 except for the entry (i, j) where we have 1. Of course the matrix $\mathcal{M}(\ell)$ associated to ℓ depends on the coordinate systems we use on X and Y. When we want to emphasize such a dependence, we write

$$\mathcal{M}_{\mathcal{E}}^{\mathcal{F}}(\ell)$$

to denote the matrix associated to $\ell : X \to Y$ using the coordinate systems $\mathcal{E}$ on the source space and $\mathcal{F}$ in the target space.

The product of composition of linear maps corresponds to the product of composition of linear maps at the level of coordinates, hence to the product row by columns of the corresponding matrices. More precisely, we have the following.

2.24 Proposition. *Let* $\varphi : X \to Y$ *and* $\psi : Y \to Z$ *be two linear maps, and let* $\mathcal{E} : X \to \mathbb{K}^n$, $\mathcal{F} : Y \to \mathbb{K}^m$, *and* $\mathcal{G} : Z \to \mathbb{K}^k$ *be three systems of coordinates on* X, Y *and* Z. *Then*

$$\mathcal{M}_{\mathcal{E}}^{\mathcal{G}}(\psi \circ \varphi) = \mathcal{M}_{\mathcal{F}}^{\mathcal{G}}(\psi) \mathcal{M}_{\mathcal{E}}^{\mathcal{F}}(\varphi)$$

rows by columns.

Proof. In fact,

$$\mathcal{G} \circ (\psi \circ \varphi) \mathcal{E}^{-1} = (\mathcal{G} \circ \psi \circ \mathcal{F}^{-1}) \circ (\mathcal{F} \circ \varphi \circ \mathcal{E}^{-1}).$$

$\square$

A special case arises if $X = Y = Z$. In this case, the space $\mathcal{L}(X, X)$ of the linear maps from X into itself, also known as the space of *endomorphisms* of X and sometimes denoted by $\mathrm{End}\,(X)$, is closed under the operations of sum, multiplication by scalars and product of composition. We say that $\mathcal{L}(X, X)$ is an *algebra* with respect to these operations and, for any coordinate system $\mathcal{E}$ on X, $\mathcal{M}_{\mathcal{E}}^{\mathcal{E}} : \mathcal{L}(X, X) \to M_{n,n}(\mathbb{K})$ is an isomorphism of algebras. The set of isomorphisms from X into itself, called the *automorphisms* of X and denoted by $\mathrm{Aut}\,(X)$, is a group with respect to the composition. If $\dim X = n$ and $\mathcal{E} : X \to \mathbb{K}^n$ is a coordinate system, then $\mathcal{M}_{\mathcal{E}}^{\mathcal{E}}(\mathrm{Aut}\,(X))$ coincides with the group $GL(n, \mathbb{K})$ of all nonsingular $n \times n$ matrices,

$$GL(n, \mathbb{K}) := \Big\{ \mathbf{L} \in M_{n,n}(\mathbb{K}) \,\Big|\, \det \mathbf{L} \neq 0 \Big\}.$$

g. Linear abstract equations

Let X, Y be two vector spaces over $\mathbb{K}$. A *linear (abstract) equation* in the unknown x is an equation of the form

$$\varphi(x) = y, \tag{2.3}$$

where $\varphi : X \to Y$ is a linear map and $y \in Y$. The equation $\varphi(x) = 0$ is called the associate *homogeneous equation* to (2.3).

Of course, we have

(i) the set of all solutions of the associate linear homogeneous equation $\varphi(x) = 0$ is $\ker \varphi$,
(ii) (2.3) is solvable iff $y \in \mathrm{Im}\,\varphi$,
(iii) (2.3) has at most a unique solution if $\ker \varphi = \{0\}$,
(iv) if $\varphi(x_0) = y$, then the set of all solutions of (2.3) is

$$\Big\{ x \in X \,\Big|\, x - x_0 \in \ker \varphi \Big\}.$$

Taking into account the rank formula, we infer the following.

2.25 Corollary. *Let* X, Y *be finite dimensional of dimension* n *and* m *respectively, and let* $\varphi : X \to Y$ *be a linear map. Then*

(i) *if $m < n$, then* $\dim \ker \varphi > 0$,

(ii) *if $m \geq n$, then φ is injective iff* $\operatorname{Rank} \varphi = n$,

(iii) *if $n = m$, then φ is injective if and only if φ is surjective.*

The claim (iii) of Corollary 2.25 is one of the forms of *Fredholm's alternative theorem: either $\varphi(x) = y$ is solvable for every $y \in Y$ or $\varphi(x) = 0$ has a nonzero solution.*

2.26 Example. A second order linear equation

$$a\, y'' + b\, y' + c\, y = f, \qquad a, b, c \in \mathbb{R}, \ f \in C^0(\mathbb{R}), \tag{2.4}$$

can be seen as an abstract linear equation $\varphi(y) = f$ by introducing the linear map

$$\varphi : C^2(\mathbb{R}) \to C^0(\mathbb{R}), \qquad y \to \varphi(y) := a\, y'' + b\, y' + c\, y. \tag{2.5}$$

Since (2.4) has a solution for every $f \in C^0(\mathbb{R})$, see [GM1], φ is onto, and the linearity yields the following

(i) the set of all solutions of the associated homogeneous equation $a\, y'' + b\, y' + c\, y = 0$ is a linear space, actually $\ker \varphi$,

(ii) if y_f is any solution of (2.4), then all solutions of (2.4) are obtained by adding to y_f a solution of the homogeneous equation $a\, y'' + b\, y' + c\, y = 0$, i.e., of $\varphi(y) = 0$. In abstract terms, the set of all solutions of $\varphi(y) = f$ is given by

$$\left\{ y \in C^2(\mathbb{R}) \,|\, y - y_f \in \ker \varphi \right\}.$$

Moreover, consider the map $\gamma : \mathbb{R}^2 \to C^2(\mathbb{R})$ that maps each $(\alpha, \beta) \in \mathbb{R}^2$ to the unique solution of the initial value problem

$$\begin{cases} a\, y'' + b\, y' + c\, y = 0, \\ y(0) = \alpha, y'(0) = \beta. \end{cases}$$

It is easy to show that $\gamma : \mathbb{R}^2 \to C^2(\mathbb{R})$ is linear, and by definition, $\operatorname{Im} \gamma = \ker \varphi$, where φ is the map in (2.5). Since γ is trivially injective, the rank formula yields $\dim \operatorname{Im} \gamma = 2 - 0 = 2$, concluding that *the space of solutions of a homogeneous second order ODE is a vector space of dimension 2.*

h. Changing coordinates

The coordinates of a vector depend on the chosen coordinate system. Let us discuss how they change.

Let X be a vector space of dimension n, and let $\mathcal{E} : X \to \mathbb{K}^n$ and $\mathcal{F} : X \to \mathbb{K}^n$ be two coordinate systems on X, that we label respectively as the *old system* and the *new system*. Denote by $(e_1, e_2, \ldots, e_n)$ and $(f_1, f_2, \ldots, f_n)$ the bases associated respectively to the old coordinate system $\mathcal{E}$ and to the new coordinate system $\mathcal{F}$. The linear map $L := \mathcal{F} \circ \mathcal{E}^{-1} : \mathbb{K}^n \to \mathbb{K}^n$ maps the *old* $\mathcal{E}$-coordinate vector of $x \in X$ to the *new* $\mathcal{F}$-coordinate vector of x, see Figure 2.5. The matrix $\mathbf{L}$ associated to L in the basis $(e_1, e_2, \ldots, e_n)$ is

$$\mathbf{L} := \mathcal{M}_{\mathcal{E}}^{\mathcal{F}}(\operatorname{Id}) = \left[\mathcal{F}(e_1) \,\middle|\, \mathcal{F}(e_2) \,\middle|\, \ldots \,\middle|\, \mathcal{F}(e_n) \right].$$

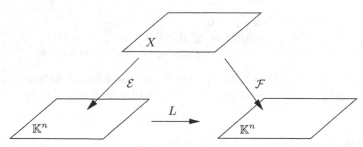

Figure 2.5. Changing the basis.

We say that L or $\mathbf{L}$ *changes coordinates from $\mathcal{E}$ to $\mathcal{F}$*. Remember that the ith column of $\mathbf{L}$ is the *new* $\mathcal{F}$-coordinate vector of the ith vector of the *old basis*, i.e, see (2.2), $\mathbf{L} = [L^i_j]$ where

$$e_j = \sum_{i=1}^{n} L^i_j f_i. \tag{2.6}$$

Let $m : X \to X$ be the linear map defined by $m(e_i) := f_i \ \forall i = 1, \ldots, m$. Then the associated matrix $\mathbf{M} = [M^i_j]$ to m using the basis $\mathcal{E}$ is

$$\mathbf{M} := \left[\mathcal{E}(f_1) \,\middle|\, \mathcal{E}(f_2) \,\middle|\, \ldots \,\middle|\, \mathcal{E}(f_n) \right]$$

or,

$$f_j = \sum_{i=1}^{n} M^i_j e_i.$$

Therefore, comparing with (2.6), $\mathbf{M} = \mathbf{L}^{-1}$.

In conclusion

(i) $\mathbf{L}$ maps the *old* coordinates to the *new* coordinates,
(ii) $\mathbf{L}^{-1}$ maps the *old* basis to the *new* basis.

Thus $\mathbf{L}$ acts differently on the basis and on the coordinates. We say that the coordinates change in a *contravariant* way. This is nothing mysterious; for instance, $7000 \, g = 7 \, Kg$: if the unit measure ϵ_1 is, say 1000 times the unit measure e_1, then we expect that the number of units e_1 associated to a measure will be 1/1000 of the number of units ϵ_1.

2.27 Example. Suppose we want to change from the canonical basis (e_1, e_2) to a new one given by the vectors $(1, 2)$ and $(3, 4)$ in $\mathbb{R}^2$. The matrix that changes the basis from $e_1 = (1, 0)$, $e_2 = (0, 1)$ into $f_1 = (e_1 + 2 e_2)$, $f_2 = (3 e_1 + 4 e_2)$ is

$$\mathbf{M} = \begin{pmatrix} 1 & 3 \\ 2 & 4 \end{pmatrix}.$$

The old coordinates of a vector P can easily be obtained from the new ones, as

$$P = x \, e_1 + y \, e_2 = \alpha \, f_1 + \beta \, f_2,$$

thus, in the old coordinates,

$$\begin{pmatrix} x \\ y \end{pmatrix} = \alpha \begin{pmatrix} 1 \\ 2 \end{pmatrix} + \beta \begin{pmatrix} 3 \\ 4 \end{pmatrix} = \mathbf{M} \begin{pmatrix} \alpha \\ \beta \end{pmatrix},$$

and, conversely, we have

$$\begin{pmatrix} \alpha \\ \beta \end{pmatrix} = \mathbf{M}^{-1} \begin{pmatrix} x \\ y \end{pmatrix}.$$

i. The associated matrix under changes of basis

Let X and Y be two vector spaces of dimension n and m. As before, the matrix associated to a linear map depends on the chosen coordinates on X and Y. Let $\mathcal{E} : X \to \mathbb{K}^n$, $\mathcal{E}' : X \to \mathbb{K}^n$ be two coordinate systems on X, and let $\mathcal{F} : X \to \mathbb{K}^n$, $\mathcal{F}' : X \to \mathbb{K}^n$ be two coordinate systems on Y. Of course, for every map ℓ we have $\ell = \mathrm{Id}_Y \circ \ell \circ \mathrm{Id}_X$, consequently

$$\mathcal{M}_{\mathcal{E}'}^{\mathcal{F}'}(\ell) = \mathcal{M}_{\mathcal{F}}^{\mathcal{F}'}(\mathrm{Id}) \mathcal{M}_{\mathcal{E}}^{\mathcal{F}}(\ell) \mathcal{M}_{\mathcal{E}'}^{\mathcal{E}}(\mathrm{Id}),$$

or, in other words, we can state the following.

2.28 Proposition. *Given the previous notation, let $\mathbf{R} \in M_{n,n}(\mathbb{K})$ be the matrix that changes coordinates from $\mathcal{E}$ to $\mathcal{E}'$, let $\mathbf{S} \in M_{m,m}(\mathbb{K})$ be the matrix that changes coordinates from $\mathcal{F}$ to $\mathcal{F}'$, let $\mathbf{A}$ $\mathbf{A}'$ be the matrices that represent ℓ respectively, in the systems of coordinates $\mathcal{E}$ and $\mathcal{F}$ and in the systems $\mathcal{E}'$ and $\mathcal{F}'$. Then*

$$\mathbf{A}' = \mathbf{S}\mathbf{A}\mathbf{R}^{-1}.$$

2.29 Corollary. *Let $\mathbf{A} \in M_{n,n}(\mathbb{K})$ and let $A : \mathbb{K}^n \to \mathbb{K}^n$ be the associated linear operator $A(\mathbf{x}) := \mathbf{A}\mathbf{x}$. Let $(\mathbf{f}_1, \mathbf{f}_2, \dots, \mathbf{f}_n)$ be a basis of $\mathbb{K}^n$. Then the matrix associated to A using the basis $(\mathbf{f}_1, \mathbf{f}_2, \dots, \mathbf{f}_n)$, both in the source and the target $\mathbb{K}^n$, is the matrix*

$$\mathbf{A}' := \mathbf{S}^{-1}\mathbf{A}\mathbf{S}$$

where

$$\mathbf{S} := \left[\mathbf{f}_1 \,\middle|\, \mathbf{f}_2 \,\middle|\, \dots \,\middle|\, \mathbf{f}_n \right].$$

Proof. Let $(\mathbf{e}_1, \mathbf{e}_2, \dots, \mathbf{e}_n)$ be the canonical basis of $\mathbb{K}^n$, then $\mathbf{S}\mathbf{e}_i = \mathbf{f}_i$ are the coordinates of $\mathbf{f}_i$ in the basis $(\mathbf{e}_1, \mathbf{e}_2, \dots, \mathbf{e}_n)$, $\mathbf{A}\mathbf{S}\mathbf{e}_i = \mathbf{A}\mathbf{f}_i$ is the coordinate vector of $A(\mathbf{f}_i)$ in canonical coordinates and, finally, $\mathbf{S}^{-1}\mathbf{A}\mathbf{S}\mathbf{e}_i$ is the coordinate vector of $A(\mathbf{f}_i)$ in the $(\mathbf{f}_1, \mathbf{f}_2, \dots, \mathbf{f}_n)$ basis. $\qquad \square$

j. The dual space $\mathcal{L}(X, \mathbb{K})$

Linear maps from X into $\mathbb{K}$ play a special role.

Let X be a vector space over $\mathbb{K}$ with $\dim X = n$. Linear maps from X into $\mathbb{K}$ are also called *linear forms* or *covectors*, and the space of linear forms, $\mathcal{L}(X, \mathbb{K})$, also denoted by X^*, is called the *dual space* of X. Suppose X is finite dimensional. Then, as we have seen, X^* has dimension n and, if $(e_1, e_2, \ldots, e_n)$ is a basis of X, then every linear form $\ell : X \to \mathbb{K}$ is represented as a $1 \times n$ matrix $\mathbf{L}$ that maps the coordinates $\mathbf{x}$ of $x \in X$ to $\ell(x)$, i.e.,

$$\mathbf{L}\mathbf{x} = \ell(x) \qquad \text{if} \qquad x = \sum_{i=1}^{n} x^i e_i, \ \mathbf{x} = (x^1, x^2, \ldots, x^n)$$

or, $\mathbf{L} = [l_1 \,|\, l_2 \,|\, \ldots \,|\, l_n]$ and

$$\ell(x) = \sum_{i=1}^{n} l_i x^i, \qquad \forall \mathbf{x} = (x^1, x^2, \ldots, x^n).$$

Consider now the linear maps $e^1, e^2, \ldots, e^n : X \to \mathbb{K}$ defined by

$$e^j(e_i) = \delta_{ij} \qquad \forall i, j = 1, \ldots, n. \tag{2.7}$$

2.30 Proposition. *We have*

(i) *for $i = 1, \ldots, n$, the map $e^i : X \to \mathbb{K}$ maps $x \in X$ to the ith coordinate of x, so that*

$$x = \sum_{i=1}^{n} e^i(x) e_i \qquad \forall x \in X,$$

(ii) *$(e^1, e^2, \ldots, e^n)$ is a basis on X^*,*
(iii) *if $x = \sum_{i=1}^{n} x^i e_i \in X$ and $\ell = \sum_{i=1}^{n} l_i e^i \in X^*$, then $\ell(x) = \sum_{i=1}^{n} l_i x^i$.*

Proof. (i) If $x = \sum_{i=1}^{n} x^i e_i$, then $e^j(x) = \sum_{i=1}^{n} x^i e^j(e_i) = x^j$.
(ii) Let $\ell := \sum_{j=1}^{n} \mu_j e^j$. Then $\ell(e_i) = \mu_i \ \forall i$. Thus, if $\ell(x) = 0 \ \forall x$ we trivially have $\mu_i = 0 \ \forall i$.
(iii) In fact,

$$\ell(x) = \Big(\sum_{j=1}^{n} l_j e^j\Big)\Big(\sum_{i=1}^{n} x^i e_i\Big) = \sum_{j=1}^{n} l_j e^j \Big(\sum_{i=1}^{n} x^i e_i\Big)$$

$$= \sum_{i=1}^{n}\sum_{j=1}^{n} l_j x^i e^j(e_i) = \sum_{i=1}^{n}\sum_{j=1}^{n} l_j x^i \delta_{ij} = \sum_{i=1}^{n} l_i x^i.$$

$\square$

The system of linear maps $(e^1, e^2, \ldots, e^n)$ characterized by (2.7) is called the *dual basis* of $(e_1, e_2, \ldots, e_n)$ in X^*.

Figure 2.6. Giuseppe Peano (1858–1932) and the frontispiece of his *Calcolo Geometrico*.

2.31 Remark. Coordinates of vectors or covectors of a vector space X of dimension n are both n-tuples. However, to distinguish them, it is useful to index the coordinates of covectors with lower indices. We can reinforce this notation even more by writing the coordinates of vectors as column vectors and coordinates of covectors as row vectors.

k. The bidual space

Of course, we may consider also the space of linear forms on X^*, denoted by X^{**} and called the *bidual* of X. Every $v \in X$ identifies a linear form on $\mathcal{L}(X^*, \mathbb{K})$, by defining $v^{**} : X^* \to \mathbb{K}$ by $v^{**}(\ell) := \ell(v)$. The map $\gamma : X \to X^{**}$, $x \to \gamma(x) := x^{**}$ we have just defined is linear and injective. Since $\dim X^{**} = \dim X^* = \dim X$, γ is surjective, hence an isomorphism, and we call it *natural* since it does not depend on other structures on X, as does the choice of a basis.

Since X^{**} and X are naturally isomorphic, there is a "symmetry" between the two spaces X and X^*. To emphasize this symmetry, it is usual to write

$$< \varphi, x > \qquad \text{instead of} \qquad \varphi(x)$$

if $\varphi \in X^*$ and $x \in X$, introducing the *evaluation map*

$$< , >: X^* \times X \to \mathbb{K}, \qquad < \varphi, x >:= \varphi(x).$$

2.32 ¶. Let X be a vector space and let X^* be its dual. A *duality* between X and X^* is a map $< , >: X^* \times X \to \mathbb{K}$ that is

(i) linear in each factor,

$$< \varphi, \alpha x + \beta y > = \alpha < \varphi, x > + \beta < \varphi, y >,$$
$$< \alpha \varphi + \beta \psi, x > = \alpha < \varphi, x > + \beta < \psi, x >,$$

for all $\alpha, \beta \in \mathbb{K}$, $x, y \in X$ and $\varphi, \psi \in X^*$,

(ii) *nondegenerate* i.e.,

$$\text{if } < \varphi, x > = 0 \ \forall x, \text{ then } \varphi = 0,$$
$$\text{if } < \varphi, x > = 0 \ \forall \varphi, \text{ then } x = 0.$$

Show that the evaluation map $(\varphi, x) \to < \varphi, x >$ that evaluates a linear map $\varphi : X \to \mathbb{K}$ at $x \in X$ is a duality.

l. Adjoint or dual maps

Let X, Y be vector spaces, X^* and Y^* their duals and $< \ , \ >_X$ and $< \ , \ >_Y$ the evaluation maps on $X^* \times X$ and $Y^* \times Y$. For every linear map $\ell : X \to Y$, one also has a map $\ell^* : Y^* \to X^*$ defined by

$$< \ell^*(y^*), x > := < y^*, \ell(x) > \qquad \forall x \in X, \forall y^* \in Y^*.$$

It turns out that ℓ^* is linear. Now if $(e_1, e_2, \ldots, e_n)$ and $(f_1, \ldots, f_m)$ are bases in X and Y respectively, and $(e^1, e^2, \ldots, e^n)$ and $(f^1, f^2, \ldots, f^m)$ are the dual bases in X^* and Y^*, then the associated matrices $\mathbf{L} = [L_j^i] \in M_{m,n}(\mathbb{K})$ and $\mathbf{M} = [M_j^i] \in M_{n,m}(\mathbb{K})$, associated respectively to ℓ and ℓ^*, are defined by

$$\ell(e_i) = \sum_{h=1}^{m} L_i^h f_h, \qquad \ell^*(f^h) = \sum_{i=1}^{n} M_h^i e^i.$$

By duality,

$$M_h^i = < \ell^*(f^h), e_i > = < f^h, \ell(e_i) > = L_i^h$$

i.e., $\mathbf{M} = \mathbf{L}^T$. Therefore we conclude that *if $\mathbf{L}$ is the matrix associated to ℓ in a given basis, then $\mathbf{L}^T$ is the matrix associated to ℓ^* in the dual basis.*

We can now discuss how coordinate changes in X reflect on the dual space.

Let X be a vector space of dimension n, X^* its dual space, $(e_1, \ldots, e_n)$, $(\epsilon_1, \epsilon_2, \ldots, \epsilon_n)$ two bases on X and $(e^1, e^2, \ldots, e^n)$ and $(\epsilon^1, \epsilon^2, \ldots, \epsilon^n)$ the corresponding dual bases on X^*.

Let $\ell : X \to X$ be the linear map defined by $\ell(e_i) := \epsilon_i \ \forall i = 1, \ldots, n$. Then by duality

$$< \ell^*(\epsilon^i), e_j > = < \epsilon^i, \ell(e_j) > = < \epsilon^i, \epsilon_j > = \delta_{ij} = < e^i, e_j > \qquad \forall i, j,$$

$\ell^*(\epsilon^i) = e^i \ \forall i = 1, \ldots, n$. If $\mathbf{L}$ and $\mathbf{L}^T$ are the associated matrices to ℓ and ℓ^*, $\mathbf{L}$ changes basis from $(e_1, e_2, \ldots, e_n)$ to $(\epsilon_1, \epsilon_2, \ldots, \epsilon_n)$ in X, and $\mathbf{L}^T$ changes basis in the dual space from $(e^1, e^2, \ldots, e^n)$ to $(\epsilon^1, \epsilon^2, \ldots, \epsilon^n)$,

$$\epsilon_i = \sum_{i=1}^{n} \mathbf{L}_j^i e_i, \qquad e^i = \sum_{j=1}^{n} (\mathbf{L}^T)_i^j \epsilon^j.$$

Now if $\varphi \in X^*$, we have $\varphi = \sum_{i=1}^{n} a_i \epsilon^i = \sum_{i=1}^{n} b_i e^i$, hence

$$\sum_{i=1}^{n} b^i e^i = \sum_{i=1}^{n} \sum_{j=1}^{n} b_i (\mathbf{L}^T)_i^j \epsilon^j = \sum_{i=1}^{n} a_i e^i.$$

Thus, if $\mathbf{a} := (a_1, a_2, \ldots, a_n)$, $\mathbf{b} := (b_1, b_2, \ldots, b_n)$, we have

$$\mathbf{a}^T = \mathbf{L}^T \mathbf{b}^T \qquad \text{or} \qquad \mathbf{a} = \mathbf{b}\, \mathbf{L}.$$

In other words, the coordinates in X^* change according to the change of basis. We say that the change of coordinates in X^* is *covariant*.

2.2 Eigenvectors and Similar Matrices

Let $A : X \to X$ be a linear operator on a vector space. How can we describe the properties of A that are invariant by isomorphisms? Since isomorphims amount to changing basis, we can put it in another way. Suppose X is finite dimensional, $\dim X = n$, then we may consider the matrix $\mathbf{A}$ associated to A using a basis (we use the same basis both in the source and the target X). But how can we catch the properties of $\mathbf{A}$ that are independent of the basis? One possibility is to try to choose an "optimal" basis in which, say, the matrix $\mathbf{A}$ takes the simplest form. As we have seen, if we choose two coordinate systems $\mathcal{E}$ and $\mathcal{F}$ on X, and $\mathbf{S}$ is the matrix that changes coordinates from $\mathcal{E}$ to $\mathcal{F}$, then the matrices $\mathbf{A}$ and $\mathbf{B}$ that represent A respectively in the basis $\mathcal{E}$ and $\mathcal{F}$ are related by

$$\mathbf{B} = \mathbf{S}\mathbf{A}\mathbf{S}^{-1}.$$

Therefore we are asking for a nonsingular matrix $\mathbf{S}$ such that $\mathbf{S}^{-1}\mathbf{A}\mathbf{S}$ has the simplest possible form: this is the problem of *reducing a matrix to a canonical form*.

Let us try to make the meaning of "simplest" for a matrix more precise.

Suppose that in X there are two supplementary invariant subspaces under A

$$X = W_1 \oplus W_2, \qquad A(W_1) \subset W_1, \; A(W_2) \subset W_2.$$

Then every $x \in X$ splits uniquely as $x = x_1 + x_2$ with $x_1 \in W_1$, $x_2 \in W_2$, and $A(x) = A(x_1) + A(x_2)$ with $A(x_1) \in W_1$ and $A(x_2) \in W_2$. In other words, A splits into two operators $A_1 : W_1 \to W_1$, $A : W_2 \to W_2$ that are the restrictions of A to W_1 and W_2. Now suppose that $\dim X = n$ and let $(e_1, e_2, \ldots, e_k)$ and $(f_1, f_2, \ldots, f_{n-k})$ be two bases respectively of W_1 and W_2. Then the matrix associated to A in the basis $(e_1, e_2, \ldots, e_k, f_1, f_2, \ldots, f_{n-k})$ of X has the form

$$\mathbf{A} = \begin{pmatrix} \boxed{\mathbf{A}_1} & 0 \\ 0 & \boxed{\mathbf{A}_2} \end{pmatrix}$$

where some of the entries are zero.

If we pursue this approach, the optimum would be the decomposition of X into n supplementary invariant subspaces $W_1, W_2, \ldots, W_n$ under A of dimension 1,

$$X = W_1 \oplus W_2 \oplus \cdots \oplus W_n, \qquad A(W_i) \subset W_i.$$

In this case, A acts on each W_i as a dilation: $A(x) = \lambda_i x \ \forall x \in W_i$ for some $\lambda_i \in \mathbb{K}$. Morever, if $(e_1, e_2, \ldots, e_n)$ is a basis of X such that $e_i \in W_i$ for each i, then the matrix associated to A in this basis is the diagonal matrix

$$\mathbf{A} = \operatorname{diag}(\lambda_1, \lambda_2, \ldots, \lambda_n).$$

2.2.1 Eigenvectors

a. Eigenvectors and eigenvalues

As usual, $\mathbb{K}$ denotes the field $\mathbb{R}$ or $\mathbb{C}$.

2.33 Definition. *Let $A : X \to X$ be a linear operator on a vector space X over $\mathbb{K}$. We say that $x \in X$ is an* eigenvector *of A if $Ax = \lambda x$ for some $\lambda \in \mathbb{K}$. If x is a nonzero eigenvector, the number λ for which $A(x) = \lambda x$ is called an* eigenvalue *of A, or more precisely, the* eigenvalue *of A relative to x. The set of eigenvalues of A is called the* spectrum *of A.*

If $\mathbf{A} \in M_{n,n}(\mathbb{K})$, we refer to eigenvalues and eigenvectors of the associated linear operator $A : \mathbb{K}^n \to \mathbb{K}^n$, $A(\mathbf{x}) := \mathbf{A}\mathbf{x}$, as the eigenvalues and the eigenvectors of $\mathbf{A}$.

From the definition, λ is an eigenvalue of A if and only if $\ker(\lambda \operatorname{Id} - A) \neq \{0\}$, equivalently, if and only if $\lambda \operatorname{Id} - A$ is not invertible. If λ is an eigenvalue, the subspace of all eigenvectors with eigenvalue λ

$$V_\lambda := \left\{ x \in X \,\middle|\, A(x) = \lambda x \right\} = \ker(\lambda \operatorname{Id} - A)$$

is called the *eigenspace* of A relative to λ.

2.34 Example. let $X = C^\infty([0, \pi])$ be the linear space of smooth functions that vanish at 0 and π and let $D^2 : X \to X$ be the linear operator $D^2(f) := f''$ that maps every function f into its second derivative. Nonzero eigenvectors of the operator D^2, that is, the nonidentically zero functions $y \in C^\infty[0, 1]$ such that $D^2 y(x) = \lambda y(x)$ for some $\lambda \in \mathbb{R}$, are called eigenfunctions.

2.35 Example. Let X be the set P_n of polynomials of degree less than n. Then, each $P_k \subset P_n \ k = 0, \ldots, n$ is an invariant subspace for the operator of differentiation. It has zero as a unique eigenvalue.

2.36 ¶. Show that the rotation in $\mathbb{R}^2$ by an angle θ has no nonzero eigenvectors if $\theta \neq 0, \pi$, since in this case there are no invariant lines.

2.37 Definition. *Let $A : X \to X$ be a linear operator on X. A subspace $W \subset X$ is invariant (under A) if $A(W) \subset W$.*

In the following proposition we collect some simple properties of eigenvectors.

2.38 Proposition. *Let $A : X \to X$ be a linear operator on X.*

(i) *$x \neq 0$ is an eigenvector if and only if $\mathrm{Span}\,\{x\}$ is an invariant subspace under A.*
(ii) *Let λ be an eigenvector of A and let V_λ be the corresponding eigenspace. Then every subspace $W \subset V_\lambda$ is an invariant subspace under A, i.e., $A(W) \subset W$.*
(iii) *$\dim \ker(\lambda \,\mathrm{Id} - A) > 0$ if and only if λ is an eigenvalue for A.*
(iv) *Let $W \subset X$ be an invariant subspace under A and let λ be an eigenvalue for $A_{|W}$. Then λ is an eigenvalue for $A : X \to X$.*
(v) *λ is an eigenvalue for A if and only if 0 is an eigenvalue for $\lambda \,\mathrm{Id} - A$.*
(vi) *Let $\varphi : X \to Y$ be an isomorphism and let $A : X \to X$ be an operator. Then $x \in X$ is an eigenvector for A if and only if $\varphi(x)$ is an eigenvector for $\varphi \circ A \circ \varphi^{-1}$, and x and $\varphi(x)$ have the same eigenvalue.*
(vii) *Nonzero eigenvectors with different eigenvalues are linearly independent.*

Proof. (i), $\ldots$, (vi) are trivial. To prove (vii) we proceed by induction on the number k of eigenvectors. For $k = 1$ the claim is trivial. Now assume by induction that the claim holds for $k - 1$ nonzero eigenvectors, and let $e_1, e_2, \ldots, e_k$ be such that $e_i \neq 0$ $\forall i = 1, \ldots, k$, $A(e_j) = \lambda_j e_j$ $\forall j = 1, \ldots, k$ with $\lambda_j \neq \lambda_i$ $\forall i \neq j$. Let

$$a_1 e_1 + a_2 e_2 + \cdots + a_k e_k = 0, \qquad (2.8)$$

be a linear combination of $e_1, e_2, \ldots, e_k$. From (2.8), multiplying by λ_1 and applying A we get

$$a_1 \lambda_1 e_1 + a_2 \lambda_1 e_2 + \cdots + a_k \lambda_1 e_k = 0,$$
$$a_1 \lambda_1 e_1 + a_2 \lambda_2 e_2 + \cdots + a_k \lambda_k e_k = 0,$$

consequently

$$\sum_{j=2}^{k} (\lambda_j - \lambda_1) a_j e_j = 0.$$

By the inductive assumption, $a_j(\lambda_j - \lambda_1) = 0$ $\forall j = 2, \ldots, n$, hence $a_j = 0$ for all $j \geq 2$. We then conclude from (2.8) that we also have $a_1 = 0$, i.e., that $e_1, e_2, \ldots, e_k$ are linearly independent. $\qquad \square$

Let $A : X \to X$ be a linear operator on X of dimension n, and let $\mathbf{A}$ be the associated matrix in a coordinate system $\mathcal{E} : X \to \mathbb{K}^n$. Then (vi) implies that $x \in X$ is an eigenvector of A if and only if $\mathbf{x} := \mathcal{E}(x)$ is an eigenvector for $\mathbf{x} \to \mathbf{A}\mathbf{x}$ and x and $\mathbf{x}$ have the same eigenvalue.

From (vii) Proposition 2.38 we infer the following.

2.39 Corollary. *Let $A : X \to X$ be a linear operator on a vector space X of dimension n. If A has n different eigenvalues, then X has a basis formed by eigenvectors of A.*

b. Similar matrices

Let $A : X \to X$ be a linear operator on a vector space X of dimension n. As we have seen, if we fix a basis, we can represent A by an $n \times n$ matrix. If $\mathbf{A}$ and $\mathbf{A}' \in M_{n,n}(\mathbb{K})$ are two such matrices that represent A in two different bases $(e_1, e_2, \ldots, e_n)$ and $(\epsilon_1, \epsilon_2, \ldots, \epsilon_n)$, then by Proposition 2.28 $\mathbf{A}' = \mathbf{S}^{-1}\mathbf{A}\mathbf{S}$ where $\mathbf{S}$ is the matrix that changes basis from $(e_1, e_2, \ldots, e_n)$ to $(\epsilon_1, \epsilon_2, \ldots, \epsilon_n)$.

2.40 Definition. *Two matrices $\mathbf{A}, \mathbf{B} \in M_{n,n}(\mathbb{K})$ are said to be similar if there exists $\mathbf{S} \in GL(n, \mathbb{K})$ such that $\mathbf{B} = \mathbf{S}^{-1}\mathbf{A}\mathbf{S}$.*

It turns out that the similarity relation is an equivalence relation on matrices, thus $n \times n$ matrices are partitioned into classes of similar matrices. Since matrices associated to a linear operator $A : X \to X$, $\dim X = n$, are similar, we can associate to A a unique class of similar matrices. It follows that if a property is preserved by similarity equivalence, then it can be taken as a property of the linear operator to which the class is referred. For instance, let $A : X \to X$ be a linear operator, and let $\mathbf{A}, \mathbf{B}$ be such that $\mathbf{B} = \mathbf{S}^{-1}\mathbf{A}\mathbf{S}$. By Binet's formula, we have

$$\det \mathbf{B} = \det \mathbf{S}^{-1} \det \mathbf{A} \det \mathbf{S} = \frac{1}{\det \mathbf{S}} \det \mathbf{A} \det \mathbf{S} = \det \mathbf{A}.$$

Thus we may define the *determinant* of the linear map $A : X \to X$ by

$$\det A := \det \mathbf{A}$$

where $\mathbf{A}$ is any matrix associated to A.

c. The characteristic polynomial

Let X be a vector space of dimension n, and let $A : X \to X$ be a linear operator. The function

$$\lambda \to p_A(\lambda) := \det(\lambda \operatorname{Id} - A), \qquad \lambda \in \mathbb{K},$$

is called the *characteristic polynomial* of A. It can be computed by representing A by a matrix $\mathbf{A}$ in a coordinate system and computing $p_A(\lambda)$ as the characteristic polynomial of any of the matrices $\mathbf{A}$ representing A,

$$p_A(\lambda) = p_{\mathbf{A}}(\lambda) = \det(\lambda \operatorname{Id} - \mathbf{A}).$$

In particular, it follows that $p_A(\ ) : \mathbb{K} \to \mathbb{K}$ is a polynomial in λ of degree n, and that the roots of $p_A(\lambda)$ are the eigenvalues of A or of $\mathbf{A}$.

Moreover, we can state

2.41 Proposition. *We have the following.*

(i) *Two similar matrices* $\mathbf{A}, \mathbf{B}$ *have the same eigenvalues and the same characteristic polynomials.*

(ii) *If* $\mathbf{A}$ *has the form*

$$\begin{pmatrix} \boxed{\mathbf{A}_1} & 0 & \cdots & 0 \\ 0 & \boxed{\mathbf{A}_2} & \cdots & 0 \\ \vdots & \vdots & \ddots & \vdots \\ 0 & 0 & \cdots & \boxed{\mathbf{A}_k} \end{pmatrix},$$

where for $i = 1, \ldots, k$, *each block* $\mathbf{A}_i$ *is a square matrix of dimension* k_i *with principal diagonal on the principal diagonal of* $\mathbf{A}$, *then*

$$p_{\mathbf{A}}(s) = p_{\mathbf{A}_1}(s) \cdot p_{\mathbf{A}_2}(s) \ldots p_{\mathbf{A}_k}(s).$$

(iii) *We have*

$$\det(s \operatorname{Id} - \mathbf{A}) = s^n - \operatorname{tr} \mathbf{A} \, s^{n-1} + \cdots + (-1)^n \det \mathbf{A}$$

$$= s^n + \sum_{k=1}^n (-1)^k a_k s^{n-k}$$

where $\operatorname{tr} \mathbf{A} := \sum_{i=1}^n \mathbf{A}_i^i$ *is the* trace *of the matrix* $\mathbf{A}$, *and* a_k *is the sum of the determinants of the* $k \times k$ *submatrices of* $\mathbf{A}$ *with principal diagonal on the principal diagonal of* $\mathbf{A}$.

Proof. (i) If $\mathbf{B} = \mathbf{SAS}^{-1}$, $\mathbf{S} \in GL(n, \mathbb{K})$, then $s \operatorname{Id} - \mathbf{B} = \mathbf{S}(s \operatorname{Id} - \mathbf{A})\mathbf{S}^{-1}$, hence $\det(s \operatorname{Id} - \mathbf{B}) = \det \mathbf{S} \det(s \operatorname{Id} - \mathbf{A})(\det \mathbf{S})^{-1} = \det(s \operatorname{Id} - \mathbf{A})$.

(ii) The matrix $s \operatorname{Id} - \mathbf{A}$ is a block matrix of the same form

$$\begin{pmatrix} \boxed{s \operatorname{Id} - \mathbf{A}_1} & 0 & \cdots & 0 \\ 0 & \boxed{s \operatorname{Id} - \mathbf{A}_2} & \cdots & 0 \\ \vdots & \vdots & \ddots & \vdots \\ 0 & 0 & \cdots & \boxed{s \operatorname{Id} - \mathbf{A}_k} \end{pmatrix},$$

hence $\det(s \operatorname{Id} - \mathbf{A}) = \prod_{i=1}^k \det(s \operatorname{Id} - \mathbf{A}_i)$.

(iii) We leave it to the reader. $\qquad\square$

Notice that there exist matrices with the same eigenvalues that are not similar, see Exercise 2.73.

d. Algebraic and geometric multiplicity

2.42 Definition. *Let $A : X \to X$ be a linear operator, and let $\lambda \in \mathbb{K}$ be an eigenvalue of A. We say that λ has* geometric multiplicity $k \in \mathbb{N}$ *if* $\dim \ker(\lambda \operatorname{Id} - A) = k$.

Let $p_A(s)$ be the characteristic polynomial of A. We say that λ has algebraic multiplicity k *if*

$$p_A(s) = (s - \lambda)^k q(s), \qquad where \ \ q(\lambda) \neq 0.$$

2.43 Proposition. *Let $A : X \to X$ be a linear operator on a vector space of dimension n and let λ be an eigenvalue of A of algebraic multiplicity m. Then $\dim \ker(\lambda \operatorname{Id} - A) \leq m$.*

Proof. Let us choose a basis $(e_1, e_2, \ldots, e_n)$ in X such that $(e_1, e_2, \ldots, e_k)$ is a basis for $V_\lambda := \ker(\lambda \operatorname{Id} - A)$. The matrix $\mathbf{A}$ associated to A in this basis has the form

$$\mathbf{A} = \left(\begin{array}{c|c} \lambda \operatorname{Id} & C \\ \hline 0 & D \end{array}\right)$$

where the first block, $\lambda \operatorname{Id}$, is a $k \times k$ matrix of dimension $k = \dim V_\lambda$. Thus Proposition 2.41 (ii) yields $p_\mathbf{A}(s) = \det(s \operatorname{Id} - \mathbf{A}) = (s - \lambda)^k p_\mathbf{D}(s)$, and the multiplicity of λ is at least k. $\square$

e. Diagonizable matrices

2.44 Definition. *We say that $\mathbf{A} \in M_{n,n}(\mathbb{K})$ is* diagonalizable, *if $\mathbf{A}$ is similar to a diagonal matrix.*

2.45 Theorem. *Let $A : X \to X$ be a linear operator on a vector space of dimension n, and let $(e_1, e_2, \ldots, e_n)$ be a basis of X. The following claims are equivalent.*

(i) *$e_1, e_2, \ldots, e_n$ are eigenvectors of A and $\lambda_1, \lambda_2, \ldots, \lambda_n$ are the relative eigenvalues.*

(ii) *We have $A(x) = \sum_{i=1}^n \lambda_i x^i e_i$ for all $x \in X$ if $x = \sum_{i=1}^n x^i e_i$.*

(iii) *The matrix that represents A in the basis $(e_1, e_2, \ldots, e_n)$ is*

$$\operatorname{diag}(\lambda_1, \lambda_2, \ldots, \lambda_n).$$

(iv) *If $\mathbf{A}$ is the matrix associated to A in the basis $(f_1, f_2, \ldots, f_n)$, then*

$$\mathbf{S}^{-1} \mathbf{A} \mathbf{S} = \operatorname{diag}(\lambda_1, \lambda_2, \ldots, \lambda_n)$$

where $\mathbf{S}$ is the matrix that changes basis from $(f_1, f_2, \ldots, f_n)$ to $(e_1, e_2, \ldots, e_n)$, i.e., the ith column of $\mathbf{S}$ is the coordinate vector of the eigenvector e_i in the basis $(f_1, f_2, \ldots, f_n)$.

Proof. (i) $\Leftrightarrow$ (ii) by linearity and (iii) $\Leftrightarrow$ (i) since (iii) is equivalent to $A(e_i) = \lambda_i e_i$. Finally (iii) $\Leftrightarrow$ (iv) by Corollary 2.29. $\square$

2.46 Corollary. *Let* $A : X \to X$ *be a linear operator on a vector space of dimension* n. *Then the following claims are equivalent.*

(i) X *splits as the direct sum of* n *one-dimensional invariant subspaces (under* A), $X = W_1 \oplus \cdots \oplus W_n$.

(ii) X *has a basis made of eigenvectors of* A.

(iii) *Let* $\lambda_1, \lambda_2, \ldots, \lambda_k$ *be all distinct eigenvalues of* A, *and let* $V_{\lambda_1}, \ldots, V_{\lambda_k}$ *be the corresponding eigenspaces. Then*

$$\sum_{i=1}^{k} \dim V_{\lambda_i} = n.$$

(iv) *If* $\mathbf{A}$ *is the matrix associated to* A *in a basis, then* $\mathbf{A}$ *is diagonizable.*

Proof. (i) implies (ii) since any nonzero vector in any of the W_i's is an eigenvector. Denoting by $(e_1, e_2, \ldots, e_n)$ a basis of eigenvectors, the spaces $W_i := \mathrm{Span}\{e_i\}$ are supplementary spaces of dimension one, hence (ii) implies (i). (iii) is a rewriting of (i) since for each eigenvalue λ, V_λ is the direct sum of the W_i's that have λ as the corresponding eigenvalue. Finally (ii) and (iii) are equivalent by Theorem 2.45. $\square$

2.47 Linear equations. The existence of a basis of X of eigenvectors of an operator $A : X \to X$ makes solving the linear equation $A(x) = y$ trivial.

Let $(e_1, e_2, \ldots, e_n)$ be a basis of X of eigenvectors of A and let $\lambda_1, \lambda_2, \ldots, \lambda_n$ be the corresponding eigenvalues. Writing $x, y \in X$ in this basis,

$$x = \sum_{i=1}^{n} x^i e_i, \qquad \sum_{i=1}^{n} y^i e_i,$$

we rewrite the equation $A(x) = y$ as

$$\sum_{i=1}^{n} (\lambda_i x^i - y^i) e_i = 0,$$

i.e., as the diagonal system

$$\begin{cases} \lambda_1 x^1 = y^1, \\ \lambda_2 x^2 = y^2, \\ \ldots, \\ \lambda_n x^n = y^n. \end{cases}$$

Therefore

(i) suppose that 0 is not an eigenvalue, then $A(x) = y$ has a unique solution

$$x = \sum_{i=1}^{n} \frac{y^i}{\lambda_i} e_i.$$

(ii) let 0 be an eigenvalue, and let $V_0 = \text{Span}\{e_1, e_2, \ldots, e_k\}$. Then $A(x) = y$ is solvable if and only if $y^1 = \cdots = y^k = 0$ and a solution of $A(x) = y$ is $x_0 := \sum_{i=k+1}^{n} \frac{y^i}{\lambda_i} e_i$. By linearity, the space of all solutions is the set

$$\left\{ x \in X \,\middle|\, x - x_0 \in \ker A = V_0 \right\}.$$

2.48 ¶. Let $A : X \to X$ be a linear operator on a finite-dimensional space. Show that A is invertible if and only if 0 is not an eigenvalue for A. In this case show that $1/\lambda$ is an eigenvalue for A^{-1} if and only if λ is an eigenvalue for A.

f. Triangularizable matrices

First, we notice that the eigenvalues of a triangular matrix are the entries of the principal diagonal. We can then state the following.

2.49 Theorem. *Let* $\mathbf{A} \in M_{n,n}(\mathbb{K})$. *If the characteristic polynomial decomposes as a product of factors of first degree, i.e., if there are (not necessarily distinct) numbers* $\lambda_1, \lambda_2, \ldots, \lambda_n \in \mathbb{K}$ *such that*

$$\det(\lambda \, \text{Id} - \mathbf{A}) = (\lambda - \lambda_1) \ldots (\lambda - \lambda_n),$$

then $\mathbf{A}$ *is similar to an upper triangular matrix.*

Proof. Let us prove the following equivalent claim. Let $A : X \to X$ be a linear operator on a vector space of dimension n. If $p_A(\lambda)$ factorizes as a product of factors of first degree, then there exists a basis $(u_1, u_2, \ldots, u_n)$ of X such that

$$\text{Span}\{u_1\}, \ \text{Span}\{u_1, u_2\}, \ \text{Span}\{u_1, u_2, u_3\}, \ \ldots, \text{Span}\{u_1, u_2, \ldots u_n\}$$

are *invariant subspaces* under A. In this case we have for the linear operator $A(\mathbf{x}) = \mathbf{A}\mathbf{x}$ associated to $\mathbf{A}$

$$\begin{cases} \mathbf{A}\mathbf{u}_1 = A(\mathbf{u}_1) = a_1^1 \mathbf{u}_1, \\ \mathbf{A}\mathbf{u}_2 = A(\mathbf{u}_2) = a_2^1 \mathbf{u}_1 + a_2^2 \mathbf{u}_2, \\ \vdots \\ \mathbf{A}\mathbf{u}_n = A(\mathbf{u}_n) = a_n^1 \mathbf{u}_1 + a_n^2 \mathbf{u}_2 + \cdots + a_n^n \mathbf{u}_n, \end{cases}$$

i.e., the matrix $\mathbf{A}$ associated to A using the basis $(u_1, u_2, \ldots, u_n)$ is upper triangular,

$$\mathbf{A} = \begin{pmatrix} a_1^1 & a_2^1 & \cdots & a_n^1 \\ 0 & a_2^2 & \cdots & a_n^2 \\ \vdots & \vdots & \ddots & \vdots \\ 0 & \cdots & 0 & a_n^n \end{pmatrix}.$$

We proceed by induction on the dimension of X. If $\dim X = 1$, the claim is trivial. Now suppose that the claim holds for any linear operator on a vector space of dimension $n - 1$, and let us prove the claim for A. From

$$p_A(\lambda) = \det(\lambda \operatorname{Id} - A) = (\lambda - \lambda_1) \ldots (\lambda - \lambda_n),$$

λ_1 is an eigenvalue of A, hence there is a corresponding nonzero eigenvalue u_1 and Span $\{u_1\}$ is an invariant subspace under A. Now we complete $\{u_1\}$ as a basis by adding vectors $v_2, \ldots v_n$, and let B be the restriction of the operator A to Span $\{v_2, \ldots v_n\}$. Let $\mathbf{B}$ be the matrix associated to B in the basis $(v_2, \ldots, v_n)$, and let $\mathbf{A}$ be the matrix associated to A in the basis $(u_1, u_2, \ldots, u_n)$. Then

$$\mathbf{A} = \begin{pmatrix} a_1^1 & \boxed{a_2^1 \ a_3^1 \ \ldots a_n^1} \\ & \\ 0 & \boxed{\mathbf{B}} \end{pmatrix}, \qquad \text{where } a_1^1 = \lambda_1.$$

Thus

$$p_A(\lambda) = p_{\mathbf{A}}(\lambda) = (\lambda - \lambda_1) p_{\mathbf{B}}(\lambda) = (\lambda - \lambda_1) p_B(\lambda).$$

It follows that the characteristic polynomial of B is $p_B(\lambda) = (\lambda - \lambda_2) \ldots (\lambda - \lambda_n)$. By the inductive hypothesis, there exists a basis $(u_2, \ldots u_n)$ of Span $\{v_2, \ldots, v_n\}$ such that

$$\text{Span } \{u_2\}, \ \text{Span } \{u_2, u_3\}, \ \ldots, \text{Span } \{u_2, \ldots, u_n\}$$

are invariant subspaces under B, hence

$$\text{Span } \{u_1\}, \ \text{Span } \{u_1, u_2\}, \ \text{Span } \{u_1, u_2, u_3\}, \ \ldots, \text{Span } \{u_1, u_2, \ldots u_n\}$$

are invariant subspaces under A.   □

2.2.2 Complex matrices

When $\mathbb{K} = \mathbb{C}$, a significant simplification arises. Because of the fundamental theorem of algebra, the characteristic polynomial $P_A(\lambda)$ of every linear operator $A : X \to X$ over a complex vector space X of dimension n, factorizes as product of n factors of first degree. In particular, A has n eigenvalues, if we count them with their multiplicities. From Theorem 2.49 we conclude the following at once.

2.50 Corollary. *Let $\mathbf{A} \in M_{n,n}(\mathbb{C})$ be a complex matrix. Then $\mathbf{A}$ is similar to an upper triangular matrix, that is, there exists a nonsingular matrix $\mathbf{S} \in M_{n,n}(\mathbb{C})$ such that $\mathbf{S}^{-1} \mathbf{A} \mathbf{S}$ is upper triangular.*

Moreover,

2.51 Corollary. *Let $\mathbf{A} \in M_{n,n}(\mathbb{C})$ be a matrix. Then $\mathbf{A}$ is diagonizable (as a complex matrix) if and only if the geometric and algebraic multiplicities of each eigenvalue agree.*

Proof. Let $\lambda_1, \lambda_2, \ldots, \lambda_k$ be the distinct eigenvalues of $\mathbf{A}$, for each $i = 1, \ldots k$, let m_i and V_{λ_i} respectively be the algebraic multiplicity and the eigenspace of λ_i.

If $\dim V_{\lambda_i} = m_i$ $\forall i$, then by the fundamental theorem of algebra

$$\sum_{i=1}^{k} \dim V_{\lambda_i} = \sum_{i=1}^{n} m_i = n.$$

Hence $\mathbf{A}$ is diagonalizable, by Corollary 2.46.

Conversely, if $\mathbf{A}$ is diagonalizable, then $\sum_{i=1}^{k} \dim V_{\lambda_i} = n$, hence by Proposition 2.43 $\dim V_{\lambda_i} \leq m_i$, hence

$$n = \sum_{i=1}^{k} m_i \geq \sum_{i=1}^{k} \dim V_{\lambda_i} = n.$$

$\square$

2.52 Remark (Real and complex eigenvalues). If $\mathbf{A} \in M_{n,n}(\mathbb{R})$, its eigenvalues are by definition the *real* solutions of the polynomial equation $\det(\lambda \operatorname{Id} - \mathbf{A}) = 0$. But $\mathbf{A}$ is also a matrix with complex entries, $\mathbf{A} \in M_{n,n}(\mathbb{C})$ and it has as *eigenvalues* which are the complex solutions of $\det(\lambda \operatorname{Id} - \mathbf{A}) = 0$. It is customary to call *eigenvalues of* $\mathbf{A}$ the complex solutions of $\det(\lambda \operatorname{Id} - \mathbf{A}) = 0$ even if $\mathbf{A}$ has real entries, while the real solutions of the same equation, which are the eigenvalues of the real matrix $\mathbf{A}$ following Definition 2.33, are called *real eigenvalues*.

The further developments we want to discuss depend on some relationships among polynomials and matrices that we now want to illustrate.

a. The Cayley–Hamilton theorem

Given a polynomial $f(t) = \sum_{k=1}^{n} a_k t^k$, to every $n \times n$ matrix $\mathbf{A}$ we can associate a new matrix $f(\mathbf{A})$ defined by

$$f(\mathbf{A}) := a_0 \operatorname{Id} + \sum_{k=1}^{n} a_k \mathbf{A}^k =: \sum_{k=0}^{n} a_k \mathbf{A}^k,$$

if we set $\mathbf{A}^0 := \operatorname{Id}$. It is easily seen that, if a polynomial $f(t)$ factors as $f(t) = p(t)q(t)$, then the matrices $p(\mathbf{A})$ and $q(\mathbf{A})$ commute, and we have $f(\mathbf{A}) = p(\mathbf{A})q(\mathbf{A}) = q(\mathbf{A})p(\mathbf{A})$.

2.53 Proposition. *Let* $\mathbf{A} \in M_n(\mathbb{C})$, *and let* $p(t)$ *be a polynomial. Then*

(i) *if* λ *is an eigenvalue of* $\mathbf{A}$, *then* $p(\lambda)$ *is an eigenvalue of* $p(\mathbf{A})$,
(ii) *if* μ *is an eigenvalue of* $p(\mathbf{A})$, *then* $\mu = p(\lambda)$ *for some eigenvalue* λ *of* $\mathbf{A}$.

Proof. (i) follows observing that λ^k, $k \in \mathbb{N}$, is an eigenvalue of $\mathbf{A}^k$ if λ is an eigenvalue of $\mathbf{A}$.

(ii) Since μ is an eigenvalue of $p(\mathbf{A})$, the matrix $p(A) - \mu \operatorname{Id}$ is singular. Let $p(t) = \sum_{i=1}^{k} a_i t^i$ be of degree k, $a_k \neq 0$. By the fundamental theorem of algebra we have

$$p(t) - \mu = a_k \prod_{i=1}^{k} (t - r_i),$$

hence $p(\mathbf{A}) - \mu \operatorname{Id} = a_k \prod_{i=1}^{k} (\mathbf{A} - r_i \operatorname{Id})$ and, since $p(\mathbf{A}) - \mu \operatorname{Id}$ is singular, at least one of its factors, say $A - r_1 \operatorname{Id}$, is singular. Consequently, r_1 is an eigenvalue of $\mathbf{A}$ and trivially, $p(r_1) - \mu = 0$. $\qquad \square$

Now consider two polynomials $\mathbf{P}(t) := \sum_j \mathbf{P}_j t^j$ and $\mathbf{Q}(t) := \sum_k \mathbf{Q}_k t^k$ with $n \times n$ matrices as coefficients. Trivially, the product polynomial $\mathbf{R}(t) := \mathbf{P}(t)\mathbf{Q}(t)$ is given by

$$\mathbf{R}(t) := \sum_{j,k} \mathbf{P}_j \mathbf{Q}_k t^{j+k}.$$

2.54 Lemma. *Using the previous notation, if $\mathbf{A} \in M_{n,n}(\mathbb{C})$ commutes with the coefficients of $\mathbf{Q}(t)$, then $\mathbf{R}(\mathbf{A}) = \mathbf{P}(\mathbf{A})\mathbf{Q}(\mathbf{A})$.*

Proof. In fact,

$$\mathbf{R}(\mathbf{A}) = \sum_{j,k} \mathbf{P}_j \mathbf{Q}_k \mathbf{A}^{j+k} = \sum_{j,k} (\mathbf{P}_j \mathbf{A}^j)(\mathbf{Q}_k \mathbf{A}^k)$$
$$= \Big(\sum_j \mathbf{P}_j \mathbf{A}^j\Big)\Big(\sum_k \mathbf{Q}_k \mathbf{A}^k\Big) = \mathbf{P}(\mathbf{A})\mathbf{Q}(\mathbf{A}).$$

$\qquad \square$

2.55 Theorem (Cayley–Hamilton). *Let $\mathbf{A} \in M_{n,n}(\mathbb{C})$ and let $p_{\mathbf{A}}(s)$ be its characteristic polynomial, $p_{\mathbf{A}}(s) := \det(s \operatorname{Id} - \mathbf{A})$. Then $p_{\mathbf{A}}(\mathbf{A}) = 0$.*

Proof. Set $\mathbf{Q}(s) := s \operatorname{Id} - \mathbf{A}$, $s \in \mathbb{C}$, and denote by $\operatorname{cof} \mathbf{Q}(s)$ the matrix of cofactors of $\mathbf{Q}(s)$. By Laplace's formulas, see (1.22),

$$\operatorname{cof} \mathbf{Q}(s)\, \mathbf{Q}(s) = \mathbf{Q}(s) \operatorname{cof} \mathbf{Q}(s) = \det \mathbf{Q}(s) \operatorname{Id} = p_{\mathbf{A}}(s) \operatorname{Id}.$$

Since $\mathbf{A}$ trivially commutes with the coefficents Id and $\mathbf{A}$ of $\mathbf{Q}(s)$, Lemma 2.54 yields

$$p_{\mathbf{A}}(\mathbf{A}) = p_{\mathbf{A}}(\mathbf{A}) \operatorname{Id} = \operatorname{cof} \mathbf{Q}(\mathbf{A})\, \mathbf{Q}(\mathbf{A}) = \operatorname{cof} \mathbf{Q}(\mathbf{A}) \cdot 0 = 0.$$

$\qquad \square$

b. Factorization and invariant subspaces

Given two polynomials P_1, P_2 with $\deg P_1 \geq \deg P_2$, we may divide P_1 by P_2, i.e., uniquely decompose P_1 as $P_1 = QP_2 + R$ where $\deg R < \deg P_2$. This allows us to define the *greatest common divisor* (g.c.d.) of two polynomials that is defined up to a scalar factor, and compute it by Euclid's algorithm. Moreover, since complex polynomials factor with irreducible factors of degree 1, the g.c.d. of two complex polynomials is a constant polynomial if and only if the two polynomials have no common root. We also have

2.56 Lemma. *Let $p(t)$ and $q(t)$ be two polynomials with no common zeros. Then there exist polynomials $a(t)$ and $b(t)$ such that $a(t)\,p(t) + b(t)\,q(t) = 1 \; \forall t \in \mathbb{C}$.*

We refer the readers to [GM2], but for their convenience we add the proof of Lemma 2.56

Proof. Let

$$\mathcal{P} := \Big\{ r(t) := a(t)p(t) + b(t)q(t) \,\Big|\, a(t), b(t) \text{ are polynomials} \Big\}$$

and let $d = \alpha p + \beta q$ be the nonzero polynomial of minimum degree in $\mathcal{P}$. We claim that d divides both p and q. Otherwise, dividing p by d we would get a nonzero polynomial $r := p - md$ and, since p and d are in $\mathcal{P}$, $r = p - md \in \mathcal{P}$ also, hence a contradiction, since r has degree strictly less than d.

Then we claim that the degree of d is zero. Otherwise, d would have a root that should be common to p and q since d divides both p and q. In conclusion, d is a nonzero constant polynomial. $\qquad\square$

2.57 Proposition. *For every polynomial p, the kernel of $p(\mathbf{A})$ is an invariant subspace for $\mathbf{A} \in M_{n,n}(\mathbb{C})$.*

Proof. Let $\mathbf{w} \in \ker p(\mathbf{A})$. Since $t\,p(t) = p(t)\,t$, we infer $\mathbf{A}p(\mathbf{A}) = p(\mathbf{A})\mathbf{A}$. Therefore

$$p(\mathbf{A})(\mathbf{A}\mathbf{w}) = (p(\mathbf{A})\,\mathbf{A})\mathbf{w} = (\mathbf{A}\,p(\mathbf{A}))\mathbf{w} = \mathbf{A}\,p(\mathbf{A})\mathbf{w} = \mathbf{A}\,0 = 0.$$

Hence $\mathbf{A}w \in \ker p(\mathbf{A})$. $\qquad\square$

2.58 Proposition. *Let p be the product of two coprime polynomials, $p(t) = p_1(t)p_2(t)$, and let $\mathbf{A} \in M_{n,n}(\mathbb{C})$. Then*

$$\ker p(\mathbf{A}) := \ker p_1(\mathbf{A}) \oplus \ker p_2(\mathbf{A}).$$

Proof. By Lemma 2.56, there exist two polynomials a_1, a_2 such that $a_1(t)p_1(t) + a_2(t)p_2(t) = 1$. Hence

$$a_1(\mathbf{A})p_1(\mathbf{A}) + a_2(\mathbf{A})p_2(\mathbf{A}) = \mathrm{Id}. \tag{2.9}$$

Set

$$W_1 := \ker p_1(\mathbf{A}), \qquad W_2 := \ker p_2(\mathbf{A}), \qquad W := \ker p(\mathbf{A}).$$

Now for every $\mathbf{x} \in W$, we have $a_1(\mathbf{A})p_1(\mathbf{A})\mathbf{x} \in W_2$ since

$$p_2(\mathbf{A})a_1(\mathbf{A})p_1(\mathbf{A})\mathbf{x} = p_2(\mathbf{A})(\,\mathrm{Id} - a_2(\mathbf{A})p_2(\mathbf{A}))\mathbf{x} = (\,\mathrm{Id} - a_2(\mathbf{A})p_2(\mathbf{A}))p_2(\mathbf{A})\mathbf{x}$$
$$= a_1(\mathbf{A})p_1(\mathbf{A})p_2(\mathbf{A})\mathbf{x} = a_1(\mathbf{A})p(\mathbf{A})\mathbf{x} = 0.$$

and, similarly, $a_2(\mathbf{A})p_2(\mathbf{A})\mathbf{x} \in W_1$. Thus $W = W_1 + W_2$. Finally $W = W_1 \oplus W_2$. In fact, if $\mathbf{y} \in W_1 \cap W_2$, then by (2.9), we have

$$\mathbf{y} = a_1(\mathbf{A})p_1(\mathbf{A})\mathbf{y} + a_2(\mathbf{A})p_2(\mathbf{A})\mathbf{y} = 0 + 0 = 0.$$

$\qquad\square$

c. Generalized eigenvectors and the spectral theorem

2.59 Definition. *Let $\mathbf{A} \in M_{n,n}(\mathbb{C})$, and let λ be an eigenvalue of $\mathbf{A}$ of multiplicity k. We call generalized eigenvectors of $\mathbf{A}$ relative to the eigenvalue λ the elements of*

$$W := \ker(\lambda\,\mathrm{Id} - \mathbf{A})^k.$$

Of course,

(i) eigenvectors relative to λ are generalized eigenvectors relative to λ,
(ii) the spaces of generalized eigenvectors are invariant subspaces for $\mathbf{A}$.

2.60 Theorem. *Let* $\mathbf{A} \in M_{n,n}(\mathbb{C})$. *Let* $\lambda_1, \lambda_2, \ldots, \lambda_k$ *be the eigenvalues of* A *with multiplicities* $m_1, m_2, \ldots, m_k$ *and let* $W_1, W_2, \ldots, W_k$ *be the subspaces of the relative generalized eigenvectors,* $W_i := \ker(\lambda_i \operatorname{Id} - A)$. *Then*

(i) *the spaces* $W_1, W_2, \ldots, W_k$ *are supplementary, consequently there is a basis of* $\mathbb{C}^n$ *of generalized eigenvectors of* $\mathbf{A}$,

(ii) $\dim W_i = m_i$.

Consequently, if we choose $\mathbf{A}' \in M_{n,n}(\mathbb{K})$ *using a basis* $(e_1, e_2, \ldots, e_n)$ *where the the first* m_1 *elements span* W_1, *the following* m_2 *elements span* W_2 *and the last* m_k *elements span* W_k. *We can then write the matrix* $\mathbf{A}'$ *in the new basis similar to* $\mathbf{A}$ *where* $\mathbf{A}'$ *has the form*

$$\mathbf{A}' = \begin{pmatrix} \boxed{\mathbf{A}_1} & 0 & \cdots & 0 \\ 0 & \boxed{\mathbf{A}_2} & \cdots & 0 \\ \vdots & \vdots & \ddots & \vdots \\ 0 & 0 & \cdots & \boxed{\mathbf{A}_k} \end{pmatrix},$$

where for every $i = 1, \ldots, k$, *the block* $\mathbf{A}_i$ *is a* $m_i \times m_i$ *matrix with* λ_i *as the only eigenvector with multiplicity* m_i *and, of course,* $(\lambda_i \operatorname{Id} - \mathbf{A}_i)^{m_i} = 0$.

Proof. (i) Clearly the polynomials $p_1(s) := (\lambda_1 - s)^{m_1}$, $p_2(s) := (\lambda_2 - s)^{m_2}$, $\ldots$, $p_k(s) := (\lambda_k - s)^{m_k}$ factorize $p_\mathbf{A}$ and are coprime. Set $\mathbf{N}_i := p_i(\mathbf{A})$ and notice that $W_i = \ker \mathbf{N}_i$. Repeatedly applying Proposition 2.58, we then get

$$\ker p_\mathbf{A}(\mathbf{A}) = \ker(\mathbf{N}_1 \mathbf{N}_2 \ldots \mathbf{N}_k) = \ker(\mathbf{N}_1) \oplus \ker(\mathbf{N}_2 \mathbf{N}_3 \ldots \mathbf{N}_k)$$
$$= \cdots = W_1 \oplus W_2 \oplus \cdots \oplus W_k.$$

(i) then follows from the Cayley–Hamilton theorem, $\ker p_\mathbf{A}(\mathbf{A}) = \mathbb{C}^n$.

(ii) It remains to show that $\dim W_i = m_i \ \forall i$. Let $(e_1, e_2, \ldots, e_n)$ be a basis such that the first h_1 elements span W_1, the following h_2 elements span W_2 and the last h_k elements span W_k. $\mathbf{A}$ is therefore similar to a block matrix

$$\mathbf{A}' = \begin{pmatrix} \boxed{\mathbf{A}_1} & 0 & \cdots & 0 \\ 0 & \mathbf{A}_2 & \cdots & 0 \\ \vdots & \vdots & \ddots & \vdots \\ 0 & 0 & \cdots & \boxed{\mathbf{A}_k} \end{pmatrix},$$

where the block $\mathbf{A}_i$ is a square matrix of dimension $h_i := \dim W_i$. On the other hand, the $q_i \times q_i$ matrix $(\lambda_i \operatorname{Id} - \mathbf{A}_i)^{m_i} = 0$ hence all the eigenvalues of $\lambda_i \operatorname{Id} - \mathbf{A}_i$ are zero. Therefore $\mathbf{A}_i$ has a unique eigenvalue λ_i with multiplicity h_i, and $p_{\mathbf{A}_i}(s) := (s - \lambda_i)^{h_i}$. We then have

$$p_\mathbf{A}(s) = p_{\mathbf{A}'}(s) = \prod_{i=1}^{k} p_{\mathbf{A}_i}(s) = \prod_{i=1}^{k} (s - \lambda)^{h_i},$$

and the uniqueness of the factorization yields $h_i = m_i$. The rest of the claim is trivial.

$\square$

Another proof of $\dim W_i = m_i$ goes as follows. First we show the following.

2.61 Lemma. *If 0 is an eigenvalue of $\mathbf{B} \in M_{n,n}(\mathbb{C})$ with multiplicity m, the 0 is an eigenvalue for $\mathbf{B}^m$ with multiplicity m.*

Proof. The function $1 - \lambda^m$, $\lambda \in \mathbb{C}$, can be factorized as $1 - \lambda^m = \prod_{i=0}^{m-1}(\omega^i - \lambda)$ where $\omega := e^{i\,2\pi/m}$ is a root of unity (the two polynomials have the same degree and take the same values at the m roots of the unity and at 0). For $z, t \in \mathbb{C}$

$$z^m - t^m = z^m\left(1 - \left(\frac{t}{z}\right)^m\right) = z^m \prod_{i=0}^{m-1}\left(\omega^i - \frac{t}{z}\right) = \prod_{i=0}^{m-1}(\omega^i z - t),$$

hence

$$z^m\,\mathrm{Id} - \mathbf{B}^m = z^m\,\mathrm{Id} - \mathbf{B}^m = \prod_{i=0}^{m-1}(\omega^i z\,\mathrm{Id} - \mathbf{B}).$$

If we set $q_0(z) := \prod_{i=0}^{m-1} q(\omega^j z)$, we have $q_0(0) \neq 0$, and

$$p_{\mathbf{B}^m}(z^m) := \det(z^m\,\mathrm{Id} - \mathbf{B}^m) = \prod_{i=0}^{m-1} p_{\mathbf{B}}(\omega^i z) = \prod_{i=0}^{m-1}(\omega^j z)^m q(\omega^j z) = z^{m^2} q_0(z).$$

$$(2.10)$$

On the other hand $p_{\mathbf{B}^m} = s^r q_1(r)$ for some q_1 with $q_1(0) \neq 0$ and some $r \geq 1$. Thus, following (2.10)

$$p_{\mathbf{B}^m}(s) = s^m q_1(s),$$

i.e., 0 is an eigenvalue of multiplicity m for $\mathbf{B}^m$.   □

Another proof that $\dim W_i = m_i$ in Theorem 2.60. Since

$$\sum_{i=1}^{n} m_i = \sum_{i=1}^{n} \dim W_i = \dim X,$$

it suffices to show that $\dim W_i \leq m_i$ $\forall i$.

Since 0 is an eigenvalue of $\mathbf{B} := \lambda_i\,\mathrm{Id} - \mathbf{A}$ of multiplicity $m := m_i$, 0 is an eigenvalue of multiplicity m for $\mathbf{B}^m$ by Lemma 2.61. Since W_i is the eigenspace corresponding to the eigenvalue 0 of $\mathbf{B}^m$, it follows from Proposition 2.43 that $\dim W_i \leq m$.   □

d. Jordan's canonical form

2.62 Definition. *A matrix $\mathbf{B} \in M_{n,n}(\mathbb{K})$ is said to be* nilpotent *if there exists $k \geq 0$ such that $\mathbf{B}^k = 0$.*

Let $\mathbf{B} \in M_{q,q}(\mathbb{C})$ be a nilpotent matrix and let k be such that $\mathbf{B}^k = 0$, but $\mathbf{B}^{k-1} \neq 0$. Fix a basis $(e_1, e_2, \dots, e_s)$ of $\ker \mathbf{B}$, and, for each $i = 1, \dots, s$, set $e_i^1 := e_i$ and define $e_i^2, e_i^3, \dots, e_i^{k_i}$ to solve the systems $\mathbf{B} e_i^j := e_i^{j-1}$ for $j = 2, 3 \dots$ as long as possible. Let $\{e_i^j\}$, $j = 1, \dots, k_i$, $i = 1, \dots, q$, be the family of vectors obtained this way.

2.63 Theorem (Canonical form of a nilpotent matrix). *Let $\mathbf{B}$ be a $q \times q$ nilpotent matrix. Using the previous notation, $\{e_j^i\}$ is a basis of $\mathbb{C}^q$. Consequently, if we write $\mathbf{B}$ with respect to this basis, we get a $q \times q$ matrix $\mathbf{B}'$ similar to $\mathbf{B}$ of the form*

$$\mathbf{B}' = \begin{pmatrix} \boxed{\mathbf{B}_1} & 0 & \cdots & 0 \\ 0 & \boxed{\mathbf{B}_2} & \cdots & 0 \\ \vdots & \vdots & \ddots & \vdots \\ 0 & 0 & \cdots & \boxed{\mathbf{B}_q} \end{pmatrix}, \qquad (2.11)$$

where each block $\mathbf{B}_i$ *has dimension* k_i *and, if* $k_i > 1$, *it has the form*

$$\mathbf{B}_i = \boxed{\begin{matrix} 0 & 1 & 0 & \cdots & 0 \\ 0 & 0 & 1 & \cdots & 0 \\ 0 & 0 & 0 & \cdots & 0 \\ \vdots & \vdots & \vdots & \ddots & 1 \\ 0 & 0 & 0 & \cdots & 0 \end{matrix}}. \qquad (2.12)$$

The reduced matrix $\mathbf{B}'$ is called the *canonical Jordan form* of the nilpotent matrix $\mathbf{B}$.

Proof. The kernels $H_j := \ker \mathbf{B}^j$ of $\mathbf{B}^j$, $j = 1, \ldots, k$, form a strictly increasing sequence of subspaces

$$\{0\} = H_0 \subset H_1 \subset H_2 \subset \cdots \subset H_{k-1} \subset H_k := \mathbb{C}^q.$$

The claim then follows by iteratively applying the following lemma. □

2.64 Lemma. *For* $j = 1, 2, \ldots, k-1$, *let* $(e_1, e_2, \ldots, e_p)$ *be a basis of* H_j *and let* $x_1, x_2, \ldots, x_r$ *be all possible solutions of* $\mathbf{B}x_j = e_j$, $j = 1, \ldots, p$. *Then* $(e_1, e_2, \ldots, e_p, x_1, x_2, \ldots, x_r)$ *is a basis for* H_{j+1}.

Proof. In fact, it is easily seen that
o the vectors $e_1, e_2, \ldots, e_p, x_1, x_2, \ldots, x_r$ are linearly independent,
o $\{e_1, e_2, \ldots, e_p, x_1, x_2, \ldots, x_r\} \subset H_{j+1}$,
o the image of H_{j+1} by $\mathbf{B}$ is contained in H_j.

Thus r, which is the number of elements e_i in the image of $\mathbf{B}$, is the dimension of the image of H_{j+1} by $\mathbf{B}$. The rank formula then yields

$$\dim H_{j+1} = \dim H_j + \dim \left(\operatorname{Im} \mathbf{B} \cap H_{j+1} \right) = p + r.$$

□

Now consider a generic matrix $\mathbf{A} \in M_{n,n}(\mathbb{C})$. We first rewrite $\mathbf{A}$ using a basis of generalized eigenvectors to get a new matrix $\mathbf{A}'$ similar to $\mathbf{A}$ of the form

$$\mathbf{A}' = \begin{pmatrix} \boxed{\mathbf{A}_1} & 0 & \cdots & 0 \\ 0 & \boxed{\mathbf{A}_2} & \cdots & 0 \\ \vdots & \vdots & \ddots & \vdots \\ 0 & 0 & \cdots & \boxed{\mathbf{A}_k} \end{pmatrix}, \qquad (2.13)$$

where each block $\mathbf{A}_i$ has the dimension of the algebraic multiplicity m_i of the eigenvalue λ_i and a unique eigenvalue λ_i. Moreover, the matrix $\mathbf{C}_i := \lambda_i \,\mathrm{Id} - \mathbf{A}_i$ is nilpotent, and precisely $\mathbf{C}_i^{m_i} = 0$ and $\mathbf{C}_i^{m_1-1} \neq 0$. Applying Theorem 2.63 to each $\mathbf{C}_i$, we then show that $\mathbf{A}_i$ is similar to $\lambda_i \,\mathrm{Id} + \mathbf{B}'$ where $\mathbf{B}'$ is as (2.11). Therefore, we conclude the following.

2.65 Theorem (Jordan's canonical form). *Let* $\lambda_1, \lambda_2, \ldots, \lambda_k$ *be all distinct eigenvalues of* $\mathbf{A} \in M_{n,n}(\mathbb{C})$. *For every* $i = 1, \ldots, k$

   (i) *let* $(u_{i,1}, \ldots, u_{i,p_i})$ *be a basis of the eigenspace* V_{λ_i} *(as we know,* $p_i \leq n_i$*),*
  (ii) *consider the generalized eigenvectors relative to* λ_i *defined as follows: for any* $j = 1, 2, \ldots, p_i$,
     a) set $e_{i,j}^1 := u_{i,j}$,
     b) set $e_{i,j}^\alpha$ *to be a solution of*

$$(\mathbf{A} - \lambda_i \,\mathrm{Id})e_{i,j}^\alpha = e_{i,j}^{\alpha-1}, \qquad \alpha = 2, \ldots, \qquad (2.14)$$

     as long as the system (2.14) *is solvable.*
     c) denote by $\alpha(i,j)$ *the number of solved systems plus 1.*

Then for every $i = 1, \ldots, k$ *the list* $(e_{i,j}^\alpha)$ *with* $j = 1, \ldots, p_i$ *and* $\alpha = 1, \ldots, \alpha(i,j)$ *is a basis for the generalized eigenspace* W_i *relative to* λ_i. *Hence the full list*

$$(e_{i,j}^\alpha) \qquad i = 1 \ldots, k, j = 1, \ldots, p_i, \alpha = 1, \ldots, a(i,j) \qquad (2.15)$$

is a basis of $\mathbb{C}^n$. *By the definition of the* $\{e_{i,j}^\alpha\}$, *if we set*

$$\mathbf{S} := \Big[e_{1,1}^1, e_{1,1}^2, \ldots, e_{1,2}^1, e_{1,2}^2, \ldots, e_{2,1}^1, e_{2,1}^2, \ldots \Big],$$

the matrix $\mathbf{J} := \mathbf{S}^{-1}\mathbf{A}\mathbf{S}$, *that represents* $\mathbf{x} \to \mathbf{A}\mathbf{x}$ *in the basis* (2.15), *has the form*

$$\mathbf{J} = \begin{pmatrix} \boxed{\mathbf{J}_{1,1}} & 0 & 0 & 0 & \ldots & 0 \\ 0 & \boxed{\mathbf{J}_{1,p_1}} & 0 & 0 & \ldots & 0 \\ 0 & 0 & \boxed{\mathbf{J}_{1,2}} & 0 & \ldots & 0 \\ \vdots & \vdots & \vdots & \vdots & \ddots & \vdots \\ 0 & 0 & 0 & 0 & \ldots & \boxed{\mathbf{J}_{k,p_k}} \end{pmatrix},$$

where $i = 1, \ldots, k$, $j = 1, \ldots, p_i$, $\mathbf{J}_{i,j}$ *has dimension* $\alpha(i,j)$ *and*

$$\mathbf{J}_{i,j} = \begin{cases} \lambda_i & \text{if } \dim \mathbf{J}_{i,j} = 1, \\ \begin{pmatrix} \lambda_i & 1 & 0 & 0 & \cdots & 0 \\ 0 & \lambda_i & 1 & 0 & \cdots & 0 \\ 0 & 0 & \lambda_i & 1 & \cdots & 0 \\ \vdots & \vdots & \vdots & \vdots & \ddots & \vdots \\ 0 & 0 & 0 & \cdots & \lambda_i & 1 \\ 0 & 0 & 0 & \cdots & 0 & \lambda_i \end{pmatrix} & \text{otherwise.} \end{cases}$$

A basis with the properties of the basis in (2.15) is called a *Jordan basis*, and the matrix $\mathbf{J}$ that represents $\mathbf{A}$ in a Jordan basis is called a *canonical Jordan form* of $\mathbf{A}$.

2.66 Example. Find a canonical Jordan form of

$$\mathbf{A} = \begin{pmatrix} 2 & 0 & 0 & 0 & 0 \\ 1 & 2 & 0 & 0 & 0 \\ 0 & 1 & 2 & 0 & 0 \\ 0 & 0 & 1 & 3 & 0 \\ 1 & 0 & 0 & 1 & 3 \end{pmatrix}.$$

$\mathbf{A}$ is lower triangular hence the eigenvalues of $\mathbf{A}$ are 2 with multiplicity 3 and 3 with multiplictiy 2.

We then have

$$\mathbf{A} - 2\,\mathrm{Id} = \begin{pmatrix} 0 & 0 & 0 & 0 & 0 \\ 1 & 0 & 0 & 0 & 0 \\ 0 & 1 & 0 & 0 & 0 \\ 0 & 0 & 1 & 1 & 0 \\ 1 & 0 & 0 & 1 & 1 \end{pmatrix}.$$

$\mathbf{A} - 2\,\mathrm{Id}$ has rank 4 since the columns of $\mathbf{A}$ of indices 1, 2, 3 and 5 are linearly independent. Therefore the eigenspace V_2 has dimension $5 - 4 = 1$ by the rank formula. We now compute a nonzero eigenvalue,

$$(\mathbf{A} - 2\,\mathrm{Id}) \begin{pmatrix} x \\ y \\ z \\ t \\ u \end{pmatrix} = \begin{pmatrix} 0 \\ x \\ y \\ z+t \\ x+t+u \end{pmatrix} = \begin{pmatrix} 0 \\ 0 \\ 0 \\ 0 \\ 0 \end{pmatrix}.$$

For instance, one eigenvector is $s_1 := (0,0,1,-1,1)^T$. We now compute the Jordan basis relative to this eigenvalue. We have $e_{1,1}^1 = s_1$ and it is possible to solve

$$\begin{pmatrix} 0 \\ x \\ y \\ z+t \\ x+t+u \end{pmatrix} = \begin{pmatrix} 0 \\ 0 \\ 1 \\ -1 \\ 1 \end{pmatrix} :$$

for instance, $s_2 := e_{1,1}^2 = (0,1,0,-1,2)^T$ is a solution. Hence we compute a solution of

$$\begin{pmatrix} 0 \\ x \\ y \\ z+t \\ x+t+u \end{pmatrix} = \begin{pmatrix} 0 \\ 1 \\ 0 \\ -1 \\ 2 \end{pmatrix}$$

hence $s_3 := e_{1,1}^3 = (1,0,0,-1,2)^T$.

Looking now at the other eigenvalue,

$$\mathbf{A} - 3\,\mathrm{Id} = \begin{pmatrix} -1 & 0 & 0 & 0 & 0 \\ 1 & -1 & 0 & 0 & 0 \\ 0 & 1 & -1 & 0 & 0 \\ 0 & 0 & 1 & 0 & 0 \\ 1 & 0 & 0 & 1 & 0 \end{pmatrix},$$

A is of rank 4 since the columns of indices 1, 2, 3 and 4 are linearly independent. Thus by the rank formula, the eigenspace relative to the eigenvalue 2 has dimension 1. We now compute an eigenvector with eigenvalue 2. We need to solve

$$(\mathbf{A} - 3\,\mathrm{Id}) \begin{pmatrix} x \\ y \\ z \\ t \\ u \end{pmatrix} = \begin{pmatrix} -x \\ -y \\ y-z \\ z \\ x+t \end{pmatrix} = \begin{pmatrix} 0 \\ 0 \\ 0 \\ 0 \\ 0 \end{pmatrix}$$

and a nonzero solution is, for instance, $s_4 := (0,0,0,0,1)^T$. Finally, we compute Jordan's basis relative to this eigenvalue. A solution of

$$\begin{pmatrix} -x \\ -y \\ y-z \\ z \\ x+t \end{pmatrix} = \begin{pmatrix} 0 \\ 0 \\ 0 \\ 0 \\ 1 \end{pmatrix}$$

is given by $s_5 = e_{2,1}^2 = (0,0,0,1,0)^T$. Thus, we conclude that the matrix

$$\mathbf{S} = \begin{bmatrix} s_1 \mid s_2 \mid s_3 \mid s_4 \mid s_5 \end{bmatrix} = \begin{pmatrix} 0 & 0 & 1 & 0 & 0 \\ 0 & 1 & 0 & 0 & 0 \\ 1 & 0 & 0 & 0 & 0 \\ -1 & -1 & -1 & 0 & 1 \\ 1 & 2 & 2 & 1 & 0 \end{pmatrix}$$

is nonsingular, since the columns are linearly independent, and by construction

$$\mathbf{S}^{-1}\mathbf{A}\mathbf{S} = \begin{pmatrix} 2 & 1 & 0 & 0 & 0 \\ 0 & 2 & 1 & 0 & 0 \\ 0 & 0 & 2 & 0 & 0 \\ 0 & 0 & 0 & 3 & 1 \\ 0 & 0 & 0 & 0 & 3 \end{pmatrix}.$$

e. Elementary divisors

As we have seen, the characteristic polynomial

$$\det(s\,\mathrm{Id} - \mathbf{A}), \qquad s \in \mathbb{K},$$

is invariant by similarity transformations. However, in general the equality of two characteristic polynomials does not imply that the two matrices be similar.

2.67 Example. The unique eigenvalue of the matrix $\mathbf{A}_\mu = \begin{pmatrix} \lambda_0 & 0 \\ \mu & \lambda_0 \end{pmatrix}$ is λ_0 and has multiplicity 2. The corresponding eigenspace is given by the solutions of the system

$$\begin{cases} 0 \cdot x^1 + 0 \cdot x^2 = 0, \\ \mu x^1 + 0 \cdot x^2 = 0. \end{cases}$$

If $\mu \neq 0$, then $V_{\lambda_0,\mu}$ has dimension 1. Notice that A_0 is diagonal, while A_μ is not diagonal. Moreover, A_0 and A_μ with $\mu \neq 0$ are not similar.

It would be interesting to find a *complete set of invariants* that characterizes the class of similarity of a matrix, without going explictly into Jordan's reduction algorithm. Here we mention a few results in this direction.

Let $\mathbf{A} \in M_{n,n}(\mathbb{C})$. The determinants of the minors of order k of the matrix $s\,\mathrm{Id} - \mathbf{A}$ form a subset $\mathcal{D}_k$ of polynomials in the s variable. Denote by $D_k(s)$ the g.c.d. of these polynomials whose coefficient of the maximal degree term is normalized to 1. Moreover set $D_0(s) := 1$. Using Laplace's formula, one sees that $D_{k-1}(s)$ divides $D_k(s)$ for all $k = 1, \ldots, n$. The polynomials

$$E_k(s) := \frac{D_k(s)}{D_{k-1}(s)}, \qquad k = 1, \ldots, n,$$

are called the *elementary divisors* of $\mathbf{A}$. They form a complete set of invariants that describe the complex similarity class of $\mathbf{A}$. In fact, the following holds.

2.68 Theorem. *The following claims are equivalent*

(i) $\mathbf{A}$ *and* $\mathbf{B}$ *are similar as complex matrices,*
(ii) $\mathbf{A}$ *and* $\mathbf{B}$ *have the same Jordan's canonical form (up to permutations of rows and columns),*
(iii) $\mathbf{A}$ *and* $\mathbf{B}$ *have the same elementary divisors.*

2.3 Exercises

2.69 ¶. Write a few 3×3 real matrices and interpret them as linear maps from $\mathbb{R}^3$ into $\mathbb{R}^3$. For each of these linear maps, choose a new basis of $\mathbb{R}^3$ and write the associate matrix with respect to the new basis both in the source and the target $\mathbb{R}^3$.

2.70 ¶. Let $V_1, V_2, \ldots, V_n$ be finite-dimensional vector spaces, and let $f_0, f_1, \ldots, f_n$ be linear maps such that

$$\{0\} \overset{f_0}{\to} V_1 \overset{f_1}{\to} V_2 \overset{f_3}{\to} \ldots \overset{f_{n-2}}{\to} V_{n-1} \overset{f_{n-1}}{\to} V_n \overset{f_n}{\to} \{0\}.$$

Show that, if $\operatorname{Im}(f_i) = \ker(f_{i+1})\ \forall i = 0, \ldots, n-1$, then $\sum_{i=1}^n (-1)^i \dim V_i = 0$.

2.71 ¶. Consider $\mathbb{R}$ as a vector space over $\mathbb{Q}$. Show that 1 and ξ are linearly independent if and only if ξ is irrational, $\xi \notin \mathbb{Q}$. Give reasons to support that $\mathbb{R}$ as a vector space over $\mathbb{Q}$ is not finite dimensional.

2.72 ¶ Lagrange multipliers. Let X, Y and Z be three vector spaces over $\mathbb{K}$ and let $f : X \to Y$, $g : X \to Z$ be two linear maps. Show that $\ker g \subset \ker f$ if and only if there exists a linear map $\ell : Z \to Y$ such that $f := \ell \circ g$.

2.73 ¶. Show that the matrices

$$\begin{pmatrix} 1 & 0 \\ 0 & 1 \end{pmatrix}, \qquad \begin{pmatrix} 1 & 0 \\ 1 & 1 \end{pmatrix},$$

have the same eigenvalues but are not similar.

2.74 ¶. Let $\lambda_1, \lambda_2, \ldots, \lambda_n$ be the eigenvalues of $\mathbf{A} \in M_{n,n}(\mathbb{C})$, possibly repeated with their multiplicities. Show that $\operatorname{tr} \mathbf{A} = \lambda_1 + \cdots + \lambda_n$ and $\det \mathbf{A} = \lambda_1 \cdot \lambda_2 \cdots \lambda_n$.

2.75 ¶. Show that $p(s) = s^n + a_{n-1}s^{n-1} + \cdots + a_0$ is the characteristic polynomial of the $n \times n$ matrix

$$\begin{pmatrix} 0 & 1 & 0 & \ldots & 0 \\ 0 & 0 & 1 & \ldots & 0 \\ \vdots & \vdots & \vdots & \ddots & \vdots \\ -a_0 & -a_1 & -a_2 & \ldots & -a_{n-1} \end{pmatrix}.$$

2.76 ¶. Let $\mathbf{A} \in M_{k,k}(\mathbb{K})$, $\mathbf{B} \in M_{n,n}(\mathbb{K})$, $\mathbf{C} \in M_{k,n}(\mathbb{K})$. Compute the characteristic polynomial of the matrix

$$\mathbf{M} := \begin{pmatrix} \mathbf{A} & \mathbf{C} \\ 0 & \mathbf{B} \end{pmatrix}.$$

2.77 ¶. Let $\ell : \mathbb{C}^n \to \mathbb{C}^n$ be defined by $\ell(\mathbf{e}_i) := \mathbf{e}_{i+1}$ if $i = 1, \ldots, n-1$ and $\ell(\mathbf{e}_n) = \mathbf{e}_1$, where $\mathbf{e}_1, \mathbf{e}_2, \ldots, \mathbf{e}_n$ is the canonical basis of $\mathbb{C}^n$. Show that the associated matrix $\mathbf{L}$ is diagonizable and that the eigenvalues are all distinct. [*Hint:* Compute the characteristic polynomial.]

2.78 ¶. Let $\mathbf{A} \in M_{n,n}(\mathbb{R})$ and suppose $\mathbf{A}^2 = \mathrm{Id}$. Show that $\mathbf{A}$ is similar to

$$\begin{pmatrix} \mathrm{Id}_k & 0 \\ 0 & -\mathrm{Id}_{n-k} \end{pmatrix}.$$

for some k, $1 \le k \le n$. [*Hint:* Consider the subspaces $V_+ := \{x \,|\, \mathbf{A}x = x\}$ and $V_- := \{x \,|\, \mathbf{A}x = -x\}$ and show that $V_+ \oplus V_- = \mathbb{R}^n$.]

2.79 ¶. Let $\mathbf{A}, \mathbf{B} \in M_{n,n}(\mathbb{R})$ be two matrices such that $\mathbf{A}^2 = \mathbf{B}^2 = \mathrm{Id}$ and $\mathrm{tr}\,\mathbf{A} = \mathrm{tr}\,\mathbf{B}$. Show that $\mathbf{A}$ and $\mathbf{B}$ are similar. [*Hint:* Use Exercise 2.78.]

2.80 ¶. Show that the diagonizable matrices span $M_{n,n}(\mathbb{R})$. [*Hint:* Consider the matrices $\mathbf{M}_{ij} = \mathrm{diag}\,(1, 2, \ldots, n) + \mathbf{E}_{i,j}$ where $\mathbf{E}_{i,j}$ has value 1 at entry (i, j) and value zero otherwise.]

2.81 ¶. Let $\mathbf{A}, \mathbf{B} \in M_{n,n}(\mathbb{R})$ and let B be symmetric. Show that the polynomial $t \to \det(\mathbf{A} + t\mathbf{B})$ has degree less than $\mathrm{Rank}\,\mathbf{B}$.

2.82 ¶. Show that any linear operator $A : \mathbb{R}^n \to \mathbb{R}^n$ has an invariant subspace of dimension 1 or 2.

2.83 ¶ Fitting decomposition. Let $f : X \to X$ be a linear operator of a finite-dimensional vector space and set $f^k := f \circ \cdots \circ f$ k-times. Show that there exists k, $1 \le k \le n$ such that
  (i) $\ker(f^k) = \ker(f^{k+1})$,
  (ii) $\mathrm{Im}\,(f^k) = \mathrm{Im}\,(f^{k+1})$,
  (iii) $f_{|\mathrm{Im}\,(f^k)} : \mathrm{Im}\,(f^k) \to \mathrm{Im}\,(f^k)$ is an isomorphism,
  (iv) $f(\ker f^k) \subset \ker(f^k)$,
  (v) $f_{|\ker(f^k)} : \ker(f^k) \to \ker(f^k)$ is nilpotent,
  (vi) $V = \ker(f^k) \oplus \mathrm{Im}\,(f^k)$.

2.84 ¶. A is nilpotent if and only if all its eigenvalues are zero.

2.85 ¶. Consider the linear operators in the linear space of polynomials

$$A(P)(t) := P'(t), \qquad B(P)(t) = tP(t).$$

Compute the operator $AB - BA$.

2.86 ¶. Let A, B be linear operators on $\mathbb{R}^n$. Show that
  (i) $\mathrm{tr}\,(AB) = \mathrm{tr}\,(BA)$,
  (ii) $AB - BA \neq \mathrm{Id}$.

2.87 ¶. Show that a linear operator $C : \mathbb{R}^2 \to \mathbb{R}^2$ can be written as $C = AB - BA$ where $A, B : \mathbb{R}^2 \to \mathbb{R}^2$ are linear operators if and only if $\mathrm{tr}\,C = 0$.

2.88 ¶. Show that the Jordan canonical form of the matrix

$$
\mathbf{A} = \begin{pmatrix}
a & a_2^1 & a_3^1 & \dots & a_n^1 \\
0 & a & a_3^2 & \dots & a_n^2 \\
0 & 0 & a & \dots & a_n^3 \\
\vdots & \vdots & \vdots & \ddots & \vdots \\
0 & 0 & 0 & \dots & a
\end{pmatrix}
$$

with $a_2^1 a_3^2 \dots a_n^{n-1} \neq 0$ is

$$
\begin{pmatrix}
a & 1 & 0 & \dots & 0 \\
0 & a & 1 & \dots & 0 \\
0 & 0 & a & \dots & 0 \\
\vdots & \vdots & \vdots & \ddots & \vdots \\
0 & 0 & 0 & \dots & a
\end{pmatrix}.
$$

3. Euclidean and Hermitian Spaces

3.1 The Geometry of Euclidean and Hermitian Spaces

Until now we have introduced several different languages, linear independence, matrices and products, linear maps that are connected in several ways to linear systems and stated some results. The structure we used is essentially linearity. A new structure, the *inner product*, provides a richer framework that we shall illustrate in this chapter.

a. Euclidean spaces

3.1 Definition. *Let X be a real vector space. An* inner product *on X is a map $(\ |\ ) : X \times X \to \mathbb{R}$ which is*

○ (BILINEAR) $(x, y) \to (x|y)$ *is linear in each factor, i.e.,*

$$(\lambda x + \mu y | z) = \lambda(x|z) + \mu(y|z),$$
$$(x|\lambda y + \mu z) = \lambda(x|y) + \mu(x|z),$$

for all $x, y, z \in X$, for all $\lambda, \mu \in \mathbb{R}$.
○ (SYMMETRIC) $(x|y) = (y|x)$ *for all $x, y \in X$.*
○ (POSITIVE DEFINITE) $(x|x) \geq 0\ \forall x$ *and $(x|x) = 0$ if and only if $x = 0$.*

The nonnegative real number

$$|x| := \sqrt{(x|x)}$$

is called the norm *of $x \in X$.*

A finite-dimensional vector space X with an inner product is called an *Euclidean vector space*, and the inner product of X is called the *scalar product* of X.

3.2 Example. The map $(\ |\ ) : \mathbb{R}^n \times \mathbb{R}^n \to \mathbb{R}$ defined by

$$(\mathbf{x}|\mathbf{y}) := \mathbf{x} \bullet \mathbf{y} = \sum_{i=1}^{n} x^i y^i, \qquad \mathbf{x} := (x^1, x^2, \dots, x^n),\ \mathbf{y} := (y^1, y^2, \dots, y^n)$$

is an inner product on $\mathbb{R}^n$, called the *standard scalar product* of $\mathbb{R}^n$, and $\mathbb{R}^n$ with this scalar product is an Euclidean space. In some sense, as we shall see later, see Proposition 3.25, it is the unique Euclidean space of dimension n.

Other examples of inner products on $\mathbb{R}^n$ can be obtained by weighing the coordinates by nonnegative real numbers. Let $\lambda_1, \lambda_2, \ldots, \lambda_n$ be positive real numbers. Then

$$(\mathbf{x}|\mathbf{y}) := \sum_{i=1}^{n} \lambda_i x^i y^i, \qquad \mathbf{x} = (x^1, x^2, \ldots, x^n), \ \mathbf{y} = (y^1, y^2, \ldots, y^n)$$

is an inner product on $\mathbb{R}^n$.

Other examples of inner products in infinite-dimensional vector spaces can be found in Chapter 10.

Let X be a vector space with an inner product. From the bilinearity of the inner product we deduce that

$$
\begin{aligned}
|x+y|^2 &= (x+y|x+y) = (x|x+y) + (y|x+y) \\
&= (x|x) + 2(x|y) + (y|y) = |x|^2 + 2(x|y) + |y|^2
\end{aligned}
\tag{3.1}
$$

from which we infer the following.

3.3 Theorem. *The following hold.*

(i) (PARALLELOGRAM IDENTITY) *We have*

$$|x+y|^2 + |x-y|^2 = 2\left(|x|^2 + |y|^2\right) \qquad \forall x, y \in X.$$

(ii) (POLARITY FORMULA) *We have*

$$(x|y) = \frac{1}{4}\left(|x+y|^2 - |x-y|^2\right) \qquad \forall x, y \in X,$$

hence we can get the scalar product of x and y by computing two norms.

(iii) (CAUCHY–SCHWARZ INEQUALITY) *The following inequality holds*

$$|(x|y)| \le |x|\,|y|, \qquad \forall x, y \in X;$$

moreover, $(x|y) = |x||y|$ if and only if either $y = 0$ or $x = \lambda y$ for some $\lambda \in \mathbb{R}$, $\lambda \ge 0$.

Proof. (i), (ii) follow trivially from (3.1). Let us prove (iii). If $y = 0$, the claim is trivial. If $y \ne 0$, the function $t \to |x + ty|^2$, $t \in \mathbb{R}$, is a second order nonnegative polynomial since

$$0 \le |x+ty|^2 = (x+ty|x+ty) = (x+ty|x) + (x+ty|ty) = |x|^2 + 2(x|y)\,t + |y|^2\,t^2;$$

hence its discriminant is nonpositive, thus $((x|y))^2 - |x|^2|y|^2 \le 0$.

If $(x|y) = |x|\,|y|$, then the discriminant of $t \to |x + ty|^2$ vanishes. If $y \ne 0$, then for some $t \in \mathbb{R}$ we have $|x + ty|^2 = 0$, i.e., $x = -ty$. Finally, $-t$ is nonnegative since $-t(y|y) = (x|y) = |x|\,|y| \ge 0$. $\qquad\square$

3.4 Definition. *Let X be a vector space with an inner product. Two vectors $x, y \in X$ are said to be* orthogonal, *and we write $x \perp y$, if $(x|y) = 0$.*

From (3.1) we immediately infer the following.

3.5 Proposition (Pythagorean theorem). *Let X be a vector space with an inner product. Then two vectors $x, y \in X$ are orthogonal if and only if*

$$|x + y|^2 = |x|^2 + |y|^2.$$

3.6 Carnot's formula. Let $\mathbf{x}, \mathbf{y} \in \mathbb{R}^2$ be two nonzero vectors of $\mathbb{R}^2$, that we think of as the plane of Euclidean geometry with an orthogonal Cartesian reference. Setting $\mathbf{x} := (a, b)$, $\mathbf{y} := (c, d)$, and denoting by θ the angle between $0\mathbf{x}$ and $0\mathbf{y}$, it is easy to see that $|\mathbf{x}|$, $|\mathbf{y}|$ are the lengths of the two segments $0\mathbf{x}$ and $0\mathbf{y}$, and that $\mathbf{x} \bullet \mathbf{y} := ac + bd = |\mathbf{x}|\,|\mathbf{y}|\cos\theta$. Thus (3.1) reads as *Carnot's formula*

$$|\mathbf{x} + \mathbf{y}|^2 = |\mathbf{x}|^2 + |\mathbf{y}|^2 + 2|\mathbf{x}|\,|\mathbf{y}|\cos\theta.$$

In general, given two vectors $\mathbf{x}, \mathbf{y} \in \mathbb{R}^n$, we have by Cauchy–Schwarz inequality $|\mathbf{x} \bullet \mathbf{y}| \leq |\mathbf{x}|\,|\mathbf{y}|$, hence there exists a $\theta \in \mathbb{R}$ such that

$$\frac{\mathbf{x} \bullet \mathbf{y}}{|\mathbf{x}|\,|\mathbf{y}|} =: \cos\theta.$$

θ is called the angle between $\mathbf{x}$ and $\mathbf{y}$ *and denoted by $\widehat{\mathbf{xy}}$.* In this way (3.1) rewrites as *Carnot's formula*

$$|\mathbf{x} + \mathbf{y}|^2 = |\mathbf{x}|^2 + |\mathbf{y}|^2 + 2|\mathbf{x}|\,|\mathbf{y}|\cos\theta.$$

Notice that the angle θ is defined up to the sign, since $\cos\theta$ is an even function.

3.7 Proposition. *Let X be a Euclidean vector space and let $(\ |\ )$ be its inner product. The norm of $x \in X$,*

$$|x| := \sqrt{(x|x)}$$

is a function $|\ | : X \to \mathbb{R}$ with the following properties

(i) $|x| \in \mathbb{R}_+ \ \forall x \in X$.
(ii) (NONDEGENERACY) $|x| = 0$ *if and only if* $x = 0$.
(iii) (1-HOMOGENEITY) $|\lambda x| = |\lambda|\,|x| \ \forall \lambda \in \mathbb{R}, \ \forall x \in X$.
(iv) (TRIANGULAR INEQUALITY) $|x + y| \leq |x| + |y| \ \forall x, y \in X$.

Proof. (i), (ii), (iii) are trivial. (iv) follows from the Cauchy–Schwarz inequality since

$$|x + y|^2 = |x|^2 + |y|^2 + 2(y|x) \leq |x|^2 + |y|^2 + 2\,|(y|x)|$$
$$\leq |x|^2 + |y|^2 + 2\,|x|\,|y| = (|x| + |y|)^2).$$

$\square$

Finally, we call the *distance* between x and $y \in X$ the number $d(x, y) := |x - y|$. It is trivial to check, using Proposition 3.7, that the *distance function* $d : X \times X \to \mathbb{R}$ defined by $d(x, y) := |x - y|$, has the following properties

(i) (NONDEGENERACY) $d(x,y) \geq 0$ $\forall x, y \in X$ and $d(x,y) = 0$ if and only if $x = y$.

(ii) (SYMMETRY) $d(x,y) = d(y,x)$ $\forall x, y \in X$.

(iii) (TRIANGULAR INEQUALITY) $d(x,y) \leq d(x,z) + d(z,y)$ $\forall x, y, z \in X$.

We refer to d as the distance in X *induced by the inner product.*

3.8 Inner products in coordinates. Let X be a Euclidean space, denote by $(\ |\ )$ its inner product, and let $(e_1, e_2, \ldots, e_n)$ be a basis of X. If $x = \sum_{i=1}^n x^i e_i$, $y = \sum_{i=1}^n y^i e_i \in X$, then by linearity

$$(x|y) = \sum_{i,j} x^i y^j (e_i|e_j).$$

The matrix

$$\mathbf{G} = [g_{ij}], \qquad g_{ij} = (e_i|e_j)$$

is called the *Gram matrix* of the scalar product in the basis $(e_1, e_2, \ldots, e_n)$. Introducing the coordinate column vectors $\mathbf{x} = (x^1, x^2, \ldots, x^n)^T$ and $\mathbf{y} = (y^1, y^2, \ldots, y^n)^T \in \mathbb{R}^n$ and denoting by $\cdot\bullet\cdot$ the standard scalar product in $\mathbb{R}^n$, we have

$$(x|y) = x \bullet \mathbf{G}y = \mathbf{x}^T \mathbf{G}\mathbf{y}$$

rows by columns. We notice that

(i) $\mathbf{G}$ is *symmetric*, $\mathbf{G}^T = \mathbf{G}$, since the scalar product is symmetric,

(ii) $\mathbf{G}$ is *positive definite*, i.e., $\mathbf{x}^T \mathbf{G}\mathbf{x} \geq 0$ $\forall \mathbf{x} \in \mathbb{R}^n$ and $\mathbf{x}^T \mathbf{G}\mathbf{x} = 0$ if and only if $\mathbf{x} = 0$, in particular, $\mathbf{G}$ is invertible.

b. Hermitian spaces

A similar structure exists on complex vector spaces.

3.9 Definition. *Let X be a vector space over $\mathbb{C}$. A Hermitian product on X is a map $(\ |\ ) : X \times X \to \mathbb{C}$ which is*

(i) (SESQUILINEAR), *i.e.,*

$$(\alpha v + \beta w | z) = \alpha(v|z) + \beta(w|z),$$
$$(v|\alpha w + \beta z) = \overline{\alpha}(v|w) + \overline{\beta}(v|z)$$

$\forall v, w, z \in X$, $\forall \alpha, \beta \in \mathbb{C}$.

(ii) (HERMITIAN) $(z|w) = \overline{(w|z)}$ $\forall w, z \in X$, *in particular $(z|z) \in \mathbb{R}$* $\forall z \in X$.

(iii) (POSITIVE DEFINITE) $(z|z) \geq 0$ *and* $(z|z) = 0$ *if and only if $z = 0$.*

The nonnegative real number $|z| := \sqrt{(z|z)}$ is called the norm *of $z \in X$.*

3.10 Definition. *A finite-dimensional complex space with a Hermitian product is called a* Hermitian space.

3.11 Example. Of course the product $(z, w) \to (z|w) := \overline{w}z$ is a Hermitian product on $\mathbb{C}$. More generally, the map $(\, | \,) : \mathbb{C}^n \times \mathbb{C}^n \to \mathbb{C}$ defined by

$$(z|w) := \mathbf{z} \bullet \mathbf{w} := \sum_{j=1}^{n} z^j \overline{w^j} \qquad \forall z = (z^1, z^2, \dots, z^n), \; w = (w^1, w^2, \dots, w^n)$$

is a Hermitian product on $\mathbb{C}^n$, called the *standard Hermitian product* of $\mathbb{C}^n$. As we shall see later, see Proposition 3.25, $\mathbb{C}^n$ equipped with the standard Hermitian product is in a sense the only Hermitian space of dimension n.

Let X be a complex vector space with a Hermitian product $(\, | \,)$. From the properties of the Hermitian product we deduce

$$\begin{aligned} |z + w|^2 &= (z + w|z + w) = (z|z + w) + (w|z + w) \\ &= (z|z) + (z|w) + (w|z) + (w|w) = |z|^2 + |w|^2 + 2\Re(z|w) \end{aligned} \qquad (3.2)$$

from which we infer at once the following.

3.12 Theorem.   (i) *We have*

$$\Re(z|w) = \frac{1}{2}\Big(|z + w|^2 - |z|^2 - |w|^2\Big) \qquad \forall z, w \in X.$$

(ii) (PARALLELOGRAM IDENTITY) *We have*

$$|z + w|^2 + |z - w|^2 = 2\left(|z|^2 + |w|^2\right) \qquad \forall z, w \in X.$$

(iii) (POLARITY FORMULA) *We have*

$$4(z|w) = \Big(|z + w|^2 - |z - w|^2\Big) + i\Big(|z + iw|^2 - |z - iw|^2\Big),$$

for all $z, w \in X$. We therefore can compute the Hermitian product of z and w by computing four norms.

(iv) (CAUCHY–SCHWARZ INEQUALITY) *The following inequality holds*

$$|(z|w)| \leq |z| \, |w|, \qquad \forall z, w \in X;$$

moreover $(z|w) = |z| \, |w|$ if and only if either $w = 0$, or $z = \lambda w$ for some $\lambda \in \mathbb{R}$, $\lambda \geq 0$.

Proof. (i), (ii), (iii) follow trivially from (3.2). Let us prove (iv). Let z, $w \in X$ and $\lambda = te^{i\theta}$, $t, \theta \in \mathbb{R}$. From (3.2)

$$0 \leq |z + \lambda w|^2 = t^2|w|^2 + |z|^2 + 2t\,\Re(e^{-i\theta}(z|w)) \qquad \forall t \in \mathbb{R},$$

hence its discriminant is nonpositive, thus

$$|\Re(e^{-i\theta}(z|w))| \leq |z| \, |w|.$$

Since θ is arbitrary, we conclude $|(z|w)| \leq |z| \, |w|$. The second part of the claim then follows as in the real case. If $(z|w) = |z| \, |w|$, then the discriminant of the real polynomial $t \to |z + tw|^2$, $t \in \mathbb{R}$, vanishes. If $w \neq 0$, for some $t \in \mathbb{R}$ we have $|z + tw|^2 = 0$, i.e., $z = -tw$. Finally, $-t$ is nonnegative since $-t(w|w) = (z|w) = |z| \, |w| \geq 0$. $\qquad \square$

3.13 ¶. Let X be a complex vector space with a Hermitian product and let $z, w \in X$. Show that $|(z|w)| = |z|\,|w|$ if and only if either $w = 0$ or there exists $\lambda \in \mathbb{C}$ such that $z = \lambda w$.

3.14 Definition. *Let X be a complex vector space with a Hermitian product $(\ |\ )$. Two vectors $z, w \in X$ are said to be* orthogonal, *and we write $z \perp w$, if $(z|w) = 0$.*

From (3.2) we immediately infer the following.

3.15 Proposition (Pythagorean theorem). *Let X be a complex vector space with a Hermitian product $(\ |\ )$. If $z, w \in X$ are orthogonal, then*

$$|z + w|^2 = |z|^2 + |w|^2.$$

We see here a difference between the real and the complex cases. Contrary to the real case, two complex vectors, such that $|z+w|^2 = |z|^2 + |w|^2$ holds, need not be orthogonal. For instance, choose $X := \mathbb{C}$, $(z|w) := \overline{w}z$, and let $z = 1$ and $w = i$.

3.16 Proposition. *Let X be a complex vector space with a Hermitian product on it. The* norm *of $z \in X$,*

$$|z| := \sqrt{(z|z)},$$

is a real-valued function $|\ | : X \to \mathbb{R}$ with the following properties

  (i) $|z| \in \mathbb{R}_+ \ \forall z \in X$.
  (ii) (NONDEGENERACY) $|z| = 0$ *if and only if $z = 0$.*
  (iii) (1-HOMOGENEITY) $|\lambda z| = |\lambda|\,|z| \ \forall \lambda \in \mathbb{C}, \ \forall z \in X$.
  (iv) (TRIANGULAR INEQUALITY) $|z + w| \le |z| + |w| \ \forall z, w \in X$.

Proof. (i), (ii), (iii) are trivial. (iv) follows from the Cauchy–Schwarz inequality since

$$|z + w|^2 = |z|^2 + |w|^2 + 2\Re(z|w) \le |z|^2 + |w|^2 + 2\,|(z|w)|$$
$$\le |z|^2 + |w|^2 + 2\,|z|\,|w| = (|z| + |w|)^2).$$

$\square$

Finally, we call *distance* between two points z, w of X the real number $d(z, w) := |z - w|$. It is trivial to check, using Proposition 3.16, that the *distance function* $d : X \times X \to \mathbb{R}$ defined by $d(z, w) := |z - w|$ has the following properties

  (i) (NONDEGENERACY) $d(z, w) \ge 0 \ \forall z, w \in X$ and $d(z, w) = 0$ if and only if $z = w$.
  (ii) (SYMMETRY) $d(z, w) = d(w, z) \ \forall z, w \in X$.
  (iii) (TRIANGULAR INEQUALITY) $d(z, w) \le d(z, x) + d(x, w) \ \forall w, x, z \in X$.

We refer to d as to the distance on X *induced by the Hermitian product.*

3.17 Hermitian products in coordinates. If X is a Hermitian space, the *Gram matrix* associated to the Hermitian product is defined by setting

$$\mathbf{G} = [g_{ij}], \qquad g_{ij} := (e_i|e_j).$$

Using linearity

$$(z|w) = \sum_{i,j=1}^{n} (e_i|e_j) z^i \overline{w^j} = \mathbf{z}^T \mathbf{G} \overline{\mathbf{w}}$$

if $\mathbf{z} = (z^1, z^2, \ldots, z^n)$, $\mathbf{w} = (w^1, w^2, \ldots, w^n) \in \mathbb{C}^n$ are the coordinate vector columns of z and w in the basis $(e_1, e_2, \ldots, e_n)$. Notice that

(i) $\mathbf{G}$ is a *Hermitian matrix*, $\overline{\mathbf{G}}^T = \mathbf{G}$,
(ii) $\mathbf{G}$ is *positive definite*, i.e., $\overline{\mathbf{z}}^T \mathbf{G} \mathbf{z} \geq 0 \ \forall \mathbf{z} \in \mathbb{C}^n$ and $\overline{\mathbf{z}}^T \mathbf{G} \mathbf{z} = 0$ if and only if $\mathbf{z} = 0$, in particular, $\mathbf{G}$ is invertible.

c. Orthonormal basis and the Gram–Schmidt algorithm

3.18 Definition. *Let X be a Euclidean space with scalar product $(\ |\ )$ or a Hermitian vector space with Hermitian product $(\ |\ )$. A system of vectors $\{e_\alpha\}_{\alpha \in \mathcal{A}} \subset X$ is called* orthonormal *if*

$$(e_\alpha|e_\beta) = \delta_{\alpha\beta} \qquad \forall \alpha, \beta \in \mathcal{A}.$$

Orthonormal vectors are linearly independent. In particular, n orthonormal vectors in a Euclidean or Hermitian vector space of dimension n form a basis, called an *orthonormal basis*.

3.19 Example. The canonical basis $(e_1, e_2, \ldots, e_n)$ of $\mathbb{R}^n$ is an orthonormal basis for the standard inner product in $\mathbb{R}^n$.

Similarly, the canonical basis $(e_1, e_2, \ldots, e_n)$ of $\mathbb{C}^n$ is an orthonormal basis for the standard Hermitian product in $\mathbb{C}^n$.

3.20 ¶. Let $(\ |\ )$ be an inner (Hermitian) product on a Euclidean (Hermitian) space X of dimension n and let $\mathbf{G}$ be the associated Gram matrix in a basis $(e_1, e_2, \ldots, e_n)$. Show that $\mathbf{G} = \mathrm{Id}_n$ if and only if $(e_1, e_2, \ldots, e_n)$ is orthonormal.

Starting from a denumerable system of linearly independent vectors, we can construct a new denumerable system of orthonormal vectors that span the same subspaces by means of the *Gram–Schmidt algorithm*.

3.21 Theorem (Gram–Schmidt). *Let X be a real (complex) vector space with inner (Hermitian) product $(\ |\ )$. Let $v_1, v_2, \ldots, v_k, \ldots$ be a denumerable set of linearly independent vectors in X. Then there exist a set of orthonormal vectors $w_1, w_2, \ldots, w_k, \ldots$ such that for each $k = 1, 2, \ldots$*

$$\mathrm{Span}\Big\{w_1, w_2, \ldots, w_k\Big\} = \mathrm{Span}\Big\{v_1, v_2, \ldots, v_k\Big\}.$$

Proof. We proceed by induction. In fact, the algorithm

$$\begin{cases} w_1' = v_1, \\ w_1 := \dfrac{w_1'}{|w_1'|}, \\ w_p' = v_p - \sum_{j=1}^{p-1}(v_p|w_j)w_j, \\ w_p := \dfrac{w_p'}{|w_p'|}, \end{cases}$$

never stops since $w_p' \neq 0 \ \forall p = 1, 2, 3, \ldots$ and produces the claimed orthonormal basis.

$\square$

3.22 Proposition (Pythagorean theorem). *Let X be a real (complex) vector space with inner (Hermitian) product $(\ |\ )$. Let $(e_1, e_2, \ldots, e_k)$ be an orthonormal basis of X. Then*

$$x = \sum_{i=1}^{k}(x|e_j)e_j \qquad x \in X,$$

that is the ith coordinate of x in the basis $(e_1, e_2, \ldots, e_n)$ is the cosine director $(x|e_i)$ of x with respect to e_i. Therefore we compute

$$(x|y) = \sum_{i=1}^{k}(x|e_i)\,(y|e_i) \qquad \text{if } X \text{ is Euclidean,}$$

$$(x|y) = \sum_{i=1}^{k}(x|e_i)\,\overline{(y|e_i)} \qquad \text{if } X \text{ is Hermitian,}$$

so that in both cases Pythagoras's theorem *holds:*

$$|x|^2 = (x|x) = \sum_{i=1}^{k}|(x|e_i)|^2.$$

Proof. In fact, by linearity, for $j = 1, \ldots, k$ and $x = \sum_{i=1}^{n} x^i e_i$ we have

$$(x|e_j) = \Big(\sum_{i=1}^{n} x^i e_i \Big| e_j\Big) = \sum_{i=1}^{n} x^i(e_i|e_j) = \sum_{i=1}^{n} x^i \delta_{ij} = x^j.$$

Similarly, using linearity and assuming X is Hermitian, we have

$$(x|y) = \Big(\sum_{i=1}^{n} x^i e_i \Big| \Big(\sum_{j=1}^{n} y^j e_j\Big) = \sum_{i,j=1}^{n} x^i \overline{y^j}(e_i|e_j)$$

$$= \sum_{i,j=1}^{n} x^i \overline{y^j}\delta_{ij} = \sum_{i=1}^{k} x^i \overline{y^i},$$

hence, by the first part,

$$(x|y) = \sum_{i=1}^{n}(x|e_i)\,\overline{(y|e_i)}.$$

$\square$

d. Isometries

3.23 Definition. *Let X, Y be two real (complex) vector spaces with inner (Hermitian) products $(\ |\ )_X$ and $(\ |\ )_Y$. We say that a linear map $A : X \to Y$ is an* isometry *if and only if*

$$|A(x)|_Y = |x|_X \qquad \forall x \in X,$$

or, equivalently, compare the polar formula, if

$$(A(x)|A(y))_Y = (x|y)_X \qquad \forall x, y \in X.$$

Isometries are trivially injective, but not surjective. If there exists a surjective isometry between two Euclidean (Hermitian) spaces, then X and Y are said to be *isometric*.

3.24 ¶. Let X, Y be two real (complex) vector spaces with inner (Hermitian) products $(\ |\ )_X$ and $(\ |\ )_Y$ and let $A : X \to Y$ be a linear map. Show that the following claims are equivalent

  (i) A is an isometry,
  (ii) $\mathcal{B} \subset X$ is an orthonormal basis if and only if $A(\mathcal{B})$ is an orthonormal basis for $A(X)$.

Let X be a real vector space with inner product $(\ |\ )$ or a complex vector space with Hermitian product $(\ |\ )$. Let $(e_1, e_2, \ldots, e_n)$ be a basis in X and $\mathcal{E} : X \to \mathbb{K}^n$, $(\mathbb{K} = \mathbb{R}$ of $\mathbb{K} = \mathbb{C})$ be the corresponding system of coordinates. Proposition 3.22 implies that the following claims are equivalent.

  (i) $(e_1, e_2, \ldots, e_n)$ *is an orthonormal basis*,
  (ii) $\mathcal{E}(x) = ((x|e_1), \ldots, (x|e_n))$,
  (iii) $\mathcal{E}$ *is an isometry between X and the Euclidean space $\mathbb{R}^n$ with the standard scalar product (or $\mathbb{C}^n$ with the standard Hermitian product).*

In this way, the Gram–Schmidt algorithm yields the following.

3.25 Proposition. *Let X be a real vector space with inner product $(\ |\ )$ (or a complex vector space with Hermitian product $(\ |\ ))$ of dimension n. Then X is isometric to $\mathbb{R}^n$ with the standard scalar product (respectively, to $\mathbb{C}^n$ with the standard Hermitian product), the isometry being the coordinate system associated to an orthonormal basis.*

In other words, using an orthonormal basis on X is the same as identifying X with $\mathbb{R}^n$ (or with $\mathbb{C}^n$) with the canonical inner (Hermitian) product.

3.26 Isometries in coordinates. Let us compute the matrix associated to an isometry $R : X \to Y$ between two Euclidean spaces of dimension n and m respectively, in an orthonormal basis (so that X and Y are respectively isometric to $\mathbb{R}^n$ ($\mathbb{C}^n$) and $\mathbb{R}^m$ ($\mathbb{C}^m$) by means of the associated coordinate system). It is therefore sufficient to discuss real isometries $R : \mathbb{R}^n \to \mathbb{R}^m$ and complex isometries $R : \mathbb{C}^n \to \mathbb{C}^m$.

Let $R : \mathbb{R}^n \to \mathbb{R}^m$ be linear and let $\mathbf{R} \in M_{m,n}(\mathbb{R})$ be the associated matrix, $R(\mathbf{x}) = \mathbf{Rx}$, $\mathbf{x} \in \mathbb{R}^n$. Denoting by $(\mathbf{e}_1, \mathbf{e}_2, \ldots, \mathbf{e}_n)$ the canonical basis of $\mathbb{R}^n$,

$$\mathbf{R} = \left[\mathbf{r}_1 \,\middle|\, \mathbf{r}_2 \,\middle|\, \ldots \,\middle|\, \mathbf{r}_n\right], \qquad \mathbf{r}_i = \mathbf{R}\mathbf{e}_i \;\forall i.$$

Since $(\mathbf{e}_1, \mathbf{e}_2, \ldots, \mathbf{e}_n)$ is orthonormal, R is an isometry if and only if $(\mathbf{r}_1, \mathbf{r}_2, \ldots, \mathbf{r}_n)$ are orthonormal. In particular, $m \geq n$ and

$$\mathbf{r}_j^T \mathbf{r}_i = \mathbf{r}_i \bullet \mathbf{r}_j = \delta_{ij}$$

i.e., the matrix $\mathbf{R}$ is an *orthogonal matrix,*

$$\mathbf{R}^T \mathbf{R} = \mathrm{Id}_n.$$

When $m = n$, the isometries $R : \mathbb{R}^n \to \mathbb{R}^n$ are necessarily surjective being injective, and form a group under composition. As above, we deduce that the group of isometries of $\mathbb{R}^n$ is isomorphic to the *orthogonal group* $O(n)$ defined by

$$O(n) := \left\{ \mathbf{R} \in M_{n,n}(\mathbb{R}) \,\middle|\, \mathbf{R}^T \mathbf{R} = \mathrm{Id}_n \right\}.$$

Observe that a square orthogonal matrix $\mathbf{R}$ is invertible with $\mathbf{R}^{-1} = \mathbf{R}^T$. If follows that $\mathbf{R}\mathbf{R}^T = \mathrm{Id}$ and $|\det \mathbf{R}| = 1$.

Similarly, consider $\mathbb{C}^n$ as a Hermitian space with the standard Hermitian product. Let $R : \mathbb{C}^n \to \mathbb{C}^m$ be linear and let $\mathbf{R} \in M_{m,n}(\mathbb{C})$ be such that $R(\mathbf{z}) = \mathbf{R}\mathbf{z}$. Denoting by $(\mathbf{e}_1, \mathbf{e}_2, \ldots, \mathbf{e}_n)$ the canonical basis of $\mathbb{R}^n$,

$$\mathbf{R} = \left[\mathbf{r}_1 \,\middle|\, \mathbf{r}_2 \,\middle|\, \ldots \,\middle|\, \mathbf{r}_n\right], \qquad \mathbf{r}_i = \mathbf{R}\mathbf{e}_i \;\forall i = 1, \ldots, m.$$

Since $(\mathbf{e}_1, \mathbf{e}_2, \ldots, \mathbf{e}_n)$ is orthonormal, R is an isometry if and only if $\mathbf{r}_1, \mathbf{r}_2, \ldots, \mathbf{r}_n$ are orthonormal. In particular, $m \geq n$ and

$$\overline{\mathbf{r}_j}^T \mathbf{r}_i = \mathbf{r}_i \bullet \mathbf{r}_j = \delta_{ij}$$

i.e., the matrix $\mathbf{R}$ is a *unitary matrix,*

$$\overline{\mathbf{R}}^T \mathbf{R} = \mathrm{Id}_n.$$

When $m = n$, the isometries $R : \mathbb{C}^n \to \mathbb{C}^n$ are necessarily surjective being injective, moreover they form a group under composition. From the above, we deduce that the group of isometries of $\mathbb{C}^n$ is isomorphic to the *unitary group* $U(n)$ defined by

$$U(n) := \left\{ \mathbf{R} \in M_{n,n}(\mathbb{C}) \,\middle|\, \overline{\mathbf{R}}^T \mathbf{R} = \mathrm{Id}_n \right\}.$$

Observe that a square unitary matrix $\mathbf{R}$ is invertible with $\mathbf{R}^{-1} = \overline{\mathbf{R}}^T$. It follows that $\mathbf{R}\overline{\mathbf{R}}^T = \mathrm{Id}$ and $|\det \mathbf{R}| = 1$.

e. The projection theorem

Let X be a real (complex) vector space with inner (Hermitian) product $(\;|\;)$ that is not necessarily finite dimensional, let $V \subset X$ be a finite-dimensional linear subspace of X of dimension k and let $(e_1, e_2, \ldots, e_k)$ be an orthonormal basis of V.

We say that $x \in X$ is *orthogonal to V* if $(x|v) = 0 \;\forall v \in V$. As $(e_1, e_2, \ldots, e_k)$ is a basis of V, $x \perp V$ if and only if $(x|e_i) = 0 \;\forall i = 1, \ldots, k$.

For all $x \in X$, the vector

$$P_V(x) := \sum_{i=1}^{k} (x|e_i)e_i \in V$$

is called the *orthogonal projection* of x in V, and the map $P_V : X \to V$, $x \to P_V(x)$, the *projection map* onto V. By Proposition 3.22, $P_V(x) = x$ if $x \in V$, hence $\operatorname{Im} P = V$ and $P^2 = P$. By Proposition 3.22 we also have $|P_V(x)|^2 = \sum_{i=1}^k |(x|e_i)|^2$. The next theorem explains the name for $P_V(x)$ and shows that in fact $P_V(x)$ is well defined as it does not depend on the chosen basis $(e_1, e_2, \ldots, e_k)$.

3.27 Theorem (of orthogonal projection). *With the previous notation, there exists a unique $z \in V$ such that $x - z$ is orthogonal to V, i.e., $(x - z|v) = 0 \ \forall v \in V$. Moreover, the following claims are equivalent.*

(i) *$x - z$ is orthogonal to V, i.e., $(x - z|v) = 0 \ \forall v \in V$,*
(ii) *$z \in V$ is the orthogonal projection of x onto V, $z = P_V(x)$,*
(iii) *z is the point in V of minimum distance from x, i.e.,*

$$|x - z| < |x - v| \qquad \forall v \in V, \ v \neq z.$$

In particular, $P_V(x)$ is well defined as it does not depend on the chosen orthonormal basis and there is a unique minimizer of the function $v \to |x - v|$, $v \in V$, the vector $z = P_V(x)$.

Proof. We first prove uniqueness. If $z_1, z_2 \in V$ are such that $(x - z_i|v) = 0$ for $i = 1, 2$, then $(z_1 - z_2|v) = 0 \ \forall v \in V$, in particular $|z_1 - z_2|^2 = 0$.

(i) $\Rightarrow$ (ii). From (i) we have $(x|e_i) = (z|e_i) \ \forall i = 1, \ldots, k$. By Proposition 3.22

$$z = \sum_{i=1}^k (z|e_i)e_i = \sum_{i=1}^k (x|e_i)e_i = P_V(x).$$

This also shows existence of a point z such that $x - z$ is orthogonal to V and that the definition of $P_V(x)$ is independent of the chosen orthonormal basis $(e_1, e_2, \ldots, e_k)$.

(ii) $\Rightarrow$ (i). If $z = P_V(x)$, we have for every $j = 1, \ldots, k$

$$(x - z|e_j) = (x|e_j) - \sum_{i=1}^k (x|e_i)(e_i|e_j) = (x|e_j) - (x|e_j) = 0,$$

hence $(x - z|v) = 0 \ \forall v$.

(i) $\Rightarrow$ (iii). Let $v \in V$. Since $(x - z|v) = 0$ we have

$$|x - v|^2 = |x - z + z - v|^2 = |x - z|^2 + |z - v|^2,$$

hence (iii).

(iii) $\Rightarrow$ (i). Let $v \in V$. The function $t \to |x - z + tv|^2$, $t \in \mathbb{R}$, has a minimum point at $t = 0$. Since

$$|x - z + tv|^2 = |x - z|^2 + 2t \Re(x - z|v) + t^2|v|^2,$$

necessarily $\Re(x - z|v) = 0$. If X is a real vector space, this means $(x - z|v) = 0$, hence (i). If X is a complex vector space, from $\Re(x - z|v) = 0 \ \forall v \in V$, we also have $\Re(e^{-i\theta}(x - z|v)) = 0 \ \forall \theta \in \mathbb{R} \ \forall v \in V$, hence $(x - z|v) = 0 \ \forall v \in V$ and thus (ii). □

We can discuss linear independence in terms of an orthogonal projection. In fact, for any finite-dimensional space $V \subset X$, $x \in V$ if and only if $x - P_V(x) = 0$, equivalently, the equation $x - P_V(x) = 0$ is an implicit equation that characterizes V as the kernel of $\operatorname{Id} - P_V$.

3.28 ¶. Let $W = \mathrm{Span}\{\mathbf{v}_1, \mathbf{v}_2, \ldots, \mathbf{v}_k\}$ be a subspace of $\mathbb{K}^n$. Describe a procedure that uses the orthogonal projection theorem to find a basis of W.

3.29 ¶. Given $\mathbf{A} \in M_{m,n}(\mathbb{R})$, describe a procedure that uses the orthogonal projection theorem in order to select a maximal system of independent rows and columns of $\mathbf{A}$.

3.30 ¶. Let $\mathbf{A} \in M_{m,n}(\mathbb{R})$. Describe a procedure to find a basis of $\ker \mathbf{A}$.

3.31 ¶. Given k linear independent vectors, choose among the vectors $(\mathbf{e}_1, \mathbf{e}_2, \ldots, \mathbf{e}_n)$ of $\mathbb{R}^n$ $(n-k)$ vectors that together with $\mathbf{v}_1, \mathbf{v}_2, \ldots, \mathbf{v}_k$ form a basis of $\mathbb{R}^n$.

3.32 Projections in coordinates. Let X be a Euclidean (Hermitian) space of dimension n and let $V \subset X$ be a subspace of dimension k. Let us compute the matrix associated to the orthogonal projection operator $P_V : X \to X$ in an orthonormal basis. Of course, it suffices to think of P_V as of the orthogonal projection on a subspace of $\mathbb{R}^n$ ($\mathbb{C}^n$ in the complex case).

Let $(\mathbf{e}_1, \mathbf{e}_2, \ldots, \mathbf{e}_n)$ be the canonical basis of $\mathbb{R}^n$ and $V \subset \mathbb{R}^n$. Let $\mathbf{v}_1, \mathbf{v}_2, \ldots, \mathbf{v}_k$ be an orthonormal basis of V and denote by $\mathbf{V} = [v_j^i]$ the $n \times k$ nonsingular matrix

$$\mathbf{V} := \left[\mathbf{v}_1 \,\middle|\, \mathbf{v}_2 \,\middle|\, \cdots \,\middle|\, \mathbf{v}_k\right]$$

so that $\mathbf{v}_j = \sum_{i=1}^{n} v_j^i \mathbf{e}_i$. Let $\mathbf{P}$ be the $n \times n$ matrix associated to the orthogonal projection onto V, $P_V(\mathbf{x}) = \mathbf{P}\mathbf{x}$, or,

$$\mathbf{P} = \left[\mathbf{p}_1 \,\middle|\, \mathbf{p}_2 \,\middle|\, \cdots \,\middle|\, \mathbf{p}_n\right], \qquad \mathbf{p}_i = \mathbf{P}\mathbf{e}_i, \; i = 1, \ldots, n.$$

Then

$$\begin{aligned}
\mathbf{p}_i = P_V(\mathbf{e}_i) &= \sum_{j=1}^{k} (\mathbf{e}_i \bullet \mathbf{v}_j)\, \mathbf{v}_j = \sum_{j=1}^{k} v_j^i \mathbf{v}_j \\
&= \sum_{j=1}^{k} \sum_{h=1}^{n} v_j^i v_j^h \mathbf{e}_h = \sum_{h=1}^{n} (\mathbf{V}\mathbf{V}^T)^{ih} \mathbf{e}_h,
\end{aligned} \tag{3.3}$$

i.e.,

$$\mathbf{P} = \mathbf{V}\mathbf{V}^T.$$

The complex case is similar. With the same notation, instead of (3.3) we have

$$\begin{aligned}
\mathbf{p}_i = P_V(\mathbf{e}_i) &= \sum_{j=1}^{k} (\mathbf{e}_i \bullet \mathbf{v}_j)\, \mathbf{v}_j = \sum_{j=1}^{k} \overline{v_j^i} \mathbf{v}_j \\
&= \sum_{j=1}^{k} \sum_{h=1}^{n} \overline{v_j^i} v_j^h \mathbf{e}_h = \sum_{h=1}^{n} (\overline{\mathbf{V}}\mathbf{V}^T)^{ih} \mathbf{e}_h,
\end{aligned} \tag{3.4}$$

i.e.,

$$\mathbf{P} = \overline{\mathbf{V}}\mathbf{V}^T.$$

f. Orthogonal subspaces

Let X be a real vector space with inner product $(\,\mid\,)$ or a complex vector space with Hermitian product $(\,\mid\,)$. Suppose X is finite dimensional and let $W \subset X$ be a linear subspace of X. The subset

$$W^{\perp} := \left\{ x \in X \,\middle|\, (x|y) = 0 \;\forall y \in W \right\}$$

is called the *orthogonal of W in X*.

3.33 Proposition. *We have*

(i) $W^\perp$ *is a linear subspace of* X,
(ii) $W \cap W^\perp = \{0\}$,
(iii) $(W^\perp)^\perp = W$,
(iv) W *and* $W^\perp$ *are supplementary, hence* $\dim W + \dim W^\perp = n$,
(v) *if* P_W *and* $P_{W^\perp}$ *are respectively, the orthogonal projections onto* W *and* $W^\perp$ *seen as linear maps from* X *into itself, then* $P_{W^\perp} = \mathrm{Id}_X - P_W$.

Proof. We prove (iv) and leave the rest to the reader. Let $(v_1, v_2, \ldots, v_k)$ be a basis of W. Then we can complete $(v_1, v_2, \ldots, v_k)$ with $n - k$ vectors of the canonical basis to get a new basis of X. Then the Gram–Schmidt procedure yields an orthonormal basis $(w_1, w_2, \ldots, w_n)$ of X such that $W = \mathrm{Span}\{w_1, w_2, \ldots, w_k\}$. On the other hand $w_{k+1}, \ldots, w_n \in W^\perp$, hence $\dim W^\perp = n - k$. $\qquad\square$

g. Riesz's theorem

3.34 Theorem (Riesz). *Let* X *be a Euclidean or Hermitian space of dimension* n. *For any* $L \in X^*$ *there is a unique* $x_L \in X$ *such that*

$$L(x) = (x|x_L) \qquad \forall x \in X. \tag{3.5}$$

Proof. Assume for instance, that X is Hermitian. Suppose $L \neq 0$, otherwise we choose $x_L = 0$, and observe that $\dim \mathrm{Im}\, L = 1$, and $V := \ker L$ has dimension $n - 1$ if $\dim X = n$. Fix $x_0 \in V^\perp$ with $|x_0| = 1$, then every $x \in X$ decomposes as

$$x = x' + \lambda x_0, \qquad x' \in \ker L, \ \lambda = (x|x_0).$$

Consequently,

$$L(x) = L(x') + \lambda L(x_0) = (x|x_0)L(x_0) = (x|\overline{L(x_0)}x_0)$$

and the claim follows choosing $x_L := \overline{L(x_0)}x_0$. $\qquad\square$

The map $\beta : X^* \to X$, $L \to x_L$ defined by the Riesz theorem is called the *Riesz map*. Notice that β is linear if X is Euclidean and *antilinear* if X is Hermitian.

3.35 The Riesz map in coordinates. Let X be a Euclidean (Hermitian) space with inner (Hermitian) product $(\ |\ )$, fix a basis and denote by $\mathbf{x} = (x^1, x^2, \ldots, x^n)$ the coordinates of x, and by $\mathbf{G}$ the Gram matrix of the inner (Hermitian) product.

Let $L \in X^*$ and let $\mathbf{L}$ be the associated matrix, $L(x) = \mathbf{Lx}$. From (3.5)

$$\mathbf{Lx} = L(x) = (x|x_L) = \mathbf{x}^T \mathbf{G} \mathbf{x}_L \qquad \text{if } X \text{ is Euclidean,}$$

$$\mathbf{Lx} = L(x) = (x|x_L) = \mathbf{x}^T \mathbf{G} \overline{\mathbf{x}_L} \qquad \text{if } X \text{ is Hermitian,}$$

i.e.,

$$\mathbf{G}\mathbf{x}_L = \mathbf{L}^T \quad \text{or} \quad \mathbf{x}_L = \mathbf{G}^{-1}\mathbf{L}^T \qquad \text{if } X \text{ is Euclidean,}$$

$$\mathbf{G}\overline{\mathbf{x}_L} = \overline{\mathbf{L}}^T \quad \text{or} \quad \mathbf{x}_L = \overline{\mathbf{G}}^{-1}\overline{\mathbf{L}}^T \qquad \text{if } X \text{ is Hermitian.}$$

In particular, if the chosen basis $(e_1, e_2, \ldots, e_n)$ is orthonormal, then $\mathbf{G} = \mathrm{Id}$ and

$$\mathbf{x}_L = \mathbf{L}^T \qquad \text{if } X \text{ is Euclidean,}$$

$$\mathbf{x}_L = \overline{\mathbf{L}}^T \qquad \text{if } X \text{ is Hermitian.}$$

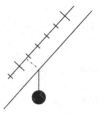

Figure 3.1. Dynamometer.

3.36 Example (Work and forces). Suppose a mass m is fixed to a dynamometer. If θ is the inclination of the dynamometer, the dynamometer shows the number

$$L = mg\cos\theta, \qquad (3.6)$$

where g is a suitable constant. Notice that we need no coordinates in $\mathbb{R}^3$ to read the dynamometer. We may model the lecture of the dynamometer as a map of the direction v of the dynamometer, that is, as a map $L : S^2 \to \mathbb{R}$ from the unit sphere $S^2 = \{x \in \mathbb{R}^3 \,\big|\, |x| = 1\}$ of the ambient space V into $\mathbb{R}$. Moreover, extending L homogeneously to the entire space V by setting $L(v) := |v|\, L(v/|v|)$, $v \in \mathbb{R}^3 \setminus \{0\}$, we see that such an extension is linear because of the simple dependence of L from the inclination. Thus we can model the *elementary work* done on the mass m, the measures made using the dynamometer, by a linear map $L : V \to \mathbb{R}$. Thinking of the ambient space V as Euclidean, by Riesz's theorem we can represent L as a scalar product, introducing a vector $F := x_L \in V$ such that

$$(v|F) = L(v) \qquad \forall v \in V.$$

We interpret such a vector as the *force* whose action on the mass produces the elementary work $L(v)$.

Now fix a basis (e_1, e_2, e_3) of V. If $F = (F^1, F^2, F^3)^T$ is the column vector of the force coordinates and $L = (L_1, L_2, L_3)$ is the 1×3 matrix of the coordinates of L in the dual basis, that is, the three readings $L_i = L(e_i)$, $i = 1, 2, 3$, of the dynamometer in the directions e_1, e_2, e_3, then, as we have seen,

$$\begin{pmatrix} F^1 \\ F^2 \\ F^3 \end{pmatrix} = \mathbf{G}^{-1} \begin{pmatrix} L_1 \\ L_2 \\ L_3 \end{pmatrix}.$$

In particular, if (e_1, e_2, e_3) is an orthonormal basis,

$$\begin{pmatrix} F^1 \\ F^2 \\ F^3 \end{pmatrix} = \begin{pmatrix} L_1 \\ L_2 \\ L_3 \end{pmatrix}.$$

h. The adjoint operator

Let X, Y be two vector spaces both on $\mathbb{K} = \mathbb{R}$ or $\mathbb{K} = \mathbb{C}$ with inner (Hermitian) products $(\ |\ )_X$ and $(\ |\ )_Y$ and let $A : X \to Y$ be a linear map. For any $y \in Y$ the map

$$x \to (A(x)|y)_Y$$

defines a linear map on X, hence by Riesz's theorem there is a unique $A^*(y) \in X$ such that

$$(A(x)|y)_Y = (y|A^*(x))_X \qquad \forall x \in X, \ \forall y \in Y. \qquad (3.7)$$

It is easily seen that the map $y \to A^*(y)$ from Y into X defined by (3.7) is linear: it is called the *adjoint* of A. Moreover,

(i) let $A, B : X \to Y$ be two linear maps between two Euclidean or Hermitian spaces. Then $(A + B)^* = A^* + B^*$,

(ii) $(\lambda A)^* = \lambda A^*$ if $\lambda \in \mathbb{R}$ and $A : X \to Y$ is a linear map between two Euclidean spaces,

(iii) $(\lambda A)^* = \overline{\lambda} A^*$ if $\lambda \in \mathbb{C}$ and $A : X \to Y$ is a linear map between two Hermitian spaces,

(iv) $(B \circ A)^* = A^* \circ B^*$ if $A : X \to Y$ and $B : Y \to Z$ are linear maps between Euclidean (Hermitian) spaces,

(v) $(A^*)^* = A$ if $A : X \to Y$ is a linear map.

3.37 ¶. Let X, Y be vector spaces. We have already defined an adjoint $\widetilde{A} : Y^* \to X^*$ with no use of inner or Hermitian products,

$$< \widetilde{A}(y^*), x >=< y^*, A(x) > \qquad \forall x \in X, \ \forall y^* \in Y^*.$$

If X and Y are Euclidean (Hermitian) spaces, denote by $\beta_X : X^* \to X$, $\beta_Y : Y^* \to Y$ the Riesz isomorphisms and by A^* the adjoint of A defined by (3.7). Show that $A^* = \beta_X \circ \widetilde{A} \circ \beta_Y^{-1}$.

3.38 The adjoint operator in coordinates. Let X, Y be two Euclidean (Hermitian) spaces with inner (Hermitian) products $(\ | \ )_X$ and $(\ | \ )_Y$. Fix two bases in X and Y, and denote the Gram matrices of the inner (Hermitian) products on X and Y respectively, by $\mathbf{G}$ and $\mathbf{H}$. Denote by $\mathbf{x}$ the coordinates of a vector x. Let $A : X \to Y$ be a linear map, A^* be the adjoint map and let $\mathbf{A}, \mathbf{A}^*$ be respectively, the associated matrices. Then we have

$$(A(x)|y)_Y = \mathbf{x}^T \mathbf{A}^T \mathbf{H} \mathbf{y}, \qquad (x|A^*(y)) = \mathbf{x}^T \mathbf{G} \mathbf{A}^* \mathbf{y},$$

if X and Y are Euclidean and

$$(A(x)|y)_Y = \mathbf{x}^T \mathbf{A}^T \mathbf{H} \overline{\mathbf{y}}, \qquad (x|A^*(y)) = \mathbf{x}^T \mathbf{G} \overline{\mathbf{A}^* \mathbf{y}},$$

if X and Y are Hermitian. Therefore

$$\mathbf{G}\mathbf{A}^* = \mathbf{A}^T \mathbf{H} \qquad \text{if } X \text{ and } Y \text{ are Euclidean,}$$

$$\mathbf{G}\overline{\mathbf{A}^*} = \mathbf{A}^T \mathbf{H} \qquad \text{if } X \text{ and } Y \text{ are Hermitian,}$$

or, recalling that $\mathbf{G}^T = \mathbf{G}$, $(\mathbf{G}^{-1})^T = \mathbf{G}^{-1}$, $\mathbf{H}^T = \mathbf{H}$ if X and Y are Euclidean and that $\overline{\mathbf{G}^T} = \mathbf{G}$, $(\overline{\mathbf{G}^{-1}})^T = \mathbf{G}^{-1}$, and $\overline{\mathbf{H}}^T = \mathbf{H}$ if X and Y are Hermitian, we find

$$\mathbf{A}^* = \mathbf{G}^{-1} \mathbf{A}^T \mathbf{H} \qquad \text{if } X \text{ and } Y \text{ are Euclidean,}$$

$$\overline{\mathbf{A}^*} = \mathbf{G}^{-1} \mathbf{A}^T \mathbf{H} \qquad \text{if } X \text{ and } Y \text{ are Hermitian.}$$

In particular,

$$\mathbf{A}^* = \mathbf{A}^T \qquad \text{in the Euclidean case,}$$

$$\mathbf{A}^* = \overline{\mathbf{A}}^T \qquad \text{in the Hermitian case} \qquad (3.8)$$

if and only if the chosen bases in X and Y are orthonormal.

3.39 Theorem. *Let $A : X \to Y$ be a linear operator between two Euclidean or two Hermitian spaces and let $A^* : Y \to X$ be its adjoint. Then* $\operatorname{Rank} A^* = \operatorname{Rank} A$. *Moreover,*

$$(\operatorname{Im} A)^{\perp} = \ker A^*, \qquad \operatorname{Im} A = (\ker A^*)^{\perp},$$
$$(\operatorname{Im} A^*)^{\perp} = \ker A, \qquad \operatorname{Im} A^* = (\ker A)^{\perp}.$$

Proof. Fix two orthonormal bases on X and Y, and let $\mathbf{A}$ be the matrix associated to A using these bases. Then, see (3.8), the matrix associated to A^* is $\mathbf{A}^T$, hence

$$\operatorname{Rank} A^* = \operatorname{Rank} \mathbf{A}^T = \operatorname{Rank} \mathbf{A} = \operatorname{Rank} A,$$

and

$$\dim(\ker A^*)^{\perp} = \dim Y - \dim \ker A^* = \operatorname{Rank} A^* = \operatorname{Rank} A = \dim \operatorname{Im} A.$$

On the other hand, $\operatorname{Im} A \subset (\ker A^*)^{\perp}$ since, if $y = A(x)$ and $A^*(v) = 0$, then $(y|v) = (A(x)|v) = (x|A^*(v)) = 0$. We then conclude that $(\ker A^*)^{\perp} = \operatorname{Im} A$. The other claims easily follow. In fact, they are all equivalent to $\operatorname{Im} A = (\ker A^*)^{\perp}$.   □

As an immediate consequence of Theorem 3.39 we have the following.

3.40 Theorem (The alternative theorem). *Let $A : X \to Y$ be a linear operator between two Euclidean or two Hermitian spaces and let $A^* : Y \to X$ be its adjoint. Then $A_{|\ker A^{\perp}} : (\ker A)^{\perp} \to \operatorname{Im} A$ and $A^*_{|\operatorname{Im} A} : \operatorname{Im} A \to (\ker A)^{\perp}$ are injective and onto, hence isomorphisms. Moreover,*

(i) *$A(x) = y$ has at least a solution if and only if y is orthogonal to $\ker A^*$,*
(ii) *y is orthogonal to $\operatorname{Im} A$ if and only if $A^*(y) = 0$,*
(iii) *A is injective if and only if A^* is surjective,*
(iv) *A is surjective if and only if A^* is injective.*

3.41 . A more direct proof of the equality $\ker A = (\operatorname{Im} A^*)^{\perp}$ is the following. For simplicity, consider the real case. Clearly, it suffices to work in coordinates and by choosing an orthonormal basis, it is enough to show that $\operatorname{Im} \mathbf{A} = (\ker \mathbf{A}^T)^{\perp}$ for every matrix $\mathbf{A} \in M_{m,n}(\mathbb{R})$.

Let $\mathbf{A} = (a^i_j) \in M_{m,n}(\mathbb{K})$ and let $\mathbf{a}^1, \mathbf{a}^2, \ldots, \mathbf{a}^m$ be the rows of $\mathbf{A}$, equivalently the columns of $\mathbf{A}^T$,

$$\mathbf{A} = \begin{pmatrix} \mathbf{a}^1 \\ \mathbf{a}^2 \\ \vdots \\ \mathbf{a}^m \end{pmatrix}, \qquad \mathbf{A}^T = \left[\mathbf{a}^1 \,\middle|\, \mathbf{a}^2 \,\middle|\, \ldots \,\middle|\, \mathbf{a}^m \right].$$

Then,

$$\mathbf{Ax} = \begin{pmatrix} a_1^1 x^1 + a_2^1 x^2 + \cdots + a_n^1 x^n \\ a_1^2 x^1 + a_2^2 x^2 + \cdots + a_n^2 x^n \\ \vdots \\ a_1^m x^1 + a_2^m x^2 + \cdots + a_n^m x^n \end{pmatrix} = \begin{pmatrix} \mathbf{a}^1 \bullet x \\ \mathbf{a}^2 \bullet x \\ \vdots \\ \mathbf{a}^m \bullet x \end{pmatrix}.$$

Consequently, $\mathbf{x} \in \ker \mathbf{A}$ if and only if $\mathbf{a}^i \bullet \mathbf{x} = 0 \ \forall i = 1, \ldots, m$, i.e.,

$$\ker \mathbf{A} = \operatorname{Span}\left\{\mathbf{a}^1, \mathbf{a}^2, \ldots, \mathbf{a}^m\right\}^\perp = (\operatorname{Im} \mathbf{A}^T)^\perp. \tag{3.9}$$

3.2 Metrics on Real Vector Spaces

In this section, we discuss bilinear forms on real vector spaces. One can develop similar considerations in the complex setting, but we refrain from doing it.

a. Bilinear forms and linear operators

3.42 Definition. *Let X be a real linear space. A bilinear form on X is a map $b : X \times X \to \mathbb{R}$ that is linear in each factor, i.e.,*

$$b(\alpha x + \beta y, z) = \alpha\, b(x, z) + \beta\, b(y, z),$$
$$b(x, \alpha y + \beta z) = \alpha\, b(x, y) + \beta\, b(z, z),$$

for all $x, y, x \in X$ and all $\alpha, \beta \in \mathbb{R}$. We denote the space of all bilinear forms on X by $\mathcal{B}(X)$.

Observe that, if $b \in \mathcal{B}(X)$, then $b(0, x) = b(0, y) = 0 \ \forall x, y \in X$. The class of bilinear forms becomes a vector space if we set

$$(b_1 + b_2)(x, y) := b_1(x, y) + b_2(x, y),$$
$$(\lambda b)(x, y) := b(\lambda x, y) = b(x, \lambda, y).$$

Suppose that X is a linear space with an inner product denoted by $(\ |\ )$. If $b \in \mathcal{B}(X)$, then for every $y \in X$ the map $x \to b(x, y)$ is linear, consequently, by Riesz's theorem there is a unique $B := B(y) \in X$ such that

$$b(x, y) = (x | B(y)) \qquad \forall x \in X. \tag{3.10}$$

It is easily seen that the map $y \to B(y)$ from Y into X defined by (3.10) is linear. Thus (3.10) defines a one-to-one correspondence between $\mathcal{B}(X)$ and the space of linear operators $\mathcal{L}(X, X)$, and it is easy to see this correspondence is a linear isomorphism between $\mathcal{B}(X)$ and $\mathcal{L}(X, X)$.

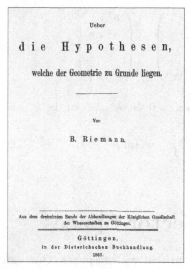

Figure 3.2. Frontispiece and a page of the celebrated dissertation of G. F. Bernhard Riemann (1826–1866).

3.43 Bilinear forms in coordinates. Let X be a finite-dimensional vector space and let $(e_1, e_2, \ldots, e_n)$ be a basis of X. Let us denote by $\mathbf{B}$ the $n \times n$ matrix, sometimes called the *Gram matrix* of b,

$$\mathbf{B} = [b_{ij}], \qquad b_{ij} = b(e_i, e_j).$$

Recall that the first index from the left is the row index. Then by linearity, if for every x, y, $\mathbf{x} = (x^1, x^2, \ldots, x^n)^T$ and $\mathbf{y} = (y^1, y^2, \ldots, y^n)^T \in \mathbb{R}^n$ are respectively, the column vectors of the coordinates of x and y, we have

$$b(x, y) = \sum_{i,j=1}^{n} b_{ij} x^i y^j = \mathbf{x}^T \bullet (\mathbf{By}) = \mathbf{x}^T \mathbf{By}.$$

In particular, a coordinate system induces a one-to-one correspondence between bilinear forms in X and bilinear forms in $\mathbb{R}^n$.

Notice that the entries of the matrix $\mathbf{B}$ have two lower indices that sum with the indices of the coordinates of the vectors x, y that have upper indices. This also reminds us that $\mathbf{B}$ is not the matrix associated to a linear operator B related to $\mathbf{B}$. In fact, if instead N is the associated linear operator to b,

$$b(x, y) = (x|N(y)) \qquad \forall x, y \in X,$$

then

$$\mathbf{y}^T \mathbf{Bx} = b(x, y) = (x|N(y)) = \mathbf{y}^T \mathbf{N}^T \mathbf{Gx}$$

where we have denoted by $\mathbf{G}$ the Gram matrix associated to the inner product on X, $\mathbf{G} = [g_{ij}]$, $g_{ij} = (e_i|e_j)$, and by $\mathbf{N}$ the $n \times n$ matrix associated to $N : X \to X$ in the basis $(e_1, e_2, \ldots, e_n)$. Thus

$$\mathbf{N}^T \mathbf{G} = \mathbf{B}$$

or, recalling that $\mathbf{G}$ is symmetric and invertible,

$$\mathbf{N} = \mathbf{G}^{-1} \mathbf{B}^T.$$

b. Symmetric bilinear forms or metrics

3.44 Definition. *Let X be a real vector space. A bilinear form $b \in \mathcal{B}(X)$ is said to be*

(i) symmetric *or a* metric, *if $b(x,y) = b(y,x)$ $\forall x, y \in X$,*
(ii) antisymmetric *if $b(x,y) = -b(y,x)$ $\forall x, y \in X$.*

The space of symmetric bilinear forms is denoted by $Sym(X)$.

3.45 ¶. Let $b \in \mathcal{B}(X)$. Show that $b_S(x,y) := \frac{1}{2}(b(x,y) + b(y,x))$, $x, y \in X$, is a symmetric bilinear form and $b_A(x,y) := \frac{1}{2}(b(x,y) - b(y,x))$, $x, y \in X$, is an antisymmetric bilinear form. In particular, one has the natural decomposition

$$b(x,y) = b_S(x,y) + b_A(x,y)$$

of b into its *symmetric* and *antisymmetric parts*. Show that b is symmetric if and only if $b = b_S$, and that b is antisymmetric if and only if $b = b_A$.

3.46 ¶. Let $b \in \mathcal{B}(X)$ be a symmetric form, and let $\mathbf{B}$ be the associated Gram matrix. Show that b is symmetric if and only if $\mathbf{B}^T = \mathbf{B}$.

3.47 ¶. Let $b \in \mathcal{B}(X)$ and let N be the associated linear operator, see (3.10). Show that N is self-adjoint, $N^* = N$, if and only if $b \in Sym(X)$. Show that $N^* = -N$ if and only if b is antisymmetric.

c. Sylvester's theorem

3.48 Definition. *Let X be a real vector space. We say that a* metric on *X, i.e., a bilinear symmetric form $g : X \times X \to \mathbb{R}$ is*

(i) nondegenerate *if $\forall x \in X$, $x \neq 0$ there is $y \in X$ such that $b(x,y) \neq 0$ and $\forall y \in X$, $y \neq 0$ there is $x \in X$ such that $b(x,y) \neq 0$,*
(ii) positively definite *if $b(x,x) > 0$ $\forall x \in X$, $x \neq 0$,*
(iii) negatively definite *if $b(x,x) < 0$ $\forall x \in X$, $x \neq 0$.*

3.49 ¶. Show that the scalar product $(x|y)$ on X is a symmetric and nondegenerate bilinear form. We shall see later, Theorems 3.52 and 3.53, that any symmetric, nondegenerate and positive bilinear form on a finite-dimensional space is actually an inner product.

3.50 Definition. *Let X be a vector space of dimension n and let $g \in Sym(X)$ be a metric on X.*

(i) *We say that a basis $(e_1, e_2, \ldots, e_n)$ is g-orthogonal if $g(e_i, e_j) = 0$ $\forall i, j = 1, \ldots, n$, $i \neq j$.*
(ii) *The radical of g is defined as the linear space*

$$\operatorname{rad}(g) := \Big\{ x \in X \mid g(x,y) = 0 \; \forall y \in X \Big\}.$$

(iii) *The range of the metric g is $r(g) := n - \dim \operatorname{rad} g$.*

Figure 3.3. Jorgen Gram (1850–1916) and James Joseph Sylvester (1814–1897).

(iv) *The* signature *of the metric g is the triplet of numbers*

$$(i_+(g), i_-(g), i_0(g))$$

where

$i_+(g) :=$ *maximum of the dimensions of the subspaces $V \subset X$*
     on which g is positive definite, $g(v, v) > 0 \ \forall v \in V, \ v \neq 0$,

$i_-(g) :=$ *maximum of the dimensions of the subspaces $V \subset X$*
     on which g is negative definite, $g(v, v) < 0 \ \forall v \in V, \ v \neq 0$,

$i_0(g) := \dim \operatorname{rad}(g)$.

One immediately sees the following.

3.51 Proposition. *We have*

(i) *The matrix associated to g in a basis $(e_1, e_2, \ldots, e_n)$ is diagonal if and only if $(e_1, e_2, \ldots, e_n)$ is g-orthogonal,*

(ii) *g is nondegenerate if and only if $\operatorname{rad}(g) = \{0\}$,*

(iii) *if $\mathbf{G}$ is the matrix associated to g in a basis, then $x \in \operatorname{rad} g$ if and only if its coordinate vector belongs to $\ker \mathbf{G}$; thus $r(g) = \operatorname{Rank} \mathbf{G}$ and g is nondegenerate if and only if $\mathbf{G}$ is not singular,*

(iv) *if X is Euclidean and $G \in \mathcal{L}(X, X)$ is the linear operator associated to g, by $g(x, y) = (x|G(y)) \ \forall x, y \in X$, then $\operatorname{rad}(G) = \ker G$, hence $r(g) = \operatorname{Rank} G$ and g is nondegenerate if and only if G is invertible.*

3.52 Theorem (Sylvester). *Let X be a finite-dimensional vector space and let $(e_1, e_2, \ldots, e_n)$ be a g-orthogonal basis for a metric g on X. Denote by n_+, n_- and n_0 the numbers of elements in the basis such that respectively, we have $g(e_i, e_i) > 0$, $g(e_i, e_i) < 0$, $g(e_i, e_i) = 0$. Then $n_+ = i_+(g)$, $n_- = i_-(g)$ and $n_0 = i_0(g)$. In particular, n_+, n_-, n_0 do not depend on the chosen g-orthogonal basis,*

$$i_+(g) + i_-(g) = r(g) \qquad and \qquad i_+(g) + i_-(g) + i_0(g) = n.$$

Proof. Suppose that $g(e_i, e_i) > 0$ for $i = 1, \ldots, n_+$. For each $v = \sum_{i=1}^{n_+} v^i e_i$, we have

$$g(v, v) = \sum_{i=1}^{n_+} |v^i|^2 g(e_i, e_i) > 0,$$

hence $\dim \mathrm{Span}\,\{e_1, e_2, \ldots, e_{n_+}\} \leq i_+(g)$. On the other hand, if $W \subset X$ is a subspace of dimension $i_+(g)$ such that $g(v, v) > 0\ \forall v \in W$, we have

$$W \cap \mathrm{Span}\,\{e_{n_++1}, \ldots, e_n\} = \{0\}$$

since $g(v, v) \leq 0$ for all $v \in \mathrm{Span}\,\{e_{n_++1}, \ldots, e_n\}$. Therefore we also have $i_+(g) \leq n - (n - n_+) = n_+$.

Similarly, one proves that $n_- = i_-(g)$. Finally, since $\mathbf{G} := [g(e_i, e_j)]$ is the matrix associated to g in the basis $(e_1, e_2, \ldots, e_n)$, we have $i_0(g) = \dim \mathrm{rad}\,(g) = \dim \ker \mathbf{G}$, and, since $\mathbf{G}$ is diagonal, $\dim \ker G = n_0$. $\qquad\square$

d. Existence of g-orthogonal bases

The Gram–Schmidt algorithm yields the existence of an orthonormal basis in a Euclidean space X. We now see that a slight modification of the Gram–Schmidt algorithm allows us to construct in a finite-dimensional space a g-orthogonal basis for a given metric g.

3.53 Theorem (Gram–Schmidt). *Let g be a metric on a finite-dimensional real vector space X. Then g has a g-orthogonal basis.*

Proof. Let r be the rank of g, $r := n - \dim \mathrm{rad}\,(g)$, and let $(w_1, w_2, \ldots, w_{n-r})$ be a basis of $\mathrm{rad}\,(g)$. If V denotes a supplementary subspace of $\mathrm{rad}\,(g)$, then V is g-orthogonal to $\mathrm{rad}\,g$ and $\dim V = r$. Moreover, for every $v \in V$ there is $z \in X$ such that $g(v, z) \neq 0$. Decomposing z as $z = w + t$, $w \in V$, $t \in \mathrm{rad}\,(g)$, we then have $g(v, w) = g(v, w) + g(v, t) = g(v, z) \neq 0$, i.e., g is nondegenerate on V. Since trivially, $(w_1, w_2, \ldots, w_{n-r})$ is g-orthogonal and V is g-orthogonal to $(w_1, w_2, \ldots, w_{n-r})$, in order to conclude it suffices to complete the basis $(w_1, w_2, \ldots, w_{n-r})$ with a g-orthogonal basis of V; in other words, it suffices to prove the claim under the further assumption that g be nondegenerate.

We proceed by induction on the dimension of X. Let $(f_1, f_2, \ldots, f_n)$ be a basis of X. We claim that there exists $e_1 \in X$ with $g(e_1, e_1) \neq 0$. In fact, if for some f_i we have $g(f_i, f_i) \neq 0$, we simply choose $e_1 := f_i$, otherwise, if $g(f_i, f_i) = 0$ for all i, for some $k \neq 0$ we must have $g(f_1, f_k) \neq 0$, since by assumption $\mathrm{rad}\,(g) = \{0\}$. In this case, we choose $e_1 := f_1 + f_k$ as

$$g(f_1 + f_k, f_1 + f_k) = g(f_1, f_1) + 2g(f_1, f_k) + g(f_k, f_k) = 0 + 2g(f_1, f_k) + 0 \neq 0.$$

Now it is easily seen that the subspace

$$V_1 := \Big\{v \in X \,\Big|\, g(e_1, v) = 0\Big\}$$

supplements $\mathrm{Span}\,\{e_1\}$, and we find a basis $(v_2, \ldots, v_n)$ of V_1 such that $g(v_j, e_1) = 0$ for all $j = 2, \ldots, n$ by setting

$$v_j := f_j - \frac{g(f_j, e_1)}{g(e_1, e_1)} e_1.$$

Since g is nondegenerate on V_1, by the induction assumption we find a g-orthogonal basis $(e_2, \ldots, e_n)$ of V_1, and the vectors $(e_1, e_2, \ldots, e_n)$ form a g-orthogonal basis of X. $\qquad\square$

A variant of the Gram–Schmidt procedure is the following one due to Carl Jacobi (1804–1851).

Let $g : X \times X \to \mathbb{R}$ be a metric on X. Let $(f_1, f_2, \ldots, f_n)$ be a basis of X, let $\mathbf{G}$ be the matrix associated to g in this basis, $\mathbf{G} = [g_{ij}]$, $g_{ij} = g(f_i, f_j)$. Set $\Delta_0 = 1$ and for $k = 1, \ldots, n$

$$\Delta_k := \det \mathbf{G}_k$$

where $\mathbf{G}_k$ is the $k \times k$ submatrix of the first k rows and k columns.

3.54 Proposition (Jacobi). *If $\Delta_k \neq 0$ for all $k = 1, \ldots, n$, there exists a g-orthogonal basis $(e_1, e_2, \ldots, e_n)$ of X; moreover*

$$g(e_k, e_k) := \frac{\Delta_{k-1}}{\Delta_k}.$$

Proof. We look for a basis $(e_1, e_2, \ldots, e_n)$ so that

$$\begin{cases} e_1 = a_1^1 f_1, \\ e_2 = a_2^1 f_1 + a_2^2 f_2, \\ \quad \vdots \\ e_n = a_n^1 f_1 + \cdots + a_n^n f_n \end{cases}$$

or, equivalently,

$$e_k := \sum_{i=1}^{k} a_k^i f_i, \qquad k = 1, \ldots, n, \tag{3.11}$$

as in the Gram–Schmidt procedure, such that $g(e_i, e_j) = 0$ for $i \neq j$. At first sight the system $g(e_i, e_j) = 0$, $i \neq j$, is a system in the unknowns a_k^i. However, if we impose that for all k's

$$g(e_k, f_i) = 0 \qquad \forall i = 1, \ldots, k-1, \tag{3.12}$$

by linearity $g(e_k, e_i) = 0$ for $i < k$, and by symmetry $g(e_k, e_i) = 0$ for $i > k$. It suffices then to fulfill (3.12) i.e., solve the system of $k-1$ equations in k unknowns $a_k^1, a_k^2, \ldots, a_k^k$

$$\sum_{j=1}^{k} g(f_j, f_i) a_k^j = 0, \qquad \forall i = 1, \ldots, k-1. \tag{3.13}$$

If we add the normalization condition

$$\sum_{j=1}^{k} g(f_j, f_k) a_k^j = 1, \tag{3.14}$$

we get a system of k equations in k unknowns of the type $\mathbf{G}_k \mathbf{x} = \mathbf{b}$, where $\mathbf{G}_k = [g_{ij}]$, $g_{ij} := g(f_i, f_j)$, $\mathbf{x} = (a_k^1, \ldots, a_k^k)^T$ and $\mathbf{b} = (0, 0, \ldots, 1)^T$. Since $\det \mathbf{G}_k = \Delta_k$ and $\Delta_k \neq 0$ by assumption, the system is solvable. Due to the arbitrariness of k, we are able to find a g-orthogonal basis of type (3.11). It remains to compute $g(e_k, e_k)$. From (3.13) and (3.14) we get

$$g(e_k, e_k) = \sum_{i,j=1}^{k} a_k^i a_k^j g(f_i, f_j) = \sum_{j=1}^{k} a_k^j \Big(\sum_{i=1}^{k} g(f_i, f_j) a_k^i \Big) = \sum_{j=1}^{k} a_k^j \delta_{jk} = a_k^k,$$

and we compute a_k^k by Cramer's formula,

$$a_k^k = \frac{\Delta_{k-1}}{\Delta_k},$$

hence $g(e_k, e_k) = \Delta_{k-1}/\Delta_k$. $\qquad\qquad\square$

3.55 Remark. Notice that Jacobi's method is a rewriting of the Gram–Schmidt procedure in the case where $g(f_i, f_i) \neq 0$ for all i's. In terms of *Gram's matrix* $\mathbf{G} := [g(e_i, e_j)]$, we have also proved that

$$\mathbf{T}^T \mathbf{G} \mathbf{T} = \operatorname{diag}\left\{ \frac{\Delta_{k-1}}{\Delta_k} \right\}$$

for a suitable triangular matrix T.

3.56 Corollary (Sylvester). *Suppose that* $\Delta_1, \ldots, \Delta_k \neq 0$. *Then the metric g is nondegenerate. Moreover,* $i_-(g)$ *equals the number of changes of sign in the sequence* $(1, \Delta_1, \Delta_2, \ldots, \Delta_n)$. *In particular, if* $\Delta_k > 0$ *for all* k's, *then g is positive definite.*

Let $(e_1, e_2, \ldots, e_n)$ be a g-orthogonal basis of X. By reordering the basis in such a way that

$$g(e_j, e_j) \begin{cases} > 0 & \text{if } j = 1, \ldots, i_+(g), \\ < 0 & \text{if } j = i_+(g) + 1, \ldots, i_+(g) + i_-(g), \\ = 0 & \text{otherwise;} \end{cases}$$

and setting

$$f_j := \begin{cases} \dfrac{e_j}{\sqrt{g(e_j, e_j)}} & \text{if } j = 1, \ldots i_+(g) + i_-(g), \\ e_j & \text{otherwise} \end{cases}$$

we get

$$g(f_j, f_j) - \begin{cases} 1 & \text{if } j = 1, \ldots, i_+(g), \\ -1 & \text{if } j = i_+(g) + 1, \ldots, i_+(g) + i_-(g), \\ 0 & \text{otherwise.} \end{cases}$$

e. Congruent matrices

It is worth seeing now how the matrix associated to a bilinear form changes when we change bases. Let $(e_1, e_2, \ldots, e_n)$ and $(f_1, f_2, \ldots, f_n)$ be two bases of X and let $\mathbf{R}$ be the matrix associated to the map $R : X \to X$, $R(e_i) := f_i$ in the basis $(e_1, e_2, \ldots, e_n)$, that is

$$\mathbf{R} := \left[\mathbf{r}_1 \,\middle|\, \mathbf{r}_2 \,\middle|\, \cdots \,\middle|\, \mathbf{r}_n \right]$$

where $\mathbf{r}_i$ is the column vector of the coordinates of f_i in the basis $(e_1, e_2, \ldots, e_n)$. As we know, if $\mathbf{x}$ and $\mathbf{x}'$ are the column vectors of the coordinates of x respectively, in the basis $(e_1, e_2, \ldots, e_n)$ and $(f_1, f_2, \ldots, f_n)$, then $\mathbf{x} = \mathbf{R}\mathbf{x}'$. Denote by $\mathbf{B}$ and $\mathbf{B}'$ the matrices associated to b respectively, in the coordinates $(e_1, e_2, \ldots, e_n)$ and $(f_1, f_2, \ldots, f_n)$. Then we have

$$b(x, y) = \mathbf{x}'^T \mathbf{B}' \mathbf{y}',$$

$$b(x, y) = \mathbf{x}^T \mathbf{B} \mathbf{y} = \mathbf{x}'^T \mathbf{R}^T \mathbf{B} \mathbf{R} \mathbf{y}',$$

hence

$$\mathbf{B}' = \mathbf{R}^T \mathbf{B} \mathbf{R}. \tag{3.15}$$

The previous argument can be of course reversed. If (3.15) holds, then $\mathbf{B}$ and $\mathbf{B}'$ are the Gram matrices of the same metric b on $\mathbb{R}^n$ in different coordinates

$$b(x, y) = \mathbf{x}^T \mathbf{b}' \mathbf{y} = (\mathbf{R}x)^T \mathbf{B}(\mathbf{R}y).$$

3.57 Definition. *Two matrices* $\mathbf{A}, \mathbf{B} \in M_{n,n}(\mathbb{R})$ *are said to be* congruent *if there exists a nonsingular matrix* $\mathbf{R} \in M_{n,n}(\mathbb{R})$ *such that* $\mathbf{B} = \mathbf{R}^T \mathbf{A} \mathbf{R}$.

It turns out that the congruence relation is an equivalence relation on matrices, thus the $n \times n$ matrices are partitioned into classes of congruent matrices. Since the matrices associated to a bilinear form in different basis are congruent, to any bilinear form corresponds a unique class of congruent matrices.

The above then reads as saying that two matrices $\mathbf{A}, \mathbf{B} \in M_{n,n}(\mathbb{R})$ are congruent if and only if they represent the same bilinear form in different coordinates. Thus, the existence of a g-orthogonal basis is equivalent to the following.

3.58 Theorem. *A symmetric matrix* $\mathbf{A} \in M_{n,n}(\mathbb{R})$ *is congruent to a diagonal matrix.*

Moreover, Sylvester's theorem reads equivalently as the following.

3.59 Theorem. *Two diagonal matrices* $\mathbf{I}, \mathbf{J} \in M_{n,n}(\mathbb{R})$ *are congruent if and only if they have the same number of positive, negative and zero entries in the diagonal. If, moreover, a symmetric matrix* $\mathbf{A} \in M_{n,n}(\mathbb{R})$ *is congruent to*

$$\mathbf{I} = \begin{pmatrix} \boxed{\mathrm{Id}_a} & 0 & 0 \\ 0 & \boxed{-\mathrm{Id}_b} & 0 \\ 0 & 0 & 0 \end{pmatrix},$$

then $(a, b, n - a - b)$ *is the signature of the metric* $\mathbf{y}^T \mathbf{A} \mathbf{x}$.

Thus the existence of a g-orthogonal matrix in conjunction with Sylvester's theorem reads as the following.

3.60 Theorem. *Two symmetric matrices* $\mathbf{A}, \mathbf{B} \in M_{n,n}(\mathbb{R})$ *are congruent if and only if the metrics* $\mathbf{y}^T \mathbf{A} \mathbf{x}$ *and* $\mathbf{y}^T \mathbf{B} \mathbf{x}$ *on* $\mathbb{R}^n$ *have the same signature* (a, b, r). *In this case,* $\mathbf{A}$ *and* $\mathbf{B}$ *are congruent to*

$$\mathbf{I} = \begin{pmatrix} \mathrm{Id}_a & 0 & 0 \\ 0 & -\mathrm{Id}_b & 0 \\ 0 & 0 & 0 \end{pmatrix}.$$

f. Classification of real metrics

Since reordering the basis elements is a linear change of coordinates, we can now reformulate Sylvester's theorem in conjunction with the existence of a g-orthonormal basis as follows. Let X, Y be two real vector spaces, and let g, h be two metrics respectively, on X and Y. We say that (X, g) and (Y, h) are *isometric* if and only if there is an isomorphism $L : X \to Y$ such that $h(L(x), L(y)) = g(x, y)\ \forall x, y \in X$. Observing that two metrics are isometric if and only if, in coordinates, their Gram matrices are congruent, from Theorem 3.60 we infer the following.

3.61 Theorem. *(X, g) and (Y, h) are isometric if and only if g and $h have the same signature,*

$$(i_+(g), i_-(g), i_0(g)) = (i_+(h), i_-(h), i_0(h)).$$

Moreover, if X has dimension n and the metric g on X has signature (a, b, r), $a + b + r = n$, then (X, g) is isometric to $(\mathbb{R}^n, h)$ where $h(\mathbf{x}, \mathbf{y}) :=$ $\mathbf{x}^T \mathbf{H} \mathbf{y}$ and

$$\mathbf{H} = \begin{pmatrix} \mathrm{Id}_a & 0 & 0 \\ 0 & -\mathrm{Id}_b & 0 \\ 0 & 0 & 0 \end{pmatrix}.$$

According to the above, the metrics over a real finite-dimensional vector space X are classified, modulus isometries, by their signature. Some of them have names:

(i) The *Euclidean metric*: $i_+(g) = n$, $i_-(g) = i_0(g) = 0$; in this case g is a scalar product.
(ii) The *pseudoeuclidean metrics*: $i_-(g) = 0$.
(iii) The *Lorenz metric* or *Minkowski metric*: $i_+(g) = n - 1$, $i_-(g) = 1$, $i_0(g) = 0$.
(iv) The *Artin metric* $i_+(g) = i_-(g) = p$, $i_0(g) = 0$.

3.62 ¶. Show that a bilinear form g on a finite-dimensional space X is an inner product on X if and only if g is symmetric and positive definite.

g. Quadratic forms

Let X be a finite-dimensional vector space over $\mathbb{R}$ and let $b \in \mathcal{B}(X)$ be a bilinear form on X. The *quadratic form* $\phi : X \to \mathbb{R}$ associated to b is defined by

$$\phi(x) = b(x, x), \qquad x \in X.$$

Observe that ϕ is fixed only by the symmetric part of b

$$b_S(x, y) := \frac{1}{2}(b(x, y) + b(y, x))$$

since $b(x, x) = b_S(x, x) \; \forall x \in X$. Moreover one can recover b_S from ϕ since b_S is symmetric,

$$b_S(x, y) = \frac{1}{2}\Big(\phi(x + y) - \phi(x) - \phi(y)\Big).$$

Another important relation between a bilinear form $b \in \mathcal{B}(X)$ and its quadratic form ϕ is the following. Let x and $v \in X$. Since

$$\phi(x + tv) = \phi(x) + t\Big(b(x, v) + b(v, x)\Big) + t^2\phi(v),$$

we have

$$\frac{d}{dt}\phi(x + tv)_{|t=0} = 2\, b_S(x, v). \tag{3.16}$$

We refer to (3.16) saying that *the symmetric part b_S of b is the first variation of the associated quadratic form.*

3.63 Homogeneous polynomials of degree two. Let $\mathbf{B} = [b_{ij}] \in M_{n,n}(\mathbb{R})$ and let

$$b(\mathbf{x}, \mathbf{y}) := \mathbf{x}^T \mathbf{B} \mathbf{y} = \sum_{i,j=1}^{n} b_{ij} x^i y^j$$

be the bilinear form defined by $\mathbf{B}$ on $\mathbb{R}^n$, $\mathbf{x} = (x^1, x^2, \ldots, x^n)$, $\mathbf{y} = (y^1, y^2, \ldots, y^n)$. Clearly,

$$\phi(\mathbf{x}) = b(\mathbf{x}, \mathbf{x}) = \mathbf{x}^T \mathbf{B} \mathbf{x} = \sum_{i,j=1}^{n} b_{ij} x^i x^j$$

is a homogeneous polynomial of degree two. Conversely, any homogeneous polynomial of degree two

$$P(x) = \sum_{\substack{i,j=1,n \\ i \le j}} b_{ij} x^i x^j = \mathbf{x}^T \mathbf{B} \mathbf{x}$$

defines a unique symmetric bilinear form in $\mathbb{R}^n$ by

$$b(\mathbf{x}, \mathbf{y}) := \frac{1}{2}\Big(P(\mathbf{x} + \mathbf{y}) - P(\mathbf{x}) - P(\mathbf{y})\Big)$$

with associated quadratic form P.

3.64 Example. Let (x, y) be the standard coordinates in $\mathbb{R}^2$. The quadratic polynomial

$$ax^2 + bxy + cy^2 = (x, y) \begin{pmatrix} a & b/2 \\ b/2 & c \end{pmatrix} \begin{pmatrix} x \\ y \end{pmatrix}$$

is the quadratic form of the metrics

$$g((x, y), (z, w)) := (z, w) \begin{pmatrix} a & b/2 \\ b/2 & c \end{pmatrix} \begin{pmatrix} x \\ y \end{pmatrix}.$$

3.65 Derivatives of a quadratic form. From (3.16) we can compute the partial derivatives of the quadratic form $\phi(x) := \mathbf{x}^T \mathbf{G} y$. In fact, choosing $v = e_h$, we have

$$\frac{\partial \phi}{\partial x^h}(x) := \frac{d}{dt}\phi(x + te_h) = 2b_S(x, e_h) = \mathbf{x}^T(\mathbf{G} + \mathbf{G}^T)e_h$$

hence, arranging the partial derivatives in a $1 \times n$ matrix, called the *Jacobian matrix* of ϕ,

$$\mathbf{D}\phi(x) := \left[\frac{\partial \phi}{\partial x^1}(x) \,\middle|\, \frac{\partial \phi}{\partial x^2}(x) \,\middle|\, \cdots \,\middle|\, \frac{\partial \phi}{\partial x^n}(x) \right]$$

we have

$$\mathbf{D}\phi(x) = \mathbf{x}^T(\mathbf{G} + \mathbf{G}^T),$$

or, taking the transpose,

$$\nabla\phi(x) := (\mathbf{D}\phi(x))^T = (\mathbf{G} + \mathbf{G}^T)\mathbf{x}.$$

h. Reducing to a sum of squares

Let g be a metric on a real vector space X of dimension n and let ϕ be the associated quadratic form. Then, choosing a basis $(e_1, e_2, \ldots, e_n)$ we have

$$\phi(x) = g(x, x) = \sum_{i=1}^{n} (x^i)^2 g(e_i, e_i)$$

if and only if $(e_1, e_2, \ldots, e_n)$ is g-orthogonal, and the number of positive, negative and zero coefficients is the signature of g. Thus, Sylvester's theorem in conjunction with the fact that we can always find a g-orthogonal basis can be rephrased as follows.

3.66 Theorem (Sylvester's law of inertia). *Let $\phi(x) = g(x, x)$ be the quadratic form associated to a metric g on an n-dimensional real vector space.*

(i) *There exists a basis $(f_1, f_2, \ldots, f_n)$ of X such that*

$$\phi(x) = \sum_{i=1}^{i_+(g)} (x^i)^2 - \sum_{i=1}^{i_-(g)} (x^i)^2, \qquad x = \sum_{i=1}^{n} x^i f_i,$$

where $(i_+(g), i_-(g), i_0(g))$ is the signature of g.

(ii) *If for some basis* $(e_1, e_2, \ldots, e_n)$

$$\phi(x) = \sum_{i=1}^{n} \phi(e_i)|x^i|^2, \qquad x := \sum_{i=1}^{n} x^i e_i, \qquad (3.17)$$

then the numbers n_+, n_- *and* n_0 *respectively, of positive, negative and zero* $\phi(e_i)$*'s are the signature* $(i_+(g), i_-(g), i_0(g))$ *of* g.

3.67 Example. In order to reduce a quadratic form ϕ to the canonical form (3.17), we may use Gram–Schmidt's algorithm. Let us repeat it focusing this time on the change of coordinates instead of on the change of basis. Suppose we want to reduce to a sum of squares by changing coordinates, the quadratic form

$$\phi(x) = \sum_{i,j=1}^{n} a_{ij} x^j x^i,$$

where at least one of the a_{ij}'s is not zero. We first look for a coefficient a_{kk} that is not zero. If we find it, we go further, otherwise if all a_{kk} vanish, at least one of the mixed terms, say a_{12}, is nonzero; the change of variables

$$\begin{cases} x^1 = y^1 + y^2, \\ x^2 = y^1 - y^2, \\ x^j = y^j \qquad \text{for } j = 3, \ldots, n, \end{cases}$$

transforms $a_{12} x^1 x^2$ into $a_{12}((y^1)^2 - (y^2)^2)$, and since $a_{11} = a_{22} = 0$, in the new coordinates $(y^1, y^2, \ldots, y^n)$ the coefficient of $(y^1)^2$ is not zero. Thus, possibly after a linear change of variables, we write ϕ as

$$\phi(x) = \frac{1}{a_{11}}(a_{11}y^1)^2 + \sum_{i=2}^{n} a_{2j} y^1 y^j + B(y^2, \ldots, y^n).$$

We now complete the square and set

$$\begin{cases} Y^1 = a_{11}y^1 + \sum_{j=2}^{n} \frac{a_{2j}}{2} y^j, \\ Y^j = y^j \qquad \text{for } j = 2, \ldots, n. \end{cases}$$

so that

$$\phi(x) = \frac{1}{a_{11}}\left(a_{11}y^1 + \sum_{j=2}^{n} \frac{a_{2j}}{2} y^j\right)^2 + C = \frac{1}{a_{11}}(Y^1)^2 + C$$

where C contains only products of $Y^2, \ldots, Y^n$. The process can then be iterated.

3.68 Example. Show that Jacobi's method in Proposition 3.54 transforms ϕ in

$$\phi(x) = \frac{\Delta_0}{\Delta_1}(x^1)^2 + \frac{\Delta_1}{\Delta_2}(x^2)^2 + \cdots + \frac{\Delta_2}{\Delta_3}(x^n)^2,$$

if $x = \sum_{i=1}^{n} x^i e_i$, for a suitable g-orthogonal basis $(e_1, e_2, \ldots, e_n)$.

3.69 Example (Classification of conics). The conics in the plane are the zeros of a second degree polynomial in two variables

$$P(x, y) := ax^2 + 2bxy + cy^2 + dx + ey + f = 0, \qquad (x, y) \in \mathbb{R}^2, \qquad (3.18)$$

where $a, b, c, d, e, f \in \mathbb{R}$. Choose a new system of coordinates (X, Y), $X = \alpha x + \beta y$, $Y = \gamma x + \delta y$ in which the quadratic part of P transforms into a sum of squares

$$ax^2 + bxy + cy^2 = pX^2 + qY^2,$$

consequently, P into

$$pX^2 + qY^2 + 2rX + 2sY + f = 0.$$

Now we can classify the conics in terms of the signs of p, q and f. If p, q are zero, the conic reduces to the straight line

$$2rX + 2sY + f = 0.$$

If $p \neq 0$ and $q = 0$, then, completing the square, the conic becomes

$$p(X - X_0)^2 + 2sY + f = 0, \qquad X_0 = \frac{r}{p},$$

i.e., a parabola with vertex in $(X_0, 0)$ and axis parallel to the axis of Y. Similarly, if $p = 0$ and $q \neq 0$, the conic is a parabola with vertex in $(0, Y_0)$, $Y_0 := s/q$, and axis parallel to the X axis. Finally, if $pq \neq 0$, completing the square, the conic is

$$p(X - X_0)^2 + q(Y - Y_0)^2 + f = 0, \qquad X_0 = r/p, \ Y_0 = s/q,$$

i.e., it is
o a hyperbola if $f \neq 0$ and $pq < 0$,
o two straight lines if $f = 0$ and $pq < 0$,
o an ellipse if $\operatorname{sgn}(f) = -\operatorname{sgn}(p)$ and $pq > 0$,
o a point if $f = 0$ and $pq > 0$,
o the empty set if $\operatorname{sgn}(f) = \operatorname{sgn}(p)$ and $pq > 0$.

Since we have operated with linear changes of coordinates that map straight lines into straight lines, ellipses into ellipses, and hyperbolas into hyperbolas, we conclude the following.

3.70 Proposition. *The conics in the plane are classified in terms of the signature of their quadratic part and of the sign of the zero term.*

3.71 ¶. The equation of a *quadric* i.e., of the zeros of a second order polynomial in n variables, see Figure 3.4 for $n = 3$, has the form

$$\phi(\mathbf{x}) := \mathbf{x}^T \mathbf{A}\mathbf{x} + 2\,\mathbf{b} \bullet x + c = 0$$

where $\mathbf{A} \in M_{n,n}(\mathbb{R})$ is symmetric, $\mathbf{b} \in X$ and $c \in \mathbb{R}$. Prove the following claims.

(i) 0 is a *center of symmetry*, i.e., $\phi(\mathbf{x}) = \phi(-\mathbf{x})$, if and only if $\mathbf{b} = 0$.
(ii) $\mathbf{x}_0$ is a center of symmetry, i.e., $\phi(\mathbf{x}_0 - \mathbf{x}) = \phi(\mathbf{x}_0 + \mathbf{x})$, if and only if $\mathbf{A}\mathbf{x}_0 = -\mathbf{b}$.
(iii) If $\det \mathbf{A} \neq 0$, then there is a center of symmetry $\mathbf{x}_0$; letting $\mathbf{x} = \mathbf{z} + \mathbf{x}_0$, we have $\phi(\mathbf{x}) = \mathbf{z}^T \mathbf{A}\mathbf{z} + c_1 = 0$ for a suitable $c_1 \in \mathbb{R}$.
(iv) By Sylvester's law of inertia, $\mathbf{z}^T \mathbf{A}\mathbf{z} + c_1$ transforms with a suitable linear change of coordinates into

$$\phi(x) = \sum_{i=1}^{p} X_i^2 - \sum_{i=p+1}^{n} X_i^2 = 1.$$

(v) Suppose $\det \mathbf{A} = 0$. Since $\mathbf{A} = \mathbf{A}^T$, we have $\ker \mathbf{A} = (\operatorname{Im}\mathbf{A})^\perp$. Choosing a basis in which the first k elements generate $\operatorname{Im}\mathbf{A}$ and the last $n - k$ $\ker \mathbf{A}$, then $\mathbf{A}$ writes as

$$\begin{pmatrix} \mathbf{A}' & 0 \\ 0 & 0 \end{pmatrix}$$

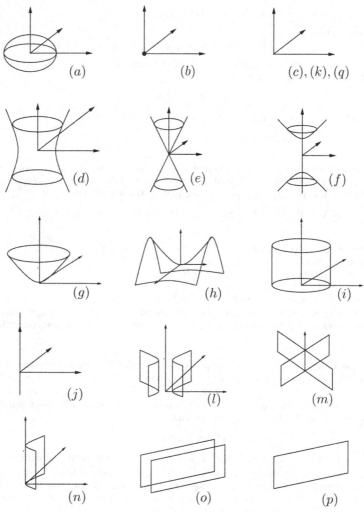

Figure 3.4. Quadrics: (a) ellipsoid: $a^2x^2+b^2y^2+c^2z^2 = 1$; (b) point: $a^2x^2+b^2y^2+c^2z^2 = 0$; (c) imaginary ellipsoid: $a^2x^2 + b^2y^2 + c^2z^2 = -1$; (d) hyperboloid of one sheet: $a^2x^2 + b^2y^2 - c^2z^2 = 1$; (e) cone: $a^2x^2 + b^2y^2 - c^2z^2 = 0$; (f) hyperboloid of two sheets: $-a^2x^2 - b^2y^2 + c^2z^2 = 1$; (g) paraboloid: $a^2x^2 + b^2y^2 - 2cz = 0$, $c > 0$; (h) saddle: $a^2x^2 - b^2y^2 - 2cz = 0$, $c > 0$; (i) elliptic cylinder: $a^2x^2 + b^2y^2 = 1$; (j) straight line: $a^2x^2 + b^2y^2 = 0$; (k) imaginary straight line: $a^2x^2 + b^2y^2 = -1$; (l) hyperbolic cylinder: $a^2x^2 - b^2y^2 = 1$; (m) nonparallel planes: $a^2x^2 - b^2y^2 = 0$; (n) parabolic cylinder: $a^2x^2 - 2cz$, $c > 0$; (o) parallel planes: $a^2x^2 = 1$; (p) plane: $a^2x^2 = 0$; (q) imaginary plane: $a^2x^2 = -1$.

in this new basis and the quadric can be written as

$$\phi(\mathbf{x}) = (\mathbf{x}')^T \mathbf{A}'\mathbf{x}' + 2(\mathbf{b}'|\mathbf{x}') + 2(\mathbf{b}''|\mathbf{x}'') + c_2 = 0$$

where $\mathbf{x}', \mathbf{b}' \in \operatorname{Im} A$, $\mathbf{x}'', \mathbf{b}'' \in \ker A$, $\mathbf{x} = \mathbf{x}' + \mathbf{x}''$, $\mathbf{b} = \mathbf{b}' + \mathbf{b}''$ and $\det \mathbf{A}' \neq 0$. Applying the argument in (iii) to

$$(\mathbf{x}')^T \mathbf{A}'\mathbf{x}' + 2\,\mathbf{b}' \bullet \mathbf{x}' + c_2,$$

we may further transform the quadric into

$$\phi(x) = (\mathbf{x}')^T \mathbf{A}'\mathbf{x}' + c_3 + 2\,\mathbf{b}'' \bullet \mathbf{x}'' = 0,$$

and, writing $y' := -2\,\mathbf{b}'' \bullet \mathbf{x}'' - c_3$, that is, by means of an affine transformation that does not change the variable $\mathbf{x}'$, we end up with $\phi(x) = (\mathbf{x}')^T \mathbf{A}'\mathbf{x}' - y' = 0$.

3.3 Exercises

3.72 ¶. Starting from specific lines or planes expressed in parametric or implicit way in $\mathbb{R}^3$, write

o the straight line through the origin perpendicular to a plane,
o the plane through the origin perpendicular to a straight line,
o the distance of a point from a straight line and from a plane,
o the distance between two straight lines,
o the perpendicular straight line to two given nonintersecting lines,
o the symmetric of a point with respect to a line and to a plane,
o the symmetric of a line with respect to a plane.

3.73 ¶. Let X, Y be two Euclidean spaces with inner products respectively, $(\ |\ )_X$ and $(\ |\ )_Y$. Show that $X \times Y$ is an Euclidean space with inner product

$$\left(\begin{pmatrix} x^1 \\ x^2 \end{pmatrix} \middle| \begin{pmatrix} y^1 \\ y^2 \end{pmatrix} \right) := (x_1|y_1) + (x_2|y_2),$$

(x_1, x_2), $(y_1, y_2) \in X \times Y$. Notice that $X \times \{0\}$ and $\{0\} \times Y$ are orthogonal subspaces of $X \times Y$.

3.74 ¶. Let $x, y \in \mathbb{R}^n$. Show that $x \perp y$ if and only if $|x - ay| \geq |x| \ \forall a \in \mathbb{R}$.

3.75 ¶. The graph of the map $A(x) := \mathbf{A}x$, $\mathbf{A} \in M_{m,n}(\mathbb{R})$ is defined as

$$G_A := \left\{ (x, y) \,\middle|\, x \in \mathbb{R}^n, \ y \in \mathbb{R}^k, \ y = A(x) \right\} \subset \mathbb{R}^n \times \mathbb{R}^k.$$

Show that G_A is a linear subspace of $\mathbb{R}^{n+k}$ of dimension n and that it is generated by the column vectors of the $(k+n) \times n$

Also show that the row vectors of the $k \times (n + k)$ matrix

$$\left(\begin{array}{c|c} \mathbf{A} & -\operatorname{Id}_k \end{array} \right)$$

generates the orthogonal to G_A.

3.76 ¶. Write in the standard basis of $\mathbb{R}^4$ the matrices of the orthogonal projection on specific subspaces of dimension 2 and 3.

3.77 ¶. Let X be Euclidean or Hermitian and let V, W be subspaces of X. Show that $V^\perp \cap W^\perp = (V + W)^\perp$.

3.78 ¶. Let $f : M_{n,n}(\mathbb{K}) \to \mathbb{K}$ be a linear map such that $f(\mathbf{AB}) = f(\mathbf{BA}) \; \forall \mathbf{A}, \mathbf{B} \in M_{n,n}(\mathbb{K})$. Show that there is $\lambda \in \mathbb{K}$ such that $f(\mathbf{X}) = \lambda \operatorname{tr} \mathbf{X}$ for all $\mathbf{X} \in M_{n,n}(\mathbb{K})$ where $\operatorname{tr} \mathbf{X} := \sum_{i=1}^{n} x_i^i$ if $\mathbf{X} = [x_j^i]$.

3.79 ¶. Show that the bilinear form $b : M_{n,n}(\mathbb{R}) \times M_{n,n}(\mathbb{R}) \to \mathbb{R}$ given by

$$b(\mathbf{A}, \mathbf{B}) := \operatorname{tr}(\mathbf{A}^T \mathbf{B}) := \sum_{i=1}^{n} (\mathbf{A}^T \mathbf{B})_i^i$$

defines an inner product on the real vector space $M_{n,n}(\mathbb{R})$. Find the orthogonal of the symmetric matrices.

3.80 ¶. Given $n + 1$ points $z_1, z_2, \ldots, z_{n+1}$ in $\mathbb{C}$, show that there exists a unique polynomial of degree at most n with prescribed values at $z_1, z_2, \ldots, z_{n+1}$. [*Hint:* If $\mathcal{P}_n$ is the set of complex polynomials of degree at most n, consider the map $\phi : \mathcal{P}_n \to \mathbb{C}^{n+1}$ given by $\phi(P) := (P(z_1), P(z_2), \ldots, P(z_n))$.]

3.81 ¶ Discrete integration. Let $t_1, t_2, \ldots, t_n$ be n points in $[a, b] \subset \mathbb{R}$. Show that there are constants $a_1, a_2, \ldots, a_n$ such that

$$\int_a^b P(t) \, dt = \sum_{j=1}^{n} a_j P(t_j)$$

for every polynomial of degree at most $n - 1$.

3.82 ¶. Let

$$Q := [0, 1]^n = \left\{ x \in \mathbb{R}^n \; \middle| \; 0 \le x_i \le 1, \; i = 1, \ldots, n \right\}$$

be the cube of side one in $\mathbb{R}^n$. Show that its diagonal has length $\sqrt{n}$. Denote by $x_1, \ldots, x_{2^n}$ the vertices of Q and by $\overline{x} := (1/2, 1/2, \ldots, 1/2)$ the center of Q. Show that the balls around $\overline{x}$ that do not intersect the balls $B(x_i, 1/2)$, $i = 1, \ldots, 2^n$, necessarily have radius at most $R_n := (\sqrt{n} - 2)/2$. Conclude that for $n > 4$, $B(\overline{x}, R_n)$ is not contained in Q.

3.83 ¶. Give a few explicit metrics in $\mathbb{R}^3$ and find the corresponding orthogonal bases.

3.84 ¶. Reduce a few explicit quadratic forms in $\mathbb{R}^3$ and $\mathbb{R}^4$ to their canonical form.

4. Self-Adjoint Operators

In this chapter, we deal with self-adjoint operators on a Euclidean or Hermitian space, and, more precisely, with the spectral theory for self-adjoint and normal operators. In the last section, we shall see methods and results of linear algebra at work on some specific examples and problems.

4.1 Elements of Spectral Theory

4.1.1 Self-adjoint operators

a. Self-adjoint operators

4.1 Definition. *Let X be a Euclidean or Hermitian space X. A linear operator $A : X \to X$ is called* self-adjoint *if $A^* = A$.*

As we can see, if $\mathbf{A}$ is the matrix associated to A in an *orthonormal* basis, then $\mathbf{A}^T$ and $\overline{\mathbf{A}}^T$ are the matrices associated to A^* in the same basis according to whether X is Euclidean or Hermitian. In particular, A is self-adjoint if and only if $\mathbf{A} = \mathbf{A}^T$ in the Euclidean case and $\overline{\mathbf{A}} = \mathbf{A}^T$ in the Hermitian case.

Moreover, as a consequence of the alternative theorem we have

$$X = \ker A \oplus \operatorname{Im} A, \qquad \ker A \perp \operatorname{Im} A$$

if $A : X \to X$ is self-adjoint. Finally, notice that the space of self-adjoint operators is a subalgebra of $\mathcal{L}(X, X)$. Typical examples of self-adjoint operators are the orthogonal projection operators. In fact, we have the following.

4.2 Proposition. *Let X be a Euclidean or Hermitian space and let $P : X \to X$ be a linear operator. P is the orthogonal projection onto its image if and only if $P^* = P$ and $P \circ P = P$.*

Proof. This follows, for instance, from 3.32. Here we present a more direct proof.

Suppose P is the orthogonal projection onto its image. Then for every $y \in X$ $(y - P(y)|z) = 0 \; \forall z \in \operatorname{Im} P$. Thus $y = P(y)$ if $y \in \operatorname{Im} P$, that is $P(x) = P \circ P(x) = P^2(x)$ $\forall x \in X$. Moreover, for $x, y \in X$

$$0 = (x - P(x)|P(y)) = (x|P(y)) - (P(x)|P(y)),$$
$$0 = (P(x)|y - P(y)) = (P(x)|y) - (P(x)|P(y)),$$

hence,

$$(P(x)|y) = (x|P(y)), \qquad \text{i.e.,} \qquad P^* = P.$$

Conversely, if $P^* = P$ and $P^2 = P$ we have

$$(x - P(x)|P(z)) = (P^*(x - P(x))|z) = (P(x) - P^2(x)|z) = (P(x) - P(x)|z) = 0$$

for all $z \in X$. $\qquad\qquad\qquad\qquad\qquad\qquad\qquad\qquad\qquad\qquad\qquad$ □

b. The spectral theorem

The following theorem, as we shall see, yields a characterization of the self-adjoint operators.

4.3 Theorem (Spectral theorem). *Let $A : X \to X$ be a self-adjoint operator on the Euclidean or Hermitian space X. Then X has an orthonormal basis made of eigenvectors of X.*

In order to prove Theorem 4.3 let us first state the following.

4.4 Proposition. *Under the hypothesis of Theorem 4.3 we have*

(i) *A has n real eigenvalues, if counted with multiplicity,*
(ii) *if $V \subset \mathbb{R}^n$ is an invariant subspace under A, then $V^\perp$ is also invariant under A,*
(iii) *eigenvectors corresponding to distinct eigenvalues are orthogonal.*

Proof. (i) Assume X is Hermitian and let $\mathbf{A} \in M_{n,n}(\mathbb{R})$ be the matrix associated to A in an orthonormal basis. Then $\overline{\mathbf{A}} = \mathbf{A}^T$, and $\mathbf{A}$ has n complex eigenvalues, if counted with multiplicity. Let $\mathbf{z} \in \mathbb{C}^n$ be an eigenvector with eigenvalue $\lambda \in \mathbb{C}$. Then

$$\overline{\mathbf{A}}\,\overline{\mathbf{z}} = \overline{\mathbf{A}\mathbf{z}} = \overline{\lambda \mathbf{z}} = \overline{\lambda}\,\overline{\mathbf{z}}.$$

Consequently, if $\mathbf{A} = (a_j^i)$, $\mathbf{z} = (z^1, z^2, \ldots, z^n)$, we have

$$\begin{cases} \lambda|\mathbf{z}|^2 = \sum_{i=1}^n \lambda z^i \, \overline{z^i} = \sum_{i=1}^n (\mathbf{A}z)^i \, \overline{z^i} = \sum_{i,j=1}^n a_j^i \, z^j \, \overline{z^i}, \\ \overline{\lambda}|\mathbf{z}|^2 = \sum_{i=1}^n \overline{\lambda} \, z^i \, \overline{z^i} = \sum_{i=1}^n z^i \, \overline{\mathbf{A}z}^i = \sum_{i,j=1}^n \overline{a_j^i \, z^j} \, z^i, = \sum_{i,j=1}^n \overline{a_i^j} \, z^j \, \overline{z^i}. \end{cases}$$

Since $a_j^i = \overline{a_i^j}$ for all $i, j = 1, \ldots, n$, we conclude that

$$(\lambda - \overline{\lambda})|\mathbf{z}|^2 = 0 \qquad \text{i.e.,} \qquad \lambda \in \mathbb{R}.$$

In the Euclidean case, $\mathbf{A}^T = \mathbf{A} = \overline{\mathbf{A}}$, also.

(ii) Let $w \in V^\perp$. For every $v \in V$ we have $(A(w)|v) = (w|A(v)) = 0$ since $A(v) \in V$ and $w \in V^\perp$. Thus $A(w) \perp V$.

(iii) Let x, y be eigenvectors of A with eigenvalues λ, μ, respectively. Then λ and μ are real and

$$(\lambda - \mu)(x|y) = (\lambda x|y) - (x|\mu y) = (A(x)|y) - (x|A(y)) = 0.$$

Thus $(x|y) = 0$ if $\lambda \neq \mu$. $\qquad\qquad\qquad\qquad\qquad\qquad\qquad\qquad\qquad$ □

Proof of Theorem 4.3. We proceed by induction on the dimension n of X. On account of Proposition 4.4 (i), the claim trivially holds if $\dim X = 1$. Suppose the theorem has been proved for all self-adjoint operators on H when $\dim H = n - 1$ and let us prove it for A. Because of (i) Proposition 4.4, all eigenvalues of A are real, hence there exists at least an eigenvector u_1 of A with norm one. Let $H := \text{Span}\{u_1\}^\perp$ and let $B := A_{|H}$ be the restriction of A to H. Because of (ii) Proposition 4.4, $B(H) \subset H$, hence $B : H \to H$ is a linear operator on H (whose dimension is $n - 1$); moreover, B is self-adjoint, since it is the restriction to a subspace of a self-adjoint operator. Therefore, by the inductive assumption, there is an orthonormal basis $(u_2, \ldots, u_n)$ of H made by eigenvectors of B, hence of A. Since $u_2, \ldots, u_n$ are orthogonal to u_1, $(u_1, u_2, \ldots, u_n)$ is an orthonormal basis of X made by eigenvectors of A. $\qquad\square$

The next proposition expresses the existence of an orthonormal basis of eigenvectors in several different ways, see Theorem 2.45. We leave its simple proof to the reader.

4.5 Proposition. *Let $A : X \to X$ be a linear operator on a Euclidean or Hermitian space X of dimension n. Let $(u_1, u_2, \ldots, u_n)$ be a basis of X and let $\lambda_1, \lambda_2, \ldots, \lambda_n$ be real numbers. The following claims are equivalent*

(i) $(u_1, u_2, \ldots, u_n)$ *is an orthonormal basis of X and each u_j is an eigenvector of A with eigenvalue λ_j, i.e.,*

$$A(u_j) = \lambda_j u_j, \qquad (u_i | u_j) = \delta_{ij} \qquad \forall i, j = 1, \ldots, n,$$

(ii) $(u_1, u_2, \ldots, u_n)$ *is an orthonormal basis and*

$$A(x) = \sum_{j=1}^{n} \lambda_j (x | u_j)\, u_j \qquad \forall x \in X,$$

(iii) $(u_1, u_2, \ldots, u_n)$ *is an orthonormal basis and for all $x, y \in X$*

$$(A(x) | y) = \begin{cases} \sum_{j=1}^{n} \lambda_j (x | u_j)\, (y | u_j) & \text{if } X \text{ is Euclidean,} \\ \sum_{j=1}^{n} \lambda_j (x | u_j)\, \overline{(y | u_j)} & \text{if } X \text{ is Hermitian.} \end{cases}$$

Moreover, we have the following, compare with Theorem 2.45.

4.6 Proposition. *Let $A : X \to X$ be a self-adjoint operator in a Euclidean or Hermitian space X of dimension n and let $\mathbf{A} \in M_{n,n}(\mathbb{K})$ be the matrix associated to A in a given orthonormal basis. Then $\mathbf{A}$ is similar to a diagonal matrix. More precisely, let $(u_1, u_2, \ldots, u_n)$ be a basis of X of eigenvectors of A, let $\lambda_1, \lambda_2, \ldots, \lambda_n \in \mathbb{R}$ be the corresponding eigenvalues and let $\mathbf{S} \in M_{n,n}(\mathbb{R})$ be the matrix that has the n-tuple of components of u_i in the given orthonormal basis as the ith column,*

$$\mathbf{S} := \left[\mathbf{u}_1 \,\middle|\, \mathbf{u}_2 \,\middle|\, \cdots \,\middle|\, \mathbf{u}_n \right].$$

Then

$$\mathbf{S}^T \mathbf{S} = \text{Id} \qquad and \qquad \mathbf{S}^T \mathbf{A} \mathbf{S} = \text{diag}\,(\lambda_1, \lambda_2, \ldots, \lambda_n)$$

if X is Euclidean, and

$$\overline{\mathbf{S}}^T \mathbf{S} = \mathrm{Id} \quad and \quad \mathbf{S}^T \mathbf{A} \mathbf{S} = \mathrm{diag}\,(\lambda_1, \lambda_2, \ldots, \lambda_n).$$

if X is Hermitian.

Proof. Since the columns of $\mathbf{S}$ are orthonormal, it follows that $\mathbf{S}^T \mathbf{S} = \mathrm{Id}$ if X is Euclidean or $\overline{\mathbf{S}}^T \mathbf{S} = \mathrm{Id}$ if X is Hermitian. The rest of the proof is contained in Theorem 2.45.
□

c. Spectral resolution

Let $A : X \to X$ be a self-adjoint operator on a Euclidean or Hermitian space X of dimension n, let $(u_1, u_2, \ldots, u_n)$ be an orthonormal basis of eigenvectors of A, let $\lambda_1, \lambda_2, \ldots, \lambda_k$ be the distinct eigenvalues of A and $V_1, V_2, \ldots, V_k$ the corresponding eigenspaces. Let $P_i : X \to V_i$ be the projector on V_i so that

$$P_i(x) = \sum_{u_j \in V_i} (x|u_j)\, u_j,$$

and by (ii) Proposition 4.4

$$A(x) = \sum_{i=1}^{k} \lambda_i P_i(x).$$

As we have seen, by (iii) Proposition 4.4, we have $V_i \perp V_j$ if $i \neq j$ and, by the spectral theorem, $\sum_{i=1}^{k} \dim V_i = n$. In other words, we can say that $\{V_i\}_i$ *is a decomposition of X in orthogonal subspaces* or state the following.

4.7 Theorem. *Let $A : X \to X$ be self-adjoint on a Euclidean or Hermitian space X of dimension n. Then there exists a unique family of projectors $P_1, P_2, \ldots, P_k$ and distinct real numbers $\lambda_1, \lambda_2, \ldots, \lambda_k$ such that*

$$P_i \circ P_j = \delta_{ij} P_j, \quad \sum_{i=1}^{k} P_i = \mathrm{Id} \quad and \quad A = \sum_{i=1}^{k} \lambda_i P_i.$$

Finally, we can easily complete the spectral theorem as follows.

4.8 Proposition. *Let X be a Euclidean or Hermitian space. A linear opertor $A : X \to X$ is self-adjoint if and only if the eigenvalues of A are real and there exists an orthonormal basis of X made of eigenvectors of A.*

d. Quadratic forms

To a self-adjoint operator $A : X \to X$ we may associate the *bilinear form* $a : X \times X \to \mathbb{K}$,

$$a(x,y) := (A(x)|y), \qquad x, y \in X,$$

which is *symmetric* if X is Euclidean and *sesquilinear*, $a(x,y) = \overline{a(y,x)}$, if X is Hermitian.

4.9 Theorem. *Let* $A : X \to X$ *be a self-adjoint operator,* $(e_1, e_2, \ldots, e_n)$ *an orthonormal basis of* X *of eigenvectors of* A *and* $\lambda_1, \lambda_2, \ldots, \lambda_n$ *be the corresponding eigenvalues. Then*

$$(A(x)|x) = \sum_{i=1}^{n} \lambda_i |(x|e_i)|^2 \qquad \forall x \in X. \tag{4.1}$$

In particular, if $\lambda_{\min}$ *and* $\lambda_{\max}$ *are respectively, the smallest and largest eigenvalues of* A, *then*

$$\lambda_{\min} |x|^2 \leq (A(x)|x) \leq \lambda_{\max} |x|^2 \qquad \forall x \in X.$$

Moreover, we have $(A(x)|x) = \lambda_{\min} |x|^2$ *(resp.* $(A(x)|x) = \lambda_{\max} |x|^2$*) if and only if* x *is an eigenvector with eigenvalue* $\lambda_{\min}$ *(resp.* $\lambda_{\max}$*).*

Proof. Proposition 4.5 yields (4.1) hence

$$\lambda_{\min} \sum_{i=1}^{n} |(x|e_i)|^2 \leq (A(x)|x) \leq \lambda_{\max} \sum_{i=1}^{n} |(x|e_i)|^2,$$

and, since $|x|^2 = \sum_{i=1}^{n} |(x|e_i)|^2 \ \forall x \in X$, the first part of the claim is proved.

Let us prove that $(A(x)|x) = \lambda_{\min}|x|^2$ if and only if x is an eigenvector with eigenvalue $\lambda_{\min}$. If x is an eigenvector of A with eigenvalue $\lambda_{\min}$, then $A(x) - \lambda_{\min} x$ hence $(A(x)|x) = (\lambda_{\min} x|x) = \lambda_{\min} |x|^2$. Conversely, suppose $(e_1, e_2, \ldots, e_n)$ is a basis of X made by eigenvectors of A and the eigenspace $V_{\lambda_{\min}}$ is spanned by $(e_1, e_2, \ldots, e_k)$. From $(A(x)|x) = \lambda_{\min}|x|^2$, we infer that

$$0 = (A(x)|x) - \lambda_{\min}|x|^2 = \sum_{i=1}^{n} (\lambda_i - \lambda_{\min})|(x|e_i)|^2$$

and, as $\lambda_i \lambda_{\min} \geq 0$, we get that $(x|e_i) = 0 \ \forall i > k$, thus $x \in V_{\lambda_{\min}}$.

We proceed similarly for $\lambda_{\max}$. $\qquad\square$

All eigenvalues can, in fact, be characterized as in Theorem 4.9. Let us order the eigenvalues, counted with their multiplicity, as

$$\lambda_1 \leq \lambda_2 \leq \cdots \leq \lambda_n$$

and let $(e_1, e_2, \ldots, e_n)$ be an orthonormal basis of corresponding eigenvectors $(e_1, e_2, \ldots, e_n)$, $A(e_i) = \lambda_i e_i \ \forall i = 1, \ldots, n$; finally, set

$$V_k := \mathrm{Span}\{e_1, e_2, \ldots, e_k\}, \qquad W_k := \{e_k, e_{k+1}, \ldots, e_n\}.$$

Since V_k, W_k are invariant subspaces under A and $V_k^\perp = W_{k+1}$, by applying Theorem 4.9 to the restriction of $(A(x)|x)$ on V_k and W_k, we find

$$\lambda_1 = \min_{|x|=1} (A(x)|x), \tag{4.2}$$

$$\lambda_k = \max\left\{ (A(x)|x) \,\Big|\, |x| = 1, \ x \in V_k \right\}$$
$$= \min\left\{ (A(x)|x) \,\Big|\, |x| = 1, \ x \in W_k \right\} \qquad \text{if } k = 2, \ldots, n-1,$$

$$\lambda_n = \max_{|x|=1}(A(x)|x).$$

Moreover, if S is a subspace of dimension $n-k+1$, we have $S \cap V_k \neq \{0\}$, then there is $x_0 \in S \cap V_k$ with $|x_0| = 1$; thus

$$\min\left\{ (A(x)|x) \,\Big|\, |x| = 1, \ x \in S \right\} \leq (A(x_0)|x_0)$$

$$\leq \max\left\{ (A(x)|x) \,\Big|\, |x| = 1, \ x \in V_k \right\} = \lambda_k.$$

Since $\dim W_k = n - k + 1$ and $\min_{x \in W_k}(A(x)|x) = \lambda_k$, we conclude with the *min-max characterization of eigenvalues* that makes no reference to eigenvectors.

4.10 Proposition (Courant). *Let A be a self-adjoint operator on a Euclidean or Hermitian space X of dimension n and let $\lambda_1 \leq \lambda_2 \leq \cdots \leq \lambda_n$ be the eigenvalues of A in nondecreasing order and counted with multiplicity. Then*

$$\lambda_k = \max_{\dim S = n-k+1} \min\left\{ (A(x)|x) \,\Big|\, |x| = 1, \ x \in S \right\}$$

$$= \min_{\dim S = k} \max\left\{ (A(x)|x) \,\Big|\, |x| = 1, \ x \in S \right\}.$$

4.11 A variational algorithm for the eigenvectors. From (4.2) we know that

$$\lambda_k := \min\left\{ (A(x)|x) \,\Big|\, |x| = 1, \ x \in V_{k-1}^\perp \right\}, \qquad k = 1 \ldots, n, \tag{4.3}$$

where $V_{-1} = \{0\}$. This yields an iterative procedure to compute the eigenvalues of A. For $j = 1$ define

$$\lambda_1 = \min_{|x|=1} (A(x)|x),$$

and for $j = 1, \ldots, n-1$ set

$$\begin{cases} V_j := \text{eigenspace of } \lambda_j, \\ W_j := (V_1 \oplus V_2 \oplus \cdots \oplus V_j)^\perp, \\ \lambda_{j+1} := \min\left\{ (A(x)|x) \,|\, |x| = 1, \ x \in W_j \right\}. \end{cases}$$

Notice that such an algorithm yields an alternative proof of the spectral theorem. We shall see in Chapter 10 that this procedure extends to certain classes of self-adjoint operators in infinite-dimensional spaces.

Finally, notice that Sylvester's theorem, Gram–Schmidt's procedure or the other variants for reducing a quadratic form to a canonical form, see Chapter 3, allow us to find the numbers of positive, negative and null eigenvalues (with multiplicity) without computing them explicitly.

e. Positive operators

A self-adjoint operator $A : X \to X$ is called *positive* (resp. *nonnegative*) if the quadratic form $\phi(x) := (Ax|x)$ is positive for $x \neq 0$ (resp. nonnegative). From the results about metrics, see Corollary 3.56, or directly from Theorem 4.9, we have the following.

4.12 Proposition. *Let $A : X \to X$ be self-adjoint. A is positive (nonnegative) if and only if all eigenvalues of A are positive (nonnegative) or iff there is $\lambda > 0$ ($\lambda \geq 0$) such that $(Ax|x) \geq \lambda|x|^2$.*

4.13 Corollary. *$A : X \to X$ is positive self-adjoint if and only if $a(x, y) = (A(x)|y)$ is an inner (Hermitian) product on X.*

4.14 Proposition (Simultaneous diagonalization). *Let $A, M : X \to X$ be linear self-adjoint operators on X. Suppose M is positive. Then there exists a basis $(e_1, e_2, \ldots, e_n)$ of X and real numbers $\lambda_1, \lambda_2, \ldots, \lambda_n$ such that*

$$(M(e_i)|e_j) = \delta_{ij}, \qquad \Lambda(c_j) = \lambda_j M e_j \ \forall i, j - 1, \ldots, n. \qquad (4.4)$$

Proof. The metric $g(x, y) := (M(x)|y)$ is a scalar (Hermitian) product on X and the linear operator $M^{-1}A : X \to X$ is self-adjoint with respect to g since

$$g(M^{-1}A(x), y) - (MM^{-1}A(x)|y) = (\Lambda(x)|y) = (x|A(y))$$
$$= (x|MM^{-1}A(y)) = (Mx|M^{-1}A(y)) = g(x, M^{-1}A(y)).$$

Therefore, $M^{-1}A$ has real eigenvalues and, by the spectral theorem, there is a g-orthonormal basis of X made of eigenvectors of $M^{-1}A$,

$$g(e_i, e_j) = (M(e_i)|e_j) = \delta_{ij}, \qquad M^{-1}A(e_j) = \lambda_j e_j \ \forall i, j = 1, \ldots, n.$$

$\square$

4.15 Remark. We cannot drop the positivity assumption in Proposition 4.14. For instance, if

$$\mathbf{M} := \begin{pmatrix} 1 & 0 \\ 0 & -1 \end{pmatrix} \quad \text{and} \quad \mathbf{A} := \begin{pmatrix} 0 & 1 \\ 1 & 0 \end{pmatrix}$$

we have $\det(\lambda \operatorname{Id} - \mathbf{M}^{-1}\mathbf{A}) = \lambda^2 + 1$, hence $\mathbf{M}^{-1}\mathbf{A}$ has no real eigenvalue.

4.16 ¶. Show the following.

Proposition. *Let X be a Euclidean space and let $g, b : X \times X \to \mathbb{R}$ be two metrics on X. Suppose g is positive. Then there exists a basis of X that is both g-orthogonal and b-orthogonal.*

4.17 ¶. Let A, M be linear self-adjoint operators and let M be positive. Then $M^{-1}A$ is self-adjoint with respect to the inner product $g(x, y) := (M(x)|y)$. Show that the eigenvalues $\lambda_1, \lambda_2, \ldots, \lambda_n$ of $M^{-1}A$ are iteratively given by

$$\lambda_1 = \min_{g(x,x)=1} g(M^{-1}A(x))x = \min_{x \neq 0} \frac{(A(x)|x)}{(M(x)|x)}$$

and for $j = 1, \ldots, n - 1$

$$\begin{cases} V_j := \text{eigenspace of } M^{-1}A \text{ relative to } \lambda_j, \\ W_j := (V_1 \oplus V_2 \oplus \cdots \oplus V_j)^{\perp}, \\ \lambda_{j+1} := \min\{(A(x)|x) \mid (M(x)|x) = 1, \ x \in W_j\}, \end{cases}$$

where $V^{\perp}$ denotes the orthogonal to V with respect to the inner product g.

4.18 ¶. Show the following.

Proposition. *Let T be a linear operator on $\mathbb{K}^n$. If $T + T^*$ is positive (nonnegative), then all eigenvalues of T have positive (nonnegative) real part.*

f. The operators A^*A and AA^*

Let $A : X \to Y$ be a linear operator between X and Y that we assume are either both Euclidean or both Hermitian. From now on we shall write Ax instead of $A(x)$ for the sake of simplicity. As usual, $A^* : Y \to X$ denotes the adjoint of A.

4.19 Proposition. *The operator $A^*A : X \to X$ is*

(i) *self-adjoint,*
(ii) *nonnegative,*
(iii) *Ax, A^*Ax and $(A^*Ax|x)$ are all nonzero or all zero, in particular A^*A is positive if and only if A is injective,*
(iv) *if $u_1, u_2, \ldots, u_n$ are eigenvectors of A^*A respectively, with eigenvalues $\lambda_1, \lambda_2, \ldots, \lambda_n$, then*

$$(Au_i|Au_j) = \lambda_i(u_i|u_j),$$

in particular, if $u_1, u_2, \ldots, u_n$ are orthogonal to each other, then $Au_1, \ldots, Au_n$ are orthogonal to each other as well.

Proof. (i) In fact, $(A^*A)^* = A^*A^{**} = A^*A$.

(ii) and (iii) If $Ax = 0$, then trivially $A^*Ax = 0$, and if $A^*Ax = 0$, then $(A^*Ax|x) = 0$. On the other hand, $(A^*Ax|x) = (Ax|Ax) = |Ax|^2$ hence $Ax = 0$ if $(A^*Ax|x) = 0$.

(iv) In fact, $(Au_i|Au_j) = (A^*Au_i|u_j) = \lambda_i(u_i|u_j) = \lambda_i|u_i|^2\delta_{ij}$. □

4.20 Proposition. *The operator* $AA^* : Y \to Y$ *is*

(i) *self-adjoint,*

(ii) *nonnegative,*

(iii) A^*x, AA^*x *and* $(AA^*x|x)$ *are either all nonzero or all zero, in particular* AA^* *is positive if and only if* $\ker A^* = \{0\}$, *equivalently if and only if* A *is surjective.*

(iv) *if* $u_1, u_2, \ldots, u_n$ *are eigenvectors of* AA^* *with eigenvalues respectively,* $\lambda_1, \lambda_2, \ldots, \lambda_n$, *then*

$$(A^*u_i|A^*u_j) = \lambda_i(u_i|u_j),$$

in particular, if $u_1, u_2, \ldots, u_n$ *are orthogonal to each other, then* $A^*u_1, \ldots, A^*u_n$ *are orthogonal to each other as well.*

Moreover, AA^* *and* A^*A *have the same nonzero eigenvalues and*

$$\operatorname{Rank} AA^* = \operatorname{Rank} A^*A = \operatorname{Rank} A = \operatorname{Rank} A^*.$$

In particular, $\operatorname{Rank} AA^* = \operatorname{Rank} A^*A \leq \min(\dim X, \dim Y)$.

Proof. The claims (i) (ii) (iii) and (iv) are proved as in Proposition 4.19.

To prove that A^*A and AA^* have the same nonzero eigenvalues, notice that if $x \in X$, $x \neq 0$, is an eigenvalue for A^*A with eigenvalue $\lambda \neq 0$, $A^*Ax = \lambda x$, then $Ax \neq 0$ by (iii) Proposition 4.19 and $AA^*(Ax) = A(A^*Ax) = A(\lambda x) = \lambda Ax$, i.e., Ax is a nonzero eigenvector for AA^* with the same eigenvalue λ. Similarly, one proves that if $y \neq 0$ is an eigenvector for AA^* with eigenvalue $\lambda \neq 0$, then by (iii) $A^*y \neq 0$ and A^*y is an eigenvector for A^*A with eigenvalue λ.

Finally, from the alternative theorem, we have

$$\operatorname{Rank} AA^* = \operatorname{Rank} A^* = \operatorname{Rank} A = \operatorname{Rank} A^*A.$$

$\square$

g. Powers of a self-adjoint operator

Let $A : X \to X$ be self-adjoint. By the spectral theorem, there is an orthonormal basis $(e_1, e_2, \ldots, e_n)$ of X and real numbers $\lambda_1, \lambda_2, \ldots, \lambda_n$ such that

$$Ax = \sum_{j=1}^{n} \lambda_j(x|e_j)e_j \qquad \forall x \in X.$$

By induction, one easily computes, using the eigenvectors $e_1, e_2, \ldots, e_n$ and the eigenvalues $\lambda_1, \lambda_2, \ldots, \lambda_n$ of A the k-power of A, $A^k := A \circ \cdots \circ A$ k times, $\forall k \geq 2$, as

$$A^k x = \sum_{i=1}^{n} (\lambda_i)^k(x|e_i) \, e_i \qquad \forall x \in X. \tag{4.5}$$

4.21 Proposition. *Let* $A : X \to X$ *be self-adjoint and* $k \geq 1$. *Then*

(i) A^k *is self-adjoint,*

(ii) λ *is an eigenvalue for* A *if and only if* λ^k *is an eigenvalue for* A^k,

(iii) $x \in X$ is an eigenvector of A with eigenvalue λ if and only if x is an eigenvector for A^k with eigenvalue λ^k. In particular, the eigenspaces of A relative to λ and of A^k relative to λ^k coincide.

(iv) If A is invertible, equivalently, if all eigenvalues of A are nonzero, then

$$A^{-1}x = \sum_{i=1}^{n} \frac{1}{\lambda_i}(x|e_i)\, e_i \qquad \forall x \in X.$$

4.22 ¶. Let $A : X \to X$ be self-adjoint. Show that

$$(A_{\ker A^\perp})^{-1}x = \sum_{\lambda_i \neq 0} \frac{1}{\lambda_i}(x|e_i)\, e_i.$$

If $p(t) = \sum_{k=1}^{m} a_k t^k$ is a polynomial of degree m, then (4.5) yields

$$p(A)(x) = \sum_{k=1}^{m} a_k A^k(x) = \sum_{k=1}^{m}\sum_{j=1}^{n} a_k \lambda_j^k (x|e_j)e_j = \sum_{j=1}^{n} p(\lambda_j)(x|e_j)e_j. \quad (4.6)$$

4.23 Proposition. *Let $A : X \to X$ be a nonnegative self-adjoint operator and let $k \in \mathbb{N}$, $k \geq 1$. There exists a unique nonnegative self-adjoint operator $B : X \to X$ such that $B^{2k} = A$. Moreover, B is positive if A is positive.*

The operator B such that $B^{2k} = A$ is called the $2k$th *root of A* and is denoted by $\sqrt[2k]{A}$.

Proof. If $A(x) = \sum_{j=1}^{n} \lambda_j (x|e_j)e_j$, (4.5) yields $B^{2k} = A$ for

$$B(x) := \sum_{j=1}^{n} \sqrt[2k]{\lambda_j}(x|e_j)e_j.$$

Uniqueness remains to be shown. Suppose B and C are self-adjoint, nonnegative and such that $A = B^{2k} = C^{2k}$. Then B and C have the same eigenvalues and the same eigenspaces by Proposition 4.21, hence $B = C$.   □

In particular, if $A : X \to X$ is nonnegative and self-adjoint, the operator *square root of A* is defined by

$$\sqrt{A}(x) := \sum_{i=1}^{n} \sqrt{\lambda_j}(x|e_j)e_j, \qquad x \in X,$$

if A has the spectral decomposition $Ax = \sum_{j=1}^{n} \lambda_j (x|e_j)e_j$.

4.24 ¶. Prove Proposition 4.14 by noticing that, if A and M are self-adjoint and M is positive, then $M^{-1/2}AM^{-1/2} : X \to X$ is well defined and self-adjoint.

4.25 ¶. Let A, B be self-adjoint and let A be positive. Show that B is positive if $S := AB + BA$ is positive. [*Hint:* Consider $A^{-1/2}BA^{-1/2}$ and apply Exercise 4.18.]

4.1.2 Normal operators

a. Simultaneous spectral decompositions

4.26 Theorem. *Let X be a Euclidean or Hermitian space. If A and B are two self-adjoint operators on X that commute,*

$$A = A^*, \qquad B = B^*, \qquad AB = BA,$$

then there exists an orthonormal basis $(e_1, e_2, \ldots, e_n)$ on X of eigenvectors of A and B, hence

$$z = \sum_{i=1}^{n} (z|e_i)e_i, \qquad Az = \sum_{i=1}^{n} \lambda_i(z|e_i)e_i, \qquad Bz = \sum_{i=1}^{n} \mu_i(z|e_i)e_i,$$

$\lambda_1, \lambda_2, \ldots, \lambda_n \in \mathbb{R}$ *and* $\mu_1, \mu_2, \ldots, \mu_n \in \mathbb{R}$ *being the eigenvalues respectively of A and B.*

This is proved by induction as in Theorem 4.3 on account of the following.

4.27 Proposition. *Under the hypoteses of Theorem 4.26, we have*

(i) *A and B have a common nonzero eigenvector,*
(ii) *if V is invariant under A and B, then $V^\perp$ is invariant under A and B as well.*

Proof. (i) Let λ be an eigenvalue of A and let V_λ be the corresponding eigenspace. For all $y \in V_\lambda$ we have $ABy = BAy = \lambda By$, i.e., $By \in V_\lambda$. Thus V_λ is invariant under B, consequently, there is an eigenvector $w \in V_\lambda$ of $B_{|V_\lambda}$, i.e., common to A and B.

(ii) For every $w \in V^\perp$ and $z \in V$, we have $Az, Bz \in V$ and $(Aw|z) = (w|Az) = 0$, $(Bw|z) = (w|Bz) = 0$, i.e., $Aw, Bw \in V^\perp$. □

4.28 ¶. Show that two symmetric matrices $\mathbf{A}, \mathbf{B}$ are simultaneously diagonizable if and only if they commute $\mathbf{AB} = \mathbf{BA}$.

b. Normal operators on Hermitian spaces

A linear operator on a Euclidean or Hermitian space is called *normal* if $NN^* = N^*N$. Of course, if we fix an orthonormal basis in X, we may represent N with an $n \times n$ matrix $\mathbf{N} \in M_{n,n}(\mathbb{C})$ and N is normal if and only if $\mathbf{N}\overline{\mathbf{N}}^T = \overline{\mathbf{N}}^T\mathbf{N}$ if X is Hermitian or $\mathbf{N}\mathbf{N}^T = \mathbf{N}^T\mathbf{N}$ if X is Euclidean.

The class of normal operators, though not trivial from the algebraic point of view (it is not closed for the operations of sum and composition), is interesting as it contains several families of important operators as subclasses. For instance, *self-adjoint* operators $N = N^*$, *anti-self-adjoint* operators $N^* = -N$, and *isometric* operators, $N^*N = \text{Id}$, are normal operators.

Moreover, *normal operators in a Hermitian space* are exactly the ones that are diagonalizable. In fact, we have the following.

4.29 Theorem (Spectral theorem). *Let X be a Hermitian space of dimension n and let $N : X \to X$ be a linear operator. Then N is normal if and only if there exists an orthonormal basis of X made by eigenvectors of N.*

Proof. Let $(e_1, e_2, \ldots, e_n)$ be an orthonormal basis of X made by eigenvectors of N. Then for every $z \in X$

$$Nz = \sum_{j=1}^{n} \lambda_j (z|e_j) e_j, \qquad N^* z = \sum_{j=1}^{n} \overline{\lambda_j} (z|e_j) e_j$$

hence $NN^* z = \sum_{j=1}^{n} |\lambda_j|^2 (z|e_j) e_j = N^* N z$.

Conversely, let

$$A := \frac{N + N^*}{2}, \qquad B := \frac{N - N^*}{2i}.$$

It is easily seen that A and B are self-adjoint and commute. Theorem 4.26 then yields a basis of orthonormal eigenvectors of A and B and therefore of eigenvectors of $N := A + iB$ and $N^* = A - iB$. $\square$

4.30 ¶. Show that $N : \mathbb{C}^n \to \mathbb{C}^n$ is normal if and only if N and N^* have the same eigenspaces.

c. Normal operators on Euclidean spaces

Let us translate the information contained in the spectral theorem for normal operators on Hermitian spaces into information about normal operators on Euclidean spaces. In order to do that, let us first make a few remarks.

As usual, in $\mathbb{C}^n$ we write $z = x + iy$, $x, y \in \mathbb{R}^n$ for $z = (x_1 + iy_1, \ldots, x_n + iy_n)$. If W is a subspace of $\mathbb{R}^n$, the subspace of $\mathbb{C}^n$

$$W \oplus iW := \Big\{ z \in \mathbb{C}^n \,\Big|\, z = x + iy, \ x, y \in W \Big\}$$

is called the *complexified of* W. Trivially, $\dim_{\mathbb{C}}(W \oplus iW) = \dim_{\mathbb{R}} W$. Also, if V is a subspace of $\mathbb{C}^n$, set

$$\overline{V} := \Big\{ z \in \mathbb{C}^n \,\Big|\, \overline{z} \in V \Big\}.$$

4.31 Lemma. *A subspace $V \subset \mathbb{C}^n$ is the complexified of a real subspace W if and only if $\overline{V} = V$.*

Proof. If $V = W \oplus iW$, trivially $\overline{V} = V$. Conversely, if $z \in V$ is such that $\overline{z} \in V$, the vectors

$$x := \frac{z + \overline{z}}{2}, \qquad y := \frac{z - \overline{z}}{2i} = \frac{(z/i) + \overline{z/i}}{2}$$

have real coordinates. Set

$$W := \Big\{ x \in \mathbb{R}^n \,\Big|\, x = \frac{z + \overline{z}}{2}, \ z \in V \Big\};$$

then it is easily seen that $V = W \oplus iW$ if $V = \overline{V}$. $\square$

For $N : \mathbb{R}^n \to \mathbb{R}^n$ we define its *complexified* as the (complex) linear operator $N_{\mathbb{C}} : \mathbb{C}^n \to \mathbb{C}^n$ defined by $N_{\mathbb{C}}(z) := Nx + iNy$ if $z = x + iy$. Then we easily see that

(i) λ is an eigenvalue of N if and only if λ is an eigenvalue of $N_{\mathbb{C}}$,
(ii) N is respectively, a self-adjoint, anti-self-adjoint, isometric or normal operator if and only if $N_{\mathbb{C}}$ is respectively, a self-adjoint, anti-self-adjoint, isometric or normal operator on $\mathbb{C}^n$,
(iii) the eigenvalues of N are either real, or pairwise complex conjugate; in the latter case the conjugate eigenvalues λ and $\overline{\lambda}$ have the same multiplicity.

4.32 Proposition. *Let $N : \mathbb{R}^n \to \mathbb{R}^n$ be a normal operator. Every real eigenvalue λ of N of multiplicity k has an eigenspace V_λ of dimension k. In particular, V_λ has an orthonormal basis made of eigenvectors.*

Proof. Let λ be a real eigenvalue for $N_{\mathbb{C}}$, $N_{\mathbb{C}} z = \lambda z$. We have

$$N_{\mathbb{C}} \overline{z} = Nx - iNy = \overline{N_{\mathbb{C}} z} = \overline{\lambda z} = \lambda \overline{z},$$

i.e., $z \in \mathbb{C}^n$ is an eigenvector of $N_{\mathbb{C}}$ with eigenvalue λ if and only if $\overline{z}$ is also an eigenvector with eigenvalue λ. The eigenspace E_λ of $N_{\mathbb{C}}$ relative to λ is then closed under conjugation and by Lemma 4.31 $E_\lambda := W_\lambda \oplus iW_\lambda$, where

$$W_\lambda := \Big\{ x \in \mathbb{R}^n \,\Big|\, x = \frac{z + \overline{z}}{2}, \; z \in E_\lambda \Big\},$$

and $\dim_{\mathbb{R}} W_\lambda = \dim_{\mathbb{C}} E_\lambda$. Since $N_{\mathbb{C}}$ is diagonalizable in $\mathbb{C}$ and $W_\lambda \subset V_\lambda$, we have

$$k = \dim_{\mathbb{C}} E_\lambda = \dim W_\lambda \le \dim_{\mathbb{R}} V_\lambda.$$

As $\dim V_\lambda \le k$, see Proposition 2.43, the claim follows. $\qquad\square$

4.33 Proposition. *Let λ be a nonreal eigenvalue of the normal operator $N : \mathbb{R}^n \to \mathbb{R}^n$ with multiplicity k. Then there exist k planes of dimension 2 that are invariant under N. More precisely, if $e_1, e_2, \ldots, e_n \in \mathbb{C}^n$ are k orthonormal eigenvectors that span the eigenspace E_λ of $N_{\mathbb{C}}$ relative to λ and we set*

$$u_{2j-1} := \frac{e_j + \overline{e_j}}{\sqrt{2}}, \qquad u_{2j} := \frac{e_j - \overline{e_j}}{\sqrt{2}i},$$

then $u_1, u_2, \ldots, u_{2k}$ are orthonormal in $\mathbb{R}^n$, and for $j = 1, \ldots, k$ the plane $\mathrm{Span}\,\{u_{2j-1}, u_{2j}\}$, is invariant under N; more precisely we have

$$\begin{cases} N(u_{2j-1}) = \alpha u_{2j-1} - \beta u_{2j}, \\ N(u_{2j}) = \beta u_{2j-1} + \alpha u_{2j} \end{cases}$$

where $\lambda =: \alpha + i\beta$.

Proof. Let $E_\lambda, E_{\overline{\lambda}}$ be the eigenspaces of $N_{\mathbb{C}}$ relative to λ and $\overline{\lambda}$. Since $N_{\mathbb{C}}$ is diagonalizable on $\mathbb{C}$, then

$$\dim_{\mathbb{C}} E_\lambda = \dim_{\mathbb{C}} E_{\overline{\lambda}} = k, \qquad E_\lambda \perp E_{\overline{\lambda}}.$$

On the other hand, for $z \in E_\lambda$

$$N_{\mathbb{C}}\bar{z} = Nx - iNy = \overline{N_{\mathbb{C}}z} = \bar{\lambda}\,\bar{z}.$$

Therefore, $z \in E_\lambda$ if and only if $\bar{z} \in E_{\bar{\lambda}}$. The complex subspace $F_\lambda := E_\lambda \oplus E_{\bar{\lambda}}$ of $\mathbb{C}^n$ has dimension $2k$ and is closed under conjugation; Lemma 4.31 then yields $F_\lambda = W_\lambda \oplus iW_\lambda$ where

$$W_\lambda := \left\{ x \in \mathbb{R}^n \,\Big|\, x = \frac{z + \bar{z}}{2},\ z \in E_\lambda \right\} \qquad \text{and} \qquad \dim_{\mathbb{R}} W_\lambda = \dim_{\mathbb{C}} E = 2k. \quad (4.7)$$

If $(e_1, e_2, \ldots, e_k)$ is an orthonormal basis of E_λ, $(\bar{e}_1, \bar{e}_2, \ldots, \bar{e}_k)$ is an orthonormal basis of $E_{\bar{\lambda}}$; since

$$\sqrt{2}e_j =: u_{2j-1} + iu_{2j}, \qquad \sqrt{2}\bar{e}_j =: u_{2j-1} - iu_{2j},$$

we see that $\{u_j\}$ is an orthonormal basis of W_λ. Finally, if $\lambda := \alpha + i\beta$, we compute

$$\begin{cases} N(u_{2j-1}) = N_{\mathbb{C}}\left(\frac{e_j + \bar{e}_j}{\sqrt{2}}\right) = \frac{\lambda e_j + \bar{\lambda}\bar{e}_j}{2} = \cdots = \alpha u_{2j-1} - \beta u_{2j}, \\ N(u_{2j}) = N\left(\frac{e_j - \bar{e}_j}{\sqrt{2}i}\right) = \frac{\lambda e_j - \bar{\lambda}\bar{e}_j}{2i} = \cdots = \beta u_{2j-1} + \alpha u_{2j}, \end{cases}$$

i.e., $\mathrm{Span}\,\{u_{2j-1}, u_{2j}\}$ is invariant under N. □

Observing that the eigenspaces of the real eigenvalues and the eigenspaces of the complex conjugate eigenvectors are pairwise orthogonal, from Propositions 4.32 and 4.33 we infer the following.

4.34 Theorem. *Let N be a normal operator on $\mathbb{R}^n$. Then $\mathbb{R}^n$ is the direct sum of 1-dimensional and 2-dimensional subspaces that are pairwise orthogonal and invariant under N. In other words, there is an orthonormal basis such that the matrix $\mathbf{N}$ associated to N in this basis has the block structure*

$$\mathbf{N}' = \begin{pmatrix} A_1 & 0 & \cdots & 0 \\ 0 & A_2 & \cdots & 0 \\ \vdots & \vdots & \ddots & \vdots \\ 0 & 0 & \cdots & A_p \end{pmatrix}.$$

To each real eigenvalue λ of multiplicity k correspond k blocks λ of dimension 1×1. To each couple of complex conjugate eigenvalues $\lambda, \bar{\lambda}$ of multiplicity k correspond k 2×2 blocks of the form

$$\begin{pmatrix} \alpha & \beta \\ -\beta & \alpha \end{pmatrix}$$

where $\alpha + i\beta := \lambda$.

4.35 Corollary. *Let $N : \mathbb{R}^n \to \mathbb{R}^n$ be a normal operator. Then*

(i) *N is self-adjoint if and only if all its eigenvalues are real,*
(ii) *N is anti-self-adjoint if and only if all its eigenvalues are purely imaginary (or zero),*
(iii) *N is an isometry if and only if all its eigenvalues have modulus one.*

4.36 ¶. Show Corollary 4.35.

4.1.3 Some representation formulas

a. The operator A^*A

Let $A : X \to Y$ be a linear operator between two Euclidean spaces or two Hermitian spaces and let $A^* : Y \to X$ be its adjoint. As we have seen, $A^*A : X \to X$ is self-adjoint, nonnegative and can be written as

$$A^*Ax = \sum_{i=1}^{n} \lambda_i(x|e_i)\, e_i$$

where $(e_1, e_2, \ldots, e_n)$ is a basis of X made of eigenvectors of A^*A and for each $i = 1, \ldots, n$ λ_i is the eigenvalue relative to e_i; accordingly, we also have

$$(A^*A)^{1/2}x := \sum_{i=1}^{n} \mu_i(x|e_i)\, e_i,$$

where $\mu_i := \sqrt{\lambda_i}$. The operator $(A^*A)^{1/2}$ and its eigenvalues $\mu_1, \ldots, \mu_n$, called the *singular values* of A, play an important role in the description of A.

4.37 ¶. Let $\mathbf{A} \in M_{m,n}(\mathbb{R})$. Show that $||\mathbf{A}|| := \sup_{|x|=1} |\mathbf{A}x|$ is the greatest singular value of $\mathbf{A}$. [*Hint:* $|\mathbf{A}x|^2 = (\mathbf{A}^*\mathbf{A}x) \bullet x$.]

4.38 Theorem (Polar decomposition). *Let $A : X \to Y$ be an operator between two Euclidean or two Hermitian spaces.*

(i) *If $\dim X \le \dim Y$, then there exists an isometry $U : X \to Y$, i.c., $U^*U = \mathrm{Id}$, such that*

$$A = U(A^*A)^{1/2}.$$

Moreover, if $A = US$ with $U^*U = \mathrm{Id}$ and $S^* = S$, then $S = (A^*A)^{1/2}$ and U is uniquely defined on $\ker S^\perp = \ker A^\perp$.

(ii) *If $\dim X \ge \dim Y$, then there exists an isometry $U : Y \to X$, i.c., $U^*U = \mathrm{Id}$ such that*

$$A = (AA^*)^{1/2}U^*.$$

Moreover, if $A = SU$ with $U^*U = \mathrm{Id}$ and $S^* = S$, then $S = (AA^*)^{1/2}$ and U is uniquely defined on $\ker S^\perp = \mathrm{Im}\, A$.

Proof. Let us show (i). Set $n := \dim X$ and $N := \dim Y$. First let us prove uniqueness. If $A = US$ where $U^*U = \mathrm{Id}$ and $S^* = S$, then $A^*A = S^*U^*US = S^*S = S^2$, i.e., $S = (A^*A)^{1/2}$. Now from $A = U(A^*A)^{1/2}$, we infer for $i = 1, \ldots, n$

$$A(e_i) = U(A^*A)^{1/2}(e_i) = U(\mu_i e_i) = \mu_i U(e_i),$$

if $(e_1, e_2, \ldots, e_n)$ is an orthonormal basis of X of eigenvectors of $(A^*A)^{1/2}$ with relative eigenvalues $\mu_1, \mu_2, \ldots, \mu_n$. Hence, $U(e_i) = \frac{1}{\mu_i}A(e_i)$ if $\mu_i \ne 0$, i.e., U is uniquely defined by A on the direct sum of the eigenspaces relative to nonzero eigenvalues of $(A^*A)^{1/2}$, that is, on the orthogonal of $\ker(A^*A)^{1/2} = \ker A$.

Now we shall exhibit U. The vectors $A(e_1), \ldots, A(e_n)$ are orthogonal and $|A(e_i)| = \mu_i$ as

$$(A(e_i)|A(e_j)) = (A^* A(e_i)|e_j) = \mu_i(e_i|e_j) = \mu_i \delta_{ij}.$$

Let us reorder the eigenvectors and the corresponding eigenvalues in such a way that for some k, $1 \le k \le n$, the vectors $A(e_1), \ldots, A(e_k)$ are not zero and $A(e_{k+1}) = \cdots = A(e_n) = 0$. For $i = 1, \ldots, k$ we set $v_i := \frac{A(e_i)}{|A(e_i)|}$ and we complete $v_1, v_2, \ldots, v_k$ to form a new orthonormal basis $(v_1, v_2, \ldots, v_N)$ of Y. Now consider $U : X \to Y$ defined by

$$U(e_i) := v_i \qquad i = 1, \ldots, n.$$

By construction $(U(e_i)|U(e_j)) = \delta_{ij}$, i.e., $U^*U = \text{Id}$, and, since $\mu_i = |A(e_i)| = 0$ for $i > k$, we conclude for every $i = 1, \ldots, n$

$$U(A^*A)^{1/2}(e_i) = U(\mu_i e_i) = \mu_i U(e_i) = \begin{cases} \mu_i \frac{A(e_i)}{|A(e_i)|} & \text{if } 1 \le i \le k \\ \mu_i v_i = 0 v_i = 0 & \text{if } k < i \le n \end{cases} = A(e_i).$$

(ii) follows by applying (i) to A^*. $\qquad\qquad\qquad\qquad\qquad\qquad\qquad\square$

b. Singular value decomposition

Combining polar decomposition and the spectral theorem we deduce the so-called *singular value decomposition* of a matrix $\mathbf{A}$. We discuss only the real case, since the complex one requires only a few straightforward changes. Let $\mathbf{A} \in M_{N,n}(\mathbb{R})$ with $n \le N$. The polar decomposition yields

$$\mathbf{A} = \mathbf{U}(\mathbf{A}^T \mathbf{A})^{1/2} \qquad \text{with} \qquad \mathbf{U}^T \mathbf{U} = \text{Id}.$$

On the other hand, since $\mathbf{A}^T \mathbf{A}$ is symmetric, the spectral theorem yields $\mathbf{S} \in M_{n,n}(\mathbb{R})$ such that

$$(\mathbf{A}^T \mathbf{A})^{1/2} = \mathbf{S}^T \text{diag}(\mu_1, \mu_2, \ldots, \mu_n) \mathbf{S}, \qquad \mathbf{S}^T \mathbf{S} = \text{Id},$$

where $\mu_1, \mu_2, \ldots, \mu_n$ are the squares of the singular values of $\mathbf{A}$. Recall that the ith column of $\mathbf{S}$ is the eigenvector of $(\mathbf{A}^* \mathbf{A})^{1/2}$ relative to the eigenvalue μ_i.

In conclusion, if we set $\mathbf{T} := \mathbf{U}\mathbf{S}^T \in M_{N,n}(\mathbb{R})$, then $\mathbf{T}^T \mathbf{T} = \text{Id}$, $\mathbf{S}^T \mathbf{S} = \text{Id}$ and

$$\mathbf{A} = \mathbf{T} \text{diag}(\mu_1, \mu_2, \ldots, \mu_n) \mathbf{S}.$$

This is the *singular value decomposition* of $\mathbf{A}$, that is implemented in most computer libraries on linear algebra. Starting from the singular value decomposition of $\mathbf{A}$, we can easily compute, of course, $(\mathbf{A}^T \mathbf{A})^{1/2}$, and the polar decomposition of $\mathbf{A}$.

4.39 . We notice that the singular value decomposition can be written in a more symmetric form if we extend $\mathbf{T}$ to a square orthogonal matrix $\mathbf{V} \in M_{N,N}(\mathbb{R})$, $\mathbf{V}^T \mathbf{V} = \text{Id}$ and extending $\text{diag}(\mu_1, \mu_2, \ldots, \mu_n)$ to a $N \times n$ matrix by adding $N - n$ null rows at the bottom. Then, again

$$A = \mathbf{V} \Delta \mathbf{S}$$

where $\mathbf{V} \in M_{N \times N}(\mathbb{R})$, $\mathbf{V}^T \mathbf{V} = \text{Id}$, $\mathbf{S} \in M_{n,n}(\mathbb{R})$, $\mathbf{S}^T \mathbf{S} = \text{Id}$ and

$$\Delta = \begin{pmatrix} \mu_1 & 0 & \cdots & 0 \\ 0 & \mu_2 & \cdots & 0 \\ \vdots & \vdots & \ddots & \vdots \\ 0 & 0 & \cdots & \mu_n \\ 0 & 0 & \cdots & 0 \\ \vdots & \vdots & \ddots & \vdots \\ 0 & 0 & \cdots & 0 \end{pmatrix}.$$

c. The Moore–Penrose inverse

Let $A : X \to Y$ be a linear operator between two Euclidean or two Hermitian spaces of dimension respectively, n and m. Denote by

$$P : X \to \ker A^\perp \qquad \text{and} \qquad Q : Y \to \operatorname{Im} A$$

the orthogonal projections operators to $\ker A^\perp$ and $\operatorname{Im} A$. Of course $Ax = Qy$ has at least a solution $x \in X$ for any $y \in Y$. Equivalently, there exists $x \in X$ such that $y - Ax \perp \operatorname{Im} A$. Since the set of solutions of $Ax = Qy$ is a translate of $\ker A$, we conclude that there exists a unique $x := A^\dagger y \in X$ such that

$$\begin{cases} y - Ax \perp \operatorname{Im} A, \\ x \in \ker A^\perp, \end{cases} \qquad \text{equivalently,} \qquad \begin{cases} Ax = Qy, \\ x = Px. \end{cases} \tag{4.8}$$

The linear map $A^\dagger : Y \to X$, $y \to A^\dagger y$, defined this way, i.e.,

$$A^\dagger := (A_{|\ker A^\perp})^{-1} \circ Q : Y \to X,$$

is called the *Moore–Penrose inverse* of $A : X \to Y$. From the definition

$$\begin{cases} AA^\dagger = Q, \\ A^\dagger A = P, \\ \ker A^\dagger = \operatorname{Im} A^\perp = \ker Q, \\ \operatorname{Im} A^\dagger = \ker A^\perp. \end{cases}$$

4.40 Proposition. $A^\dagger$ *is the unique linear map* $B : Y \to X$ *such that*

$$AB = Q, \qquad BA = P \quad \text{and} \quad \ker B = \ker Q; \tag{4.9}$$

moreover we have

$$A^* A A^\dagger = A^\dagger A A^* = A^*. \tag{4.10}$$

Proof. We prove that $B = A^\dagger$ by showing for all $y \in Y$ the vector $x := By$ satisfies (4.8). The first equality in (4.9) yields $Ax = ABy = Qy$ and the last two imply $x = By = BQy = BAx = Px$.

Finally, from $AA^\dagger = Q$ and $A^\dagger A = P$, we infer that

$$A^* AA^\dagger = A^* Q = A^*, \qquad A^\dagger AA^* = PA^* = A^*,$$

using also that $A^* Q = A^*$ and $PA^* = A^*$ since A and A^* are such that $\text{Im}\, A = (\ker A^*)^\perp$ and $\text{Im}\, A^* = \ker A^\perp$. □

The equation (4.10) allow us to compute $A^\dagger$ easily when A is injective or surjective.

4.41 Corollary. *Let $A : X \to Y$ be a linear map between Euclidean or Hermitian spaces of dimension n and m, respectively.*

(i) *If $\ker A = \{0\}$, then $n \le m$, $A^* A$ is invertible and*

$$A^\dagger = (A^* A)^{-1} A^*;$$

moreover, if $A = U(A^* A)^{1/2}$ is the polar decomposition of A, then $A^\dagger = (A^* A)^{-1/2} U^*$.

(ii) *If $\ker A^* = \{0\}$, then $n \ge m$, AA^* is invertible, and*

$$A^\dagger = A^* (AA^*)^{-1};$$

moreover, if $A = (AA^*)^{1/2} U^*$ is the polar decomposition of A, then $A^\dagger = U(AA^*)^{-1/2}$.

For more on the Moore-Penrose inverse, see Chapter 10.

4.2 Some Applications

In this final section, we illustrate methods of linear algebra in a few specific examples.

4.2.1 The method of least squares

a. The method of least squares

Suppose we have m experimental data $y_1, y_2, \ldots, y_m$ when performing an experiment of which we have a mathematical model that imposes that the data should be functions, $\phi(x)$, of a variable $x \in X$. Our problem is that of finding $x \in X$ in such a way that the theoretical data $\phi(x)$ be as close as possible to the data of the experiment.

We can formalize our problem as follows. We list the experimental data as a vector $y = (y_1, y_2, \ldots, y_m) \in \mathbb{R}^m$ and represent the mathematical model as a map $\phi : X \to \mathbb{R}^m$. Then, we introduce a *cost function* $C = C(\phi(x), y)$ that evaluates the error between the expected result when the parameter is x, and the experimental data. Our problem then becomes that of finding a minimizer of the cost function C.

If we choose

  (i) the model of the data to be *linear*, i.e., X is a vector space of dimension n and $\phi = A : X \to \mathbb{R}^m$ is a linear operator,
  (ii) as cost function, the function square distance between the expected and the experimental data,

$$C(x) = |Ax - y|^2 = (Ax - y | Ax - y), \qquad (4.11)$$

we talk of the (*linear*) *least squares problem*.

4.42 Theorem. *Let X and Y be Euclidean spaces, $A : X \to Y$ a linear map, $y \in Y$ and $C : X \to \mathbb{R}$ the function*

$$C(x) := |Ax - y|_Y^2, \qquad x \in X.$$

The following claims are equivalent

  (i) *x is a minimizer of C,*
  (ii) *$y - Ax \perp \operatorname{Im} A$,*
  (iii) *x solves the canonical equation*

$$A^*(Ax - y) = 0. \qquad (4.12)$$

Consequently C has at least a minimizer in X and the space of minimizers of C is a translate of $\ker A$.

Proof. Clearly minimizing is equivalent to finding $z = Ax \in \operatorname{Im} A$ of least distance from y. By the orthogonal projection theorem, x is a minimizer if and only if Ax is the orthogonal projection of y onto $\operatorname{Im} A$. We therefore deduce that a minimizer $x \in X$ for C exists, that for two minimizers x_1, x_2 of C we have $Ax_1 = Ax_2$, i.e., $x_1 - x_2 \in \ker A$ and that (i) and (ii) are equivalent. Finally, since $\operatorname{Im} A^\perp = \ker A^*$, (ii) and (iii) are clearly equivalent. $\qquad \square$

4.43 Remark. The equation (4.12) expresses the fact that the function $x \to |Ax - b|^2$ is stationary at a minimizer. In fact, compare 3.65, since $\nabla_x(z|x) = z$ and $\nabla_x(\mathbf{L}x|x) = 2\mathbf{L}x$ if $\mathbf{L}$ is self-adjoint, we have

$$|\mathbf{A}x - b|^2 = |b|^2 - 2(b|\mathbf{A}x) + |\mathbf{A}x|^2,$$
$$\nabla(b|\mathbf{A}x) = \nabla(\mathbf{A}^*b|x) = \mathbf{A}^*b,$$
$$\nabla_x|\mathbf{A}x|^2 = \nabla_x(\mathbf{A}^*\mathbf{A}x|x) = 2\mathbf{A}^*\mathbf{A}x$$

hence

$$\nabla_x|b - \mathbf{A}x|^2 = 2\mathbf{A}^*(\mathbf{A}x - b).$$

As a consequence of Theorem 4.42 on account of (4.8) we can state the following

4.44 Corollary. *The unique minimizer of $C(x) = |Ax - y|_Y^2$ in $\operatorname{Im} A^* = \ker A^\perp$ is $x = A^\dagger y$.*

b. The function of linear regression

Given m vectors $x_1, x_2, \ldots, x_m$ in a Euclidean space X and m corresponding numbers $y_1, y_2, \ldots, y_m$, we want to find a linear map $L : X \to \mathbb{R}$ that minimizes the quantity

$$\mathcal{F}(L) := \sum_{i=1}^{m} |y_i - L(x_i)|^2.$$

This is in fact a dual formulation of the linear least squares problem.

By Riesz's theorem, to every linear map $L : X \to \mathbb{R}$ corresponds a unique vector $w_L \in X$ such that $L(y) := (y|w_L)$, and conversely. Therefore, we need to find $w \in X$ such that

$$C(w) := \sum_{i=1}^{m} |y_i - (x_i|w)|^2 \to \min.$$

If $y := (y_1, y_2, \ldots, y_m) \in \mathbb{R}^m$ and $A : X \to \mathbb{R}^m$ is the linear map

$$Aw := \Big((x_1|w), (x_2|w), \ldots (x_n|w) \Big),$$

we are again seeking a minimizer of $C : X \to \mathbb{R}$

$$C(w) := |y - Aw|^2, \qquad w \in X.$$

Theorem 4.42 tells us that the set of minimizers is nonempty, it is a translate of $\ker A$ and the unique minimizer of C in $\ker A^\perp = \operatorname{Im} A^*$ is $w := A^\dagger y$. Notice that

$$A^* \alpha = \sum_{i=1}^{n} \alpha_i x_i, \qquad \alpha = (\alpha^1, \alpha^2, \ldots, \alpha^m) \in \mathbb{R}^m$$

hence, $w \in \operatorname{Im} A^* = \ker A^\perp$ if and only if w is a linear combination of $x_1, x_2, \ldots, x_m$. We therefore conclude that $A^\dagger y$ *is the unique minimizer of the cost function* C *that is a linear combination of* $x_1, x_2, \ldots, x_m$. The corresponding linear map $L(x) := (x|A^\dagger y)$ is called the *function of linear regression*.

4.2.2 Trigonometric polynomials

Let us reconsider in the more abstract setting of vector spaces some of the results about trigonometric polynomials, see e.g., Section 5.4.1 of [GM2].

Let $\mathcal{P}_{n,2\pi}$ be the class of trigonometric polynomials of degree m with complex coefficients

$$\mathcal{P}_{n,2\pi} := \Big\{ P(x) = \sum_{k=-n}^{n} c_k e^{i\,kx} \,\Big|\, c_k \in \mathbb{C}, \ k = -n, \ldots, n \Big\}.$$

Recall that the vector $(c_{-n}, \ldots, c_n) \in \mathbb{C}^{2n+1}$ is called the spectrum of $P(x) = \sum_{k=-n}^{n} c_k e^{ikx}$. Clearly, $\mathcal{P}_{n,2\pi}$ is a vector space over $\mathbb{C}$ of dimension at most $2n + 1$. The function $(P|Q) : \mathcal{P}_{n,2\pi} \times \mathcal{P}_{n,2\pi} \to \mathbb{C}$ defined by

$$(P|Q) := \frac{1}{2\pi} \int_{-\pi}^{\pi} P(t)\overline{Q(t)}\, dt$$

is a Hermitian product on $\mathcal{P}_{n,2\pi}$ that makes $\mathcal{P}_{n,2\pi}$ a Hermitian space. Since

$$(e^{ikx}|e^{ihx}) = \frac{1}{2\pi} \int_{-\pi}^{\pi} e^{i(k-h)x}\, dx = \delta_{hk},$$

see Lemma 5.45 of [GM2], we have the following.

4.45 Proposition. *The trigonometric polynomials* $\{e^{ikx}\}_{k=-n,n}$ *form an orthonormal set of* $2n + 1$ *vectors in* $\mathcal{P}_{n,2\pi}$ *and we have the following.*

(i) $\mathcal{P}_{n,2\pi}$ *is a Hermitian space of dimension* $2n + 1$.

(ii) *The map* $\hat{\ } : \mathcal{P}_{n,2\pi} \to \mathbb{C}^{2n+1}$, *that maps a trigonometric polynomial to its spectrum is well defined since it is the coordinate system in* $\mathcal{P}_{n,2\pi}$ *relative to the orthonormal basis* $\{e^{ikx}\}$. *In particular* $\hat{\ } : \mathcal{P}_{n,2\pi} \to \mathbb{C}^{2n+1}$ *is a (complex) isometry.*

(iii) (FOURIER COEFFICIENTS) *For* $k = -n, \ldots, n$ *we have*

$$c_k = (P|e^{ikx}) = \frac{1}{2\pi} \int_{-\pi}^{\pi} P(t)e^{-ikt}\, dt.$$

(iv) (ENERGY IDENTITY)

$$\frac{1}{2\pi} \int_{-\pi}^{\pi} |P(t)|^2\, dt = ||P||^2 := (P|P) = \sum_{k=-n}^{n} |(P|e^{ikx})|^2 = \sum_{k=-n}^{n} |c_k|^2.$$

a. Spectrum and products

Let $P(x) = \sum_{k=-n}^{n} c_k e^{ikx}$ and $Q(x) = \sum_{k=-n}^{n} d_k e^{ikx}$ be two trigonometric polynomials of order n. Their product is the trigonometric polynomial of order $2n$

$$P(x)Q(x) = \sum_{k=-n}^{n} c_k e^{ikx} \sum_{k=-n}^{n} d_k e^{ikx} = \sum_{h,k=-n}^{n} c_h d_k e^{i(h+k)x}$$

$$= \sum_{p=-2n}^{2n} \left(\sum_{h+k=p} c_h d_k \right) e^{ipx}.$$

If we denote by $\{c_k\} * \{d_k\}$ the product in the sense of Cauchy of the spectra of P and Q, we can state the following.

4.46 Proposition. *The spectrum of $P(x)Q(x)$ is the product in the sense of Cauchy of the spectra of P and Q*

$$(\widehat{PQ})_k = \widehat{P}_k * \widehat{Q}_k.$$

4.47 Definition. *The* convolution product *of P and Q is defined by*

$$P * Q(x) := \frac{1}{2\pi} \int_{-\pi}^{\pi} P(x+t)\overline{Q(t)} \, dt.$$

Notice that the operation $(P, Q) \to P * Q$ is linear in the first factor and antilinear in the second one. We have

4.48 Proposition. *$P*Q$ is a trigonometric polynomial of degree n. Moreover the spectrum of $P * Q$ is the term-by-term product of the spectra of P and Q,*

$$(\widehat{P * Q})_k := \widehat{P}_k \, \widehat{Q}_k.$$

Proof. In fact for $h, \ k = -n, \ldots, n$, we have

$$e^{i \, hx} * e^{i \, kx}(x) = \frac{1}{2\pi} \int_{-\pi}^{\pi} e^{i \, h(x+t)} e^{-i \, kt} \, dt = e^{i \, hx} \frac{1}{2\pi} \int_{-\pi}^{\pi} e^{i \, (h-k)t} \, dt = \delta_{hk} e^{i \, kx}$$

hence, if $P(x) = \sum_{k=-n}^{n} c_k e^{i \, kx}$ and $Q(x) = \sum_{k=-n}^{n} d_k e^{i \, kx}$, we have

$$P * Q(x) = \sum_{h=-n}^{n} \sum_{k=-n}^{n} c_h \overline{d_k} \delta_{hk} e^{i \, kx} = \sum_{k=-n}^{n} c_k \overline{d_k} e^{i \, kx}.$$

$\square$

b. Sampling of trigonometric polynomials

A trigonometric polynomial of degree n can be reconstructed from its values on a suitable choice of its values on $2n + 1$ points, see Section 5.4.1 of [GM2]. Set $x_j := \frac{2\pi}{2n+1} j$, $j = -n, \ldots, n$, then the sampling map

$$C : \mathcal{P}_{n,2\pi} \to \mathbb{C}^{2n+1}, \qquad C(P) := (P(x_{-n}), \ldots, P(x_n))$$

is invertible, in fact, see Theorem 5.49 of [GM2],

$$P(x) := \frac{1}{2n + 1} \sum_{j=-n}^{n} P(x_j) D_n(x - x_j)$$

where $D_n(t)$ is the *Dirichlet kernel* of order n

$$D_n(t) := \sum_{k=-n}^{n} e^{i \, kt} = 1 + 2 \sum_{k=1}^{n} \cos kt.$$

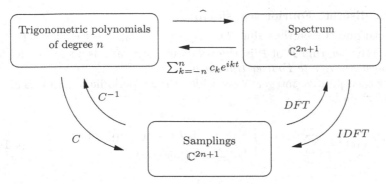

Figure 4.1. The scenario of trigonometric polynomials, spectra and samples.

4.49 Proposition. $\frac{1}{\sqrt{2n+1}}C$ *and its inverse* $\sqrt{2n+1}C^{-1}$: $\mathbb{C}^{2n+1}$ $\rightarrow$ $\mathcal{P}_{n,2\pi}$ *given by*

$$\sqrt{2n+1}C^{-1}(\mathbf{z})(x) := \frac{1}{\sqrt{2n+1}} \sum_{j=-n}^{n} z_j D_n(x - x_j)$$

are isometries between $\mathcal{P}_{n,2\pi}$ *and* $\mathbb{C}^{2n+1}$.

Proof. In fact, C maps $e^{i\,kt}$, $k = -n, \ldots, n$, to an orthonormal basis of $\mathbb{C}^{2n+1}$:

$$(C(e^{i\,ht})|C(e^{i\,kt})) = \sum_{j=-n}^{n} \exp\left(i\,\frac{2\pi(h-k)}{2n+1}j\right) = D_n\left(\frac{2\pi}{2n+1}(h-k)\right) = (2n+1)\delta_{hk}.$$

$\square$

From the samples, we can directly compute the spectrum of P.

4.50 Proposition. *Let* $P(x) \in \mathcal{P}_{n,2\pi}$ *and* $x_j := \frac{2\pi}{2n+1}j$, $j = -n, \ldots, n$. *Then*

$$\frac{1}{2\pi} \int_{-\pi}^{\pi} P(t)e^{-ikt}\,dt = \frac{1}{2n+1} \sum_{k=-n}^{n} P(x_j)e^{-ikx_j}. \qquad (4.13)$$

Proof. Since (4.13) is linear in P, it suffices to prove it when $P(x) = e^{i\,ht}$, $h = -n, \ldots, n$. In this case, we have $\frac{1}{2\pi}\int_{-\pi}^{\pi} P(t)e^{-i\,kt}\,dt = \delta_{hk}$ and

$$\frac{1}{2n+1} \sum_{j=-n}^{n} e^{i(h-k)x_j} = \frac{1}{2n+1} \sum_{j=-n}^{n} D_n(x_{h-k}) = \delta_{hk},$$

since $D_n(x_j) = 0$ for $j \neq 0$, $j \in [-n, n]$ and $D_n(0) = 2n+1$. $\square$

c. The discrete Fourier transform

The relation between the values $\{P(t_j)\}$ of $P \in \mathcal{P}_{n,2\pi}$ at the $2n+1$ points t_j and the spectrum $\widehat{P}$ of P in the previous paragraph is a special case of the so-called *discrete Fourier transform*.

For each positive integer N, consider the 2π-periodic function $E_N(t) :$ $\mathbb{R} \to \mathbb{C}$ given by

$$E_N(t) := \sum_{k=0}^{N-1} e^{i\,kt} = \begin{cases} N & \text{if } t \text{ is a multiple of } 2\pi, \\ \frac{1-e^{i\,Nt}}{1-e^{i\,t}} & \text{otherwise.} \end{cases} \qquad (4.14)$$

Let $\omega = e^{i\frac{2\pi}{N}}$ and let $1, \omega, \omega^2, \ldots, \omega^{N-1}$ be the Nth roots of one. For $h \in \mathbb{Z}$ we have

$$\frac{1}{N} \sum_{k=0}^{N-1} \omega^{hk} = \begin{cases} 1 & \text{if } h \text{ is a multiple of } N, \\ 0 & \text{otherwise,} \end{cases} \qquad (4.15)$$

in particular,

$$\frac{1}{N} \sum_{k=0}^{N-1} \omega^{hk} = \delta_{hk} \qquad \text{if } -N < h < N. \qquad (4.16)$$

The *discrete Fourier transform of order* N, $DFT_N : \mathbb{C}^N \to \mathbb{C}^N$, is defined by $DFT_N(\mathbf{y}) := \mathbf{U}\mathbf{y}$ rows by column, where

$$\mathbf{U} = [U_j^i], \qquad U_j^i := \frac{1}{N}\omega^{-ij} \qquad \forall i, j = 0, \ldots, N-1.$$

The *inverse discrete Fourier transform of order* N, $IDFT_N : \mathbb{C}^N \to \mathbb{C}^N$, is defined by $IDFT_N(\mathbf{z}) := \mathbf{V}\mathbf{z}$ where

$$\mathbf{V} = [V_j^i], \qquad V_j^i = \omega^{ij}, \qquad \forall i, j = 0, N-1.$$

4.51 Proposition. $IDFT_N$ *is the inverse of* DFT_N. *Moreover, the operators* $\sqrt{N}\,DFT_N$ *and* $\frac{1}{\sqrt{N}}\,IDFT_N$ *are isometries of* $\mathbb{C}^N$.

Proof. In fact, by (4.16)

$$(\mathbf{U}\mathbf{V})_j^i = \frac{1}{N} \sum_{k=0}^{N-1} \omega^{-ik}\omega^{kj} = \frac{1}{N} \sum_{k=0}^{N-1} \omega^{(j-i)k} = \delta_j^i,$$

i.e., $\mathbf{V} = \mathbf{U}^{-1}$ and, by the definition of $\mathbf{U}$ and $\mathbf{V}$, $\overline{\mathbf{U}}^T = \frac{1}{N}\mathbf{V}$, hence $\overline{\mathbf{U}}^T = \frac{1}{N}\mathbf{V} = \frac{1}{N}\mathbf{U}^{-1}$. $\square$

Notice that, according to their definitions, we need N^2 multiplications to compute DFT_N (or $IDFT_N$). There is an algorithm, that we shall not describe here, called the *Fast Fourier Transform* that, using the redundance of some multiplications, computes DFT_N (or $IDFT_N$) with only N multiplications with a performance of $O(N \log N)$.

Let $P(t) = \sum_{k=-n}^{n} c_k e^{ikx} \in \mathcal{P}_{n,2\pi}$ and let $N \geq 2n+1$. A computation similar to the one in Proposition 4.50 shows that

$$\widehat{P}_k = \frac{1}{2\pi} \int_{-\pi}^{\pi} P(t) e^{-kt}\, dt = \frac{1}{N} \sum_{j=0}^{N-1} P(x_j) e^{-ikx_j} = DFT_N y \qquad (4.17)$$

where $y := (P(x_0), \dots, P(x_N))$ and $x_j := \frac{2\pi}{N} j$, $-N < j < N$. Thus *the spectrum of P is the DFT_N of its values at x_j if $n < N/2$.*

On the other hand, if $z := (z_0, \dots, z_{N-1})$ is the vector defined by

$$z_k := \begin{cases} \widehat{P}_k & \text{if } 0 \leq k \leq n, \\ 0 & \text{if } n < k < N/2, \\ \widehat{P}_{k-N} & \text{if } N/2 \leq k \leq N/2+n, \\ 0 & \text{if } N/2+n < k < N \end{cases}$$

and we recall that $IDFT_N$ is the inverse of DFT_N, we have

$$P(x_j) := (IDFT_N z)_j,$$

i.e., the values of P at x_j are the $IDFT_N$ of the spectrum of P.

4.52 Frequency spectrum. In applications, the DFT_N and $IDFT_N$ may appear slightly differently. If f is a T_0-periodic function, one lets T_0/N be the period of sampling, so that $t_j := \frac{T_0}{N} j = jT$, $j = 0, 1, \dots, N-1$, are the sampling points and DTF_N produces the sequence

$$c_k := \frac{1}{N} \sum_{j=0}^{N-1} f(jT) e^{i\frac{2\pi}{N} kj}.$$

In other words, the values of $\{c_k\}$ are regarded as the values of the component of frequency $\nu_k := k\frac{1}{T_0} = \frac{k}{NT}$, i.e., as the samples of the so-called *frequency spectrum* $\widehat{f} : \mathbb{R} \to \mathbb{C}$ of f, defined by

$$\widehat{f}(\nu) := \begin{cases} c_k & \text{if } \nu = \nu_k = \frac{k}{NT}, \\ 0 & \text{otherwise.} \end{cases}$$

The discrete Fourier transform and its inverse then rewrite as

$$\hat{f}\left(\frac{k}{NT}\right) = \frac{1}{N}\sum_{j=0}^{N-1} f(jT)e^{-i\frac{2\pi}{N}jk},$$

$$f(kT) = \sum_{j=0}^{N-1} \hat{f}\left(\frac{j}{NT}\right)e^{i\frac{2\pi}{N}jk}.$$

4.2.3 Systems of difference equations

Linear difference equations of first and second order are discussed e.g., in [GM2]. Here we shall discuss systems of linear difference equations.

a. Systems of linear difference equations

First let us consider systems of first order. Let $\mathbf{A} \in M_{k,k}(\mathbb{C})$. The homogeneous linear recurrence for the sequence in $\mathbb{C}^k$

$$\begin{cases} X_{n+1} = \mathbf{A}X_n, & n \geq 0, \\ X_0 \text{ given} \end{cases}$$

has the unique solution $X_n := \mathbf{A}^n X_0 \ \forall n$, as one can easily check.

4.53 Proposition. *Given $\{F_n\}$ in $\mathbb{C}^k$, the recurrence*

$$\begin{cases} X_{n+1} = \mathbf{A}X_n + F_{n+1}, & n \geq 0, \\ X_0 \text{ given} \end{cases}$$

has the solution

$$X_n := \mathbf{A}^n X_0 + \sum_{j=0}^{n} \mathbf{A}^{n-j}F_j \qquad \forall n \geq 0$$

where we assume $F_0 := 0$.

Proof. In fact, for $n \geq 0$ we have

$$X_{n+1} = \mathbf{A}^{n+1}X_0 + \sum_{j=0}^{n+1}\mathbf{A}^{n+1-j}F_j = \mathbf{A}^{n+1}X_0 + \sum_{j=0}^{n}\mathbf{A}^{n+1-j}F_j + F_{n+1}$$

$$= \mathbf{A}\,\mathbf{A}^n X_0 + \mathbf{A}\sum_{j=0}^{n}\mathbf{A}^{n-j}F_j + F_{n+1} = \mathbf{A}\left(\mathbf{A}^n X_0 + \sum_{j=0}^{n}\mathbf{A}^{n-j}F_j\right) + F_{n+1}$$

$$= \mathbf{A}X_n + F_{n+1}.$$

$\square$

4.54 Higher order linear difference equations. Every equation

$$x_{n+k} + a_{k-1}x_{n+k-1} + \cdots + a_0 x_n = f_{n+1}, \qquad n \geq 0 \qquad (4.18)$$

can be transformed into a $k \times k$ system of difference equations of first order. In fact, if

$$X_n := (x_n, x_{n+1}, \ldots, x_{n+k-1})^T \in \mathbb{C}^k,$$
$$F_n := (0, 0, \ldots, 0, f_n)^T \in \mathbb{C}^k,$$

and $\mathbf{A}$ is the $k \times k$ matrix

$$\mathbf{A} := \begin{pmatrix} 0 & 1 & 0 & \cdots & 0 \\ 0 & 0 & 1 & \cdots & 0 \\ \vdots & \vdots & \vdots & \ddots & \vdots \\ 0 & 0 & 0 & \cdots & 1 \\ -a_0 & -a_1 & -a_2 & \cdots & -a_{k-1} \end{pmatrix}, \qquad (4.19)$$

it is easily seen that

$$X_{n+1} = \mathbf{A}X_n + F_{n+1} \qquad (4.20)$$

and conversely, if $\{X_n\}$ solves (4.20), then $\{x_n\}$, $x_n := X_n^1$ $\forall n$, solves (4.18). In this way the theory of higher order linear difference equations is subsumed to that of first order systems. In this respect, one computes for the matrix $\mathbf{A}$ in (4.19) that

$$\det(\lambda \operatorname{Id} - \mathbf{A}) = \lambda^k + \sum_{j=0}^{k-1} a_j \lambda^j.$$

This polynomial in λ is called the *characteristic polynomial* of the difference equation (4.18).

b. Power of a matrix

Let us compute the power of $\mathbf{A}$ in an efficient way. To do this we remark the following.

(i) If $\mathbf{B}$ is similar to $\mathbf{A}$, $\mathbf{A} = \mathbf{S}^{-1}\mathbf{BS}$ for some $\mathbf{S}$ with $\det \mathbf{S} \neq 0$, then

$$\mathbf{A}^2 = \mathbf{S}^{-1}\mathbf{BSS}^{-1}\mathbf{BS} = \mathbf{S}^{-1}\mathbf{B}^2\mathbf{S}$$

and, by induction,

$$\mathbf{A}^n = \mathbf{S}^{-1}\mathbf{B}^n\mathbf{S} \qquad \forall n.$$

(ii) If $\mathbf{B}$ is a block matrix with square blocks in the principal diagonal

$$\mathbf{B} = \begin{pmatrix} \boxed{\mathbf{B}_1} & 0 & \cdots & 0 \\ 0 & \boxed{\mathbf{B}_2} & \cdots & 0 \\ \vdots & \vdots & \ddots & \vdots \\ 0 & \cdots & 0 & \boxed{\mathbf{B}_r} \end{pmatrix},$$

then

$$\mathbf{B}^n = \begin{pmatrix} \boxed{\mathbf{B}_1^n} & 0 & \cdots & 0 \\ 0 & \boxed{\mathbf{B}_2^n} & \cdots & 0 \\ \vdots & \vdots & \ddots & \vdots \\ 0 & \cdots & 0 & \boxed{\mathbf{B}_r^n} \end{pmatrix}.$$

Let $\lambda_1, \lambda_2, \ldots, \lambda_k$ be the distinct eigenvalues of $\mathbf{A}$ with multiplicities $m_1, m_2, \ldots, m_k$. For every k, let p_k be the dimension of the eigenspace relative to λ_k (the geometric multiplicity). Then, see Theorem 2.65, there exists a nonsingular matrix $\mathbf{S} \in M_{k,k}(\mathbb{C})$ such that $\mathbf{J} := \mathbf{S}^{-1}\mathbf{A}\mathbf{S}$ has the Jordan form

$$\mathbf{J} = \begin{pmatrix} \boxed{\mathbf{J}_{1,1}} & 0 & 0 & \cdots & 0 \\ 0 & \boxed{\mathbf{J}_{1,2}} & 0 & \cdots & 0 \\ \vdots & \vdots & \vdots & \ddots & \vdots \\ 0 & 0 & 0 & \cdots & \boxed{\mathbf{J}_{k,p_k}} \end{pmatrix}$$

where $i = 1, \ldots, k$, $j = 1, \ldots, p_i$ and

$$\mathbf{J}_{i,j} = \begin{cases} \lambda_i & \text{if } \mathbf{J}_{i,j} \text{ has dimension 1,} \\ \begin{pmatrix} \lambda_i & 1 & 0 & 0 & \cdots & 0 \\ 0 & \lambda_i & 1 & 0 & \cdots & 0 \\ \vdots & \vdots & \vdots & \vdots & \ddots & \vdots \\ 0 & 0 & 0 & \cdots & \lambda_i & 1 \\ 0 & 0 & 0 & \cdots & 0 & \lambda_i \end{pmatrix} & \text{otherwise.} \end{cases}$$

Consequently $\mathbf{A}^n = \mathbf{S}\mathbf{J}^n\mathbf{S}^{-1}$, and

$$
\mathbf{J}^n = \begin{pmatrix}
\boxed{\mathbf{J}^n_{1,1}} & 0 & 0 & \cdots & 0 \\
0 & \boxed{\mathbf{J}^n_{1,2}} & 0 & \cdots & 0 \\
\vdots & \vdots & \vdots & \ddots & \vdots \\
0 & 0 & 0 & \cdots & \boxed{\mathbf{J}^n_{k,p_k}}
\end{pmatrix}.
$$

It remains to compute the power of each Jordan block. If $\mathbf{J}' = \mathbf{J}_{i,j} = (\lambda)$ has dimension one, then $\mathbf{J}'^n = \lambda^n$. If instead $\mathbf{J}' = \mathbf{J}_{i,j}$ is a block of dimension q at least two,

$$
\mathbf{J}' = \begin{pmatrix}
\lambda & 1 & 0 & \cdots & 0 \\
0 & \lambda & 1 & \cdots & 0 \\
\vdots & \vdots & \vdots & \ddots & \vdots \\
0 & \cdots & 0 & \lambda & 1 \\
0 & 0 & \cdots & 0 & \lambda
\end{pmatrix},
$$

then
$$
\mathbf{J}' = \lambda \operatorname{Id} + \mathbf{B}, \qquad \mathbf{B}^i_j := \delta_{i+1,j}.
$$

Since
$$
(\mathbf{B}^r)^i_j = \begin{cases} \delta_{r+i,j} & \text{if } r < q, \\ 0 & \text{if } r \geq q, \end{cases}
$$

we have $\mathbf{B}^q = 0$. Thus Newton's binomial formula yields

$$
\mathbf{J}'^n = (\lambda \operatorname{Id} + \mathbf{B})^n = \sum_{j=0}^n \binom{n}{j} \lambda^{n-j} \mathbf{B}^j
$$

$$
= \sum_{j=0}^{q-1} \binom{n}{j} \lambda^{n-j} \mathbf{B}^j = \lambda^n \sum_{j=0}^{q-1} \binom{n}{j} \frac{1}{\lambda^j} \mathbf{B}^j,
$$

i.e.,

$$
\mathbf{J}'^n = \lambda^n \begin{pmatrix}
1 & n\frac{1}{\lambda} & \binom{n}{2}\frac{1}{\lambda^2} & \cdots & \binom{n}{q-1}\frac{1}{\lambda^{q-1}} \\
0 & 1 & n\frac{1}{\lambda} & \cdots & \binom{n}{q-2}\frac{1}{\lambda^{q-2}} \\
\vdots & \vdots & \vdots & \ddots & \vdots \\
0 & 0 & \cdots & 1 & n\frac{1}{\lambda} \\
0 & 0 & \cdots & 0 & 1
\end{pmatrix}.
$$

Notice that each element of $\mathbf{A}^n = \mathbf{S}\mathbf{J}^n\mathbf{S}^{-1}$ has the form

$$\sum_{j=1}^{k} \lambda_j^n p_j(n)$$

where $\lambda_1, \lambda_2, \ldots, \lambda_n$ are the eigenvalues of $\mathbf{A}$ and $p_j(t)$ is a polynomial of degree at most $p_j - 1$, where p_j is the algebraic multiplicity of λ_j. It follows that for $\rho > \max_i(|\lambda_i|)$ there is a constant C_ρ such that every solution of $X_{n+1} = \mathbf{A}X_n$ satisfies

$$|X_n| = \sup \leq C_\rho \rho^n |X_0| \qquad \forall n.$$

In particular we have the following.

4.55 Theorem. *If all eigenvalues of $\mathbf{A}$ have modulus less than one, then every solution of $X_{n+1} = \mathbf{A}X_n$ converges to zero as $n \to +\infty$.*

Proof. Fix $\sigma > 0$ such that $\max_{i=1,n} |\lambda_i| < \sigma < 1$. As we have seen, there exists a constant C_σ such that if X_n is a solution of $X_{n+1} = \mathbf{A}X_n \ \forall n$, then

$$|X_n| \leq C_\sigma \sigma^n |X_0|, \qquad \forall n.$$

Since $0 < \sigma < 1$, $\sigma^n \to 0$, and the claim is proved. $\qquad\square$

4.56 Example (Fibonacci numbers). Consider the sequence of Fibonacci numbers

$$\begin{cases} f_{n+2} = f_{n+1} + f_n & n \geq 0, \\ f_0 = 0, \ f_1 = 1, \end{cases} \qquad (4.21)$$

that is given by

$$f_n := \frac{1}{\sqrt{5}}\left(\left(\frac{1+\sqrt{5}}{2}\right)^n - \left(\frac{1-\sqrt{5}}{2}\right)^n\right), \qquad n \geq 0, \qquad (4.22)$$

see e.g., [GM2]. Let us find it again as an application of the above. Set

$$F_n := \begin{pmatrix} f_n \\ f_{n+1} \end{pmatrix}$$

then

$$\begin{cases} F_{n+1} = \begin{pmatrix} f_{n+1} \\ f_{n+2} \end{pmatrix} = \begin{pmatrix} f_{n+1} \\ f_n + f_{n+1} \end{pmatrix} = \begin{pmatrix} 0 & 1 \\ 1 & 1 \end{pmatrix} F_n, \\ F_0 = \begin{pmatrix} 0 \\ 1 \end{pmatrix} \end{cases}$$

and,

$$F_n = \mathbf{A}^n \begin{pmatrix} 0 \\ 1 \end{pmatrix}$$

where

$$\mathbf{A} = \begin{pmatrix} 0 & 1 \\ 1 & 1 \end{pmatrix}.$$

The characteristic polynomial of $\mathbf{A}$ is $\det(\lambda \operatorname{Id} - \mathbf{A}) = \lambda(\lambda - 1) - 1$, hence $\mathbf{A}$ has two distinct eigenvalues

$$\lambda := \frac{1 + \sqrt{5}}{2}, \qquad \mu := \frac{1 - \sqrt{5}}{2}.$$

An eigenvalue relative to λ is $(1, \lambda)$ and an eigenvector relative to μ is $(1, \mu)$. The matrix $\mathbf{A}$ is diagonalizable as $\mathbf{A} = \mathbf{S} \Delta \mathbf{S}^{-1}$ where

$$\mathbf{S} := \begin{pmatrix} 1 & 1 \\ \lambda & \mu \end{pmatrix}, \qquad \mathbf{S}^{-1} = \frac{1}{\lambda - \mu} \begin{pmatrix} -\mu & 1 \\ \lambda & -1 \end{pmatrix} \qquad \Delta = \begin{pmatrix} \lambda & 0 \\ 0 & \mu \end{pmatrix}.$$

It follows that

$$\mathbf{F}_n = \mathbf{A}^n \begin{pmatrix} 0 \\ 1 \end{pmatrix} = \mathbf{S} \Delta^n \mathbf{S}^{-1} \begin{pmatrix} 0 \\ 1 \end{pmatrix} = \frac{1}{\lambda - \mu} \mathbf{S} \begin{pmatrix} \lambda^n & 0 \\ 0 & \mu^n \end{pmatrix} \begin{pmatrix} 1 \\ -1 \end{pmatrix}$$

$$= \frac{1}{\lambda - \mu} \begin{pmatrix} 1 & 1 \\ \lambda & \mu \end{pmatrix} \begin{pmatrix} \lambda^n \\ -\mu^n \end{pmatrix}.$$

Consequently,

$$f_n = \frac{1}{\lambda - \mu} \left(\lambda^n - \mu^n \right) = \frac{1}{\sqrt{5}} \left(\left(\frac{1 + \sqrt{5}}{2} \right)^n - \left(\frac{1 - \sqrt{5}}{2} \right)^n \right).$$

4.2.4 An ODE system: small oscillations

Let $\mathbf{x}_1, \mathbf{x}_2, \ldots, \mathbf{x}_N$ be N point masses in $\mathbb{R}^3$ each respectively, to a nonzero mass $m_1, m_2, \ldots, m_N$. Assume that each point exerts a force on the other points according to *Hooke's law*, i.e., the force exerted by the mass at $\mathbf{x}_j$ on $\mathbf{x}_i$ is proportional to the distance of $\mathbf{x}_j$ from $\mathbf{x}_i$ and directed along the line through $\mathbf{x}_i$ and $\mathbf{x}_j$,

$$\mathbf{f}_{ij} := -k_{ij}(\mathbf{x}_j - \mathbf{x}_i), \qquad j \neq i.$$

By Newton's reaction law, the force exerted by $\mathbf{x}_i$ on $\mathbf{x}_j$ is equal and opposite in direction, $f_{ji} = -f_{ij}$, consequently the elastic constants k_{ij}, $i \neq j$, satisfy the symmetry condition $k_{ij} = k_{ji}$. In conclusion, the total force exerted by the system on the mass at $\mathbf{x}_i$ is

$$\mathbf{f}_i = - \sum_{\substack{j=1,N \\ i \neq j}} k_{ij}(\mathbf{x}_j - \mathbf{x}_i) = - \sum_{\substack{j=1,N \\ i \neq j}} k_{ij}\mathbf{x}_j + \left(\sum_{\substack{j=1,N \\ i \neq j}} k_{ij} \right) \mathbf{x}_i = - \sum_{j=1}^{N} k_{ij}\mathbf{x}_j$$

where we set $k_{ii} := - \sum_{\substack{j=1,N \\ j \neq i}} k_{ij}$. Newton's equation then takes the form

$$m_i \mathbf{x}_i'' - \mathbf{f}_i = 0, \qquad i = 1, \ldots, N, \tag{4.23}$$

with the particularity that the jth component of the force depends only on the jth component of the mass. The system then splits into 3 systems of N equations of second order, one for each coordinate. If we use the matrix notation, things simplify. Denote by $\mathbf{M} := \operatorname{diag}\{m_1, m_2, \ldots, m_N\}$

the positive diagonal matrix of masses, by $\mathbf{K} := (k_{ij}) \in M_{N,N}(\mathbb{R})$ the symmetric matrix of elastic contants, and by $X(t) \in \mathbb{R}^N$ the jth coordinates of the points $\mathbf{x}_1, \ldots, \mathbf{x}_N$

$$X = (x_1^j, \ldots, x_N^j), \qquad \mathbf{x}_i =: (x_i^1, x_i^2, x_i^3),$$

i.e., the columns of the matrix

$$\mathbf{X}(t) := \left[\mathbf{x}_1(t) \,\middle|\, \mathbf{x}_2(t) \,\middle|\, \ldots \,\middle|\, \mathbf{x}_N(t) \right]^T \in M_{N,3}(\mathbb{R}).$$

Then (4.23) transforms into the system of equations

$$\mathbf{M}X''(t) + \mathbf{K}X(t) = 0 \tag{4.24}$$

where the product is the product rows by columns. Finally, if $\mathbf{X}''(t)$ denotes the matrix of second derivatives of the entries of $\mathbf{X}(t)$, the system (4.23) can be written as

$$\mathbf{X}''(t) + \mathbf{M}^{-1}\mathbf{K}\mathbf{X}(t) = 0, \tag{4.25}$$

in the unknown $\mathbf{X} : \mathbb{R} \to M_{N,n}(\mathbb{R})$.

Since $\mathbf{M}^{-1}\mathbf{K}$ is symmetric, there is an orthonormal basis of $\mathbb{R}^N$ made by eigenvalues of $\mathbf{M}^{-1}\mathbf{K}$ and real numbers $\lambda_1, \lambda_2, \ldots, \lambda_N$ such that

$$(u_i | u_j) = \delta_{ij} \qquad \text{and} \qquad \mathbf{M}^{-1}\mathbf{K}u_j := \lambda_j u_j;$$

notice that $u_1, u_2, \ldots, u_N$ are pairwise orthonormal vectors since $\mathbf{M}$ is diagonal. Denoting by P_j the projection operator onto Span $\{u_j\}$ we also have

$$\text{Id} = \sum_{j=1}^{N} P_j, \qquad \mathbf{M}^{-1}\mathbf{K} = \sum_{j=1}^{N} \lambda_j P_j.$$

Thus, projecting (4.25) onto Span $\{u_j\}$ we find

$$0 = P_j(0) = P_j(\mathbf{X}'' + \mathbf{M}^{-1}\mathbf{K}\mathbf{X}) = (P_j\mathbf{X})'' + \lambda_j(P_j\mathbf{X}), \qquad \forall j = 1, \ldots, N,$$

i.e., the system (4.25) splits into N second order equations each in the unknowns of the matrix $P_j\mathbf{X}(t)$.

Since $\mathbf{K}$ is positive, the eigenvalues are positive, consequently each element of the matrix $P_j\mathbf{X}(t)$ is a solution of the harmonic oscillator

$$y(t) = \cos(\sqrt{\lambda_j}t)y(0) + \frac{\sin(\sqrt{\lambda_j}t)}{\sqrt{\lambda_j}}y'(0),$$

hence

$$P_j\mathbf{X}(t) = \cos(\sqrt{\lambda_j}t)P_j\mathbf{X}(0) + \frac{\sin(\sqrt{\lambda_j}t)}{\sqrt{\lambda_j}}P_j\mathbf{X}'(0).$$

In conclusion, since $\text{Id} = \sum_{j=1}^{N} P_j$, we have

$$\mathbf{X}(t) = \sum_{j=1}^{N} P_j \mathbf{X}(t)$$

$$= \Big(\sum_{j=1}^{N} \cos(\sqrt{\lambda_j}t) P_j \Big) \mathbf{X}(0) + \Big(\sum_{j=1}^{N} \frac{\sin(\sqrt{\lambda_j}t)}{\sqrt{\lambda_j}} P_j \Big) \mathbf{X}'(0). \tag{4.26}$$

The numbers $\sqrt{\lambda_1}/(2\pi), \dots \sqrt{\lambda_n}/(2\pi)$ are called the *proper frequencies* of the system.

We may also use a functional notation

$$\sin \mathbf{A} := \sum_{k=0}^{\infty} (-1)^n \frac{\mathbf{A}^{2n+1}}{(2n+1)!}, \qquad \cos \mathbf{A} := \sum_{k=0}^{\infty} (-1)^n \frac{\mathbf{A}^{2n}}{(2n)!}$$

and we can write (4.26) as

$$\mathbf{X}(t) = \cos(t\sqrt{\mathbf{A}})\mathbf{X}(0) + \frac{\sin(t\sqrt{\mathbf{A}})}{\sqrt{\mathbf{A}}} \mathbf{X}'(0),$$

where $\mathbf{A} := \mathbf{M}^{-1}\mathbf{K}$.

4.3 Exercises

4.57 ¶. Let $\mathbf{A}$ be an $n \times n$ matrix and let λ be its eigenvalue of greatest modulus. Show that $|\lambda| \leq \sup_i(|a_1^i| + |a_2^i| + \cdots + |a_n^i|)$.

4.58 ¶ Gram matrix. Let $\{f_1, f_2, \dots, f_m\}$ be m vectors in $\mathbb{R}^n$. The matrix $\mathbf{G} = [g_{ij}] \in M_{m,n}(\mathbb{R})$ defined by $g_{ij} = (f_i|f_j)$ is called *Gram's matrix*. Show that $\mathbf{G}$ is nonnegative and it is positive if and only if $f_1, f_2, \dots, f_m$ are linearly independent.

4.59 ¶. Let $A, B : \mathbb{C}^n \to \mathbb{C}^n$ be self-adjoint and let A be positive. Show that the eigenvalues of $A^{-1}B$ are real. Show also that $A^{-1}B$ is positive if B is positive.

4.60 ¶. Let $\mathbf{A} = [a_i^j] \in M_{n,n}(\mathbb{K})$ be self-adjoint and positive. Show that $\det \mathbf{A} \leq (\operatorname{tr} \mathbf{A}/n)^n$ and deduce $\det \mathbf{A} \leq \prod_{i=1}^n a_i^i$. [*Hint:* Use the inequality between geometric and aritmethic means, see [GM1].]

4.61 ¶. Let $\mathbf{A} \in M_{n,n}(\mathbb{K})$ and let $\mathbf{a}_1, \mathbf{a}_2, \dots, \mathbf{a}_n \in \mathbb{K}^n$ be the columns of $\mathbf{A}$. Prove *Hadamard's formula* $\det \mathbf{A} \leq \prod_{i=1}^n |\mathbf{a}_i|$. [*Hint:* Consider $\mathbf{H} = \mathbf{A}^*\mathbf{A}$.]

4.62 ¶. Let $\mathbf{A}, \mathbf{B} \in M_{n,n}(\mathbb{R})$ be symmetric and suppose that $\mathbf{A}$ is positive. Then the number of positive, negative and zero eigenvalues, counted with their multiplicity, of $\mathbf{AB}$ and of $\mathbf{B}$ coincide.

4.63 ¶. Show that $||N^*N|| = ||N||^2$ if N is normal.

4.64 ¶ Discrete Fourier transform. Let $T : \mathbb{C}^N \to \mathbb{C}^N$ be the cycling forward shifting operator $T((z_0, z_1, \ldots, z_{N-1})) := (z_1, z_2, \ldots, z_{N-1}, z_0)$. Show that

(i) T is self-adjoint,
(ii) the N eigenvalues of T are the Nth roots of 1,
(iii) the vectors

$$\mathbf{u}_k := \frac{1}{\sqrt{N}}(1, \omega^k, \omega^{2k}, \ldots, \omega^{k(N-1)}), \qquad \omega := e^{i\frac{2\pi}{N}}, \ k = 0, \ldots, N-1,$$

form an orthonormal basis of $\mathbb{C}^N$ of eigenvectors of T; finally the cosine directions $(\mathbf{z}|\mathbf{u}_k)$ of $\mathbf{z} \in \mathbb{C}^N$ with respect to the basis $(\mathbf{u}_0, \ldots, \mathbf{u}_{N_1})$ are given by the Discrete Fourier transform of z.

4.65 ¶. Let $A, B : X \to X$ be two self-adjoint operators on a Euclidean or Hermitian space. Suppose that all eigenvalues of $A - B$ are strictly positive. Order the eigenvalues $\lambda_1, \lambda_2, \ldots, \lambda_n$ of A and $\mu_1, \mu_2, \ldots, \mu_n$ of B in a nondecreasing order. Show that $\lambda_i < \mu_i$ $\forall i = 1, \ldots, n$. [*Hint:* Use the variational characterization of the eigenvalues.]

4.66 ¶. Let $A : X \to X$ be self-adjoint on a Euclidean or Hermitian space. Let $\lambda_1, \lambda_2, \ldots, \lambda_n$ and $\mu_1, \mu_2, \ldots, \mu_n$ be respectively, the eigenvalues and the singular values of A that we think of as ordered as $|\lambda_1| \leq |\lambda_2| \leq \cdots \leq |\lambda_n|$ and $\mu_1 \leq \mu_2 \leq \cdots \leq \mu_n$. Show that $|\lambda_i| = \mu_i$ $\forall i = 1, \ldots, n$. [*Hint:* $A^*A = A^2$.]

4.67 ¶. Let $A : X \to X$ be a linear operator on a Euclidean or Hermitian space. Let m, M be respectively the smallest and the greatest singular value of A. Show that $m \leq |\lambda| \leq M$ for any eigenvalue λ of A.

4.68 ¶. Let $A : X \to Y$ be a linear operator between two Euclidean or two Hermitian spaces. Show that

(i) $(A^*A)^{1/2}$ maps $\ker A$ to $\{0\}$,
(ii) $(A^*A)^{1/2}$ is an isomorphism from $\ker A^\perp$ onto itself,
(iii) $(AA^*)^{1/2}$ is an isomorphism from $\operatorname{Im} A$ onto itself.

4.69 ¶. Let $A : X \to Y$ be a linear operator between two Euclidean or two Hermitian spaces. Let $(u_1, u_2, \ldots, u_n)$ and $\mu_1, \mu_2, \ldots, \mu_n$, $\mu_i \geq 0$ be such that $(u_1, u_2, \ldots, u_n)$ is an orthonormal basis of X and $(A^*A)^{1/2}x = \sum_i \mu_i(x|u_i)u_i$. Show that

(i) $AA^*y = \sum_{\mu_i \neq 0} \mu_i(y|Au_i)Au_i$ $\forall y \in Y$,
(ii) If B denotes the restriction of $(A^*A)^{1/2}$ to $\ker A^\perp$, see Exercise 4.68, then

$$B^{-1}x = \sum_{\mu_i \neq 0} \frac{1}{\mu_i}(x|u_i)u_i \qquad \forall x \in \ker A^\perp,$$

(iii) If C denotes the restriction of $(AA^*)^{1/2}$ to $\operatorname{Im} A$, see Exercise 4.68, then

$$C^{-1}y = \sum_{\mu_i \neq 0} \frac{1}{\mu_i}(y|Au_i)Au_i \qquad \forall y \in \operatorname{Im} A.$$

4.70 ¶. Let $\mathbf{A} \in M_{N,n}(\mathbb{K})$, $N \geq n$, with $\operatorname{Rank} \mathbf{A} = n$. Select n vectors $u_1, u_2, \ldots, u_n \in \mathbb{K}^n$ such that $\mathbf{A}u_1, \ldots, \mathbf{A}u_n \in \mathbb{K}^N$ are orthonormal. [*Hint:* Find $\mathbf{U} \in M_{n,n}(\mathbb{K})$ such that $\mathbf{A}\mathbf{U}$ is an isometry.]

4.71 ¶. Let $\mathbf{A} \in M_{N,n}(\mathbb{R})$ and $\mathbf{A} = \mathbf{U}\Delta\mathbf{V}$, where $\mathbf{U} \in O(N)$, $\mathbf{V} \in O(n)$. According to 4.39, show that $\mathbf{A}^\dagger = \mathbf{V}^T \Delta' \mathbf{U}^T$ where

$$
\Delta' = \begin{pmatrix}
\frac{1}{\mu_1} & 0 & \cdots & 0 & 0 \\
0 & \frac{1}{\mu_2} & \cdots & 0 & 0 \\
\vdots & \vdots & \ddots & \vdots & \vdots \\
0 & 0 & \cdots & \frac{1}{\mu_k} & 0 \\
0 & 0 & \cdots & 0 & 0 \\
\vdots & \vdots & \ddots & \vdots & \vdots \\
0 & 0 & \cdots & 0 & 0
\end{pmatrix},
$$

$\mu_1, \mu_2, \ldots, \mu_k$ being the nonzero singular values of $\mathbf{A}$.

4.72 ¶. For $\mathbf{u} : \mathbb{R} \to \mathbb{R}^2$, discuss the system of equations

$$
\frac{d^2\mathbf{u}}{dt} + \begin{pmatrix} 2 & -1 \\ -1 & 2 \end{pmatrix} \mathbf{u} = 0.
$$

4.73 ¶. Let $\mathbf{A} \in M_{n,n}(\mathbb{R})$ be a symmetric matrix. Discuss the following systems of ODEs

$$
\mathbf{x}'(t) + \mathbf{A}\mathbf{x}(t) = 0,
$$
$$
- i\mathbf{x}'(t) + \mathbf{A}\mathbf{x}(t) = 0,
$$
$$
\mathbf{x}''(t) + \mathbf{A}\mathbf{x}(t) = 0, \qquad \text{where } \mathbf{A} \text{ is positive definite}
$$

and show that the solutions are given respectively by

$$
e^{-t\mathbf{A}}\mathbf{x}(0), \quad e^{-it\mathbf{A}}\mathbf{x}(0), \qquad \cos(t\sqrt{\mathbf{A}})\mathbf{x}(0) + \frac{\sin(t\sqrt{\mathbf{A}})}{\sqrt{\mathbf{A}}}\mathbf{x}'(0).
$$

4.74 ¶. Let $\mathbf{A}$ be symmetric. Show that for the solutions of $\mathbf{x}''(t) + \mathbf{A}\mathbf{x}(t) = 0$ the energy is conserved. Assuming $\mathbf{A}$ positive, show that $|\mathbf{x}(t)| \leq E/\lambda_1$ where E is the energy of $\mathbf{x}(t)$ and λ the smallest eigenvalue of $\mathbf{A}$.

4.75 ¶. Let $\mathbf{A}$ be a Hermitian matrix. Show that $|\mathbf{x}(t)| = \text{const}$ if $\mathbf{x}(t)$ solves the Schrödinger equation $i\mathbf{x}' + \mathbf{A}\mathbf{x} = 0$.

Part II

Metrics and Topology

Felix Hausdorff (1869–1942), Maurice Fréchet (1878–1973) and René-Louis Baire (1874–1932).

5. Metric Spaces and Continuous Functions

The rethinking process of infinitesimal calculus, that was started with the definition of the limit of a sequence by Bernhard Bolzano (1781–1848) and Augustin-Louis Cauchy (1789–1857) at the beginning of the XIX century and was carried on with the introduction of the system of real numbers by Richard Dedekind (1831–1916) and Georg Cantor (1845–1918) and of the system of complex numbers with the parallel development of the theory of functions by Camille Jordan (1838–1922), Karl Weierstrass (1815–1897), J. Henri Poincaré (1854–1912), G. F. Bernhard Riemann (1826–1866), Jacques Hadamard (1865–1963), Emile Borel (1871–1956), René-Louis Baire (1874–1932), Henri Lebesgue (1875–1941) during the whole of the XIX and beginning of the XX century, led to the introduction of new concepts such as open and closed sets, the point of accumulation and the compact set. These notions found their natural collocation and their correct generalization in the notion of a metric space, introduced by Maurice Fréchet (1878–1973) in 1906 and eventually developed by Felix Hausdorff (1869–1942) together with the more general notion of topological space.

The intuitive notion of a "continuous function" probably dates back to the classical age. It corresponds to the notion of deformation without "tearing". A function from X to Y is more or less considered to be continuous if, when x varies slightly, the target point $y = f(x)$ also varies slightly. The critical analysis of this intuitive idea also led, with Bernhard Bolzano (1781–1848) and Augustin-Louis Cauchy (1789–1857), to the correct definition of continuity and the limit of a function and to the study of the properties of continuous functions. We owe the theorem of intermediate values to Bolzano and Cauchy, around 1860 Karl Weierstrass proved that continuous functions take on maximum and minimum values in a closed and limited interval, and in 1870 Eduard Heine (1821–1881) studied uniform continuity. The notion of a continuous function also appears in the work of J. Henri Poincaré (1854–1912) in an apparently totally different context, in the so-called *analysis situs*, which is today's topology and algebraic topology. For Henri Poincare, *analysis situs* is the science that enables us to know the qualitative properties of geometrical figures. Poincaré referred to the properties that are preserved when geometrical figures undergo any kind of deformation except those that introduce tearing and glueing of points. An intuitive idea for some of these aspects may be provided by the following examples.

Figure 5.1. Frontispieces of *Les éspaces abstraits* by Maurice Fréchet (1878–1973) and of the *Mengenlehre* by Felix Hausdorff (1869–1942).

○ Let us draw a disc on a rubber sheet. No matter how one pulls at the rubber sheet, without tearing it, the disc stays whole. Similarly, if one draws a ring, any way one pulls the rubber sheet without tearing or glueing any points, the central hole is preserved. Let us think of a loop of string that surrounds an infinite pole. In order to separate the string from the pole one has to break one of the two. Even more, if the string is wrappped several times around the pole, the *linking number* between string and pole is constant, regardless of the shape of the coils.

○ We have already seen Euler's formula for polyhedra in [GM1]. It is a remarkable formula whose context is not classical geometry. It was Poincaré who extended it to all surfaces of the type of the sphere, i.e., surfaces that can be obtained as continuous deformations of a sphere without tearing or glueing.

○ $\mathbb{R}$, $\mathbb{R}^2$, $\mathbb{R}^3$ are clearly different objects as linear vectorspaces. As we have seen, they have the same cardinality and are thus undistinguishable as sets. Therefore it is impossible to give meaning to the concept of dimension if one stays inside the theory of sets. One can show, instead, that their algebraic dimension is preserved by deformations without tearing or glueing.

At the core of this analysis of geometrical figures we have the notion of a continuous deformation that corresponds to the notion of of a continuous one-to-one map whose inverse is also continuous, called *homeomorphisms*. We have already discussed some relevant properties of continuous functions $f : \mathbb{R} \to \mathbb{R}$ e $f : \mathbb{R}^2 \to \mathbb{R}$ in [GM1] and [GM2]. Here we shall discuss continuity in a sufficiently general context, though not in the most general.

Poincaré himself was convinced of the enormous importance of extending the methods and ideas of his *analysis situs* to more than three dimensions.

> ... *L'analysis situs à plus de trois dimensions présente des difficultés énormes; il faut pour tenter de les surmonter être bien persuadé de l'extrême importance de cette science. Si cette importance n'est pas bien comprise de tout le monde, c'est que tout le monde n'y a pas suffisamment réfléchi.*[1]

In the first twenty years of this century with the contribution, among others, of David Hilbert (1862–1943), Maurice Fréchet (1878–1973), Felix Hausdorff (1869–1942), Pavel Alexandroff (1896–1982) and Pavel Urysohn (1898–1924), the fundamental role of the notion of an *open set* in the study of continuity was made clear, and *general topology* was developed as the study of the properties of geometrical figures that are invariant with respect to homeomorphisms, thus linking back to Euler who, in 1739, had solved the famous problem of Königsberg's bridges with a topological method. There are innumerable successive applications, so much so that continuity and the structures related to it have become one of the most pervasive languages of mathematics.

In this chapter and in the next, we shall discuss topological notions and continuity in the context of *metric spaces*.

5.1 Metric Spaces

5.1.1 Basic definitions

a. Metrics

5.1 Definition. *Let X be a set. A* distance *or* metric *on X is a map* $d : X \times X \to \mathbb{R}_+$ *for which the following conditions hold:*

(i) (IDENTITY) $d(x, y) > 0$ if $x \neq y \in X$, and $d(x, x) = 0 \ \forall x \in X$.
(ii) (SYMMETRY) $d(x, y) = d(y, x) \ \forall x, y \in X$.
(iii) (TRIANGLE INEQUALITY) $d(x, y) \leq d(x, z) + d(z, y), \ \forall \ x, y, z \in X$.

A metric space *is a set X with a distance d. Formally we say that (X, d) is a metric space if X is a set and d is a distance on X.*

The properties (i), (ii) and (iii) are often called *metric axioms*.

---

[1] The analysis situs in more than three dimensions presents enormous difficulties; in order to overcome them one has to be strongly convinced of the extreme importance of this science. If its importance is not well understood by everyone, it is because they have not sufficiently thought about it.

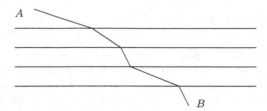

Figure 5.2. Time as distance.

5.2 Example. The Euclidean distance $d(x, y) := |x - y|$, $x, y \in \mathbb{R}$, is a distance on $\mathbb{R}$.
   On $\mathbb{R}^2$ and $\mathbb{R}^3$ distances are defined by the *Euclidean distance*, given for $n = 2, 3$
by

$$d_E(\mathbf{x}, \mathbf{y}) := \left\{ \sum_{i=1}^{n} (x_i - y_i)^2 \right\}^{1/2},$$

where $\mathbf{x} := (x_1, x_2), \mathbf{y} := (y_1, y_2)$ if $n = 2$, or $\mathbf{x} := (x_1, x_2, x_3), \mathbf{y} := (y_1, y_2, y_3)$ if $n = 3$.
In other words, $\mathbb{R}, \mathbb{R}^2, \mathbb{R}^3$ are metric spaces with the Euclidean distance.

5.3 Example. Imagine $\mathbb{R}^3$ as a union of strips $\Sigma_n := \{(x_1, x_2, x_3) \mid n \le x_3 < n+1\}$,
made by materials of different indices of refractions ν_n. The time $t(A, B)$ needed for a
light ray to go from A to B in $\mathbb{R}^3$ defines a distance on $\mathbb{R}^3$, see Figure 5.2.

5.4 Example. In the infinite cylinder $C = \{(x, y, z) \mid x^2 + y^2 = 1\} \subset \mathbb{R}^3$, we may
define a distance between two points P and Q as the minimal length of the line on C,
or *geodesic*, connecting P and Q. Observe that we can always cut the cylinder along a
directrix in such a way that the curve is not touched. If we unfold the cut cylinder to a
plane, the distance between P and Q is the Euclidean distance of the two image points.

5.5 ¶. Of course $100|x - y|$ is also a distance on $\mathbb{R}$, only the scale factor has changed.
More generally, if $f : \mathbb{R} \to \mathbb{R}$ is an injective map, then $d(x, y) := |f(x) - f(y)|$ is again
a distance on $\mathbb{R}$.

5.6 Definition. *Let (X, d) be a metric space. The* open ball *or* spherical
open neighborhood *centered at $x_0 \in X$ of radius $\rho > 0$ is the set*

$$B(x_0, \rho) := \Big\{ x \in X \,\Big|\, d(x, x_0) < \rho \Big\}.$$

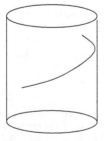 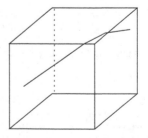

Figure 5.3. Metrics on a cylinder and on the boundary of a cube.

Notice the strict inequality in the definition of $B(x, r)$. In $\mathbb{R}$, $\mathbb{R}^2$, $\mathbb{R}^3$ with the Euclidean metric, $B(x_0, r)$ is respectively, the open interval $]x_0 - r, x_0 + r[$, the open disc of center $x_0 \in \mathbb{R}^2$ and radius $r > 0$, and the ball bounded by the sphere of $\mathbb{R}^3$ of center $x_0 \in \mathbb{R}^3$ and radius $r > 0$.

We say that a subset $E \subset X$ of a metric space is *bounded* if it is contained in some open ball. The *diameter* of $E \subset X$ is given by

$$\text{diam } E := \sup\Big\{ d(x, y) \,\Big|\, x, y \in E \Big\},$$

and, trivially, E is bounded iff $\text{diam } E < +\infty$.

Despite the suggestive language, the open balls of a metric space need not be either round nor convex; however they have some of the usual properties of discs in $\mathbb{R}^2$. For instance

(i) $B(x_0, r) \subset B(x_0, s) \; \forall x_0 \in X$ and $0 < r < s$,
(ii) $\cup_{r>0} B(x_0, r) = X \; \forall x_0 \in X$,
(iii) $\cap_{r>0} B(x_0, r) = \{x_0\} \; \forall x_0 \in X$,
(iv) $\forall x_0 \in X$ and $\forall z \in B(x_0, r)$ the open ball centered at z and radius $\rho := r - d(x_0, z) > 0$ is contained in $B(x_0, r)$,
(v) for every couple of balls $B(x, r)$ and $B(y, s)$ with a nonvoid intersection and $\forall z \in B(x, r) \cap B(y, s)$, there exists $t > 0$ such that $B(z, t) \subset B(x, r) \cap B(y, s)$, in fact $t := \min(r - d(x, z), s - d(y, z))$,
(vi) for every $x, y \in X$ with $x \neq y$ the balls $B(x, r_1)$ and $B(y, r_2)$ are disjoint if $r_1 + r_2 \leq d(x, y)$.

5.7 ¶. Prove the previous claims. Notice how essential the strict inequality in the definition of $B(x_0, \rho)$ is.

b. Convergence

A distance d on a set X allow us to define the notion of convergent sequence in X in a natural way.

5.8 Definition. *Let (X, d) be a metric space. We say that the sequence $\{x_n\} \subset X$ converges to $x \in X$, and we write $x_n \to x$, if $d(x_n, x) \to 0$ in $\mathbb{R}$, that is , if for any $r > 0$ there exists $\overline{n}$ such that $d(x_n, x) < r$ for all $n \geq \overline{n}$.*

The metric axioms at once yield that the basic facts we know for limits of sequences of real numbers also hold for limits of sequences in an arbitrary metric space. We have

(i) the limit of a sequence $\{x_n\}$ is unique, if it exists,
(ii) if $\{x_n\}$ converges, then $\{x_n\}$ is bounded,
(iii) computing the limit of $\{x_n\}$ consists in having a candidate $x \in X$ and then showing that the sequence of nonnegative real numbers $\{d(x_n, x)\}$ converges to zero,
(iv) if $x_n \to x$, then any subsequence of $\{x_n\}$ has the same limit x.

Thus, the choice of a distance on a given set X suffices to *pass to the limit in X* (in the sense specified by the metric d). However, given a set X, there is no distance on X that is reasonably absolute (even in $\mathbb{R}$), but we may consider different distances in X. The corresponding convergences have different meanings and can be suited to treat specific problems. They all use the same general language, but the exact meaning of what convergence means is hidden in the definition of the distance. This flexibility makes the language of metric spaces useful in a large context.

5.1.2 Examples of metric spaces

Relevant examples of distances are provided by linear vector spaces on the fields $\mathbb{K} = \mathbb{R}$ or $\mathbb{C}$ in which we have defined a *norm*.

5.9 Definition. *Let X be a linear space over $\mathbb{K} = \mathbb{R}$ or $\mathbb{C}$. A norm on X is a function $|| \; || : X \to \mathbb{R}_+$ satisfying the following properties*

   (i) (FINITENESS) $||x|| \in \mathbb{R} \; \forall x \in X$.
  (ii) (IDENTITY) $||x|| \geq 0$ *and* $||x|| = 0$ *if and only if* $x = 0$.
 (iii) (1-HOMOGENEITY) $||\lambda x|| = |\lambda| \, ||x|| \; \forall \, x \in X, \; \forall \lambda \in \mathbb{K}$.
  (iv) (TRIANGLE INEQUALITY) $||x + y|| \leq ||x|| + ||y|| \; \forall \, x, y \in X$.

If $|| \cdot ||$ is a norm on X, we say that $(X, || \; ||)$ is a linear normed space *or simply that X is a* normed space *with norm $|| \; ||$.*

Let X be a linear space with norm $|| \; ||$. It is easy to show that the function $d : X \times X \to \mathbb{R}_+$ given by

$$d(x, y) := ||x - y||, \qquad x, y \in X,$$

satisfies the metric axioms, hence defines a distance on X, called the *natural distance* in the normed space $(X, || \; ||)$. Obviously, such a distance is *translation invariant*, i.e., $d(x + z, y + z) = d(x, y) \; \forall x, y, z \in X$.

Trivial examples of metric spaces are provided by the nonempty subsets of a metric space. If A is a subset of a metric space (X, d), then the restriction of d to $A \times A$ is trivially, a distance on A. We say that A is a metric space with the induced distance from X.

5.10 ¶. For instance, the cylinder $C := \{(x, y, z) \in \mathbb{R}^3 \mid x^2 + y^2 = 1\}$ is a metric space with the Euclidean distance that, for $\mathbf{x}, \mathbf{y} \in C$, yields $d(\mathbf{x}, \mathbf{y}) :=$length of the chord joining $\mathbf{x}$ and $\mathbf{y}$. The *geodesic distance* d_g of Example 5.4, that is the length of the shortest path in C between $\mathbf{x}$ and $\mathbf{y}$, defines another distance. C with the geodesic distance d_g has to be considered as another metric space different from C with the Euclidean distance. A simple calculation shows that

$$||\mathbf{x} - \mathbf{y}|| \leq d_g(\mathbf{x}, \mathbf{y}) \leq \frac{\pi}{2} ||\mathbf{x} - \mathbf{y}||.$$

We shall now illustrate a few examples of metric spaces.

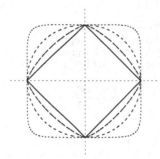

Figure 5.4. The ball centered at $(0,0)$ of radius 1 in $\mathbb{R}^2$ respectively, for the metrics d_1, $d_{1.3}$, d_2 and d_7. The unit ball centered at $(0,0)$ of radius one for the metric d_∞ is the square $]-1,1[\times]-1,1[$.

a. Metrics on finite-dimensional vector spaces

5.11 ¶. As we have already seen, $\mathbb{R}^n$ with the Euclidean distance $|\mathbf{x} - \mathbf{y}|$ is a metric space. More generally, any Euclidean or Hermitian vector space X is a normed space with norm given by

$$||x|| := \sqrt{(x|x)}$$

cf. Chapter 3. X is therefore a metric space with the induced distance

$$d(x,y) := ||x - y||,$$

called the *Euclidean distance* of X.

5.12 ¶ ∞-distance. Set for $\mathbf{x} = (x_1, x_2, \ldots, x_n) \in \mathbb{R}^n$

$$||\mathbf{x}||_\infty := \max(|x_1|, |x_2|, \ldots, |x_n|).$$

Show that $x \to ||x||_\infty$ is a norm on $\mathbb{R}^n$. Hence, $\mathbb{R}^n$, equipped with the distance

$$d_\infty(\mathbf{x}, \mathbf{y}) := ||x - y||_\infty,$$

is a metric space different from the standard $\mathbb{R}^n$ with the Euclidean distance of Exercise 5.11. In $\mathbb{R}^2$, the unit ball centered at $(0,0)$ of radius one for the metric d_∞ is the square $]-1,1[\times]-1,1[$, see Figure 5.4.

5.13 ¶ p-distance. Given a real number $p \geq 1$, we set for $\mathbf{x} \in \mathbb{R}^n$

$$||\mathbf{x}||_p := \Big(\sum_{i=1}^{n} |x_i|^p \Big)^{1/p}.$$

Show that $||\mathbf{x}||_p$ is a norm on $\mathbb{R}^n$, hence

$$d_p(\mathbf{x}, \mathbf{y}) := ||x - y||_p$$

is a distance on $\mathbb{R}^n$. Observe that $|| \ ||_2$ and d_2 are the Euclidean norm and distance in $\mathbb{R}^n$. In $\mathbb{R}^2$, the unit ball centered at $(0,0)$ of radius one for the metric d_p for some values of p is shown in Figure 5.4. [*Hint:* The triangle inequality for the p-norm is called *Minkowski's discrete inequality*

$$||\mathbf{a} + \mathbf{b}||_p \le ||\mathbf{a}||_p + ||\mathbf{b}||_p, \qquad \forall \mathbf{a} := (a_1, a_2, \ldots, a_n), \ \mathbf{b} = (b_1, b_2, \ldots, b_n),$$

which follows for instance from Minkowski's inequality for integral norms, see [GM1]. Alternatively, we can proceed as follows. Suppose $\mathbf{a}$ and $\mathbf{b}$ are nonzero, otherwise the inequality is trivial, apply the convexity inequality $f(\lambda x + (1-\lambda)y) \le \lambda f(x) + (1-\lambda)f(y)$ to $f(t) = t^p$ with $x := a_i/||\mathbf{a}||_p$, $y := b_i/||\mathbf{b}||_p$, $\lambda := ||\mathbf{a}||_p/(||\mathbf{a}||_p + ||\mathbf{b}||_p)$, and sum on i from 1 to n, to get

$$\frac{||\mathbf{a} + \mathbf{b}||_p}{||\mathbf{a}||_p + ||\mathbf{b}||_p} \le 1.]$$

5.14 ¶ Product spaces. Let $(X_1, d^{(1)})$, $(X_2, d^{(2)})$, ... $(X_n, d^{(n)})$ be n metric spaces and let $Y = X_1 \times X_2 \times \cdots \times X_n$ be the Cartesian product of $X_1, \ldots, X_n$. Show that each of the functions defined on $X \times Y$ by

$$\begin{cases} d_p(\mathbf{x}, \mathbf{y}) := \left(\sum_{i=1}^n d^{(i)}(x_i, y_i)^p \right)^{1/p}, & \text{if } p > 1, \\ d_\infty(\mathbf{x}, \mathbf{y}) := \max\left\{ d^{(i)}(x_i, y_i) \,\middle|\, i = 1, \ldots, n \right\} \end{cases}$$

for $\mathbf{x} = (x^1, x^2, \ldots, x^n)$, $\mathbf{y} = (y^1, y^2, \ldots, y^n) \in Y$, are distances on Y. Notice that if $X_1 = \cdots = X_n = \mathbb{R}$ with the Euclidean distance, Y is $\mathbb{R}^n$, then the distances $d_p(x, y)$ are just the distances in Exercises 5.13 and 5.12. Also show that if $\{\mathbf{x}_k\} \subset Y$, $\mathbf{x}_k := (x_k^1, x_k^2, \ldots x_k^n) \ \forall k$, and $\mathbf{x} = (x^1, x^2, \ldots, x^n)$, then the following claims are equivalent.

   (i)   There exists $p \ge 1$ such that $d_p(\mathbf{x}_k, \mathbf{x}) \to 0$,
   (ii) $d_p(\mathbf{x}_k, \mathbf{x}) \to 0 \ \forall p \ge 1$,
   (iii) $d_\infty(\mathbf{x}_k, \mathbf{x}) \to 0 \ \forall p \ge 1$,
   (iv) $\forall i = 1, \ldots, n \ d_i(x_k^i, x^i) \to 0$.

5.15 ¶ Discrete distance. Let X be any set. The *discrete distance on X* is given by

$$d(x, y) = \begin{cases} 1 & \text{if } x \ne y, \\ 0 & \text{if } x = y. \end{cases}$$

Show that the balls for the discrete distance are

$$B(x, r) = \left\{ y \in X \,\middle|\, d(x, y) < r \right\} = \begin{cases} \{x\} & \text{if } r \le 1, \\ X & \text{if } r \ge 1, \end{cases}$$

and that convergent sequences with respect to the discrete distance reduce to sequences that are definitively constant.

5.16 ¶ Codes distance. Let X be a set that we think of as a set of symbols, and let $X^n = X \times X \times \cdots \times X$ the space of ordered words on n symbols. Given two words $x = (x_1, x_2, \ldots, x_n)$ and $y = (y_1, y_2, \ldots, y_n) \in X^n$, let

$$d(x, y) := \#\left\{ i \,\middle|\, x_i \ne y_i \right\}$$

be the number of bits in x and y that are different. Show that $d(x, y)$ defines a distance in X^n. Characterize the balls of X^n relative to that distance. [*Hint:* Write $d(x, y) = \sum_{i=1}^n d(x_i, y_i)$ where d is the discrete distance in X.]

Figure 5.5. The first page of the *Thèse* at the Faculty of Sciences of Paris by Maurice Fréchet (1878–1973), published on the *Rendiconti del Circolo Matematico di Palermo*.

b. Metrics on spaces of sequences

We now introduce some distances and norms on infinite-dimensional vector spaces.

5.17 Example (ℓ_∞ space). Consider the space of all real (or complex) sequences $\mathbf{x} := (x_1, \dots)$. For $\mathbf{x} = \{x_n\}$, $\mathbf{y} := \{y_n\}$, set

$$||\mathbf{x}||_\infty := \sup_n |x_n|, \qquad d_\infty(\mathbf{x}, \mathbf{y}) := ||\mathbf{x} - \mathbf{y}||_\infty.$$

It is easy to show that $x \to ||x||_\infty$ satisfies the axioms of a norm apart from the fact that it may take the value $+\infty$. Thus $x \to ||x||_\infty$ is a norm in the space

$$\ell_\infty := \Big\{ \mathbf{x} = \{x_n\} \,\Big|\, ||\mathbf{x}||_\infty < +\infty \Big\},$$

that is, on the linear vector space of *bounded sequences*. Consequently,

$$d_\infty(x, y) := ||\mathbf{x} - \mathbf{y}||_\infty, \qquad \mathbf{x}, \mathbf{y} \in \ell_\infty,$$

is a distance on ℓ_∞, called the *uniform distance*. Convergence of $\{\mathbf{x}_k\} \subset \ell_\infty$ to $\mathbf{x} \in \ell_\infty$ in the uniform norm, called the *uniform convergence*, amounts to

$$||\mathbf{x}_k - \mathbf{x}||_\infty = \sup_i |x_k^i - x^i| \to 0 \qquad \text{as } k \to \infty, \tag{5.1}$$

where $\mathbf{x}_k = (x_k^1, x_k^2, \dots)$ and $\mathbf{x} = (x^1, x^2, \dots)$. Notice that the uniform convergence in (5.1) is stronger than the *pointwise convergence*

$$\forall i \quad x_k^i \to x^i \qquad \text{as } k \to \infty.$$

For instance, let $\varphi(t) := te^{-t}$, $t \in \mathbb{R}^+$, and consider the sequence of sequences $\{\mathbf{x}_k\}$ where $\mathbf{x}_i := \{x_i^n\}_n$, $x_i^n := \varphi\big(\frac{k}{n}\big)$. Then $\forall i$ we have $x_k^i = \frac{i}{k}e^{-i/k} \to 0$ as $k \to \infty$, while

$$||\mathbf{x}_k - 0||_\infty = \sup\Big\{ \frac{i}{k}e^{-i/k} \,\Big|\, i = 0, 1, \dots \Big\} = \frac{1}{e} \neq 0.$$

Of course $\mathbb{R}^n$ with the metric d_∞ in Exercise 5.12 is a subset of ℓ_∞ endowed with the induced metric d_∞. This follows from the identification

$$(x^1, \dots, x^n) \leftrightarrow (x^1, \dots, x^n, 0, \dots, 0, \dots).$$

5.18 Example (ℓ_p spaces, $p \geq 1$). Consider the space of all real (or complex) sequences $\mathbf{x} := (x_1, \dots)$. For $1 \leq p \leq \infty$, $\mathbf{x} = \{x_n\}$ and $\mathbf{y} := \{y_n\}$ set

$$||\mathbf{x}||_{\ell_p} := \Big(\sum_{k=1}^{\infty} |x_n|^p \Big)^{1/p}, \qquad d_{\ell_p}(\mathbf{x}, \mathbf{y}) := ||\mathbf{x} - \mathbf{y}||_{\ell_p}.$$

Trivially, $||\mathbf{x}||_{\ell_p} = 0$ if and only if any element of the sequence $\mathbf{x}$ is zero, moreover *Minkowski's inequality*

$$||\mathbf{x} + \mathbf{y}||_{\ell_p} \leq ||\mathbf{x}||_{\ell_p} + ||\mathbf{y}||_{\ell_p},$$

holds as it follows from Exercise 5.13 (passing to the limit as $n \to \infty$ in Minkowski's inequality in $\mathbb{R}^n$). Thus $|| \ ||_{\ell_p}$ satisfies the metric axioms apart from the fact that it may take the value $+\infty$. Hence, $|| \ ||_{\ell_p}$ is a norm in the linear space of sequences

$$\ell_p := \Big\{ \mathbf{x} = \{x_n\} \,\Big|\, ||\mathbf{x}||_{\ell_p} < +\infty \Big\}.$$

Consequently, $d_{\ell_p}(\mathbf{x}, \mathbf{y}) := ||x - y||_{\ell_p}$ is a distance on ℓ_p.

Convergence of $\{\mathbf{x}_k\} \subset \ell_p$ to $\mathbf{x} \in \ell_p$ amounts to

$$\sum_{i=1}^{\infty} |x_k^i - x^i|^p \to 0 \qquad \text{as } k \to \infty,$$

where $\mathbf{x}_k = (x_k^1, x_k^2, \dots)$ and $\mathbf{x} = (x^1, x^2, \dots)$. Notice that $\mathbb{R}^n$ with the metric d_p in Exercise 5.13 is a subset of ℓ_p endowed with the induced metric d_{ℓ_p}. This follows for instance from the identification

$$(x^1, \dots, x^n) \leftrightarrow (x^1, \dots, x^n, 0, \dots, 0, \dots).$$

Finally, observe that $||\mathbf{x}||_{\ell_q} \leq ||\mathbf{x}||_{\ell_p} \ \forall \mathbf{x}$ if $1 \leq p \leq q$, hence

$$\ell_1 \subset \ell_p \subset \ell_q. \tag{5.2}$$

Since there exist sequences $\mathbf{x} = \{x_n\}$ such that $||\mathbf{x}||_{\ell_q} < +\infty$ while $||\mathbf{x}||_{\ell_p} = +\infty$ if $p < q$, as for instance

$$\mathbf{x} := \Big\{ \frac{1}{n} \Big\}^{1/p},$$

the inclusions (5.2) are strict if $1 < p < q$.

The case $p = 2$ is particularly relevant since the ℓ_2 norm is induced by the *scalar product*

$$(\mathbf{x}|\mathbf{y})_{\ell_2} := \sum_{i=1}^{\infty} x^i y^i, \qquad ||\mathbf{x}||_{\ell_2} = \sqrt{(\mathbf{x}|\mathbf{x})_{\ell_2}}.$$

ℓ_2 is called the *Hilbert coordinate space*, and the set

$$\Big\{ \mathbf{x} = (x^1, x^2, \dots,) \in \ell_2 \,\Big|\, |x_i| \leq \frac{1}{i} \ \forall i \Big\}$$

the *Hilbert cube*.

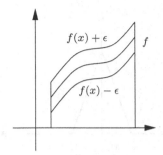

$f(x) + \epsilon$ f

$f(x) - \epsilon$

Figure 5.6. Tubular neighborhood of the graph of f.

c. Metrics on spaces of functions

The language of metric spaces is particularly relevant in dealing with different types of convergences of functions. As examples of metric spaces of functions, we then introduce a few normed spaces that are relevant in the sequel.

5.19 Example (Continuous functions). Denote by $C^0([0,1])$ the space of all continuous functions $f : [0,1] \to \mathbb{R}$. For $f : [0,1] \to \mathbb{R}$ set

$$||f||_{\infty,[0,1]} := \sup_{x \in [0,1]} |f(x)|.$$

We have

(i) $||f||_{\infty,[0,1]} < +\infty$ by Weierstrass's theorem,
(ii) $||f||_{\infty,[0,1]} = 0$ iff $f(x) = 0 \ \forall x$,
(iii) $||\lambda f||_{\infty,[0,1]} = |\lambda| ||f||_{\infty,[0,1]}$,
(iv) $||f + g||_{\infty,[0,1]} \le ||f||_{\infty,[0,1]} + ||g||_{\infty,[0,1]}$.

To prove (iv) for instance, observe that for all $x \in [0,1]$, we have

$$|f(x) + g(x)| \le |f(x)| + |g(x)| \le ||f||_{\infty,[0,1]} + ||g||_{\infty,[0,1]}$$

hence the right-hand side is an upperbound for the values of $f + g$.

The map $f \to ||f||_{\infty,[0,1]}$ is then a *norm* on $C^0([0,1])$, called the *uniform* or *infinity norm*. Consequently $C^0([0,1])$ is a normed space and a metric space with the *uniform distance*

$$d_\infty(f,g) := ||f - g||_{\infty,[0,1]} = \max_{t \in [0,1]} |f(t) - g(t)|, \qquad f,g \in C^0([0,1]).$$

In this space, the ball $B(f,\epsilon)$ of center f and radius $\epsilon > 0$ is the set of all continuous functions $g \in C^0([0,1])$ such that

$$|g(x) - f(x)| < \epsilon \qquad \forall \, x \in [0,1]$$

or the family of all continuous functions with graphs in the *tubular neighborhood* of radius ϵ of the graph of f

$$U(f,\epsilon) := \Big\{ (x,y) \, \Big| \, x \in [0,1], \ y \in \mathbb{R}, |y - f(x)| < \epsilon \Big\}, \tag{5.3}$$

see Figure 5.6.

The *uniform convergence* in $C^0([0,1])$, that is the convergence in the uniform norm, of $\{f_k\} \subset C^0([0,1])$ to $f \in C^0([0,1])$ amounts to computing

$$M_k := ||f_k - f||_{\infty,[0,1]} = \max_{t \in [0,1]} |f_k(t) - f(t)|$$

for every $k = 1, 2, \ldots$ and to showing that $M_k \to 0$ as $k \to +\infty$.

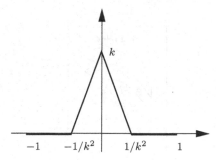

Figure 5.7. The function f_k in (5.4).

5.20 Example (Functions of class $C^1([0,1])$). Denote by $C^1([0,1])$ the space of all functions $f : [0,1] \to \mathbb{R}$ of class C^1, see [GM1]. For $f \in C^1([0,1])$, set

$$||f||_{C^1([0,1])} := \sup_{x \in [0,1]} |f(x)| + \sup_{x \in [0,1]} |f'(x)| = ||f||_{\infty,[0,1]} + ||f'||_{\infty,[0,1]}.$$

It is easy to check that $f \to ||f||_{C^1([0,1])}$ is a norm in the vector space $C^1([0,1])$. Consequently,

$$d_{C^1([0,1])}(f,g) := ||f - g||_{C^1([0,1])}$$

defines a distance in $C^1([0,1])$. In this case, a function $g \in C^1$ has a distance less than ϵ from f if $||f - g||_{\infty,[0,1]} + ||f' - g'||_{\infty,[0,1]} < \epsilon$; equivalently, if the graph of g is in the tubular neighborhood $U(f, \epsilon_1)$ of the graph of f, *and the graph of g' is in the tubular neighborhood $U(f', \epsilon_2)$ of f'* with $\epsilon_1 + \epsilon_2 = \epsilon$, see (5.3). Moreover, convergence in the $C^1([0,1])$-norm of $\{f_k\} \subset C^1([0,1])$ to $f \in C^1([0,1])$, $||f_k - f||_{C^1([0,1])} \to 0$, amounts to

$$\begin{cases} f_k \to f & \text{uniformly in } [0,1], \\ f'_k \to f' & \text{uniformly in } [0,1]. \end{cases}$$

Figures 5.8 and 5.9 show graphs of Lipschitz functions and functions of class $C^1([0,1])$ that are closer and closer to zero in the uniform norm, but with uniform norm of the derivatives larger than one.

5.21 Example (Integral metrics). Another norm and corresponding distance in $C^0([0,1])$ is given by the *distance in the mean*

$$||f||_{L^1([0,1])} := \int_0^1 |f(t)| \, dt, \qquad d_{L^1([0,1])}(f,g) := ||f - g||_{L^1(0,1)} := \int_0^1 |f - g| \, dx.$$

5.22 ¶. Show that the L^1-norm in $C^0([0,1])$ satisfies the norm axioms.

Convergence with respect to the L^1-distance differs from the uniform one. For instance, for $k = 1, 2, \ldots$ set

$$f_k(x) = \begin{cases} -k^3 (|x| - \frac{1}{k^2}) & \text{if } 0 < |x| < \frac{1}{k^2}, \\ 0 & \text{if } \frac{1}{k^2} \leq |x| < 1. \end{cases} \tag{5.4}$$

We have $||f_k||_{\infty,[0,1]} = f(0) = k \to +\infty$ while $||f_k||_{L^1(0,1)} = 1/(2k) \to 0$, cf. Figure 5.7.

More generally, the $L^p([0,1])$-norm, $1 \leq p < \infty$, on $C^0([0,1])$, is defined by

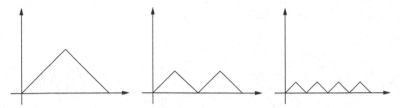

Figure 5.8. The Lebesgue example.

$$\|f\|_{L^p(0,1)} := \left(\int_0^1 |f(x)|^p \, dx \right)^{1/p}.$$

It turns out that $f \to \|f\|_{L^p(0,1)}$ satisfies the axioms of a norm, hence

$$d_{L^p([0,1])}(f,g) := \|f - g\|_{L^p([0,1])} := \left(\int_0^1 |f - g|^p \, dx \right)^{1/p}$$

is a distance in $C^0([0,1])$; it is called the $L^p([0,1])$-*distance*.

5.23 ¶. Show that the $L^p([0,1])$-norm in $C^0([0,1])$ satisfies the norm axioms. [*Hint:* The triangle inequality is in fact Minkowski's inequality, see [GM1].]

5.1.3 Continuity and limits in metric spaces

a. Lipschitz-continuous maps between metric spaces

5.24 Definition. *Let* (X, d_X) *and* (Y, d_Y) *be two metric spaces and let* $0 < \alpha \le 1$. *We say that a function* $f : X \to Y$ *is* α-*Hölder-continuous if there exists* $L > 0$ *such that*

$$d_Y(f(x), f(y)) \le L \, d_X(x,y)^\alpha, \qquad \forall \, x, y \in X. \tag{5.5}$$

1-Hölder-continuous functions are also called Lipschitz continuous. The smallest constant L for which (5.5) *holds is called the* α-*Hölder constant of* f, *often denoted by* $[f]_\alpha$. *When* $\alpha = 1$, *the 1-Hölder constant is also called the* Lipschitz *constant of* f *and denoted by* $[f]_1$, Lip_f *or* $\text{Lip}(f)$.

5.25 Example (The distance function). Let (X, d) be a metric space. For any $x_0 \in X$, the function $f(x) := d(x, x_0) : X \to \mathbb{R}$ is a Lipschitz-continuous function with $\text{Lip}(f) = 1$. In fact, from the triangle inequality, we get

$$|f(y) - f(x)| = |d(y, x_0) - d(x, x_0)| \le d(x,y) \qquad \forall x, y \in X,$$

hence f is Lipschitz continuous with $\text{Lip}(f) \le 1$. Choosing $x = x_0$, we have

$$|f(y) - f(x_0)| = |d(y, x_0) - d(x_0, x_0)| = d(y, x_0),$$

thus $\text{Lip}(f) \ge 1$.

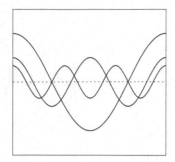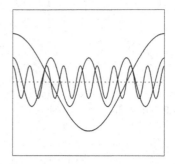

Figure 5.9. On the left, the sequence $f_k(x) := k^{-1}\cos(kx)$ that converges uniformly to zero with slopes equibounded by one. On the right, $g_k(x) := k^{-1}\cos(k^2 x)$, that converges uniformly to zero, but with slopes that diverge to infinity. Given any function $f \in C^1([0,1])$, a similar phenomenon occurs for the sequences $f_k(x) := f(kx)/k$, $g_k(x) = f(k^2 x)/k$.

5.26 ¶ Distance from a set. Let (X, d) be a metric space. The distance function from $x \in X$ to a nonempty subset $A \subset X$ is defined by

$$d(x, A) := \inf\left\{ d(x, y) \,\middle|\, y \in A \right\}.$$

It is easy to show that $f(x) := d(x, A) : X \to \mathbb{R}$ is a Lipschitz-continuous function with

$$\mathrm{Lip}\,(f) = \begin{cases} 0 & \text{if } d(x, A) = 0 \ \forall x, \\ 1 & \text{otherwise.} \end{cases}$$

If $d(x, A)$ is identically zero, then the claim is trivial. On the other hand, for any $x, y \in X$ and $z \in A$ we have $d(x, z) \leq d(x, y) + d(y, z)$ hence, taking the infimum in z,

$$d(x, A) - d(y, A) \leq d(x, y)$$

and interchanging x and y,

$$|d(x, A) - d(y, A)| \leq d(x, y),$$

that is, $x \to d(x, A)$ is Lipschitz continuous with Lipschitz constant less than one. Since there exists a $x \notin A$ such that $d(x, A) > 0$, there exists a sequence $\{z_n\} \subset A$ such that $\frac{d(x, z_n)}{d(x, A)} < 1 + \frac{1}{n}$. Therefore,

$$|d(x, A) - d(x_n, A)| = d(x, A) \geq \frac{n}{n+1} d(x, x_n),$$

from which we infer that the Lipschitz constant of $x \to d(x, A)$ must not be smaller than one.

b. Continuous maps in metric spaces

The notion of continuity that we introduced in [GM1], [GM2] for functions on one real variable can be extended in the context of the abstract *metric structure*. In fact, by paraphrasing the definition of continuity of functions $f : \mathbb{R} \to \mathbb{R}^+$ we get

5.27 Definition. *Let (X, d_X) and (Y, d_Y) be two metric spaces. We say that $f : X \to Y$ is* continuous at x_0 *if $\forall \epsilon > 0$ there exists $\delta > 0$ such that $d_Y(f(x), f(x_0)) < \epsilon$ whenever $d_X(x, x_0) < \delta$, i.e.,*

$$\forall \epsilon > 0 \; \exists \delta > 0 \text{ such that } f(B_X(x_0, \delta)) \subset B_Y(f(x_0), \epsilon). \qquad (5.6)$$

We say that $f : X \to Y$ is continuous in $E \subset X$ *if f is continuous at every point $x_0 \in E$. When $E = X$ and $f : X \to Y$ is continuous at any point of X, we simply say that $f : X \to Y$ is* continuous.

5.28 ¶. Show that α-Hölder-continuous functions, $0 < \alpha \leq 1$, in particular Lipschitz-continuous functions, between two metric spaces are continuous.

Let (X, d_X) and (Y, d_Y) be two metric spaces and $E \subset X$. Since E is a metric space with the induced distance of X, Definition 5.27 also applies to the function $f : E \to Y$. Thus $f : E \to Y$ is continuous at $x_0 \in E$ if

$$\forall \epsilon > 0 \; \exists \delta > 0 \text{ such that } f(B_X(x_0, \delta) \cap E) \subset B_Y(f(x_0), \epsilon) \qquad (5.7)$$

and we say that $f : E \to Y$ is continuous if $f : E \to Y$ is continuous at any point $x_0 \in E$.

5.29 Remark. As in the case of functions of one real variable, the domain of the function f is relevant in order to decide if f is continuous or not. For instance, $f : X \to Y$ is continuous in $E \subset X$ if

$$\forall x_0 \in E \; \forall \epsilon > 0 \; \exists \delta > 0 \text{ such that } f(B_X(x_0, \delta)) \subset B_Y(f(x_0), \epsilon), \qquad (5.8)$$

while the restriction $f_{|E} : E \to Y$ of f to E is continuous in E if

$$\forall x_0 \in E \; \forall \epsilon > 0 \; \exists \delta > 0 \text{ such that } f(B_X(x_0, \delta) \cap E) \subset B_Y(f(x_0), \epsilon). \qquad (5.9)$$

We deduce that the *restriction property* holds: *if $f : X \to Y$ is continuous in E, then its restriction $f_{|E} : E \to Y$ to E is continuous.* The opposite is in general false, as simple examples show.

5.30 Proposition. *Let X, Y, Z be three metric spaces, and $x_0 \in X$. If $f : X \to Y$ is continuous at x_0 and $g : Y \to Z$ is continuous at $f(x_0)$, then $g \circ f : X \to Z$ is continuous at x_0. In particular, the composition of two continuous functions is continuous.*

Proof. Let $\epsilon > 0$. Since g is continuous at $f(x_0)$, there exists $\sigma > 0$ such that $g(B_Y(f(x_0), \sigma)) \subset B_Z(g(f(x_0)), \epsilon)$. Since f is continuous at x_0, there exists $\delta > 0$ such that $f(B_X(x_0, \delta)) \subset B_Y(f(x_0), \sigma)$, consequently

$$g \circ f(B_X(x_0, \delta)) \subset g(B_Y(f(x_0), \sigma)) \subset B_Z(g \circ f(x_0), \epsilon).$$

$\square$

Continuity can be expressed in terms of convergent sequences. As in the proof of Theorem 2.46 of [GM2], one shows

5.31 Theorem. *Let (X, d_X) and (Y, d_Y) be two metric spaces. $f : X \to Y$ is continuous at $x_0 \in X$ if and only if $f(x_n) \to f(x_0)$ in (Y, d_Y) whenever $\{x_n\} \subset X$, $x_n \to x_0$ in (X, d_X).*

c. Limits in metric spaces

Related to the notion of continuity is the notion of the limit. Again, we want to rephrase $f(x) \to y_0$ as $x \to x_0$. For that we need f to be defined near x_0, but not necessarily at x_0. For this purpose we introduce the

5.32 Definition. *Let X be a metric space and $A \subset X$. We say that $x_0 \in X$ is an* accumulation point *of A if each ball centered at x_0 contains at least one point of A distinct from x_0,*

$$\forall r > 0 \quad B(x_0, r) \cap A \setminus \{x_0\} \neq \emptyset.$$

Accumulation points are also called cluster points.

5.33 ¶. Consider $\mathbb{R}$ with the Euclidean metric. Show that
  (i) the set of points of accumulation of $A :=]a, b[$, $B = [a, b]$, $C = [a, b[$ is the closed interval $[a, b]$,
  (ii) the set of points of accumulation of $A :=]0, 1[\cup \{2\}$, $B = [0, 1] \cup \{2\}$, $C = [0, 1[\cup \{2\}$ is the closed interval $[0, 1]$,
  (iii) the set of points of accumulation of the rational numbers and of the irrational numbers is the whole $\mathbb{R}$.

We shall return to this notion, but for the moment the definition suffices.

5.34 Definition. *Let (X, d_X) and (Y, d_Y) be two metric spaces, let $E \subset X$ and let $x_0 \in X$ be a point of accumulation of E. Given $f : E \setminus \{x_0\} \to Y$, we say that $y_0 \in Y$ is the* limit *of $f(x)$ as $x \to x_0$, $x \in E$, and we write*

$$f(x) \to y_0 \ as \ x \to x_0, \qquad or \qquad \lim_{\substack{x \to x_0 \\ x \in E}} f(x) = y_0$$

if for any $\epsilon > 0$ there exists $\delta > 0$ such that $d_Y(f(x), y_0) < \epsilon$ whenever $x \in E$ and $0 < d_X(x, x_0) < \delta$. Equivalently,

$$\forall \epsilon > 0 \ \exists \delta > 0 \ such \ that \ f(B_X(x_0, \delta) \cap E \setminus \{x_0\}) \subset B_Y(y_0, \epsilon).$$

Notice that, while in order to deal with the continuity of f at x_0 we only need f to be defined at x_0; when we deal with the notion of limit we only need that x_0 be a point of accumulation of E. These two requirements are unrelated, since not all points of E are points of accumulation and not all points of accumulation of E are in E, see, e.g., Exercise 5.33. Moreover, the condition $0 < d_X(x, x_0)$ in the definition of limit expresses the fact that we can disregard the value of f at x_0 (in case f is defined at x_0). Also notice that *the limit is unique if it exists*, and that limits are preserved by restriction. To be precise, we have

5.35 Proposition. *Let (X, d_X) and (Y, d_Y) be two metric spaces. Suppose $F \subset E \subset X$ and let $x_0 \in X$ be a point of accumulation for F. If $f(x) \to y$ as $x \to x_0$, $x \in E$, then $f(x) \to y$ as $x \to x_0$, $x \in F$.*

5.36 ¶. As for functions of one variable, the notions of limit and continuity are strongly related. Show the following.

Proposition. *Let X and Y be two metric spaces, $E \subset X$ and $x_0 \in X$.*

(i) *If x_0 belongs to E and is not a point of accumulation of E, then every function $f : E \to Y$ is continuous at x_0.*

(ii) *Suppose that x_0 belongs to E and is a point of accumulation for E. Then*

    a) *$f : E \to Y$ is continuous at x_0 if and only if $f(x) \to f(x_0)$ as $x \to x_0$, $x \in E$,*

    b) *$f(x) \to y$ as $x \to x_0$, $x \in E$, if and only if the function $g : E \cup \{x_0\} \to Y$ defined by*

$$g(x) := \begin{cases} f(x) & \text{if } x \in E \setminus \{x_0\}, \\ y & \text{if } x = x_0 \end{cases}$$

is continuous at x_0.

We conclude with a change of variable theorem for limits, see e.g., Proposition 2.27 of [GM1] and Example 2.49 of [GM2].

5.37 Proposition. *Let X, Y, Z be metric spaces, $E \subset X$ and let x_0 be a point of accumulation for E. Let $f : E \to Y$, $g : f(E) \to Z$ be two functions and suppose that $f(x_0)$ is an accumulation point of $f(E)$. If*

(i) *$g(y) \to L$ as $y \to y_0$, $y \in f(E)$,*

(ii) *$f(x) \to y_0$ as $x \to x_0$, $x \in E$,*

(iii) *either $f(x_0) = y_0$, or $f(x) \neq y_0$ for all $x \in E$ and $x \neq x_0$,*

then $g(f(x)) \to L$ as $x \to x_0$, $x \in E$.

d. The junction property

A property we have just hinted at in the case of real functions is the *junction property*, see Section 2.1.2 of [GM1], which is more significant for functions of several variables.

Let X be a set. We say that a family $\{U_\alpha\}$ of subsets of a metric space is *locally finite* at a point $x_0 \in X$ if there exists $r > 0$ such that $B(x_0, r)$ meets at most a finite number of the U_α's.

5.38 Proposition. *Let (X, d_X), (Y, d_Y) be metric spaces, $f : X \to Y$ a function, $x_0 \in X$, and let $\{U_\alpha\}$ be a family of subsets of X locally finite at x_0.*

(i) *Suppose that x_0 is a point of accumulation of U_α and that $f(x) \to y$ as $x \to x_0$, $x \in U_\alpha$, for all α. Then $f(x) \to y$ as $x \to x_0$, $x \in X$.*

(ii) *If $x_0 \in \cap_\alpha U_\alpha$ and $f : U_\alpha \subset X \to Y$ is continuous at x_0 for all α, then $f : X \to Y$ is continuous at x_0.*

5.39 ¶. Prove Proposition 5.38.

5.40 Example. An assumption on the covering is necessary in order that the conclusions of Proposition 5.38 hold. Set $A := \{(x, y) \mid x^2 < y < 2x^2\} \subset \mathbb{R}^2$ and

$$f(x, y) := \begin{cases} 1 & \text{if } (x, y) \in A, \\ 0 & \text{otherwise.} \end{cases}$$

The function f is discontinuous at $x_0 := (0,0)$, since its oscillation is one in every ball centered at x_0. Denote by U_m the straight line through the origin

$$U_m := \{(x,y) \,|\, y = mx\}, \quad m \in \mathbb{R}, \qquad U_\infty := \{(x,y) \,|\, x = 0\}.$$

The U_α's, $\alpha \in \mathbb{R} \cup \{\infty\}$ form a covering of $\mathbb{R}^2$ that is not locally finite at x_0 and for any $\alpha \in \mathbb{R} \cup \infty$, the restriction of f to each U_α is zero near the origin. In particular, each restriction $f_{|U_\alpha} : U_\alpha \to \mathbb{R}$ is continuous at the origin.

5.1.4 Functions from $\mathbb{R}^n$ into $\mathbb{R}^m$

It is important to be acquainted with the limit notion we have just introduced in an abstract context. For this purpose, in this section, we shall focus on mappings between Euclidean spaces and illustrate with a few examples some of the abstract notions previously introduced.

a. The vector space $C^0(A, \mathbb{R}^m)$

Denote by $e^i : \mathbb{R}^n \to \mathbb{R}$ the linear map that maps $x = (x^1, x^2, \ldots, x^n) \in \mathbb{R}^n$ into its ith component, $e^i(x) := x^i$. Any map $f : X \to \mathbb{R}^n$ from a set X into $\mathbb{R}^n$ writes as an n-tuple of real-valued functions $f(x) = (f^1(x), \ldots, f^n(x))$, where for any $i = 1, \ldots, n$ the function $f^i : X \to \mathbb{R}$ is given by $f^i(x) := e^i(f(x))$.

From

$$|y_1|, |y_2|, \ldots, |y_n| \leq |\mathbf{y}| \leq \sum_{i=1}^{n} |y_i| \qquad \mathbf{y} \in \mathbb{R}^n$$

we readily infer the following.

5.41 Proposition. *The following claims hold.*

(i) *The maps $e^i : \mathbb{R}^n \to \mathbb{R}$, $i = 1, \ldots, n$, are Lipschitz continuous.*

(ii) *Let (X, d) be a metric space. Then*

    a) $f : X \to \mathbb{R}^n$ is continuous at $x_0 \in X$ if and only if all its components $f^1, f^2, \ldots, f^n$ are continuous at x_0,

    b) if $f, g : X \to \mathbb{R}^n$ are continuous at x_0, then $f + g : X \to \mathbb{R}^n$ is continuous at x_0,

    c) if $f : X \to \mathbb{R}^n$ and $\lambda : X \to \mathbb{R}$ are continuous at x_0 then the map $\lambda f : X \to \mathbb{R}^n$ defined by $\lambda f(x) := \lambda(x)f(x)$, is continuous at x_0.

5.42 Example. The function $f : \mathbb{R}^3 \to \mathbb{R}$, $f(x, y, x) := \sin(x^2 y) + x^2$ is continuous at $\mathbb{R}^3$. In fact, if $\mathbf{x}_0 := (x_0, y_0, z_0)$, then the coordinate functions $\mathbf{x} = (x, y, z) \to x$, $\mathbf{x} \to y$, $\mathbf{x} \to z$ are continuous at $\mathbf{x}_0$ by Proposition 5.41. By Proposition 5.41 (iii), $\mathbf{x} \to x^2 y$ and $\mathbf{x} \to z^2$ are continuous at $\mathbf{x}_0$, and by (ii) Proposition 5.41, $\mathbf{x} \to x^2 y + z^2$ is continuous at $\mathbf{x}_0$. Finally $\sin(x^2 y + x^2)$ is continuous since sin is continuous.

5.43 Definition. *Let X and Y be two metric spaces. We denote by $C^0(X, Y)$ the class of all continuous function $f : X \to Y$.*

As a consequence of Proposition 5.41 $C^0(X, \mathbb{R}^m)$ is a vector space. Moreover, if $\lambda \in C^0(X, \mathbb{R})$ and $f \in C^0(X, \mathbb{R}^m)$, then $\lambda f : X \to \mathbb{R}^n$ given by $\lambda f(x) := \lambda(x) f(x)$, $x \in X$, belongs to $C^0(X, \mathbb{R}^m)$. In particular,

5.44 Corollary. *Polynomials in n variables belong to $C^0(\mathbb{R}^n, \mathbb{R})$. Therefore, maps $f : \mathbb{R}^n \to \mathbb{R}^m$ whose components are polynomials of n variables are continuous. In particular, linear maps $L \in \mathcal{L}(\mathbb{R}^n, \mathbb{R}^m)$ are continuous.*

It is worth noticing that in fact

5.45 Proposition. *Let $L : \mathbb{R}^n \to \mathbb{R}^m$ be linear. Then L is Lipschitz continuous in $\mathbb{R}^n$.*

Proof. As L is linear, we have

$$\mathrm{Lip}(f) := \sup_{\substack{x, y \in \mathbb{R}^n \\ x \neq y}} \frac{||L(x) - L(y)||_{\mathbb{R}^m}}{||x - y||_{\mathbb{R}^n}}$$

$$= \sup_{\substack{x, y \in \mathbb{R}^n \\ x \neq y}} \frac{||L(x - y)||_{\mathbb{R}^m}}{||x - y||_{\mathbb{R}^n}} = \sup_{0 \neq z \in \mathbb{R}^n} \frac{||L(z)||_{\mathbb{R}^m}}{||z||_{\mathbb{R}^n}} =: ||L||.$$

Let us prove that $||L|| < +\infty$. Since L is continuous at zero by Corollary 5.44, there exists $\delta > 0$ such that $||L(w)|| < 1$ whenever $||w|| < \delta$. For any nonzero $z \in \mathbb{R}^n$, set $w := \frac{\delta z}{2 ||z||}$. Since $||w|| < \delta$, we have $||L(w)|| < 1$. Therefore, writing $z = \frac{2 ||z||}{\delta} w$ and using the linearity of L

$$||L(z)|| = \left|\left| \frac{2 ||z||}{\delta} L(w) \right|\right| = \frac{2 ||z||}{\delta} ||L(w)|| < \frac{2}{\delta} ||z||$$

hence

$$||L|| \leq \frac{2}{\delta} < +\infty.$$

$\square$

For a more detailed description of linear maps in normed spaces, see Chapters 9 and 10.

b. Some nonlinear continuous transformations from $\mathbb{R}^n$ into $\mathbb{R}^N$

We now present a few examples of nonlinear continuous transformations between Euclidean spaces.

5.46 Example. For $k = 0, 1, \ldots$ consider the map $u_k :\,] - 1, 1[\to \mathbb{R}^2$ given by

$$u_k(t) = \begin{cases} (\cos kt, \sin kt) & \text{if } t \in]0, 2\pi/k[, \\ (1, 0) & \text{otherwise.} \end{cases}$$

This is a Lipschitz function whose graph is given in Figure 5.10. Notice that the graph of $u_k = \{(t, u_k(t))\}$ is a curve that "converges" as $k \to \infty$ to a horizontal line plus a vertical circle at 0. Compare with the function $\mathrm{sgn}\, x$ from $\mathbb{R}$ to $\mathbb{R}$.

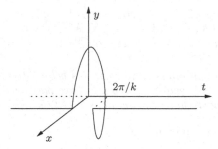

Figure 5.10. The function u_k in Example 5.46.

5.47 Example (Stereographic projection). Let

$$S^n := \left\{ x \in \mathbb{R}^{n+1} \,\middle|\, |x| = 1 \right\}$$

be the unit sphere in $\mathbb{R}^{n+1}$. If $x = (x_1, \ldots x_n, x_{n+1}) \in \mathbb{R}^{n+1}$, let us denote the coordinates of x by (y, z) where $y = (x_1, x_2, \ldots, x_n) \in \mathbb{R}^n$ and $z = x_{n+1} \in \mathbb{R}$. With this notation, $S^n = \{(y, z) \in \mathbb{R}^n \times \mathbb{R} \mid |y|^2 + z^2 = 1\}$. Furthermore, denote by $P_S = (0, -1) \in S^n$ the South pole of S^n. The *stereographic projection* (from the South pole) is the map that projects from the South pole the sphere onto the $\{z = 0\}$ plane,

$$\sigma : S^n \setminus \{P_S\} \subset \mathbb{R}^{n+1} \to \mathbb{R}^n, \qquad (y, z) \to \frac{y}{1 + z}.$$

It is easily seen that σ is injective, surjective and continuous with a continuous inverse given by

$$\sigma^{-1} : \mathbb{R}^n \to S^n \setminus \{P_S\}, \qquad \sigma^{-1}(x) := \left(\frac{2}{1 + |x|^2} x, \frac{1 - |x|^2}{1 + |x|^2} \right)$$

that maps $x \in \mathbb{R}^n$ into the point of S^n lying in the segment joining the South pole of S^n with x, see Figure 5.11.

5.48 Example (Polar coordinates). The transformation

$$\sigma : \Sigma := \left\{ (\rho, \theta) \,\middle|\, \rho > 0, \ 0 \le \theta < 2\pi \right\} \to \mathbb{R}^2, \qquad (\rho, \theta) \to (\rho \cos \theta, \rho \sin \theta)$$

defines a map that is injective and continuous with range $\mathbb{R}^2 \setminus \{0\}$.

The extension of the map to the third coordinate

$$\widetilde{\sigma} : \Sigma \times \mathbb{R} \to \mathbb{R}^2 \times \mathbb{R} \sim \mathbb{R}^3, \qquad (\rho, \theta, z) \to (\rho \cos \theta, \rho \sin \theta, z)$$

defines the so-called *cylindrical coordinates* in $\mathbb{R}^3$.

5.49 Example (Spherical coordinates). The representation of points $(x, y, z) \in \mathbb{R}^3$ as

$$\begin{cases} x = \rho \sin \varphi \cos \theta, \\ y = \rho \sin \varphi \sin \theta, \\ z = \rho \cos \varphi, \end{cases}$$

see Figure 5.12, defines the *spherical coordinates in* $\mathbb{R}^3$. This in turn defines a continuous transformation $(\rho, \theta, \varphi) \to (x, y, z)$ from

$$\Sigma := \left\{ (\rho, \theta, \varphi) \,\middle|\, \rho > 0, \ 0 \le \theta < 2\pi, \ 0 \le \varphi \le \pi \right\} =]0, +\infty[\times [0, 2\pi[\times [0, \pi]$$

into $\mathbb{R}^3 \setminus \{0\}$.

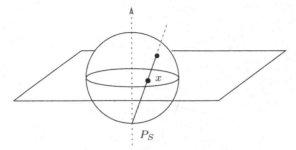

Figure 5.11. The stereographic projection from the South pole.

Complex-valued functions of one complex variable provide examples of tranformations of the plane.

5.50 Example ($w = z^n$). The map $z \to z^n$ defines a continuous transformation of $\mathbb{C}$ to $\mathbb{C}$. The inverse image of each nonzero $w \in \mathbb{C}$ is made by n distinct points, given by the n roots of w; those n points collapse to zero when $w = 0$. If we write the transformation $w = z^n$ as

$$|w| = |z|^n \qquad \text{Arg}\, w = n \text{Arg}\, z,$$

we see, identifyng $\mathbb{C}$ with $\mathbb{R}^2$, that the circle of radius r and center 0 is mapped into the circle of radius r^n and center 0. Moreover, if a point goes clockwise along the circle, then the normalized point image $\frac{z^n}{|z|^n}$ goes along the unit circle clockwise n times.

5.51 ¶. The map $z \to w = z^n$ restricted to $\varphi_0 < \text{Arg}\, z < \varphi_1$ with $0 < \varphi_1 - \varphi_0 \le 2\pi/n$ is injective and continuous.

5.52 ¶. Show that the map $z \to w = z^2$ maps the family of parallel lines to the axes (but the axes themselves) into two families of parabolas with the common axis as the real axis and the common foci at the origin, see Figure 5.13.

5.53 Example (The Joukowski function). This is the map

$$\lambda(z) := \frac{1}{2}\left(z + \frac{1}{z}\right), \qquad z \ne 0,$$

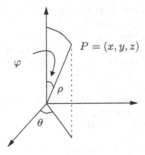

Figure 5.12. Spherical coordinates.

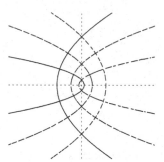

Figure 5.13. The transformation $w = z^2$ maps families of lines parallel to the axes, except for the axes, into two families of parabolas with the common axis as the real axis and the common foci at the origin.

which appears in several problems of aerodynamics. It is a continuous function defined in $\mathbb{C} \setminus \{0\}$. Since $\lambda(z) = \lambda(1/z)$, every point $w \neq \pm 1, 0$ has at most, and, in fact, exactly two distinct inverse images z_1, z_2 satisfying $z_1 z_2 = 1$.

5.54 ¶. Show that $\lambda(z) = 1/2(z + 1/z)$ is one-to-one from $\{|z| < 1,\ z \neq 0\}$ or $\{|z| > 1\}$ into the complement of the segment $\{w \mid -1 \leq \Re w \leq 1\}$. λ maps the family of circles $\{z \mid |z| = r\}_r,\ 0 < r < 1$, into a family of co-focal ellipses and maps the diameters $z = t e^{i\alpha},\ -1 < t < 1,\ 0 < \alpha < \pi$, in a family of co-focal hyperbolas, see Figure 5.14.

5.55 Example (The Möbius transformations). These maps, defined by

$$L(z) := \frac{az + b}{cz + d}, \qquad ad - bc \neq 0 \tag{5.10}$$

are continuous and injective from $\mathbb{C} \setminus \{-d/c\}$ into $\mathbb{C} \setminus \{a/c\}$ and have several relevant properties that we list below, asking the reader to show that they hold.

5.56 ¶. Show the following.

(i) $L(z) \to a/c$ as $|z| \to \infty$ and $|L(z)| \to \infty$ as $z \to -d/c$. Because of this, we write $L(\infty) = a/c$, $L(-d/c) = \infty$ and say that L is *continuous* from $\mathbb{C} \cup \{\infty\}$ into itself.

(ii) Show that every rational function, i.e., the quotient of two complex polynomials, defines a continuous transformation of $\mathbb{C} \cup \{\infty\}$ into itself, as in (i).

(iii) The Möbius transformations $L(z)$ in (5.10) are the only rational functions from $\mathbb{C} \cup \{\infty\}$ into itself that are injective.

(iv) The Möbius transformations $(a_i z + b_i)/(c_i z + d_i)$, $i = 1, 2$, are identical if and only if (a_1, b_1, c_1, d_1) is a nonzero multiple of (a_2, b_2, c_2, d_2).

(v) The Möbius transformations form a group G with respect to the composition of maps; the subset $H \subset G$, $H := \{z,\ 1 - z,\ 1/z,\ 1/(1 - z),\ (z - 1)/z\}$ is a subgroup of G.

(vi) A Möbius transformation maps straight lines and circles into straight lines and circles (show this first for the map $1/z$, taking into account that the equations for straight lines and circles have the form $A(x^2 + y^2) + 2Bx + 2Cy + D = 0$ if $z = x + iy$).

(vii) The map in (5.10) maps circles and straight lines through $-c/d$ into a straight line and any other straight line or circle into a circle.

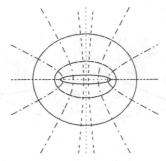

Figure 5.14. The Joukowski function maps circles $|z| = r$, $0 < r < 1$, and diameters $z = \pm e^{\pm i\alpha}$, $0 \leq t < 1$, $0 < \alpha < 2\pi$, respectively into a family of ellipses and of cofocal hyperbolas.

(viii) The only Möbius transformation with at least two fixed points is z. Two Möbius transformations are equal if they agree at three distinct points. There is a unique Möbius transformation that maps three distinct points $z_1, z_2, z_3 \in \mathbb{C} \cup \{\infty\}$ into three distinct points $w_1, w_2, w_3 \in \mathbb{C} \cup \{\infty\}$.

5.57 Example (Exponential and logarithm). The complex function $z \to \exp z$, see [GM2], is continuous from $\mathbb{C} \to \mathbb{C}$, periodic of period $2\pi i$ with image $\mathbb{C} \setminus \{0\}$. In particular e^z does not vanish, and every nonzero w has infinitely many preimages.

5.58 ¶. Taking into account what we have proved in [GM2], show the following.

(i) $w = e^z$ is injective with a continuous inverse in every strip parallel to the real axis of width $h \leq 2\pi$, and has an image as the interior of an angle of radiants h and vertex at the origin;

(ii) $w = e^z$ maps every straight line which is not parallel to the axes into a logarithmic spiral, see Chapter 7.

c. The calculus of limits for functions of several variables

Though we may have appeared pedantic, we have always insisted in specifying the domain $E \subset X$ in which the independent variables varied. This is in fact particularly relevant when dealing with limits and continuity of functions of several variables, as in this case there are several reasonable ways of approaching a point x_0. Different choices may and, in general, do lead to different answers concerning the existence and/or the equality of the limits

$$\lim_{\substack{x \to x_0 \\ x \in E}} f(x) \quad \text{and} \quad \lim_{\substack{x \to x_0 \\ x \in F}} f(x).$$

Let (X, d_X) and (Y, d_Y) be two metric spaces, $f : X \to Y$ and $x_0 \in X$ a point of accumulation of X.

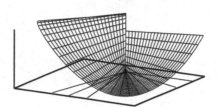

Figure 5.15. The function in Example 5.59.

(i) If we find two sets E_1, E_2 such that x_0 is an accumulation point of both E_1 and E_2, and the restrictions $f : E_1 \subset X \to Y$ and $f : E_2 \subset X \to Y$ of f have different limits, then f has no limit when $x \to x_0$, $x \in E_1 \cup E_2$.

(ii) if we want to show that $f(x)$ has limit as $x \to x_0$, we may
   a) guess a possible limit $y_0 \in Y$, for instance computing the limit y_0 of a suitable restriction of f,
   b) show that the real-valued function $x \to d_Y(f(x), y_0)$ converges to zero as $x \to x_0$, for instance proving that

$$d_Y(f(x), y_0) \le h(x) \text{ for all } x \in X, \ x \ne x_0,$$

where $h : X \to \mathbb{R}$ is such that $h(x) \to 0$ as $x \to x_0$.

5.59 Example. Let $f : \mathbb{R}^2 \setminus \{(0,0)\} \to \mathbb{R}$ be defined by $f(x, y) := xy/(x^2 + y^2)$ for $(x, y) \ne (0, 0)$. Let us show that f has no limit as $(x, y) \to (0, 0)$. By contradiction, suppose that $f(x, y) \to L \in \mathbb{R}$ as $(x, y) \to (0, 0)$. Then for any sequence $\{(x_n, y_n)\} \subset \mathbb{R}^2 \setminus \{(0, 0)\}$ converging to $(0, 0)$ we find $f(x_n, y_n) \to L$. Choosing $(x_n, y_n) := (1/n, k/n)$, we have

$$f\left(\frac{1}{n}, \frac{k}{n}\right) = \frac{k}{1 + k^2}$$

hence, as $n \to \infty$, $L = k/(1 + k^2)$. Since k is arbitrary, we have a contradiction.

This is even more evident if we observe that f is *positively homogeneous of degree* 0, i.e., $f(\lambda x, \lambda y) = f(x, y)$ for all $\lambda > 0$, i.e., f is constant along half-lines from the origin, see Figure 5.15. It is then clear that f has limit at $(0, 0)$ if and only if f is constant, which is not the case. Notice that from the inequality $2xy \le x^2 + y^2$ we can easily infer that $|f(x, y)| \le 1/2 \ \forall (x, y) \in \mathbb{R}^2 \setminus \{(0, 0)\}$, i.e., that f is a bounded function.

5.60 Example. Let $f(x, y) := \sin(x^2 y)/(x^2 + y^2)$ for $(x, y) \ne (0, 0)$. In this case $(1/n, 0) \to (0, 0)$ and $f(1/n, 0) = 0 \to 0$. Thus 0 is the only possible limit as $(x, y) \to (0, 0)$; and, in fact it is, since

$$|f(x, y) - 0| = \frac{|\sin(x^2 y)|}{x^2 + y^2} \le \frac{|x| \, |y|}{x^2 + y^2} |x| \le \frac{1}{2} |x| \to 0$$

as $(x, y) \to (0, 0)$. Here we used $|\sin t| \le |t| \ \forall t$, $2|x| \, |y| \le x^2 + y^2 \ \forall (x, y)$ and that $(x, y) \to |x|$ is a continuous map in $\mathbb{R}^2$, see Proposition 5.41.

We can also consider the restriction of f to *continuous paths from x_0*, i.e., choose a map $\varphi : [0,1] \to \mathbb{R}^2$ that is continuous at least at 0 with $\varphi(0) = x_0$ and $\varphi(t) \neq x_0$ for $t \neq 0$ and compute, if possible

$$\lim_{t \to 0^+} f(\varphi(t)).$$

Such limits may or may not exist and their values depend on the chosen path, for a fixed f. Of course, if

$$\lim_{\substack{x \to x_0 \\ x \in E}} f(x) = L,$$

then, on account of the restriction property and of the change of variable theorem,

$$\lim_{\substack{x \to x_0 \\ x \in F}} f(x) = L \qquad \text{and} \qquad \lim_{t \to 0^+} f(\varphi(t)) = L$$

respectively for any $F \subset E$ of which x_0 remains a point of accumulation and for any continuous path in E, $\varphi([0,1]) \subset E$.

5.61 Example. Let us reconsider the function

$$f : \mathbb{R}^2 \setminus \{(0,0)\} \to \mathbb{R}, \qquad f(x,y) := \frac{xy}{x^2 + y^2}$$

which is continuous in $\mathbb{R}^2 \setminus \{(0,0)\}$. Suppose that we move from zero along the straight line $\{(x,y) \,|\, y = mx, x \in \mathbb{R}\}$ that we parametrize by $x \to (x, mx)$. Then

$$f(\varphi(x)) = f(x, mx) = \frac{m}{1 + m^2} \to \frac{m}{1 + m^2}, \quad \text{as } x \to 0,$$

in particular, the previous limit depends on m, hence $f(x,y)$ has no limit as $(x,y) \to (0,0)$.

Set $E := \{(x,y) \,|\, x \in \mathbb{R}, \ 0 < y < \lambda x^2\}$. We instead have

$$\lim_{\substack{(x,y) \to (0,0) \\ (x,y) \in E}} f(x,y) = 0$$

In fact, in this case

$$0 \le \frac{|xy|}{x^2 + y^2} \le \frac{\lambda |x|^3}{x^2} = \lambda |x| \quad \text{in } E.$$

5.62 Example. The function

$$f(x,y) = \begin{cases} \frac{x^2 y}{x^4 + y^2} & \text{if } (x,y) \in \mathbb{R}^2 \setminus \{(0,0)\}, \\ 0 & \text{if } (x,y) = (0,0) \end{cases}$$

is continuous in $\mathbb{R}^2 \setminus \{(0,0)\}$ but is not continuous at $(0,0)$. Restricting f to a straight line through $(0,0)$ parametrized as $\varphi(t) = (ta, tb)$, $t \in \mathbb{R}$, gives

$$f(\varphi(t)) = f(at, bt) = \frac{a^2 b t^3}{a^2 t^2 + b^2 t^2} = \frac{a^2 b}{a^2 + b^2} t \to 0 \qquad \text{as } t \to 0.$$

However, restricting f along the graph of the function $y = \alpha x^2$ parametrized as $\varphi(x) := (x, \alpha x^2)$, gives

$$f(x, \alpha x) = \frac{\alpha x^4}{x^4 + \alpha^2 x^4} \to \frac{\alpha}{1 + \alpha^2}, \qquad \text{as } x \to 0,$$

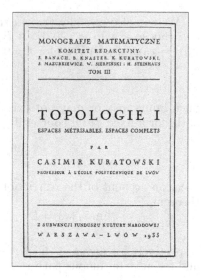

MONOGRAFJE MATEMATYCZNE
KOMITET REDAKCYJNY:
S. BANACH. B. KNASTER. K. KURATOWSKI.
S. MAZURKIEWICZ. W. SIERPIŃSKI : H. STEINHAUS
TOM III

TOPOLOGIE I

ESPACES MÉTRISABLES. ESPACES COMPLETS

P A R

CASIMIR KURATOWSKI
PROFESSEUR À L'ÉCOLE POLYTECHNIQUE DE LWÓW

Z SUBWENCJI FUNDUSZU KULTURY NARODOWEJ
W A R S Z A W A – L W Ó W 1933

Figure 5.16. Kazimierz Kuratowski (1896–1980) and the frontispiece of the first volume of his *Topologie*.

thus f has no limit as $(x, y) \to (0,0)$.

Let us now consider the restriction of f to the set

$$E := \left\{ (x,y) \,\Big|\, x \geq 0, \ |y| \leq x^3 \right\}.$$

We have

$$\lim_{\substack{(x,y) \to (0,0) \\ (x,y) \in E}} f(x,y) = 0.$$

In fact,

$$\left| \frac{x^2 y}{x^4 + y^2} - 0 \right| = \frac{|x|^2 |y|}{x^4 + y^2} \leq \frac{|x|^5}{x^4 + y^2} \leq |x| \to 0,$$

since $(x,y) \in E$.

We conclude by observing that for functions $f : \mathbb{R}^n \to \mathbb{R}$ the expression

$$\lim_{x \to x_0} f(x) = +\infty$$

means that $\forall M \in \mathbb{R}$ there is $\delta > 0$ such that $f(x) > M \ \forall x \in B(x_0, \delta) \setminus \{x_0\}$.

5.2 The Topology of Metric Spaces

In this section we introduce some families of subsets of a metric space X that are defined by the metric structure, namely the families of *open* and *closed sets*. Recall that if X is a set, $\mathcal{P}(X)$ denotes the set of all subsets of X: $A \in \mathcal{P}(X)$ if and only if $A \subset X$.

5.2.1 Basic facts

a. Open sets

5.63 Definition. *A subset A of a metric space (X, d) is called an* open set *if for all $x \in A$ there exists a ball centered at x contained in A, i.e.,*

$$\forall x \in A \; \exists r_x > 0 \text{ such that } B(x, r_x) \subset A. \qquad (5.11)$$

5.64 Proposition. *A subset A of a metric space X is open if and only if either A is empty or is a union of open balls.*

Proof. Let A be open. Then either A is empty or A is trivially a union of open balls, $A = \cup_{x \in A} B(x, r_x)$. Conversely, (5.11) trivially holds if $A = \emptyset$. If instead $x \in A \neq \emptyset$, since we assume that A is union of balls, there is $y \in X$ and $\rho > 0$ such that $x \in B(y, \rho) \subset A$. Thus $y \in A$ and, setting $r := \rho - d(x, y)$, we have $r > 0$ and by the triangle inequality $B(x, r) \subset B(y, \rho) \subset A$. $\qquad \square$

In particular,

5.65 Corollary. *The open balls of a metric space X are open sets.*

5.66 ¶. Let (X, d) be a metric space and $r \geq 0$. Show that $\{y \in X \mid d(y, x) > r\}$ is an open set in X.

5.67 ¶. Let (X, d) be a metric space. Show that $\{x_n\} \subset X$ converges to $x \in X$ if and only if, for any open set A such that $x \in A$, there exists $\overline{n}$ such that $x_n \in A$ for all $n \geq \overline{n}$.

The following is also easily seen.

5.68 Proposition. *Let (X, d) be a metric space. Then*

(i) *$\emptyset$ and X are open sets,*
(ii) *if $\{A_\alpha\}$ is a family of open sets, then $\cup_\alpha A_\alpha$ is an open set, too,*
(iii) *if $A_1, A_2, \ldots, A_n$ are finitely many open sets, then $\cap_{i=1}^n A_i$ is open.*

5.69 ¶. By considering the open sets $\{] - \frac{1}{n}, \frac{1}{n}[\mid n \in \mathbb{N}\}$, show that the intersection of infinitely many open sets needs not be an open set.

b. Closed sets

Recall that the complement of $A \subset X$ is the set

$$A^c := X \setminus A.$$

5.70 Definition. *Let X be a metric space. $F \subset X$ is called a closed set if $F^c = X \setminus F$ is an open set.*

The de Morgan formulas

$$(\cup_\alpha A_\alpha)^c = \bigcap_\alpha (A_\alpha)^c \qquad (\cap_\alpha A_\alpha)^c = \bigcup_\alpha (A_\alpha)^c$$

together with Proposition 5.68 yield at once the following.

5.71 Proposition. *Let X be a metric space. Then*

(i) *$\emptyset$ and X are closed sets,*
(ii) *the intersection of any family of closed sets is a closed set,*
(iii) *the union of finitely many closed sets is a closed set.*

5.72 ¶. Show that $[a, b]$, $[a, +\infty[$ and $[-a, +\infty[$ are closed sets in $\mathbb{R}$, while $[0, 1[$ is neither closed nor open.

5.73 ¶. Show that the set $\{\frac{1}{n} \mid n = 1, 2, \dots\}$ is neither closed nor open.

5.74 ¶. Show that any finite subset of a metric space is a closed set.

5.75 ¶. Show that the closed ball $\{x \in X \mid d(x, x_0) \leq r\}$, and its boundary $\{x \in X \mid d(x, x_0) = r\}$ are closed sets.

One may characterize closed sets in terms of convergent sequences.

5.76 Proposition. *Let (X, d) be a metric space. A set $F \subset X$ is a closed set if and only if every* convergent *sequence with values in F converges to a point of F.*

Proof. Suppose that F is closed and that $\{x_k\} \subset F$ converges to $x \in X$. Let us prove that $x \in F$. Assuming on the contrary, that $x \notin F$, there exists $r > 0$ such that $B(x, r) \cap F = \emptyset$. As $\{x_n\} \subset F$, we have $d(x_n, x) \geq r$ $\forall n$, a contradiction since $d(x_n, x) \to 0$.

Conversely, suppose that, whenever $\{x_k\} \subset F$ and $x_k \to x$, we have $x \in F$, but F is not closed. Thus $X \setminus F$ is not open, hence there exists a point $x \in X \setminus F$ such that $\forall r > 0$ $B(x, r) \cap F \neq \emptyset$. Choosing $r = 1, \frac{1}{2}, \frac{1}{3}, \dots$, we inductively construct a sequence $\{x_n\} \subset F$ such that $d(x_n, x) < \frac{1}{n}$, hence converging to x. Thus $x \in F$ by assumption, but $x \in X \setminus F$ by construction, a contradiction.   □

c. Continuity

5.77 Theorem. *Let (X, d_X) and (Y, d_Y) be two metric spaces and $f : X \to Y$. Then the following claims are equivalent*

(i) *f is continuous,*
(ii) *$f^{-1}(B)$ is an open set in X for any open ball B of Y,*
(iii) *$f^{-1}(A)$ is an open set in X for any open set A in Y,*
(iv) *$f^{-1}(F)$ is a closed set in X for any closed set F in Y.*

Proof. (i) $\Rightarrow$ (ii). Let B be an open ball in Y and let x be a point in $f^{-1}(B)$. Since $f(x) \in B$, there exists a ball $B_Y(f(x), \epsilon) \subset B$. Since f is continuous at x, there exists $\delta > 0$ such that $f(B_X(x, \delta)) \subset B_Y(f(x), \epsilon) \subset B$ that is $B_X(x, \delta) \subset f^{-1}(B)$. As x is arbitrary, $f^{-1}(B)$ is an open set in X.

(ii) $\Rightarrow$ (i) Suppose $f^{-1}(B)$ is open for any open ball B of Y. Then, given x_0, $f^{-1}(B_Y(f(x_0, \epsilon)))$ is open, hence there is $\delta > 0$ such that $B_X(x_0, \delta)$ is contained in $f^{-1}(B_Y(f(x_0, \epsilon)))$, i.e., $f(B_X(x_0, \delta)) \subset B_Y(f(x_0), \epsilon)$, hence f is continuous at x_0.

(ii) and (iii) are equivalent since $f^{-1}(\cup_i A_i) = \cup_i f^{-1}(A_i)$ for any family $\{A_i\}$ of subsets of X.

(iii) and (iv) are equivalent on account of the de Morgan formulas. $\square$

5.78 ¶. Let $f, g : X \to Y$ be two continuous functions between metric spaces. Show that the set $\{x \in X \mid f(x) = g(x)\}$ is closed.

5.79 ¶. It is convenient to set

Definition. *Let (X, d) be a metric space. $U \subset X$ is said to be a* neighborhood *of $x_0 \in X$ if there exists an open set A of X such that $x_0 \in A \subset U$.*

In particular

o $B(x_0, r)$ is a neighborhood of any $x \in B(x_0, r)$,
o A is open if and only if A is a neighborhood of any point of A.

Let (X, d), (Y, d) be two metric spaces let $x_0 \in X$ and let $f : X \to Y$. Show that f is continuous at x_0 if and only if the inverse image of an open neighborhood of $f(x_0)$ is an open neighborhood of x_0.

Finally, we state a junction rule for continuous functions, see Proposition 5.38.

5.80 Proposition. *Let (X, d) be a metric space, and let $\{U_\alpha\}$ be a covering of X. Suppose that either all U_α's are open sets or all U_α's are closed and for any $x \in X$ there is an open ball that intersects only finitely many U_α. Then*

(i) *$A \subset X$ is an open (closed) set in X if and only if each $A \cap U_\alpha$ is an open (closed) set in U_α,*

(ii) *Let Y be another metric space and let $f : X \to Y$. Then f is continuous if and only if all the restrictions $\phi_{|U_\alpha} : U_\alpha \to Y$ are continuous.*

5.81 ¶. Some kind of assumption on the covering in Proposition 5.80 is necessary. If $X := [a, b]$, $x_0 \in [a, b]$, $U_x := \{x\}$ for all $x \in [a, b]$, show that the claims in Proposition 5.80 are false.

d. Continuous real-valued maps

Let (X, d) be a metric space and $f : X \to \mathbb{R}$. From Theorem 5.77 we find that $f : X \to \mathbb{R}$ is continuous if and only if $f^{-1}(]a, b[)$ is an open set for every bounded interval $]a, b[\subset \mathbb{R}$. Moreover,

5.82 Corollary. *Let $f : X \to \mathbb{R}$ be a function defined on a metric space X and let $t \in \mathbb{R}$. Then*

(i) $\{x \in X \mid f(x) > t\}$, $\{x \in X \mid f(x) < t\}$ *are open sets,*

(ii) $\{x \in X \mid f(x) \geq t\}$, $\{x \in X \mid f(x) \leq t\}$ *and* $\{x \in X \mid f(x) = t\}$ *are closed sets.*

5.83 Proposition. *Let (X, d) be a metric space. Then $F \subset X$ is a closed set of X if and only if $F = \{x \mid d(x, F) = 0\}$.*

Proof. By Corollary 5.82, $\{x \mid d(x, F) = 0\}$ is closed, $x \to d(x, F)$ being Lipschitz continuous, see Example 5.25. Therefore $F = \{x \mid d(x, F) = 0\}$ implies that F is closed. Conversely assume that F is closed and that there exists $x \notin F$ such that $d(x, F) = 0$. Since F is closed by assumption, there exists $r > 0$ such that $B(x, r) \cap F = \emptyset$. But then $d(x, F) \geq r > 0$, a contradiction.   □

5.84 ¶. Prove the following

Proposition. *Let (X, d) be a metric space. Then*

(i) *$F \subset X$ is a closed set if and only if there exists a continuous function $f : X \to \mathbb{R}$ such that $F = \{x \in X \mid f(x) \leq 0\}$,*

(ii) *$A \subset X$ is an open set if and only if there exists a continuous function $f : X \to \mathbb{R}$ such that $A = \{x \in X \mid f(x) < 0\}$.*

Actually f can be chosen to be a Lipschitz-continuous function.

[*Hint:* If F is closed, choose $f(x) := d(x, F)$, while if A is an open set, choose $f(x) = -d(x, X \setminus A)$.]

e. The topology of a metric space

5.85 Definition. *The* topology *of a metric space X is the family $\tau_X \subset \mathcal{P}(X)$ of its open sets.*

It may happen that different distances d_1 and d_2 on the same set X that define different families of balls produce the same family of open sets for the same reason that a ball is union of infinitely many squares and a square is union of infinitely many balls. We say that the two distances are *topologically equivalent* if (X, d_1) and (X, d_2) have the same topology, i.e., the same family of open sets. The following proposition yields necessary and sufficient conditions in order that two distances be topologically equivalent.

5.86 Proposition. *Let d_1, d_2 be two distances in X and let $B_1(x, r)$ and $B_2(x, r)$ be the corresponding balls of center x and radius r. The following claims are equivalent*

(i) *d_1 and d_2 are topologically equivalent,*

(ii) *every ball $B_1(x, r)$ is open for d_2 and every ball $B_2(x, r)$ is open for d_1.*

(iii) *$\forall x \in X$ and $r > 0$ there are $r_x, \rho_x > 0$ such that $B_2(x, r_x) \subset B_1(x, r)$ and $B_1(x, \rho_x) \subset B_2(x, r)$,*

(iv) *the identity map $i : X \to X$ is a homeomorphism between the metric spaces (X, d_1) and (X, d_2).*

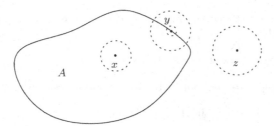

Figure 5.17. x is an interior point to A, y is a boundary point to A and z is an exterior point to A. x and y are adherent points to A and z is not.

5.87 ¶. Show that the distances in $\mathbb{R}^n$ d_∞ and d_p $\forall p \geq 1$, see Exercise 5.13, are all topologically equivalent to the Euclidean distance d_2. If we substitute $\mathbb{R}^n$ with the infinitely-dimensional vector space of sequences ℓ_1, the three distances give rise to different open sets.

We say that a property of X is a *topological property* of X if it can be expressed only in terms of set operations and open sets. For instance, being an open or closed set, the closure of or the boundary of, or a convergent sequence are topological properties of X, see Section 5.2.2 for more.

As we have seen, *f is continuous if and only if the inverse image of open sets is open.* A trivial consequence, for instance, is that *the composition of continuous functions is continuous*, see Proposition 5.30. Also we see that the continuity of $f : X \to Y$ is strongly related to the topologies

$$\tau_X := \{A \subset X \,|\, A \text{ open in } X\}, \qquad \tau_Y := \{A \subset Y \,|\, A \text{ open in } Y\},$$

respectively on X and Y, and in fact it depends on the metrics only through τ_X and τ_Y. In other words being a continuous function $f : X \to Y$ is a topological property of X and Y.

f. Interior, exterior, adherent and boundary points

5.88 Definition. *Let X be a metric space and $A \subset X$. We say that $x_0 \in X$ is* interior *to A if there is an open ball $B(x_0, r)$ such that $B(x_0, r) \subset A$; we say that x_0 is* exterior *to A if x_0 is interior to $X \setminus A$; we say that x_0 is* adherent *to A if it is not interior to $X \setminus A$; finally, we say that x_0 is a* boundary point *of A if x_0 is neither interior to A nor interior to $X \setminus A$.*

The set of *interior points to A* is denoted by $\overset{\circ}{A}$ or by $\mathrm{int}\, A$, the set of adherent points of A, called also the *closure of A*, is denoted by $\overline{A}$ or by $\mathrm{cl}\,(A)$, and finally the set of boundary points to A is called the *boundary of A* and is denoted by ∂A.

5.89 ¶. Let (X, d) be a metric space and $B(x_0, r)$ be an open ball of X. Show that

(i) every point of $B(x_0, r)$ is interior to $B(x_0, r)$, i.e., $\mathrm{int}\, B(x_0, r) = B(x_0, r)$,
(ii) every point x such that $d(x, x_0) = r$ is a boundary point to $B(x_0, r)$, i.e., $\partial B(x_0, r) = \{x \,|\, d(x, x_0) = r\}$,
(iii) every point x with $d(x, x_0) > r$ is exterior to $B(x_0, r)$,

(iv) every point x such that $d(x, x_0) \leq r$ is adherent to $B(x_0, r)$, i.e., $\mathrm{cl}\,(B(x_0, r)) = \{x \mid d(x, x_0) \leq r\}$.

5.90 ¶. Let X be a metric space and $A \subset X$. Show that
(i) int $A \subset A$,
(ii) int A is an open set and actually the largest open set contained in A,

$$\text{int } A = \cup \Big\{ U \,\Big|\, U \text{ open } U \subset A \Big\},$$

(iii) A is open if and only if $A = \text{int } A$.

5.91 ¶. Let X be a metric space and $A \subset X$. Show that
(i) $A \subset \overline{A}$,
(ii) $\overline{A}$ is closed and actually the smallest closed set that contains A, i.e.,

$$\mathrm{cl}\,(A) = \cap \Big\{ F \,\Big|\, F \text{ closed }, \ F \supset A \Big\},$$

(iii) A is closed if and only if $A = \overline{A}$,
(iv) $\overline{A} = \{x \in X \mid d(x, A) = 0\}$.

5.92 ¶. Let X be a metric space and $A \subset X$. Show that
(i) $\partial A = \partial (X \setminus A)$,
(ii) $\partial A \cap \text{int } A = \emptyset$, $\overline{A} = \partial A \cup \text{int } A$, $\partial A = \overline{A} \setminus \text{int } A$,
(iii) $\partial A = \overline{A} \cap \overline{A^c}$, in particular ∂A is a closed set,
(iv) $\partial \partial A = \emptyset$, $\overline{\overline{A}} = \overline{A}$, int int $A = \text{int } A$,
(v) A is closed if and only if $\partial A \subset A$,
(vi) A is open if and only if $\partial A \cap A = \emptyset$.

5.93 ¶. Let (X, d_X) and (Y, d_Y) be metric spaces and $f : X \to Y$. Show that the following claims are equivalent
(i) $f : X \to Y$ is continuous,
(ii) $f(\overline{A}) \subset \overline{f(A)}$ for all $A \subset X$,
(iii) $\overline{f^{-1}(B)} \subset f^{-1}(\overline{B})$ for all $B \subset Y$.

g. Points of accumulation

Let $A \subset X$ be a subset of a metric space. The set of points of accumulation, or cluster points, of A, denoted by $\mathcal{D}A$, is sometimes called the *derived of* A. Trivially $\mathcal{D}A \subset \overline{A}$, and the set of adherent points to A that are not points of accumulation of A, $\mathcal{I}(A) := \overline{A} \setminus \mathcal{D}A$, are the points $x \in \overline{A}$ such that $B(x, r) \cap A = \{x\}$ for some $r > 0$. These points are contained in A,

$$\mathcal{I}(A) = \overline{A} \setminus \mathcal{D}A \subset A$$

and are called *isolated points* of A.

5.94 ¶. Show that $\mathcal{D}A \subset \overline{A}$ and that A is closed if and only if $\mathcal{D}A \subset A$.

5.95 Proposition. *Let (X, d) be a metric space, $F \subset X$ and $x \in X$. We have*

(i) x *is adherent to F if and only if there exists a sequence $\{x_n\} \subset F$ that converges to x,*

(ii) x *is an accumulation point for F if and only if there exists a sequence $\{x_n\} \subset F$ taking distinct values in F that converges to x; in particular,*

    a) x is an accumulation point for F if and only if there exists a sequence $\{x_n\} \subset F \setminus \{x\}$ that converges to x,

    b) in every open set containing an accumulation point for F there are infinitely many distinct points of F.

Proof. (i) If there is a sequence $\{x_n\} \subset F$ that converges to $x \in X$, in every neighborhood of x there is at least a point of F, hence x is adherent to F. Conversely, if x is adherent to F, there is a $x_n \in B(x, \frac{1}{n}) \cap F$ for each n, hence $\{x_n\} \subset F$ and $x_n \to x$.

(ii) If moreover x is a point of accumulation of F, we can choose $x_n \in F \setminus \{x\}$ and moreover $x_n \in B(x, r_n)$, $r_n := \min(d(x, x_{n-1}), \frac{1}{n})$. The sequence $\{x_n\}$ has the desired properties. $\qquad\square$

h. Subsets and relative topology

Let (X, d) be a metric space and $Y \subset X$. Then (Y, d) is a metric space, too. The family of open sets in Y induced by the distance d is called the *relative topology* of Y. We want to compare the topology of X and the relative topology of Y.

The open ball in Y with center $x \in Y$ and radius $r > 0$ is

$$B_Y(x, r) := \left\{ y \in Y \,\middle|\, d(y, x) < r \right\} = B_X(x, r) \cap Y.$$

5.96 Proposition. *Let (X, d) be a metric space and let $Y \subset X$. Then*

(i) B *is open in Y if and only if there exists an open set $A \subset X$ in X such that $B = A \cap Y$,*

(ii) B *is closed in Y if and only if there exists a closed set A in X such that $B = A \cap Y$.*

Proof. Since (ii) follows at once from (i), we prove (i). Suppose that A is open in X and let x be a point in $A \cap Y$. Since A is open in X, there exists a ball $B_X(x, r) \subset A$, hence $B_Y(x, r) = B_X(x, r) \cap Y \subset A \cap Y$. Thus $A \cap Y$ is open in Y.

Conversely, suppose that B is open in Y. Then for any $x \in B$ there is a ball $B_Y(x, r_x) = B_X(x, r_x) \cap B \subset B$. The set $A := \cup\{B_X(x, r_x) \,|\, x \in B\}$ is an open set in X and $A \cap Y = B$. $\qquad\square$

Also the notions of interior, exterior, adherent and boundary points, in (Y, d) are related to the same notions in (X, d), and whenever we want to emphasize the dependence on Y of the interior, closure, derived and boundary sets we write $\mathrm{int}_Y(A)$, $\overline{A}_Y$, $\mathcal{D}_Y A$, $\partial_Y A$ instead of $\mathrm{int}\,A$, $\overline{A}$, $\mathcal{D}A$, ∂A.

5.97 Proposition. *For any $A \subset Y$ we have*

(i) $\mathrm{int}_Y(A) = \mathrm{int}_X(A) \cap Y$,

(ii) $\overline{A}_Y = \overline{A}_X \cap Y$,

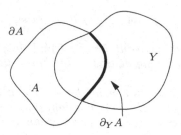

Figure 5.18. ∂A and $\partial_Y A$.

(iii) $\mathcal{D}_Y A = \mathcal{D}_X A \cap Y$,
(iv) $\partial_Y A = \partial_X A \setminus \partial_X Y$.

5.98 ¶. Let $Y := [0,1[\subset \mathbb{R}$. The open balls of Y are the subsets of the type $\{y \in [0,1[\,|\, |y - x| < r\}$. If x is not zero and r is sufficiently small, $\{y \,|\, |y - x| < r\} \cap [0,1[$ is again an open interval with center x, $]x - r, x + r[$. But, if $x = 0$, then for $r < 1$

$$B_Y(0,r) := [0, r[.$$

Notice that $x = 0$ is an interior point of Y (for the relative topology of Y), but it is a boundary point for the topology of X. This is in agreement with the intuition: in the first case we are considering Y as a space in itself and nothing exists outside it, every point is an interior point and $\partial_Y Y = \emptyset$; in the second case Y is a subset of $\mathbb{R}$ and 0 is at the frontier between Y and $\mathbb{R} \setminus Y$.

5.99 ¶. Prove the claims of this paragraph that we have not proved.

5.2.2 A digression on general topology

a. Topological spaces

As a further step of abstraction, following Felix Hausdorff (1869–1942) and Kazimierz Kuratowski (1896–1980), we can separate the topological structure of open sets from the metric structure, giving a set-definition of open sets in terms of their properties.

5.100 Definition. *Let X be a set. A* topology *in X is a distinct family of subsets $\tau \subset \mathcal{P}(X)$, called* open sets, *such that*

o $\emptyset, X \in \tau$,
o *if $\{A_\alpha\} \subset \tau$, then $\cup_\alpha A_\alpha \in \tau$,*
o *if $A_1, A_2, \ldots, A_n \in \tau$, then $\cap_{k=1}^n A_k \in \tau$.*

A set X endowed with a topology is called a topological space. *Sometimes we write it as (X, τ).*

5.101 Definition. *A function $f : X \to Y$ between topological spaces (X, τ_X) and (Y, τ_Y) is said to be* continuous *if $f^{-1}(B) \in \tau_X$ whenever*

$B \in \tau_Y$. $f : X \to Y$ *is said to be a* homeomorphism *if f is both injective and surjective and both f and f^{-1} are continuous, or, in other words $A \in \tau_X$ if and only if $f(A) \in \tau_Y$.*

Two topological spaces are said to be homeomorphic *if and only if there exists a homeomorphism between them.*

Proposition 5.68 then reads as follows.

5.102 Proposition. *Let (X, d) be a metric space. Then the family formed by the empty set and by the sets that are the union of open balls of X is a topology on X, called the topology on X induced by the metric d.*

The topological structure is more flexible than the metric structure, and allows us to greatly enlarge the notion of the space on which we can operate with continuous deformations. This is in fact necessary if one wants to deal with qualitative properties of geometric figures, in the old terminology, with *analysis situs*. We shall not dwell on these topics nor with the systematic analysis of different topologies that one can introduce on a set, i.e., on the study of *general topology*. However, it is proper to distinguish between *metric properties* and *topological properties*.

According to Felix Klein (1849–1925) a *geometry* is the study of the properties of figures or spaces that are *invariant under the action of a certain set of transformations*. For instance, Euclidean plane geometry is the study of the plane figures and of their properties that are invariant under *similarity transformations*.

Given a metric space (X, d), a property of an object defined in terms of the set operations in X and of the metric of X is a *metric property* of X, for instance whether $\{x_n\} \subset X$ is convergent or not is a metric property of X. More generally, in the class of metric spaces, the natural transformations are those $h : (X, d_X) \to (Y, d_Y)$ that are one-to-one and do not change the distances $d_Y(h(x), h(x)) - d_X(x, y)$. Also two metric spaces (X, d) and (Y, d) are said to be *isometric* if there exists an isometry between them. A *metric invariant* is a predicate defined on a class of metric spaces that is true (respectively, false) for all spaces isometric with (X, d) whenever it is true (false) for (X, d). With this languange, the metric properties that make sense for a class of metric spaces, being evidently preserved by isometries, are metric invariants. And the *Geometry of Metric Spaces*, that is the study of metric spaces, of their metric properties, is in fact the study of *metric invariants*.

5.103 ¶. Let (X, d_1) and (Y, d_2) be two metric spaces and denote them respectively, by $B_1(x, r)$ and $B_2(x, r)$ the ball centered at x and radius r respectively, for the metrics d_1 and d_2. Show that a one-to-one map $h : X \to Y$ is an isometry if and only if the action of h preserves the balls, i.e.,

$$h(B_1(x, r)) = B_2(h(x), r) \qquad \forall x \in X, \forall r > 0.$$

Similarly, given a topological space (X, τ_X), a property of an object defined in terms of the set operations and open sets of X is called a *topological property* of X, for instance being an open or closed subset, being the closure or boundary of a subset, or being a convergent sequence in X are topological properties of X.

In the class of topological spaces, the natural group of transformations is the group of *homeomorphisms*, that are precisely all the one-to-one maps whose actions preserve the open sets. Two topological spaces are said *homeomorphic* if there is a homeomorphism from one to the other. A *topological invariant* is a predicate defined on a class of topological spaces that is true (false) in any topological space that is homeomorphic to X whenever it is true (false) on X. With this language, topological properties that make sense for a class of topological spaces, being evidently preserved by the homeomorphims, are topological invariants. And the *topology*, that is the study of objects and of their properties that are preserved by the action of homeomorphisms, is in fact the study of *topological invariants*.

b. Topologizing a set

On a set X we may introduce several topologies, that is subsets of $\mathcal{P}(X)$. Since such subsets are ordered by inclusion, topologies are partially ordered by inclusion. On one hand, we may consider the *indiscrete topology* $\tau = \{\emptyset, X\}$ in which no other sets than $\emptyset$ and X are open, thus there are no "small" neighborhoods. On the other hand, we can consider the *discrete topology* in which any subset is an open set, $\tau = \mathcal{P}(X)$, thus any point is an open set. There is a kind of general procedure to introduce a topology in such a way that the sets of a given family $\mathcal{E} \subset \mathcal{P}(X)$ are all open sets. Of course we can take the discrete topology but what is significant is the smallest family of subsets τ that contains $\mathcal{E}$ and is closed with respect to finite intersections and arbitrary unions. This is called the *coarser topology* or the *weaker topology* for which $\mathcal{E} \subset \tau$. It is unique and can be obtained adding possibly to $\mathcal{E}$ the empty set, X, the finite intersections of elements of $\mathcal{E}$ and the arbitrary union of these finite intersections. This previous construction is necessary, but in general it is quite complicated and $\mathcal{E}$ loses control on τ, since τ builds up from finite intersections of elements of $\mathcal{E}$. However, if the family $\mathcal{E}$ has the following property, as for instance it happens for the balls of a metric space, this can be avoided.

A *basis* $\mathcal{B}$ of X is a family of subsets of X with the following property: for every couple U_α and $U_\beta \in \mathcal{B}$ there is $U_\gamma \in \mathcal{B}$ such that $U_\gamma \subset U_\alpha \cap U_\beta$. We have the following.

5.104 Proposition. *Let* $\mathcal{B} = \{U_\alpha\}$ *be a basis for* X. *Then the family* τ *consisting of* $\emptyset$, X *and all the unions of members of* $\mathcal{B}$ *is the weaker topology in* X *containing* $\mathcal{B}$.

c. Separation properties

It is worth noticing that several separation properties that are trivial in a metric space do not hold, in general, in a topological space. The following claims,

○ sets consisting of a single points are closed,

- for any two distinct points x and $y \in X$ there exist disjoint open sets A and B such that $x \in A$ and $y \in B$,
- for any $x \in X$ and closed set $F \subset X$ there exist disjoint open sets A and B such that $x \in A$ and $F \subset B$.
- for any pair of disjoint closed sets E and F there exist disjoint open sets A and B such that $E \subset A$ and $F \subset B$,

are all true in a metric space, but do not hold in the indiscrete topology. A topological space is called a *Hausdorff topological space* if (ii) holds, *regular* if (iii) holds and *normal* if (iv) holds. It is easy to show that (i) and (iv) imply (iii), (i) and (iii) imply (ii), and (ii) implies (i).

We conclude by stating a theorem that ensures that a topological space be *metrizable*, i.e., when we can introduce a metric on it so that the topology is the one induced by the metric.

5.105 Theorem (Uryshon). *A topological space X with a countable basis is metrizable if and only if it is regular.*

5.3 Completeness

a. Complete metric spaces

5.106 Definition. *A sequence $\{x_n\}$ with values in a metric space (X, d) is a* Cauchy sequence *if*

$$\forall \epsilon > 0 \quad \exists \, \nu \text{ such that } d(x_n, x_m) < \epsilon \quad \forall n, m \geq \nu.$$

It is easily seen that

5.107 Proposition. *In a metric space*

(i) *every convergent sequence is a Cauchy sequence,*
(ii) *any subsequence of a Cauchy sequence is again a Cauchy sequence,*
(iii) *if $\{x_{k_n}\}$ is a subsequence of a Cauchy sequence $\{x_n\}$ such that $x_{k_n} \to x_0$, then $x_n \to x_0$.*

5.108 Definition. *A metric space (X, d) is called* complete *if every Cauchy sequence converges in X.*

By definition, a Cauchy sequence and a complete metric space are metric invariants.

With Definition 5.108, Theorems 2.35 and 4.23 of [GM2] read as $\mathbb{R}$, $\mathbb{R}^2$, $\mathbb{C}$ *are complete metric spaces.* Moreover, since

$$|x_1|, |x_2|, \ldots, |x_n| \leq ||x|| \leq \sum_{i=1}^{n} |x_i| \qquad \forall x = (x_1, x_2, \ldots, x_n),$$

$\{x_k\} \subset \mathbb{R}^n$ or $\mathbb{C}^n$ is a convergent sequence (respectively, Cauchy sequence) if and only if the sequences of coordinates $\{x_k^i\}$, $i = 1, \ldots, n$ are convergent sequences (Cauchy sequences). Thus

5.109 Theorem. *For all $n \geq 1$, $\mathbb{R}^n$ and $\mathbb{C}^n$ endowed with the Euclidean metric are complete metric spaces.*

b. Completion of a metric space

Several useful metric spaces are complete. Notice that closed sets of a complete metric space are complete metric spaces with the induced distance. However, there are noncomplete metric spaces. The simplest significant examples are of course the open intervals of $\mathbb{R}$ and the set of rational numbers with the Euclidean distance.

Let X be a metric space. A complete metric space X^* is called a *completion of X* if

(i) X is isometric to a subspace $\widetilde{X}$ of X^*,
(ii) $\widetilde{X}$ is dense in X^*, i.e., $\operatorname{cl} \widetilde{X} = X^*$.

We have the following.

5.110 Theorem (Hausdorff). *Every metric space X has a completion and any two completions of X are isometric.*

Though every noncomplete metric space can be regarded as a subspace of its completion, it is worth remarking that from an effective point of view the real problem is to realize a suited handy model of this completion. For instance, the Hausdorff model, when applied to rationals, constructs the completion as equivalence class of Cauchy sequences of rationals, instead of real numbers. In the same way, the Hausdorff procedure applied to a metric space X of functions produces a space of equivalence classes of Cauchy sequences. It would be desirable to obtain another class of functions as completion, instead. But this can be far from trivial. For instance a space of functions that is the completion of $C^0([0,1])$ with the $L^1([0,1])$ distance can be obtained by the Lebesgue integration theory.

5.111 ¶. Show that a closed set F of a complete metric space is complete with the induced metric.

5.112 ¶. Let (X, d) be a metric space and $A \subset X$. Show that the closure of A is a completion of A.

Proof of Theorem 5.110. In fact we outline the main steps leaving to the reader the task of completing the arguments.

(i) We consider the family of all Cauchy sequences of X and we say that two Cauchy sequences $\{y_n\}$ and $\{z_n\}$ are equivalent if $d(y_n, z_n) \to 0$ (i.e., if, a posteriori, $\{y_n\}$ and $\{z_n\}$ "have the same limit"). Denote by $\overline{X}$ the set of equivalence classes obtained this way. Given two classes of equivalence Y and Z in $\overline{X}$, let $\{y_n\}$ and $\{z_n\}$ be two representatives respectively of Y and Z. Then one sees

Figure 5.19. Felix Hausdorff (1869–1942) and René-Louis Baire (1874–1932).

(i) $\{d(z_n, y_n)\}$ is a Cauchy sequence of real numbers, hence converges to a real number. Moreover, such a limit does not depend on the representatives $\{y_n\}$ e $\{z_n\}$ of Y and Z, so that

$$d(z_n, y_n) \longrightarrow d(Z, Y).$$

(ii) $d(Y, Z)$ is a distance in $\overline{X}$.

(ii) Let $\widetilde{X}$ be the subspace of $\overline{X}$ of the equivalence classes of the constant sequences with values in X. It turns out that X is isometric to $\widetilde{X}$. Let $Y \in \overline{X}$ and let $\{y_n\}$ be a representative of Y. Denote by Y_ν the class of all Cauchy sequences that are equivalent to the constant sequence $\{z_n\}$ where $z_n := y_\nu \ \forall n$. Then it is easily seen that $Y_\nu \to Y$ in $\overline{X}$ and that $\widetilde{X}$ is dense in $\overline{X}$.

(iii) Let $\{Y_\nu\}$ be a Cauchy sequence in $\overline{X}$. For all ν we choose $Z_\nu \in \widetilde{X}$ such that $d(Y_\nu, Z_\nu) < 1/\nu$ and we let $z_\nu \in X$ be a representative of Z_ν. Then we see that $\{z_\nu\}$ is a Cauchy sequence in X and, if Z is the equivalence class of $\{z_\nu\}$, then $Y_\nu \longrightarrow Z$. This proves that $\overline{X}$ is complete.

(iv) It remains to prove that any two completions are isometric.

Suppose that $\overline{X}$ and $\overline{\overline{X}}$ are two completions of X. With the above notation, we find $\widetilde{X} \subset \overline{X}$ and $\widetilde{\widetilde{X}} \subset \overline{\overline{X}}$ that are isometric and one-to-one with X. Therefore $\widetilde{X}$ and $\widetilde{\widetilde{X}}$ are isometric and in a one-to-one correspondence. Because $\widetilde{X}$ is dense in $\overline{X}$ and $\widetilde{\widetilde{X}}$ are dense respectively in $\overline{X}$ and $\overline{\overline{X}}$ it is not difficult to extend the isometry $i : \widetilde{X} \to \widetilde{\widetilde{X}}$ to an isometry between $\overline{X}$ and $\overline{\overline{X}}$. □

c. Equivalent metrics

Completeness is a metric invariant and not a topological invariant. This means that isometric spaces are both complete or noncomplete and that there exist metric spaces X and Y that are homeomorphic, but X is complete and Y is noncomplete. In fact, homeomorphisms preserve convergent sequences but not Cauchy sequences.

5.113 Example. Consider $X := \mathbb{R}$ endowed with the Euclidean metric and $Y := \mathbb{R}$ endowed with the distance

$$d(x, y) = \left| \frac{x}{1 + |x|} - \frac{y}{1 + |y|} \right|.$$

X and Y are homeomorphic, a homeomorphism being given by the map $h(x) := \frac{x}{1+|x|}$, $x \in \mathbb{R}$. In particular both distances give rise to the same converging sequences. However the sequence $\{n\}$ is not a Cauchy sequence for the Euclidean distance, but it is a Cauchy sequence for the metric d since for $n, m \in \mathbb{N}$, $m \geq n$

$$d(m,n) = \left| \frac{n}{1+n} - \frac{m}{1+m} \right| \leq 1 - \frac{n}{1+n} \to 0 \qquad \text{per } \nu \to \infty.$$

Since $\{n\}$ does not converge in $(\mathbb{R}, d)$, $Y = (\mathbb{R}, d)$ is not complete.

Homeomorphic, but nonisometric spaces can sometimes have the same Cauchy sequences. A *sufficient condition* ensuring that Cauchy sequences with respect to different metrics on the same set X are the same, is that the two metrics be equivalent, i.e., there exist constants $\lambda_1, \lambda_2 > 0$ such that

$$\lambda_1 d_1(x,y) \leq d_2(x,y) \leq \lambda_2 d_1(x,y).$$

5.114 ¶. Show that two metric spaces which are equivalent are also topologically equivalent, compare Proposition 5.86.

d. The nested sequence theorem

An extension of Cantor's principle or the nested intervals theorem in $\mathbb{R}$, see [GM2], holds in a *complete* metric space.

5.115 Proposition. *Let $\{E_k\}$ be a monotone-decreasing sequence of nonempty sets, i.e., $\emptyset \neq E_{k+1} \subset E_k \ \forall k = 0, 1, \ldots$, in a complete metric space X. If $\operatorname{diam}(E_k) \to 0$, then there exists one and only one point $\overline{x} \in X$ with the following property: any ball centered at $\overline{x}$ contains one, and therefore infinitely many of the E_k's. Moreover, if all the E_k are closed, then $\cap_k E_k = \{\overline{x}\}$.*

As a special case we have the following.

5.116 Corollary. *In a complete metric space a sequence of nested closed balls with diameters that converge to zero have a unique common point.*

Notice that the conclusion of Corollary 5.116 does not hold if the diameters do not converge to zero: for $E_k := [k, +\infty[\subset \mathbb{R}$ we have $\emptyset \neq E_{k+1} \subset E_k$ and $\cap_k E_k = \emptyset$.

5.117 ¶. Prove Proposition 5.115.

e. Baire's theorem

5.118 Theorem (Baire). *Let X be a complete metric space that can be written as a denumerable union of closed sets $\{E_i\}$, $X = \cup_{i=1}^{\infty} E_i$. Then at least one of the E_i's contains a ball of X.*

Proof. Suppose that none of the E_i's contains a ball of X and let $x_1 \notin E_1$; Since E_i is closed, there is r_1 such that $\mathrm{cl}\,(B(x_1,r_1)) \cap E_1 = \emptyset$. Inside $\mathrm{cl}\,(B(x_1,r_1/2))$ there is now $x_2 \notin E_2$ (otherwise $\mathrm{cl}\,(B(x_1,r_1/2)) \subset E_2$ which is a contradiction) and r_2 such that $\mathrm{cl}\,(B(x_2,r_2)) \cap E_2 = \emptyset$, also we may choose $r_2 < r_1/2$. Iterating this procedure we find a monotonic-decreasing family of closed balls $\{B(x_k,r_k)\}$, $\mathrm{cl}\,(B(x_1,r_1)) \supset \mathrm{cl}\,(B(x_2,r_2)) \supset \cdots$ such that $\mathrm{cl}\,(B(x_n,r_n)) \cap E_n = \emptyset$. Thus the common point to all these balls, that exists by Corollary 5.116, would not belong to any of the E_n, a contradiction. □

An equivalent formulation is the following.

5.119 Proposition. *In a complete metric space, the denumerable intersection of open dense sets of a complete metric space X is dense in X.*

5.120 Definition. *A subset A of a metric space X is said* nowhere dense *if its closure has no interior point,*

$$\mathrm{int}\,\mathrm{cl}\,(A) = \emptyset,$$

equivalently, if $X \setminus \overline{A}$ is dense in X. A set is called meager *or of the first category if it can be written as a countable union of nowhere dense sets. If a set is not of the first category, then we say that it is of the* second category.

5.121 Proposition. *In a complete metric space a meager set has no interior point, or, equivalently, its complement is dense.*

Proof. Let $\{A_n\}$ be a family such that $\mathrm{int}\,\mathrm{cl}\,A_n = \emptyset$. Suppose there is an open set U with $U \subset \cup_n A_n$. From $U \subset \cup_n A_n \subset \cup_n \overline{A_n}$ we deduce $\cap_n \overline{A_n}^c \subset U^c$. Baire's theorem, see Proposition 5.119, then implies that U^c is dense. Since U^c is closed, we conclude that $U^c = X$ i.e., $U = \emptyset$. □

5.122 Corollary. *A complete metric space is a set of second category.*

This form of Baire's theorem is often used to prove existence results or to show that a certain property is *generic*, i.e., holds for "almost all points" in the sense of the category, i.e., that the set

$$X \setminus \{x \in X \,|\, p(x) \text{ holds}\}$$

is a meager set. In this way one can show[2], see also Chapters 9 and 10, the following.

5.123 Proposition. *The class of continuous functions on the interval $[0,1]$ which have infinite right-derivative at every point, are of second category in $C^0([0,1])$ with the uniform distance; in particular, there exist continuous functions that are nowhere differentiable.*

Finally we notice that, though for a meager set A we have $\mathrm{int}\,A = \emptyset$, we may have $\mathrm{int}\,\mathrm{cl}\,A \neq \emptyset$: consider $A := \mathbb{Q} \subset \mathbb{R}$.

---

[2] See, e.g., J. Dugundji, *Topology*, Allyn and Bacon Inc., Boston.

5.4 Exercises

5.124 ¶. Show that $|x - y|^2$ is not a distance in $\mathbb{R}$.

5.125 ¶. Let (X, d) be a metric space and $M > 0$. Show that the functions

$$d_1(x, y) := \min(M, d(x, y)), \qquad d_2(x, y) := d(x, y)/(1 + d(x, y))$$

are also distances in (X, d) that give rise to the same topology.

5.126 ¶. Plot the balls of the following metric in $\mathbb{C}$

$$d(z, w) = \begin{cases} |z - w| & \text{if } \arg z = \arg w \text{ or } z = w, \\ |z| + |w| & \text{otherwise.} \end{cases}$$

5.127 ¶. Let (X, d_X) be a metric space. Show that, if $f : [0, +\infty[\to [0 + \infty[$ is concave and $f(0) = 0$, then

$$d(x, y) := f(d_X(x, y))$$

is a distance on X, in particular $d^\alpha(x, y)$ is a distance for any α, $0 < \alpha \le 1$. Notice that instead $\| \ \|^\alpha$, $0 < \alpha < 1$, is not a norm, if $\| \ \|$ is a norm.

5.128 ¶. Let $f : (X, d_X) \to (Y, d_Y)$ be α-Hölder continuous. Show that $f : (X, d_X^\alpha) \to (Y, d_Y)$ is Lipschitz continuous.

5.129 ¶. Let S be the space of all sequences of real numbers. Show that the function $d : S \times S \to \mathbb{R}$ given by

$$d(x, y) := \sum_{n=1}^{\infty} \frac{1}{2^n} \frac{|x_n - y_n|}{1 + |x_n - y_n|}$$

if $x = \{x_n\}$, $y = \{y_n\}$, is a distance on S.

5.130 ¶ Constancy of sign. Let X and Y be metric spaces and F be a closed set of Y, let x_0 be a point of accumulation for X and $f : X \to Y$. If $f(x) \to y$ as $x \to x_0$ and $y \notin F$, then there exists $\delta > 0$ such that $f(B(x_0, \delta) \setminus \{x_0\}) \cap F = \emptyset$.

5.131 ¶. Let (X, d) and (Y, δ) be two metric spaces and let $X \times Y$ be the product metric space with the metric

$$\rho((x_1, y_1), (x_2, y_2)) := \sqrt{d(x_1, x_2)^2 + \delta(y_1, y_2)^2}.$$

Show that the projection maps $\pi(x, y) = x$, $\widehat{\pi}(x, y) = y$, are

(i) continuous,
(ii) map open sets into open sets,
(iii) but, in general, do not map closed sets into closed sets.

5.132 ¶ Continuity of operations on functions. Let $* : Y \times Z \to W$ be a map which we think of as an operation. Given $f : X \to Z$ and $g : X \to Y$, we may then define the map $f * g : Y \times Z \to W$ by $f * g(x) = f(x) * g(x)$, $x \in X$. Suppose that X, Y, Z, W are metric spaces, consider $Y \times Z$ as the product metric space with the distance as in Exercise 5.131. Show that if f, g are continuous at x_0, and $*$ is continuous at $(f(x_0), g(x_0))$, then $f * g$ is continuous at x_0.

5.133 ¶. Show that

(i) the parametric equation of straight lines in $\mathbb{R}^n$, $t \to \mathbf{a}t + \mathbf{b}$, $\mathbf{a}, \mathbf{b} \in \mathbb{R}^n$, is a continuous function,

(ii) the parametric equation of the helix in $\mathbb{R}^3$, $t \to (\cos t, \sin t, t)$, $t \in \mathbb{R}$, is a continuous function.

5.134 ¶. Let (X, d_X) and (Y, d_Y) be two metric spaces, $E \subset Y$, x_0 be a point of accumulation of E and $f : E \subset X \to Y$. Show that $f(x) \to y_0$ as $x \to x_0$, $x \in E$, if and only if $\forall \, \epsilon > 0 \; \exists \, \delta > 0$ such that $f(E \cap B(x_0, \delta) \setminus \{x_0\}) \subset B(y_0, \epsilon)$.

5.135 ¶. Show that the *scalar product* in $\mathbb{R}^n$, $(\mathbf{x}|\mathbf{y}) := \sum_{i=1}^n x_i y_i$, $\forall \mathbf{x} = x^1, x^2, \ldots, x^n$, $\mathbf{y} = y^1, y^2, \ldots, y^n \in \mathbb{R}^n$, is a continuous function of the $2n$ variables $(\mathbf{x}, \mathbf{y}) \in \mathbb{R}^{2n}$.

5.136 ¶. Find the maximal domain of definition of the following functions and decide whether they are continuous there:

$$\frac{xz}{1 + y^2}, \qquad \frac{xy^2}{x - \log y}, \qquad \sqrt{xe^y - ye^x}.$$

5.137 ¶. Decide whether the following functions are continuous

$$\begin{cases} \frac{x^3 y}{x^4 + y^2} & \text{if } (x, y) \neq (0, 0), \\ 0 & \text{if } (x, y) = (0, 0), \end{cases} \qquad \begin{cases} \frac{xy^3}{x^4 + y^2} & \text{if } (x, y) \neq (0, 0), \\ 0 & \text{if } (x, y) = (0, 0), \end{cases}$$

$$\begin{cases} \frac{x^2 \sin y}{x^2 + y^2} & \text{if } (x, y) \neq (0, 0), \\ 0 & \text{if } (x, y) = (0, 0), \end{cases} \qquad \begin{cases} \frac{x}{y - x^2} & \text{if } y - x^2 > \sqrt{|x|} \text{ and } (x, y) \neq (0, 0), \\ 0 & \text{if } (x, y) = (0, 0). \end{cases}$$

5.138 ¶. Compute, if they exist, the limits as $(x, y) \to (0, 0)$ of

$$\frac{\log^2(1 + xy)}{x^2 + y^2}, \qquad \frac{x \sin(x^2 + 3y^2)}{x^2 + y^2},$$

$$\frac{\sin x (1 - \cos x)}{x^2 + y^2}, \qquad \frac{x^2 \sin^2 y}{\sin(x^2 + y^2)}.$$

5.139 ¶. Consider $\mathbb{R}$ with the Euclidean topology. Show that

(i) if $A :=]a, b[$, we have int $A = A$, $\overline{A} = [a, b]$ and $\partial A = \{a, b\}$,

(ii) if $A := [a, b[$, we have int $A =]a, b[$, $\overline{A} = [a, b]$ and $\partial A = \{a, b\}$,

(iii) if $A := [a, +\infty[$, we have int $A =]a, +\infty[$, $\overline{A} = A$ and $\partial A = \{a\}$,

(iv) if $A := \mathbb{Q} \subset \mathbb{R}$, we have int $A = \emptyset$, $\overline{A} = \mathbb{R}$ and $\partial A = \mathbb{R}$,

(v) if $A := \{(x, y) \in \mathbb{R}^2 \mid x = y\}$, we have int $A = \emptyset$, $\overline{A} = A$ and $\partial A = A$,

(vi) if $A := \mathbb{N} \subset \mathbb{R}$, we have int $A = \emptyset$, $\overline{A} = A$ and $\partial A = A$.

5.140 ¶. Let (X, d) be a metric space and $\{A_i\}$ be a family of subsets of X. Show that

$$\cup_i \overline{A_i} \subset \overline{\cup_i A_i}, \qquad \overline{\cap_i A_i} \subset \cap_i \overline{A_i}.$$

5.141 ¶. Prove the following

Theorem. *Any open set A of $\mathbb{R}$ is either empty or a finite or denumerable union of disjoint open intervals with endpoints that do not belong to A.*

[*Hint:* Show that (i) $\forall x \in A$ there is an interval $]\xi, \eta[$ with $x \in]\xi, \eta[$ and $\xi, \eta \notin A$, (ii) if two such intervals $]\xi_1, \eta_1[$ and $]\xi_2, \eta_2[$ have a common point in A and endpoints not in A, then they are equal, (iii) since each of those intervals contains a rational, then they are at most countable many.]

Show that the previous theorem does not hold in $\mathbb{R}^n$. Show that we instead have the following.

Theorem. *Every open set $A \subset \mathbb{R}^n$ is the union of a finite or countable union of cubes with disjoint interiors.*

5.142 ¶. Prove the following theorem, see Exercise 5.141,

Theorem. *Every closed set $F \subset \mathbb{R}$ can be obtained by taking out from $\mathbb{R}$ a finite or countable family of disjoint open intervals.*

5.143 ¶. Let X be a metric space.
  (i) Show that $x_0 \in X$ is an interior point of $A \subset X$ if and only if there is an open set U such that $x_0 \in U \subset A$.
  (ii) Using only open sets, express that x_0 is an exterior point to A, an adherent point to A and a boundary point to A.

5.144 ¶. Let X be a metric space. Show that A is open if and only if any sequence $\{x_n\}$ that converges to $x_0 \in A$ is *definitively* in A, i.e., $\exists \overline{n}$ such that $x_n \in A \ \forall n \geq \overline{n}$.

5.145 ¶. Let (X, d_X) and (Y, d_Y) be two metric spaces and let $f : X \to Y$ be a continuous map. Show that
  (i) if $y_0 \in Y$ is an interior point of $B \subset Y$ and if $f(x_0) = y_0$, then x_0 is an interior point of $f^{-1}(B)$.
  (ii) if $x_0 \in X$ is adherent to $A \subset X$, then $f(x_0)$ is adherent to $f(A)$,
  (iii) if $x_0 \in X$ is a boundary point of $A \subset X$, then $f(x_0)$ is a boundary point for $f(A)$,
  (iv) if $x_0 \in X$ is a point of accumulation of $A \subset X$ and f is injective, then $f(x_0)$ is a point of accumulation of $f(A)$.

5.146 ¶. Let X be a metric space and $A \subset X$. Show that $x_0 \in X$ is an accumulation point for A if and only if for every open set U with $x_0 \in U$ we have $U \cap A \setminus \{x_0\} \neq \emptyset$. Show also that being an accumulation point for a set is a *topological notion*. [*Hint:* Use (iv) of Exercise 5.145.]

5.147 ¶. Let X be a metric space and $A \subset X$. Show that x is a point of accumulation of A if and only if $x \in \overline{A \setminus \{x\}}$.

5.148 ¶. Let X be a metric space. A set $A \subset X$ without points of accumulation in X, $\mathcal{D}(A) = \emptyset$, is called *discrete*. A set without isolated points, $\mathcal{I}(A) = \emptyset$, is called *perfect*. Of course every point of a discrete set is isolated, since $A \subset \overline{A} = \mathcal{I}(A) \subset A$. Show that the converse is false: a set of isolated points, $A = \mathcal{I}(A)$, needs not be necessarily discrete. We may only deduce that $\mathcal{D}A = \partial A$.

5.149 ¶. Let X be a metric space. Recall that a set $E \subset X$ is *dense* in X if $\overline{E} = X$. Show that the following claims are equivalent
  (i) D is dense in X,
  (ii) every nonempty open set intersects D,
  (iii) $D^c = X \setminus D$ has no interior points,
  (iv) every open ball $B(x, r)$ intersects D.

5.150 ¶. $\mathbb{Q}$ is dense in $\mathbb{R}$, i.e., $\overline{\mathbb{Q}} = \mathbb{R}$, and $\partial\mathbb{Q} = \mathbb{R}$. Show that $\mathbb{R} \setminus \mathbb{Q}$ is dense in $\mathbb{R}$. Show that the set E of points of $\mathbb{R}^n$ with rational coordinates and its complement are dense in $\mathbb{R}^n$.

5.151 ¶. Let Γ be an additive subgroup of $\mathbb{R}$. Show that either Γ is dense in $\mathbb{R}$ or Γ is the subgroup of integer multiples of a fixed real number.

5.152 ¶. Let X be a metric space. Show $x_n \to x$ if and only if for every open set A with $x \in A$ there is $\overline{n}$ such that $x_n \in A \ \forall n \geq \overline{n}$. In particular, the notion of convergence is a topological notion.

5.153 ¶. The notion of a convergent sequence makes sense in a topological space. One says that $\{x_n\} \subset X$ converge to $x \in X$ if for every open set A with $x \in A$ there is $\overline{n}$ such that $x_n \in A \ \forall n \geq \overline{n}$. However, in this generality limits are not unique. If in X we consider the indiscrete topology $\tau = \{\emptyset, X\}$, every sequence with values in X converges to any point in X. Show that limits of converging sequences are unique in a Hausdorff topological space. Finally, let us notice that in an arbitrary topological space, closed sets cannot be characterized in terms of limits of sequences, see Proposition 5.76.

5.154 ¶. Let (X, τ) be a topological space. A set $F \subset X$ is called *sequentially closed with respect to τ* if every convergent sequence with values in F has limit in F. Show that the family of sequentially closed sets satisfies the axioms of closed sets. Consequently there is a topology (a priori different from τ) for which the closed sets are the family of sequentially closed sets.

5.155 ¶. Let X be a metric space. Show that $\operatorname{diam} \overline{A} = \operatorname{diam} A$, but in general $\operatorname{diam} \operatorname{int} A \leq \operatorname{diam} A$.

5.156 ¶. Let $\emptyset \neq E \subset \mathbb{R}$ be bounded from above. Show that $\sup E \in \overline{E}$; if $\sup E \notin E$, then $\sup E$ is a point of accumulation of E; finally, show that there exist $\max E$ and $\min E$, if E is nonempty, bounded and closed.

5.157 ¶. Let X be a metric space. Show that $\partial A = \emptyset$ iff A is both open and closed. Show that in $\mathbb{R}^n$ we have $\partial A = \emptyset$ iff $A = \emptyset$ or $A = \mathbb{R}^n$.

5.158 ¶. Let X be a metric space. Show that $\partial \operatorname{int} A \subset \partial A$, and that it may happen that $\partial \operatorname{int} A \neq \partial A$.

5.159 ¶. Sometimes one says that A is a *regular open set* if $A = \operatorname{int} \overline{A}$, and that C is a *regular closed set* if $C = \overline{\operatorname{int} C}$. Show examples of regular and nonregular open and closed sets in $\mathbb{R}^2$ and $\mathbb{R}^3$. Show the following:
  (i) The interior of a closed set is a regular open set, the closure of an open set is a regular closed set.
  (ii) The complement of a regular open (closed) set is a regular closed (open) set.
  (iii) If A and B are regular open sets, then $A \cap B$ is a regular open set; if C and D are regular closed sets then $C \cup D$ is a regular closed set.

5.160 ¶. Let X be a metric space. A subset $D \subset X$ is dense in X if and only if for every $x \in X$ we can find a sequence $\{x_n\}$ with values in D such that $x_n \to x$.

5.161 ¶. Let (X, d) and (Y, d) be two metric spaces. Show that
  (i) if $f : X \to Y$ is continuous, then $f : E \subset X \to Y$ is continuous in E with the induced metric,

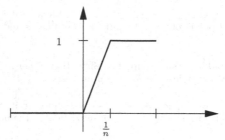

Figure 5.20. A Cauchy sequence in $C^0([0,1])$ with the L^1-metric, with a noncontinuous "limit".

(ii) $f : X \to Y$ is continuous if and only if $f : X \to f(X)$ is continuous.

5.162 ¶. Let X and Y be two metric spaces and let $f : X \to Y$. Show that f is continuous if and only if $\partial f^{-1}(A) \subset f^{-1}(\partial A) \ \forall \ A \subset X$.

5.163 ¶ Open and closed maps. Let (X, d_X) and (Y, d_Y) be two metric spaces. A map $f : X \to Y$ is called *open* (respectively, *closed*) if the image of an open (respectively, closed) set of X is an open (respectively, closed) set in Y. Show that
  (i) the coordinate maps $\pi_i : \mathbb{R}^n \to \mathbb{R}$, $\mathbf{x} = (x_1, x_2, \dots, x_n) \to x_i$, $i = 1, \dots, n$, are open maps but not closed maps,
  (ii) similarly the coordinate maps of a product $\pi_x : X \times Y \to X$, $\pi_y : X \times Y \to Y$ given by $\pi_x(x, y) = x$, $\pi_y(x, y) = y$ are open but in general not closed maps,
 (iii) $f : X \to Y$ is an open map if and only if $f(\text{int } A) \subset \text{int } f(A) \ \forall A \subset X$,
 (iv) $f : X \to Y$ is a closed map if and only if $\overline{f(A)} \subset f(\overline{A}) \ \forall A \subset X$.

5.164 ¶. Let $f : X \to Y$ be injective. Show that f is an open map if and only if it is a closed map.

5.165 ¶. A metric space (X, d_X) is called *topologically complete* if there exists a distance d in X topologically equivalent to d_X for which (X, d) is complete. Show that being topologically complete is a topological invariant.

5.166 ¶. Let (X, d) be a metric space. Show that the following two claims are equivalent
  (i) (X, d) is a complete metric space,
  (ii) If $\{F_\alpha\}$ is a family of closed sets of X such that
    a) any finite subfamily of $\{F_\alpha\}$ has nonempty intersection,
    b) $\inf\{\text{diam } F_\alpha\} = 0$,
  then $\cap_\alpha F_\alpha$ is nonempty and consists of exactly one point.

5.167 ¶. Show that the irrational numbers in $[0, 1]$ cannot be written as countable union of closed sets in $[0, 1]$. [*Hint:* Suppose they are, so that $[0, 1] = \cup_{r \in \mathbb{Q}}\{r\} \cup \cup_i E_i$ and use Baire's theorem.]

5.168 ¶. Show that a complete metric space made of countably many points has at least an isolated point. In particular, a complete metric space without isolated points is not countable. Notice that, if $x_n \to x_\infty$ in $\mathbb{R}$, then $A := \{x_n \mid n = 1, 2 \dots\} \cup \{x_\infty\}$ with the induced distance is a countable complete metric space.

5.169 ¶. Show that $C^0([0,1])$ with the L^1-metric is not complete. [*Hint:* Consider the sequence in Figure 5.20.]

5.170 ¶. Show that $X = \{n \,|\, n = 0, 1, 2, \ldots\}$ and $Y = \{1/n \,|\, n = 1, 2, \ldots\}$ are homeomorphic as subspaces of $\mathbb{R}$, but X is complete, while Y is not complete.

6. Compactness and Connectedness

In this chapter we shall discuss, still in the metric context, two important topological invariants: *compactness* and *connectedness*.

6.1 Compactness

Let E be a subset of $\mathbb{R}^2$. We ask ourselves whether there exists a point $x_0 \in E$ of maximal distance from the origin. Of course E needs to be bounded, $\sup_{x \in E} d(0, x) < +\infty$, if we want a positive answer, and it is easily seen that if E is not closed, our question may have a negative answer, for instance if $E = B(0, 1)$. Assuming E bounded and closed, how can we prove existence? We can find a maximizing sequence, i.e., a sequence $\{x_n\} \subset E$ such that

$$d(0, x_k) \longrightarrow \sup_{x \in E} d(0, x),$$

and our question has a positive answer if $\{x_k\}$ is converging or, at least, if $\{x_k\}$ has a subsequence that converges to some point $x_0 \in E$. In fact, in this case, $d(0, x_{k_n}) \to d(0, x_0)$, $x \to d(0, x)$ is continuous, and $d(0, x_{n_k}) \to \sup_{x \in E} d(0, x)$, too, thus concluding that

$$d(0, x_0) = \sup_{x \in E} d(0, x).$$

6.1.1 Compact spaces

a. Sequential compactness

6.1 Definition. *Let (X, d) be a metric space. A subset $K \subset X$ is said to be* sequentially compact *if every sequence $\{x_k\} \subset K$ has a subsequence $\{x_{n_k}\}$ that converges to a point of K.*

Necessary conditions for compactness are stated in the following

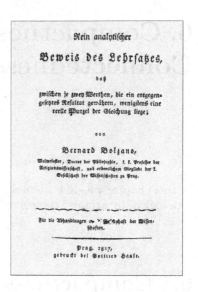

Figure 6.1. Bernhard Bolzano (1781–1848) and the frontispiece of the work where Bolzano–Weierstrass theorem appears.

6.2 Proposition. *We have*

(i) *Any sequentially compact metric space (X, d) is complete;*
(ii) *Any sequentially compact subset of a metric space (X, d) is bounded, closed, and complete with the induced metric.*

Proof. (i) Let $\{x_k\} \subset X$ be a Cauchy sequence. Sequential compactness allows us to extract a convergent subsequence; since $\{x_k\}$ is Cauchy, the entire sequence converges, see Proposition 5.107.

(ii) Let K be sequentially compact. Every point $x \in \overline{K}$ is the limit of a sequence with values in K; by assumption $x \in K$, thus $K = \overline{K}$ and K is closed. Suppose that K is not bounded. Then there is a sequence $\{x_n\} \subset K$ such that $d(x_i, x_j) > 1 \ \forall i, j$. Such a sequence has no convergent subsequences, a contradiction. Finally, K is complete by (i). $\qquad\square$

b. Compact sets in $\mathbb{R}^n$

In general, bounded and closed sets of a metric space are not sequentially compact. However we have

6.3 Theorem. *In $\mathbb{R}^n$, $n \geq 1$, a set is sequentially compact if and only if it is closed and bounded.*

This follows from

6.4 Theorem (Bolzano–Weierstrass). *Any infinite and bounded subset E of $\mathbb{R}^n$, $n \geq 1$, has at least a point of accumulation.*

Proof. Since E is bounded, there is a cube C_0 of side L, so that

$$E \subset C_0 := [a_1^{(0)}, b_1^{(0)}] \times \cdots \times [a_n^{(0)}, b_n^{(0)}], \qquad b_i^{(0)} - a_i^{(0)} = L.$$

Since E is infinite, if we divide C_0 in 2^n equal subcubes, one of them

$$C_1 := [a_1^{(1)}, b_1^{(1)}] \times \cdots \times [a_n^{(1)}, b_n^{(1)}], \qquad b_i^{(1)} - a_i^{(1)} = L/2,$$

contains infinitely many elements of E. By induction, we divide C_i in 2^n equal subcubes with no common interiors, and choose one of them, C_{i+1}, that contains infinitely many elements of E. If

$$C_i := [a_1^{(i)}, b_1^{(i)}] \times \cdots \times [a_n^{(i)}, b_n^{(i)}] \qquad b_i^{(i)} - a_i^{(i)} = L/2^i,$$

the vertices of C_i converge,

$$a_k^{(i)}, b_k^{(i)} \longrightarrow a_k^\infty, b_k^\infty \qquad \text{and} \qquad a_k^\infty = b_k^\infty$$

since for each $k = 1, \ldots, n$ the sequences $\{a^{(i)}{}_k\}$ and $\{b^{(i)}{}_k\}$ are real-valued Cauchy sequences. The point $a := (a_1^\infty, \ldots, a_n^\infty)$ is then an accumulation point for E, since for any $r > 0$, $C_i \subset B(a, r)$ for i sufficiently large. $\qquad \square$

Another useful consequence of Bolzano–Weierstrass theorem is

6.5 Theorem. *Any bounded sequence $\{x_k\}$ of $\mathbb{R}^n$ has a convergent subsequence.*

Proof. If $\{x_k\}$ takes finitely many values, then at least one of them, say α, is taken infinitely often. If $\{p_k\}_{k \in \mathbb{N}}$ are the indices such that $x_{p_k} = \alpha$, then $\{x_{p_k}\}$ converges, since it is constant. Assume now that $\{x_k\}$ takes infinitely many values. The Bolzano–Weierstrass theorem yields a point of accumulation x_∞ for these values. Now we choose ρ_1 as the first index for which $|x_{\rho_1} - x_\infty| < 1$, ρ_2 as the first index greater than ρ_1 such that $|x_{\rho_2} - x_\infty| < 1/2$ and so on: then $\{x_{\rho_k}\}$ is a subsequence of $\{x_n\}$ and $x_{\rho_k} \to x_\infty$. $\qquad \square$

c. Coverings and ϵ-nets

There are other ways to express compactness.

Let A be a subset of a metric space X. A *covering of A* is a family $\mathcal{A} = \{A_\alpha\}$ of subsets of X such that $A \subset \cup_\alpha A_\alpha$. We have already said that $\mathcal{A} = \{A_\alpha\}$ is an *open covering* of A if each A_α is an open set, and that $\{A_\alpha\}$ is a *finite covering* of A if the number of the A_α's is finite. A *subcovering* of a covering $\mathcal{A}$ of A is a subfamily of $\mathcal{A}$ that is still a covering of $\mathcal{A}$.

6.6 Definition. *We say that a subset A of a metric space X is totally bounded if for any $\epsilon > 0$ there is a finite number of balls $B(x_i, \epsilon)$, $i = 1, 2, \ldots, N$ of radius ϵ, each centered at $x_i \in X$, such that $A \subset \cup_{i=1}^N B(x_i, \epsilon)$.*

For a given $\epsilon > 0$, the corresponding balls are said to form an *ϵ-covering* of A, and their centers, characterized by the fact that each point of A has distance less than ϵ from some of the x_i's, form a set $\{x_i\}$ called an *ϵ-net* for A. With this terminology A *is totally bounded iff for every $\epsilon > 0$ there exists an ϵ-net for A.* Notice also that $A \subset X$ is totally bounded if and only if for every $\epsilon > 0$ there exists a finite covering $\{A_i\}$ of X with sets having diam $A_i < \epsilon$.

6.7 Definition. *We say that a subset K of a metric space is* compact *if every open covering of K contains a finite subcovering.*

We have the following.

6.8 Theorem. *Let X be a metric space. The following claims are equivalent.*

(i) X *is sequentially compact.*
(ii) X *is complete and totally bounded.*
(iii) X *is compact.*

The implication (ii) $\Rightarrow$ (i) is known as the *Hausdorff criterion* and the implication (i) $\Rightarrow$ (iii) as the *finite covering lemma*.

Proof. (i) $\Rightarrow$ (ii) By Proposition 6.2, X is complete. Suppose X is not totally bounded. Then for some $r > 0$ no finite family of balls of radius r can cover X. Start with $x_1 \in X$; since $B(x_1, r)$ does not cover X, there is $x_2 \in X$ such that $d(x_2, x_1) \geq r$. Since $\{B(x_1, r), B(x_2, r)\}$ does not cover X either, there is $x_3 \in X$ such that $d(x_3, x_1) \geq r$ and $d(x_3, x_2) \geq r$. By induction, we construct a sequence $\{x_i\}$ such that $d(x_i, x_j) \geq r$ $\forall i > j$, hence $d(x_i, x_j) \geq r$ $\forall i, j$. Such a sequence has no convergent subsequence, but this contradicts the assumption.

(ii) $\Rightarrow$ (iii) By contradiction, suppose that X has an open covering $\mathcal{A} = \{A_\alpha\}$ with no finite subcovering. Since X is totally bounded, there exists a finite covering $\{C_i\}$ of K,

$$\bigcup_{i=1}^{n} C_i = X, \qquad \text{such that} \qquad \operatorname{diam} C_i < 1, \qquad i = 1, \ldots, n.$$

By the assumption, there exists at least k_1 such that $\mathcal{A}$ has no finite subcovering for C_{k_1}. Of course $X_1 := C_{k_1}$ is a metric space which is totally bounded; therefore we can cover C_{k_1} with finitely many open sets with diameter less than $1/2$, and $\mathcal{A}$ has no finite subcovering for one of them that we call X_2. By induction, we construct a sequence $\{X_i\}$ of subsets of X with

$$X_1 \supset X_2 \supset \cdots, \qquad \operatorname{diam} X_i \leq 1/2^i,$$

such that none of them can be covered by finitely many open sets of $\mathcal{A}$. Now we choose for each k a point $x_k \in X_k$. Since $\{x_k\}$ is trivially a Cauchy sequence and X is complete, $\{x_k\}$ converges to some $x_0 \in X$. Let $A_0 \in \mathcal{A}$ be such that $x_0 \in A$ and let r be such that $B(x_0, r) \subset A$ (A is an open set). For k sufficiently large we then have $d(x_k, x_0) < r$ for all $x \in X_k$, i.e., $X_k \subset B(x_0, r) \subset A_0$. In conclusion, X_k is covered by one open set in $\mathcal{A}$, a contradiction since by construction no finite subcovering of $\mathcal{A}$ could cover X_k.

(iii) $\Rightarrow$ (i) If, by contradiction, $\{x_k\}$ has no convergent subsequence, then $\{x_k\}$ is an infinite set without points of accumulation in X. For every $x \in X$ there is a ball $B(x, r_x)$ centered at x that contains at most one point of $\{x_k\}$. The family of these balls $\mathcal{J} := \{B(x, r_x)\}_{x \in X}$ is an open covering of X with no finite subcovering of $\{x_k\}$ hence of X, contradicting the assumption. $\square$

6.9 Remark. Clearly the notions of compactness and sequential compactness are topological notions. They have a meaning in the more general setting of *topological spaces*, while the notion of totally bounded sets is just a metric notion. We shall not deal with compactness in topological spaces. We only mention that compactness and sequential compactness are not equivalent in the context of topological spaces.

6.10 ¶. Let X be a metric space. Show that any closed subset of a compact set is compact.

6.11 ¶. Let X be a metric space. Show that finite unions and generic intersections of compact sets are compact.

6.12 ¶. Show that a finite set is compact.

6.1.2 Continuous functions and compactness

a. The Weierstrass theorem

As in [GM2], continuity of $f : K \to \mathbb{R}$ and compactness of K yield existence of a minimizer.

6.13 Definition. *Let $f : X \to \mathbb{R}$. Points $x_-, x_+ \in X$ such that*

$$f(x_-) = \inf_{x \in X} f(x), \qquad f(x_+) = \sup_{x \in X} f(x)$$

are called respectively, a minimum point *or a* minimizer *and a* maximum point *or a* maximizer *for $f : X \to \mathbb{R}$.*

A sequence $\{x_k\} \subset X$ such that $f(x_k) \to \inf_{x \in X} f(x)$ (resp. $f(x_k) \to \sup_{x \in X} f(x)$) is called a minimizing sequence *(resp. a* maximizing sequence*).*

Notice that any function $f : X \to \mathbb{R}$ defined on a set X has a minimizing and a maximizing sequence. In fact, because of the properties of the infimum, there exists a sequence $\{y_k\} \subset f(X)$ such that $y_k \to \inf_{x \in X} f(x)$ (that may be $-\infty$), and for each k there exists a point $x_k \in X$ such that $f(x_k) = y_k$, hence $f(x_k) \to \inf_{x \in X} f(x)$.

6.14 Theorem (Weierstrass). *Let $f : X \to \mathbb{R}$ be a continuous real-valued function defined in a compact metric space. Then f achieves its maximum and minimum values, i.e., there exists $x_-, x_+ \in X$ such that*

$$f(x_-) = \inf_{x \in X} f(x), \qquad f(x_+) = \sup_{x \in X} f(x).$$

Proof. Let us prove the existence of a minimizer. Let $\{x_k\} \subset K$ be a minimizing sequence. Since X is compact, is has a subsequence $\{x_{n_k}\}$ that converges to some $x_- \in X$. By continuity of f, $f(x_{n_k}) \to f(x_-)$, while by restriction

$$y_{n_k} := f(x_{n_k}) \to \inf_{x \in X} f(x).$$

The uniqueness of the limit yields $\inf_{x \in X} f(x) = f(x_-)$.   □

In fact, we proved that, if $f : X \to \mathbb{R}$ is continuous and X is compact, then any minimizing (resp. maximizing) sequence has a subsequence that converges to a minimum (resp. maximum) point.

b. Continuity and compactness

Compactness and sequential compactness are topological invariants. In fact, we have the following.

6.15 Theorem. *Let $f : X \to Y$ be a continuous function between two metric spaces. If X is compact, then $f(X)$ is compact.*

Proof. Let $\{V_\alpha\}$ be an open covering of $f(X)$. Since f is continuous, $\{f^{-1}(V_\alpha)\}$ is an open convering of X. Consequently, there are indices $\alpha_1, \ldots, \alpha_N$ such that

$$X \subset f^{-1}(V_{\alpha_1}) \cup \ldots \cup f^{-1}(V_{\alpha_N}),$$

hence

$$f(X) \subset V_{\alpha_1} \cup \ldots \cup V_{\alpha_N},$$

i.e., $f(X)$ is compact. □

Another proof of Theorem 6.15. Let us prove that $f(X)$ is sequentially compact whenever X is sequentially compact. If $\{y_n\} \subset f(X)$ and $\{x_n\} \subset X$ is such that $f(x_n) = y_n$ $\forall n$, since X is sequentially compact, a subsequence $\{x_{k_n}\}$ of $\{x_n\}$ converges to a point $x_0 \in X$. By continuity, the subsequnce $\{f(x_{k_n})\}$ of $\{y_n\}$ converges to $f(x_0) \in f(X)$. Then Theorem 6.8 applies. □

6.16 ¶. Infer Theorem 6.14 from Theorem 6.15.

6.17 ¶. Suppose that E is a noncompact metric space. Show that there exist

(i) $f : E \to \mathbb{R}$ continuous and unbounded,
(ii) $f : E \to \mathbb{R}$ continuous and bounded without maximizers and/or minimizers.

c. Continuity of the inverse function

Compactness also plays an important role in dealing with the continuity of the inverse function of invertible maps.

6.18 Theorem. *Let $f : X \to Y$ be a continuous function between two metric spaces. If X is compact, then f is a closed function. In particular, if f is injective, then the inverse funcion $f^{-1} : f(X) \to X$ is continuous.*

Proof. Let $F \subset X$ be a closed set. Since X is compact, F is compact. From Theorem 6.15 we then infer that $f(F)$ is compact, hence closed.

Suppose f injective and let $g : f(X) \to X$ be the inverse of f. We then have $g^{-1}(E) = f(E) \ \forall E \subset X$, hence $g^{-1}(F)$ is a closed set if F is a closed set in X. □

6.19 Corollary. *Let $f : X \to Y$ be a one-to-one, continuous map between two metric spaces. If X is compact, then f is a homeomorphism.*

6.20 Example. The following example shows that the assumption of compactness in Theorem 6.18 cannot be avoided. Let $X = [0, 2\pi[$, Y be the unit circle of $\mathbb{C}$ centered at the origin and $f(t) := e^{it}$, $t \in X$. Clearly $f(t) = \cos t + i \sin t$ is continuous and injective, but its inverse function f^{-1} is not continuous at the point $(1, 0) = f(0)$.

6.1.3 Semicontinuity and the Fréchet–Weierstrass theorem

Going through the proof of Weierstrass's theorem we see that a weaker assumption suffices to prove existence of a minimizer. In fact, if instead of the continuity of f we assume[1]

$$f(x_-) \leq \liminf_{k \to \infty} f(x_k), \qquad \text{whenever } \{x_k\} \text{ is such that } x_k \to x_-, \quad (6.1)$$

then for any convergent subsequence $\{x_{n_k}\}$ of a minimizing sequence,

$$x_{n_k} \to x_0, \qquad f(x_{n_k}) \to \inf_{x \in X} f(x),$$

we have

$$\inf_{x \in X} f(x) \leq f(x_0) \leq \liminf_{k \to \infty} f(x_{n_k}) = \lim_{k \to \infty} f(x_{n_k}) = \inf_{x \in X} f(x),$$

i.e., again $f(x_0) = \inf_{x \in X} f(x)$. We therefore introduce the following definitions.

6.21 Definition. *We say that a function $f : X \to \mathbb{R}$ defined on a metric space X is* sequentially lower semicontinuous *at $x \in X$, s.l.s.c. for short, if*

$$f(x) \leq \liminf_{k \to \infty} f(x_k) \qquad \text{whenever } \{x_k\} \subset X \text{ is such that } x_k \to x.$$

6.22 Definition. *We say that a subset E of a metric space X is* relatively compact *if its closure $\overline{E}$ is compact.*

6.23 Definition. *Let X be a metric space. We say that $f : X \to \mathbb{R}$ is* coercive *if for all $t \in \mathbb{R}$ the level sets of f,*

$$\left\{ x \in X \,\middle|\, f(x) \leq t \right\}$$

are relatively compact.

Then we can state the following.

6.24 Theorem (Fréchet–Weierstrass). *Let X be a metric space and let $f : X \to \mathbb{R}$ be bounded from below, coercive and sequentially lower semicontinuous. Then f takes its minimum value.*

---

[1] See Exercises 6.26 and 6.28 for the definition of $\liminf$ and related information.

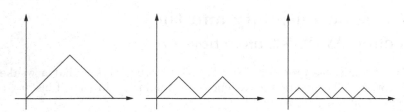

Figure 6.2. Lebesgue's example of a sequence of curves of length $\sqrt{2}$ that converges in the uniform distance to a curve of length 1.

6.25 Example. There are many interesting examples of functions that are semicontinuous but not continuous: a typical example is the length of a curve. Though we postpone details, Lebesgue's example in Figure 6.2 shows that the function length, defined on the space of piecewise linear curves with the uniform distance, is not continuous. In fact length$(f_k) = \sqrt{2}$, $f_k(x) \to f_\infty(x) := 0$ uniformly in $[0, 1]$, and length$(f_\infty) = 1 < 2\pi$. We shall prove later that in fact the length functional is sequentially lower semicontinuous.

6.26 ¶. We say that $f : X \to \mathbb{R}$ is *lower semicontinuous*, for short *l.s.c.*, if for all $t \in \mathbb{R}$ the sets $\{x \in X \mid f(x) \le t\}$ are closed. Sequential lower semicontinuity and lower semicontinuity are topological concepts; they turn out to be different, in general. Show that if X is a metric space, then f is lower semicontinuous if and only if f is sequentially semicontinuous.

6.27 ¶. Let X be a metric space. We recall, see e.g., [GM2], that $\ell \in \overline{\mathbb{R}}$ is the lim inf of $f : X \to \mathbb{R}$ as $y \to x$,
$$\ell = \liminf_{y \to x} f(y),$$
if x is a point of accumulation of X and
(i) $\forall m < \ell \,\exists\, \delta$ such that $f(y) > m$ if $y \in B(x, \delta) \setminus \{x_0\}$,
(ii) $\forall m > \ell \,\forall\, \delta > 0 \,\exists\, y_\delta \in B(x, \delta) \setminus \{x\}$ such that $f(y_\delta) < m$.
Show that the lim inf always exists and is given by
$$\liminf_{y \to x} f(y) = \sup_{r > 0} \inf_{B(x,r) \setminus \{x\}} f(y) = \lim_{r \to 0} \inf_{B(x,r) \setminus \{x\}} f(y).$$
Similarly we can define the lim sup of $f : X \to \mathbb{R}$, so that
$$\limsup_{y \to x} f(y) = - \liminf_{y \to x} (-f(x)).$$
Explicitly define it and show that
$$\limsup_{y \to x} f(y) = \lim_{r \to 0^+} \sup_{B(x,r) \setminus \{x\}} f(y).$$
Finally, show that $f : X \to \mathbb{R}$ is sequentially lower semicontinuous if and only if $\forall x \in X$
$$f(x) \le \liminf_{y \to x} f(y).$$

6.28 ¶. Let $f : X \to \mathbb{R}$ be defined on a metric space X. Show that
(i) $\liminf_{y \to x} f(y) \le \limsup_{y \to x} f(y)$,
(ii) $f(x) \le \liminf_{y \to x} f(y)$ if and only if $-f(x) \ge \limsup_{y \to x} -f(y)$, hence f is lower semicontinuous if and only if $-f$ is upper semicontinuous,
(iii) $f(y) \to \ell$ as $y \to x$ if and only if $\liminf_{y \to x} f(y) = \limsup_{y \to x} f(y) = \ell$,
(iv) $\liminf_{y \to x} f + \liminf_{y \to x} g \le \liminf_{y \to x} (f + g)$,
(v) $\liminf_{x \to x_0} (f(g(x))) = f\Big(\liminf_{x \to x_0} g(x)\Big)$, if either f is continuous at $L := \liminf_{x \to x_0} g(x)$ or $f(x) \ne L$ for any $x \ne x_0$,
(vi) f is bounded from below in a neighborhood of x if and only if $\liminf_{y \to x} f > -\infty$.

6.2 Extending Continuous Functions

6.2.1 Uniformly continuous functions

6.29 Definition. *Let (X, d_X) and (Y, d_Y) be two metric spaces. We say that $f : X \to Y$ is* uniformly continuous *in X if for any $\epsilon > 0$ there exists $\delta > 0$ such that $d_Y(f(x), f(y)) < \epsilon$ for all $x, y \in X$ with $d_X(x, y) < \delta$.*

6.30 Remark. Uniform continuity is a *global* property, in contrast with continuity (at all points) which is a *local* property. A comparison is worthwhile

(i) $f : X \to Y$ is continuous if $\forall x_0 \in X$, $\forall\, \epsilon > 0 \,\exists\, \delta > 0$ (in principle δ depends on ϵ and x_0) $d_Y(f(x), f(x_0)) < \epsilon$ whenever $d_X(x, x_0) < \delta$.

(ii) $f : X \to Y$ is *uniformly continuous* in X if $\forall \epsilon > 0 \,\exists \delta > 0$ (in this case δ depends on ϵ but not on x_0) $d_Y(f(x), f(x_0)) < \epsilon$ whenever $d_X(x, x_0) < \delta$.

Of course, if f is uniformly continuous in X, f is continuous in X and uniformly continuous on any subset of X. Moreover if $\{U_\alpha\}$ is a *finite* partition of X and each $F_{|U_\alpha} : U_\alpha \to Y$ is uniformly continuous in U_α, then $f : X \to Y$ is uniformly continuous in X.

6.31 ¶. Show that Lipschitz-continuous and more generally Hölder-continuous functions, see Definition 5.24, are uniformly continuous functions.

6.32 ¶. Show that $f : X \to Y$ is not uniformly continuous in X if and only if there exist two sequences $\{x_n\}$, $\{y_n\} \subset X$ and $\epsilon_0 > 0$ such that $d_X(x_n, y_n) \to 0$ and $d_Y(f(x_n), f(y_n)) \geq \epsilon_0 \;\forall n$.

6.33 ¶. Show that

(i) x^2 and $\sin x^2$, $x \in \mathbb{R}$, are not uniformly continuous in $\mathbb{R}$,

(ii) $1/x$ is not uniformly continuous in $]0, 1]$,

(iii) $\sin x^2$, $x \in \mathbb{R}$, is not uniformly continuous in $\mathbb{R}$.

Using directly Lagrange's theorem, show that

(iii) x^2, $x \in [0, 1]$, is uniformly continuous in $[0, 1]$,

(iv) e^{-x}, $x \in \mathbb{R}$, is uniformly continuous in $[0, +\infty[$.

6.34 ¶. Let X, Y be two metric spaces and let $f : X \to Y$ be uniformly continuous. Show that the image of a Cauchy sequence is a Cauchy sequence on Y.

6.35 Theorem (Heine–Cantor–Borel). *Let $f : X \to Y$ be a continuous map between metric spaces. If X is compact, then f is uniformly continuous.*

Proof. By contradiction, suppose that f is not uniformly continuous. Then there is $\epsilon_0 > 0$, and two sequences $\{x_n\}$, $\{y_n\} \subset X$ such that

$$d_X(x_n, y_n) < \frac{1}{n}, \quad \text{and} \quad d_Y(f(x_n), f(y_n)) > \epsilon_0 \ \forall n. \qquad (6.2)$$

Since X is compact, $\{x_n\}$ has a convergent subsequence, $x_{k_n} \to x$, $x \in X$. The first inequality in (6.2) yields that $\{y_{k_n}\}$ converges to x, too. On account of the continuity of f,

$$d_Y(f(x_{k_n}), f(x)) \to 0, \qquad d_Y(f(y_{k_n}), f(x)) \to 0,$$

hence $d_Y(f(x_{k_n}), f(y_{k_n})) \to 0$: this contradicts the second inequality in (6.2). $\qquad \square$

6.2.2 Extending uniformly continuous functions to the closure of their domains

Let X, Y be metric spaces, $E \subset X$ and $f : E \to Y$ be a continuous function. Under which conditions is there a *continuous extension* of f over $\overline{E}$, i.e., a continuous $g : \overline{E} \to Y$ such that $g = f$ in E? Notice that we do not want to change the target Y. Of course, such an extension may not exist, for instance if $E =]0, 1]$ and $f(x) = 1/x$, $x \in]0, 1]$. On the other hand, if it exists, it is unique. In fact, if g_1 and $g_2 : \overline{E} \to Y$ are two continuous extensions, then $\Delta := \{x \in \overline{E} \,|\, g_1(x) = g_2(x)\}$ is closed and contains E, hence $\Delta = \overline{E}$.

6.36 Theorem. *Let X and Y be two metric spaces. Suppose that Y is complete and that $f : E \subset X \to Y$ is a uniformly continuous map. Then f extends uniquely to a continuous function on $\overline{E}$; moreover the extension is uniformly continuous in $\overline{E}$.*

Proof. First we observe

(i) since f is uniformly continuous in E, if $\{x_n\}$ is a Cauchy sequence in E, then $\{f(x_n)\}$ is a Cauchy sequence in Y, hence it converges in Y,
(ii) since f is uniformly continuous, if $\{x_n\}$ $\{y_n\} \subset E$ are such that $x_n \to x$ and $y_n \to x$ for some $x \in X$, then the Cauchy sequences $\{f(x_n)\}$ and $\{f(y_n)\}$ have the same limit.

Define $F : \overline{E} \to Y$ as follows. For any $x \in \overline{E}$, let $\{x_n\} \subset E$ be such that $x_n \to x$. Define

$$F(x) := \lim_{x_n \to x} f(x_n).$$

We then leave to the reader the task of proving that

(i) F is welldefined, i.e., its definition makes sense, since for any x the value $F(x)$ is independent of the chosen sequence $\{x_n\}$ converging to x,
(ii) $F(x) = f(x) \ \forall x \in E$,
(iii) F is uniformly continuous in $\overline{E}$,
(iv) F extends f, i.e., $F(x) = f(x) \forall x \in E$.

$\qquad \square$

6.37 ¶. As a special case of Theorem 6.36, we notice that a function $f : E \subset X \to Y$, which is uniformly continuous on a dense subset $E \subset X$, extends to a uniformly continuous function defined on the whole X.

6.2.3 Extending continuous functions

Let X, Y be metric spaces, $E \subset X$ and $f : E \to Y$ be a continuous function. Under which conditions can f be extended to a *continuous* function $F : X \to Y$? This is a basic question for continuous maps.

a. Lipschitz-continuous functions

We first consider real-valued Lipschitz-continuous maps, $f : E \subset X \to \mathbb{R}$.

6.38 Theorem (McShane). *Let (X, d) be a metric space, $E \subset X$ and let $f : E \to \mathbb{R}$ be a Lipschitz map. Then there exists a Lipschitz-continuous map $F : X \to \mathbb{R}$ with the same Lipschitz constant as f, which extends f.*

Proof. Let $L := \operatorname{Lip}(f)$. For $x \in X$ let us define

$$F(x) := \inf_{y \in E} \Big(f(y) + L\, d(x, y) \Big)$$

and show that it has the required properties. For $x \in E$ we clearly have $F(x) \le f(x)$ while, f being Lipschitz, gives

$$f(x) \le f(y) + L\, d(x, y) \qquad \forall y \in E,$$

i.e., $f(x) \le F(x)$, thus concluding that $F(x) = f(x) \,\forall x \in E$.

Moreover, we have

$$F(x) \le \inf_{z \in E} \Big(f(z) + L\, d(y, z) \Big) + L\, d(x, y) = F(y) + L\, d(x, y)$$

and similarly

$$F(y) \le F(x) + L\, d(x, y).$$

Hence F is Lipschitz continuous with $\operatorname{Lip}(F) \le L$. As F is an extension of f, we trivially have $\operatorname{Lip}(F) \ge L$, thus concluding $\operatorname{Lip}(F) = L$.   □

The previous theorem allows us to extend vector-valued Lipschitz-continuous maps $f : E \subset X \to \mathbb{R}^m$, but the Lipschitz extension will have, in principle, a Lipschitz constant less than $\sqrt{m}\operatorname{Lip}(f)$. Actually, a more elaborated argument allows us to prove the following.

6.39 Theorem (Kirszbraun). *Let $f : E \subset \mathbb{R}^n \to \mathbb{R}^m$ be a Lipschitz-continuous map. Then f has an extension $F : \mathbb{R}^n \to \mathbb{R}^m$ such that $\operatorname{Lip} F = \operatorname{Lip} f$.*

In fact there exist several extensions of Kirszbraun's theorem that we will not discuss. We only mention that it may fail if either $\mathbb{R}^n$ or $\mathbb{R}^m$ is remetrized by some norm not induced by an inner product.

6.40 ¶ (Federer). Let X be $\mathbb{R}^2$ with the infinity norm $||x||_\infty = \sup(|x^1| + |x^2|)$ and the map $f : A \subset X \to \mathbb{R}^2$, where $A := \{(-1, 1), (1, -1), (1, 1)\}$ and

$$f(-1, 1) := (-1, 0), \quad f(1, -1) := (1, 0), \quad f(1, 1) := (0, \sqrt{3}).$$

Show that $\operatorname{Lip}(f) = 1$, but f has no 1-Lipschitz extension to $A \cup \{(0, 0)\}$.

6.2.4 Tietze's theorem

An extension of Theorem 6.38 holds for continuous functions in a closed domain.

6.41 Theorem (Tietze). *Let X be a metric space, $E \subset X$ be a closed subset of X, and f a continuous function from E into $[-1, 1]$ (respectively, $\mathbb{R}$). Then f has a continuous extension from X into $[-1, 1]$ (respectively, $\mathbb{R}$).*

Actually we have the following.

6.42 Theorem (Dugundji). *Let X be a metric space, E a closed subset of X and let f be a continuous function from E into $\mathbb{R}^n$. Then f has a continuous extension from X into $\mathbb{R}^n$; moreover the range of f is contained in the convex hull of $f(E)$.*

We recall that the *convex hull* of a subset $E \subset \mathbb{R}^n$ is the intersection of all convex sets that contain E.

Proof of Tietze's theorem. First assume that f is bounded. Then it is not restrictive to assume that $\inf_E f = 1$ and $\sup_E f = 2$. We shall prove that the function

$$F(x) := \begin{cases} f(x) & \text{if } x \in E, \\ \dfrac{\inf_{y \in E}(f(y)d(x,y))}{d(x,E)} & \text{if } x \notin E \end{cases}$$

is a continuous extension of f and $1 \leq F(x) \leq 2 \ \forall x \in X$.

Since the last claim is trivial, we need to prove that F is continuous in X. Decompose $X = \text{int } E \cup (X \setminus E) \cup \partial E$. If $x_0 \in \text{int } E$, then F is continuous at x_0 by assumption. Let $x_0 \in X \setminus \overline{E}$. In this case $x \to d(x, E)$ is continuous and strictly positive in an open neighborhood of x_0, therefore it suffices to prove that that $h(x) := \inf_{y \in E}(f(y)d(x,y))$ is continuous at x_0. We notice that for $y \in E$ and $x, x_0 \in X$ we have

$$f(y)\, d(x, y) \leq f(y)\, d(x_0, y) + f(y)\, d(x, x_0) \leq f(y)\, d(x_0, y) + 2\, d(x, x_0),$$

hence

$$h(x) \leq h(x_0) + 2d(x, x_0)$$

and, exchanging x with x_0,

$$|h(x) - h(x_0)| \leq 2d(x, x_0).$$

This proves continuity of h at x_0.

Let $x_0 \in \partial E$. For $\epsilon > 0$ let $r > 0$ be such that $|f(y) - f(x_0)| < \epsilon$ provided $d(y, x_0) < r$ and $y \in E$. For $x \in B(x_0, r/4)$ we have

$$\inf_{E \cap B(x_0, r/4)} (f(y)d(x, y)) \leq f(x_0)d(x, x_0) \leq 2\frac{r}{4} = \frac{r}{2}$$

and

$$\inf_{E \setminus B(x_0, r/4)} (f(y)d(x, y)) \geq d(x_0, y) - d(x, x_0) \geq \frac{3}{4}r.$$

Therefore we find for x with $d(x_0, x) < r/4$,

$$h(x) = \inf_{y \in E}(f(y)d(x, y)) = \inf_{E \cap B(x_0, r/4)} f(y)d(x, y)$$

and

$$d(x, E) = d(x, E \cap B(x_0, r/4)).$$

On the other hand, for $y \in E \cap B(x_0, r)$ we have $|f(x_0) - f(y)| < \epsilon$ hence

$$(f(x_0) - \epsilon)d(x, E) \leq h(x) \leq (f(x_0) + \epsilon)d(x, E)$$

if $x \in B(x_0, r/4)$, i.e., $h(x)$ is continuous at x_0.

Finally, if f is not bounded, we apply the above to $g := \varphi \circ f$, φ being the homeomorphism $\varphi : \mathbb{R} \to]0, 2[$, $\varphi(x) = \frac{1}{2} + \frac{x}{2(1 + |x|)}$. If G extends continuously g, then $F := \varphi^{-1} \circ G$ continuously extends f. $\qquad \square$

6.43 Remark. The extension $F : X \to \mathbb{R}$ of $f : \underline{E} \subset X \to \mathbb{R}$ provided by Tietze's theorem is Lipschitz continuous outside $\overline{E}$.

Sketch of the proof of Theorem 6.42, assuming $X = \mathbb{R}^n$ and $E \subset X$ compact. Choose a countable dense set $\{e_k\}_k$ in E and for $x \notin E$ and $k = 1, 2, \ldots$, and set

$$\varphi_k(x) := \max\left\{2 - \frac{|x - e_k|}{d(x, E)}, 0\right\}.$$

The function

$$\widetilde{f}(x) := \begin{cases} f(x), & x \in E, \\ \dfrac{\sum_{k \geq 1} 2^{-k} \varphi_k(x) f(e_k)}{\sum_{i \geq 1} 2^{-k} \varphi_k(x)}, & x \notin E, \end{cases}$$

defines a continuous extension $\widetilde{f}$, moreover $\widetilde{f}(\mathbb{R}^n)$ is contained in the convex-hull of $f(E)$. $\qquad \square$

6.44 ¶. Let E and F be two disjoint nonempty closed sets of a metric space (X, d). Check that the function $f : X \to [0, 1]$ given by

$$f(x) = \frac{d(x, E)}{d(x, E) + d(x, F)}$$

is continuous in X, has values in $[0, 1]$, $f(x) = 0 \ \forall x \in E$ and $f(x) = 1 \ \forall x \in F$.

6.45 ¶. Let E and F be two disjoint nonempty closed sets of a metric space (X, d). Using the function f in Exercise 6.44 show that there exist two open sets $A, B \subset X$ with $A \cap B = \emptyset$, $A \supset E$ and $B \supset F$.

Indeed Exercise 6.45 has an inverse.

6.46 Lemma (Uryshon lemma). *Let X be a topological space such that each couple of disjoint closed sets can be separated by two open disjoint sets. Then, given a couple of disjoint closed sets E and F, there exists a continuous function $f : X \to [0, 1]$ such that $f(x) = 1 \ \forall x \in E$ and $f(x) = 0 \ \forall x \in F$.*

This lemma answers the problem of finding nontrivial continuous functions in a topological space and is a basic step in the construction of the so-called *partition* or *decomposition of unity*, a means that allow us to pass from local to global constructions and vice versa.

Since we shall not need these results in a general context, we refrain from further comments and address the reader to one of the treatises in general topology.

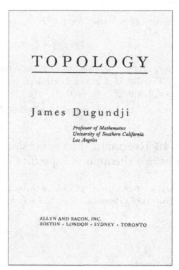

PRÉFACE AU VOLUME I.

La Topologie traite des propriétés des ensembles de points,
invariantes par rapports aux transformations bicontinues.
Une transformation (univoque) $y = f(x)$ est dite *continue*,
lorsque la condition $x = \lim x_n$ entraîne $f(x) = \lim f(x_n)$. Elle
est dite *bicontinue* ou une *homéomorphie*, lorsqu'elle admet, en
outre, une transformation inverse $x = f^{-1}(y)$ continue.
Le terme „ensemble de *points*" exige quelques explications:
on peut notamment se demander quel est *l'espace* dont on consi-
dère les points.
Comme on sait, la notion de point de l'espace euclidien à 3
dimensions a été étendue dans la Géométrie analytique sur l'espace
à un nombre arbitraire des dimensions: un point p de l'espace
euclidien C^k (à k dimensions) est par définition un système de k
nombres réels $p^{(1)}, p^{(2)}, ..., p^{(k)}$; la convergence $\lim p_n = p$ signifie
que l'on a $\lim p_n^{(l)} = p^{(l)}$, quel que soit $l \leqslant k$.
Le développement récent de la Topologie et des autres bran-
ches des mathématiques modernes (surtout celui de la Théorie
générale des fonctions et du Calcul fonctionnel) a montré que cette
conception de l'espace était encore trop étroite: dans un grand
nombre des problèmes on est conduit à considérer, outre l'espace
C^k, l'espace C^{k_0} à une *infinité de dimensions* (nommé aussi „espace
$C_{ω}$ de Fréchet") et dont les points p sont des suites infinies
$p^{(1)}, p^{(2)}, ..., p^{(l)}, ...$ de nombres réels; la convergence $\lim p_n = p$
y signifie que l'on a $\lim p_n^{(l)} = p^{(l)}$, quel que soit l.
Or, c'est précisément l'étude des invariants des homéomor-
phies entre sous-ensembles de l'espace C^{k_0} qui constitue le vrai
domaine de la Topologie à l'état actuel de cette science. Ajoutons

TOPOLOGY

James Dugundji

Professor of Mathematics
University of Southern California
Los Angeles

ALLYN AND BACON, INC.
BOSTON · LONDON · SYDNEY · TORONTO

Figure 6.3. The first page of the *Preface of Topologie* by Kazimierz Kuratowski (1896–1980) and the frontispiece of a classical in general topology.

6.3 Connectedness

Intuitively, a space is *connected* if it does not consist of two separate pieces.

6.3.1 Connected spaces

6.47 Definition. *A metric space X is said to be* connected *if it is not the union of two nonempty disjoint open sets. A subset $E \subset X$ is connected if it is connected as subspace of X.*

This can be formulated in other ways. For example we say that two sets A and B of a metric space X are *separated* if both $A \cap \overline{B}$ and $\overline{A} \cap B$ are empty, i.e., no point of A is adherent to B and no point of B is adherent to A.

6.48 Proposition. *Let X be a metric space. The following properties are equivalent.*

(i) *X is connected.*
(ii) *There are no closed sets F, G in X such that $F \cap G = \emptyset$ and $X = F \cup G$.*
(iii) *The only subsets of X both open and closed are $\emptyset$ and X.*
(iv) *X is not the union of two nonempty and separated subsets.*

Proof. Trivially (i) ⇔ (ii) ⇔ (iii). Let us prove that (i) ⇒ (iv). By contradiction, suppose $X = A \cup B$ where A and B are nonempty and separated. From $A \cap \overline{B} = \emptyset$ and $A \cup B = X$ we infer $A \subset \overline{B}^c$ and $\overline{B}^c \subset A$, hence $A = \overline{B}^c$, i.e., A is an open set. Similarly we infer that B is open, concluding that X is not connected, a contradiction. Finally, let us prove that (iv) ⇒ (ii). By contradiction, assume that X is not connected. Then $X = A \cup B$ with A, B closed, disjoint and nonempty, thus $(A \cap \overline{B}) = (\overline{A} \cap B) = A \cap B = \emptyset$. Thus X is separated, a contradiction.   □

a. Connected subsets

6.49 Theorem. *$E \subset \mathbb{R}$ is connected if and only if E is an interval.*

Proof. If $E \subset \mathbb{R}$ is not an interval, there exist $x, y \in E$ and $z \notin E$ with $x < z < y$. Thus the sets $E_1 := E \cap] -\infty, z[$ and $E_2 := E \cap]z, +\infty[$ are nonempty and separated. Since $E = E_1 \cup E_2$, E is not connected, a contradiction.

Conversely, if E is not connected, then $E = A \cup B$ with A and B nonempty and separated. Let $x \in A$ and $y \in B$ and, without loss of generality, suppose $x < y$. Define

$$z := \sup(A \cap [x, y]).$$

We have $z \in \overline{A}$ hence $z \notin B$; in particular $x \leq z < y$. If $z \notin A$ then $x < z < y$ and $z \notin E$, i.e., E is not an interval. Otherwise, if $z \in A$, then $z \notin \overline{B}$ and there exists z_1 such that $z < z_1 < y$ and $z_1 \notin B$ but then $x < z_1 < y$ with $z_1 \notin E$, thus once again, E is not an interval.   □

6.50 ¶. Show that the closure of a connected set is connected.

6.51 ¶. Show that any subset of $\mathbb{Q}$ having more than one point is not connected.

6.52 ¶. Let A and B be nonempty, disjoint and open sets of a metric space X such that $X = A \cup B$ and let Y be a connected subset of X. Then either $Y \subset A$ or $Y \subset B$.

6.53 ¶. Let $\{Y_\alpha\}$ be a family of connected subsets of X such that $\cap_\alpha Y_\alpha \neq \emptyset$. Show that $Y := \cup_\alpha Y_\alpha$ is connected. [*Hint:* Argue by contradiction.]

6.54 ¶. Let $\{Y_n\}$ be a sequence of connected subsets such that $Y_n \cap Y_{n+1} \neq \emptyset \ \forall \ n$. Then $Y := \cup_n Y_n$ is connected.

b. Connected components

Because of Exercise 6.53, the following definition makes sense.

6.55 Definition. *Let X be a metric space. The* connected component *of X containing $x_0 \in X$ is the largest connected subset C_{x_0} of X such that $x_0 \in C_{x_0}$.*

6.56 Proposition. *Let X be a metric space. We have the following.*

(i) *The distinct connected components of the points of X form a partition of X.*
(ii) *Each connected component $C \subset X$ is a closed set.*
(iii) *If $y \in C_x$, then $C_x = C_y$.*

(iv) *If $Y \subset X$ is a nonempty open and closed subset of X, then Y is a connected component of X.*

Observe that the connected components are not necessarily open. For instance, consider $X = \mathbb{Q}$ for which $C_x := \{x\} \ \forall x \in \mathbb{Q}$.

Of particular interest are the *locally connected metric spaces*, i.e., spaces X for which for every $x \in X$ there exists $r_x > 0$ such that $B(x, r_x)$ is connected.

6.57 Proposition. *Let X be metric space. The following claims are equivalent.*

(i) *Each connected component is open.*
(ii) *X is locally connected.*

Proof. Each point in X has a connected open neighborhood by (i), hence (ii) holds. Let C be a connected component of X, let $x \in C$ and, by assumption, let $B(x, r_x)$ be a connected ball centered at x. As $B(x, r_x)$ is connected, trivially $B(x, r_x) \subset C$, i.e., C is open.   □

6.58 Corollary. *Every convex set of $\mathbb{R}^n$ is connected.*

Proof. In fact every convex set $K \subset \mathbb{R}^n$ is the union of all segments joining a fixed point $x_0 \in K$ to points $x \in K$. Then Exercise 6.53 applies.   □

The class of all connected sets of a metric space X is a topological invariant. This follows at once from the following.

6.59 Theorem. *Let $f : X \to Y$ be a continuous map between metric spaces. If X is connected, then $f(X) \subset Y$ is connected.*

Proof. Assume by contradiction that $f(X)$ is not connected. Then there exist nonempty open sets $C, D \subset Y$ such that $C \cap D \cap f(X) = \emptyset$, $(C \cup D) \cap f(X) = f(X)$. Since f is continuous, $A := f^{-1}(C)$, $B := f^{-1}(D)$ are nonempty open sets in X, such that $A \cap B = \emptyset$ and $A \cup B = X$. A contradiction, since X is connected.   □

Since the intervals are the only connected subsets of $\mathbb{R}$, we again find the intermediate value theorem of [GM1] and, more generally,

6.60 Corollary. *Let $f : X \to \mathbb{R}$ be a continuous function defined on a connected metric space. Then f takes all values between any two that it assumes.*

c. Segment-connected sets in $\mathbb{R}^n$

In $\mathbb{R}^n$ we can introduce a more restrictive notion of connectedness that in some respect is more natural. If $x, y \in \mathbb{R}^n$, a *polyline* joining x to y is the union of finitely many segments of the type

$$[x, x_1], [x_1, x_2], \ldots, [x_{N-1}, y]$$

where $x_i \in \mathbb{R}^n$ and $[x_i, x_{i+1}]$ denotes the *segment* joining x_i with x_{i+1}. It is easy to check that a polyline joining x to y can be seen as the image or *trajectory* of a piecewise linear function $\gamma : [0, 1] \to \mathbb{R}^n$. Notice that piecewise linear functions are Lipschitz continuous.

6.61 Definition. *We say that $A \subset \mathbb{R}^n$ is* segment-connected *if each pair of its points can be joined by a polyline that lies entirely in A.*

If $A[x]$ denotes the set of all points that can be joined to x by a polyline in A, we see that A is segment-connected *if and only if $A = A[x]$.* Moreover we have the following.

6.62 Proposition. *Any segment-connected $A \subset \mathbb{R}^n$ is connected.*

Of course, not every connected set is segment-connected, indeed a circle in $\mathbb{R}^2$ is connected but not segment-connected. However, we have the following.

6.63 Theorem. *Let A be an nonempty open set of $\mathbb{R}^n$. Then A is connected if and only if A is segment-connected.*

Proof. Let $x_0 \in A$, let $B := A[x]$ be the set of all points that can be connected with x_0 by a polyline and let $C := A \setminus A[x]$. We now prove that both B and C are open. Since A is connected, we infer $A = A[x]$ hence, A is segment-connected.

Let $x \in B$. Since A is open, there exists $B(x, r) \subset A$. Since x is connected with x_0 by a polyline, adding a further segment we can connect each point of $B(x, r)$ with x_0 by a polyline. Therefore $B(x, r) \subset B$ if $x \in B$, i.e., B is open. Similarly, if $x \in C$, let $B(x, r) \subset A$. No points in $B(x, r)$ can be connected with x_0 by a polyline since on the contrary adding a further segment, we can connect x with x_0. So $B(x, r) \subset C$ if $x \in C$, i.e., C is open. $\square$

d. Path-connectedness

Another notion of connection that makes sense in a topological space is *joining by paths*.

Let X be a metric space. A *path* or a *curve* in X joining x with y is a continuous function $f : [0, 1] \to X$ with $f(0) = x$ and $f(1) = y$. The image of the path is called the *trajectory* of the path.

6.64 Definition. *A metric space X is said* path-connected *if any two points in X can be joined by a path.*

Evidently $\mathbb{R}^n$ is path-connected. We have, as in Theorem 6.63, the following.

6.65 Proposition. *Any path-connected metric space X is connected. The converse is however false in general.*

6.66 ¶. Consider the set $A \subset \mathbb{R}^2$, $A = G \cup I$ where G is the graph of $f(x) := \sin 1/x$, $0 < x < 1$, and $I = \{0\} \times [-1, 1]$. Show that A is connected but not path-connected.

Similarly to connected sets, if $\{A_\alpha\} \subset X$ are path-connected with $\cap_\alpha A_\alpha \neq \emptyset$, then $A := \cup_\alpha A_\alpha$ is path-connected. Because of this, one can define the *path-connected component of X containing a given $x_0 \in X$* as the maximal subset of X containing x_0 that is path-connected. However, examples show that the path-connected components are not closed, in general. But we have the following.

6.67 Proposition. *Let X be metric space. The following claims are equivalent.*

(i) *Each path-connected component is open (hence closed).*
(ii) *Each point of x has a path-connected open neighborhood.*

Proof. (ii) follows trivially from (i). Let C be a path-connected component of X, let $x \in C$ and by assumption let $B(x, r_x)$ be a path-connected ball centered at x. Then trivially $B(x, r_x) \subset C$, i.e., C is open. Moreover C is also closed since $X \setminus C$ is the union of the other path-connected components that are open sets, as we have proved.   □

6.68 Corollary. *An open set A of $\mathbb{R}^n$ is connected if and only if it is path-connected.*

Proof. Suppose that A is connected and let $U \subset A$ be a nonempty open set. Each point $x \in U$ then has a ball $B(x, r) \subset U$ that is path-connected. By Proposition 6.57 any path-connected component C of A is open and closed in A. Since A is connected, $C = A$.   □

6.3.2 Some applications

Topological invariants can be used to prove that two given spaces are *not* homeomorphic.

6.69 Proposition. $\mathbb{R}$ *and* $\mathbb{R}^n$, $n > 1$, *are not homeomorphic.*

Proof. Assume, by contradiction, that $h : \mathbb{R}^n \to \mathbb{R}$ is a homeomorphism, and let x_0 be a point of $\mathbb{R}^n$. Then clearly $\mathbb{R}^n \setminus \{x_0\}$ is connected, but $h(\mathbb{R}^n \setminus \{x_0\}) = \mathbb{R} \setminus \{h(x_0)\}$ is not connected, a contradiction.   □

Much more delicate is proving that

6.70 Theorem. $\mathbb{R}^n$ *and* $\mathbb{R}^m$, $n \neq m$, *are not homeomorphic.*

The idea of the proof remains the same. It will be sufficient to have a topological invariant that distinguishes between different $\mathbb{R}^n$.

Similarly, one shows that $[0, 1]$ and $[0, 1]^n$, $n > 1$, are not homeomorphic even if one-to-one correspondence exists.

6.71 ¶. Show that for any one-to-one mapping $h : [0, 1]^n \to [0, 1]$ neither h nor h^{-1} is continuous.

6.72 ¶. Show that the unit circle S^1 of $\mathbb{R}^2$ is not homeomorphic to $\mathbb{R}$. [*Hint:* Suppose $h : S^1 \to \mathbb{R}$ is a homeomorphism and let $x_0 \in S^1$. Then $S^1 \setminus \{x_0\}$ is connected, while $\mathbb{R} \setminus \{h(x_0)\}$ is not connected.]

6.73 Theorem. *In $\mathbb{R}$ each closed interval is homeomorphic to $[-1,1]$, each open interval is homeomorphic to $]-1,1[$ and each half-open interval is homeomorphic to $]-1,1]$. Moreover, no two of these three intervals are homeomorphic.*

Proof. The first part is trivial. To prove the second part, it suffices to argue as in Proposition 6.69 removing respectively, 2, 0 or 1 points to one of the three standard intervals, thus destroying connectedness. □

6.74 ¶. Show that the unit ball $S^n := \{x \in \mathbb{R}^{n+1} \,|\, |x| = 1\}$ in $\mathbb{R}^{n+1}$ is connected and that S^1 and S^n, $n > 1$, are not homeomorphic.

6.75 ¶. Let $A \subset \mathbb{R}^n$ and let $C \subset \mathbb{R}^n$ be a connected set containing points of both A and $\mathbb{R}^n \setminus A$. Show that C contains points of ∂A.

6.76 ¶. Show that the numbers of connected components and of path-connected components are topological invariants.

Theorem. *Let $f : X \to Y$ be a continuous function. The image of each connected (path-connected) component of X must lie in a connected component of Y. Moreover, if f is a homeomorphism, then f induces a one-to-one correspondence between connected (path-connected) components of X and Y.*

6.77 ¶. In set theory, the following theorem of Cantor–Bernstein holds, see Theorem 3.58 of [GM2].

Theorem. *If there exist injective maps $X \to Y$ and $Y \to X$, then there exists a one-to-one map between X and Y.*

This theorem becomes false if we require also continuity.

Theorem (Kuratowski). *There may exist continuous and one-to-one maps $f : X \to Y$ and $g : Y \to X$ between metric spaces and yet X and Y are not homeomorphic.*

[*Hint:* Let $X, Y \subset \mathbb{R}$ be given by

$$X =]0, 1[\cup \{2\} \cup]3, 4[\cup \{5\} \cup \ldots \cup]3n, 3n + 1[\cup \{3n + 2\} \cup \ldots$$
$$Y =]0, 1] \cup]3, 4[\cup \{5\} \cup \ldots \cup]3n, 3n + 1[\cup \{3n + 2\} \cup \ldots$$

By Exercise 6.76, X and Y are not homeomorphic, since the component $]0, 1]$ of Y is not homeomorphic to any component of X, but the maps $f : X \to Y$ and $g : Y \to X$ given by

$$f(x) := \begin{cases} x & \text{if } x \neq 2, \\ 1 & \text{if } x = 2, \end{cases} \quad \text{and} \quad g(x) := \begin{cases} x/2 & \text{if } x \in]0, 1[, \\ \dfrac{x-2}{2} & \text{if } x \in]3, 4[, \\ x - 3 & \text{otherwise} \end{cases}$$

are continuous and one-to-one.]

6.4 Exercises

6.78 ¶. Show that a continuous map between compact spaces needs not be an open map, i.e., needs not map open sets into open sets.

6.79 ¶. Show that an open set in $\mathbb{R}^n$ has at most countable many connected components. Show that this is no longer true for closed sets.

6.80 ¶. The *distance* between two subsets A and B of a metric space is defined by

$$d(A, B) := \inf_{\substack{a \in A \\ b \in B}} d(a, b).$$

Of course, the distance between two disjoint open sets or closed sets may be zero. Show that, if A is closed and B is compact, then $d(A, B) > 0$. [*Hint:* Suppose $\exists\, a_n, b_n$ such that $d(a_n, b_n) \to 0 \ldots$]

6.81 ¶. Let (X, d_X) and (Y, d_Y) be metric spaces, and let $(X \times Y, d_{X \times Y})$ be their Cartesian product with one of the equivalent distances in Exercise 5.14. Let $\pi : X \times Y$ be the projection map onto the first component, $\pi(x, y) := x$. π is an open map, see Exercise 5.131. Assuming Y compact, show that π is a closed map, i.e., maps closed sets into closed sets.

6.82 ¶. Let $f : X \to Y$ be a map between two metric spaces and suppose Y is compact. Show that f is continuous if and only if its graph

$$G_f := \Big\{ (x, y) \in X \times Y \,\Big|\, x \in X, \ y = f(x) \Big\}$$

is closed in $X \times Y$ endowed with one of the distances in Exercise 5.14. Show that, in general, the claim is false if Y is not compact.

6.83 ¶. Let K be a compact set in $\mathbb{R}^2$, and for every $x \in \mathbb{R}$ set $K_x := \{y \in \mathbb{R} \,|\, (x, y) \in K\}$ and $f(x) := \operatorname{diam} K_x$, $x \in \mathbb{R}$. Show that f is upper semicontinuous.

6.84 ¶. A map $f : X \to Y$ is said to be *proper* if the inverse image of any compact set $K \subset Y$ is a compact set in X. Show that f is a closed map if it is continuous and proper.

6.85 ¶. Show Theorem 6.35 using the finite covering property of X. [*Hint:* $\forall \epsilon > 0$ to every $x \in X$ we can associate a $\delta(x) > 0$ such that $d_Y(f(x), f(y)) < \epsilon/2$ whenever $y \in X$ and $d_X(x, y) < \delta(x)$. From the open covering $\{B(x, \delta(x))\}$ of X we can extract a finite subcovering $\{B(x_i, r_{x_i})\}_{i=1,\ldots,N}$ such that $X \subset B(x_1, \delta(x_1)) \cup \ldots \cup B(x_n, \delta(x_N))$. Set $\delta := \min\{\delta(x_1), \ldots, \delta(x_N)\}$.]

6.86 ¶. Let $f : E \to \mathbb{R}^m$ be uniformly continuous on a bounded set E. Show that $f(E)$ is bounded. [*Hint:* The closure of E is a compact set $\ldots$]

6.87 ¶. Show that
   (i) if $f : X \to \mathbb{R}^n$ and $g : X \to \mathbb{R}^n$ are uniformly continuous, then $f + g$ and λf, $\lambda \in \mathbb{R}$, are uniformly continuous,
   (ii) if $f : X \to Y$ is uniformly continuous in $A \subset X$ and $B \subset X$, then f is uniformly continuous in $A \cup B$.

6.88 ¶. Let $f, g : X \to \mathbb{R}$ be uniformly continuous. Give conditions such that fg is uniformly continuous.

6.89 ¶. Show that the composition of uniformly continuous functions is uniformly continuous.

6.90 ¶. Concerning maps $f : [0, +\infty[\to \mathbb{R}$, show the following.

  (i) If f is continuous and $f(x) \to \lambda \in \mathbb{R}$ as $x \to +\infty$, then f is uniformly continuous in $[0, +\infty[$.
 (ii) If f is continuous and has an asymptote, then f is uniformly continuous in $[0, +\infty[$.
(iii) If $f : [0, +\infty[\to \mathbb{R}$ is uniformly continuous in $[0, +\infty[$, then there exists constants A and B such that $|f(x)| \leq A|x| + B \ \forall x \geq 0$.
 (iv) If f is bounded, then there exists a *concave* function $\omega(t)$, $t \geq 0$, such that $|f(x) - f(x)| \leq \omega(|x - y|) \ \forall x, y \geq 0$.

6.91 ¶. Let $K \subset X$ be a compact subset of a metric space X and $x \in X \setminus K$. Show that there exists $y \in K$ such that $d(x, y) = d(x, K)$.

6.92 ¶. Let X be a metric compact space and $f : X \to X$ be an isometry. Show that $f(X) = X$. [*Hint:* $f^2, f^3, \ldots$, are isometries.]

6.93 ¶¶. Show that the set of points of $\mathbb{R}^2$ whose coordinates are not both rational, is connected.

6.94 ¶. Let B be a, at most, countable subset of $\mathbb{R}^n$, $n > 1$. Show that $C := \mathbb{R}^n \setminus B$ is segment-connected. [*Hint:* Assume that $0 \in C$ and show that each $x \in C$ can be connected with the origin by a path contained in C, thus C is path-connected. Now if the segment $[0, x]$ is contained in C we have reached the end of our proof, otherwise consider any segment R transversal to $[0, x]$ and show that there is $z \in R$ such that the polyline $[0, z] \cup [z, x]$ does not intersect B.]

6.95 ¶. Let $f : \mathbb{R}^n \to \mathbb{R}$, $n > 1$, be continuous. Show that there are at most two points $y \in \mathbb{R}$ for which $f^{-1}(y)$ is at most countable. [*Hint:* Take into account Exercise 6.94.]

7. Curves

The intuitive idea of a curve is that of the motion of a point in space. This idea summarizes the heritage of the ancient Greeks who used to think of curves as geometric figures characterized by specific geometric properties, as the conics, and the heritage of the XVIII century, when, with the development of mechanics, curves were thought of as the trace of a moving point.

7.1 Curves in $\mathbb{R}^n$

7.1.1 Curves and trajectories

From a mathematical point of view, it is convenient to think of a *curve* as of a continuous map γ from an interval I of $\mathbb{R}$ into $\mathbb{R}^n$, $\gamma \in C^0(I, \mathbb{R}^n)$. The *image* $\gamma(I)$ of a curve $\gamma \in C^0(I, \mathbb{R}^n)$, is called the *trace* or the *trajectory* of γ. We say that $\gamma : I \to \mathbb{R}^n$ is a *parametrization* of Γ if $\gamma(I) = \Gamma$, intuitively, a curve is a (continuous) way to travel on Γ. If $x, y \in \mathbb{R}^n$, a curve $\gamma \in C^0([a, b], \mathbb{R}^n)$ such that $\gamma(a) = x$, $\gamma(b) = y$, is often called a *path* joining x and y. A curve is what in kinematics is called the *motion law of a material point*, and its image or trajectory is the "line" in $\mathbb{R}^n$ spanned when the point moves.

If the basis in $\mathbb{R}^n$ is understood, —as we shall do from now on, fixing the standard basis of $\mathbb{R}^n$— a curve $\gamma \in C^0(I, \mathbb{R}^n)$ writes as an n-tuple of continuous real-valued functions of one variable, $\gamma(t) = (\gamma^1(t), \gamma^2(t), \ldots, \gamma^n(t))$, $\gamma^i : I \to \mathbb{R}$, $\gamma^i(t)$ being the component of $\gamma(t)$ $\forall t \in I$.

Let $k = 1, 2, \ldots,$ or ∞. We say that a curve $\gamma \in C^k(I, \mathbb{R}^n)$ if all the components of γ are real-valued functions respectively, of class $C^k(I, \mathbb{R})$, and that γ is a curve of class C^k if $\gamma \in C^k(I, \mathbb{R}^n)$ We also say that $\gamma : [a, b] \to \mathbb{R}^n$ is a *closed curve* of class C^k if γ is closed, $\gamma \in C^k(I, \mathbb{R}^n)$ and moreover, the derivatives of order up to k of each component of γ at a and b coincide,

$$D_j \gamma^i(a) = D_j \gamma^i(b) \qquad \forall i = 1, \ldots, n, \ \forall j = 1, \ldots, k.$$

If $\gamma : I \to \mathbb{R}^n$ is of class C^1, the vector

$$\gamma'(t_0) := ((\gamma^1)'(t_0), \ldots, (\gamma^n)'(t_0))$$

is the *derivative* or the *velocity vector* of γ at $t_0 \in I$, and its modulus $|\gamma'(t_0)|$ is the *(scalar) velocity* of γ at t_0. We also call $\gamma'(t_0)$ the *tangent vector* to γ at t_0. Finally, if $\gamma \in C^2(I, \mathbb{R}^n)$, the vector

$$\gamma''(t_0) := ((\gamma^1)''(t_0), \ldots, (\gamma^n)''(t_0))$$

is called the *acceleration vector* of γ at t_0.

7.1 Example (Segment). Let x and y be two distinct points in $\mathbb{R}^n$. The curve $s(t)$: $\mathbb{R} \to \mathbb{R}^n$ given by

$$s(t) = (1-t)x + ty = x + t(y-x), \qquad t \in [0,1],$$

is an affine map, called the *parametric equation* of the line through x in the direction of y. Thus its trajectory is the line $L \subset \mathbb{R}^n$ through $s(0) = x$ and $s(1) = y$ with constant vector velocity $s'(t) = y - x$. In kinematics, $s(t)$ is the position of a point traveling on the straight line $s(\mathbb{R})$ with constant velocity $|y - x|$ assuming $s(0) = x$ and $s(1) = y$.

Therefore the restriction $s_{|[0,1]}$ of s,

$$s(t) = (1-t)x + ty, \qquad 0 \le t \le 1,$$

describes the uniform motion of a point starting from x at time $t = 0$ and arriving in y at time $t = 1$ with constant speed $|y - x|$ and is called the *parametric equation of the segment joining y to x*.

7.2 Example (Uniform circular motion). The curve $\gamma : \mathbb{R} \to \mathbb{R}^2$ given by $\gamma(t) = (\cos t, \sin t)$ has as its trajectory the unit circle of $\mathbb{R}^2$ $\{(x, y) \,|\, x^2 + y^2 = 1\}$ with velocity one. In fact, $\gamma'(t) = (-\sin t, \cos t)$ thus $|\gamma'(t)| = 1$ $\forall t$. γ describes the uniform circular motion of a point on the unit circle that starts at time $t = 0$ at $(1, 0)$ and moves counterclockwise with angular velocity one, cf. [GM1]. Notice that $\gamma' \perp \gamma$ and $\gamma'' \perp \gamma'$ since

$$(\gamma'(t)|\gamma(t)) = \frac{1}{2}\frac{d|\gamma|^2}{dt}(t) = 0,$$

$$(\gamma''(t)|\gamma'(t)) = \frac{1}{2}\frac{d|\gamma'|^2}{dt}(t) = 0.$$

Finally, observe that the restriction of γ to $[0, 2\pi[$ runs on the unit circle once, since $\gamma_{|[0,2\pi[}$ is injective.

The uniform circular motion is better described looking at $\mathbb{R}^2$ as the Gauss plane of complex numbers, see [GM2]. Doing so, we substitute $\gamma(t)$ with $t \to e^{it}$, $t \in \mathbb{R}$, since we have $e^{it} = \cos t + i \sin t$.

7.3 Example (Graphs). Let $f \in C^0(I, \mathbb{R}^n)$ be a curve. The graph of f,

$$G_f := \Big\{(x, y) \in I \times \mathbb{R}^n \,|\, x \in I, \; y = f(x)\Big\} \subset \mathbb{R}^{n+1},$$

has the standard parametrization, still denoted by G_f, $G_f : I \to \mathbb{R}^{n+1}$, $G_f(t) := t \to (t, f(t))$, called the *graph-curve* of f. Observe that G_f is an injective map, in particular G_f is never a closed curve, G_f is of class C^k if f is of class C^k, $k = 1 \ldots, \infty$, and

$$G_f'(t) = (1, f'(t))$$

if f is of class C^1. A point that moves with the graph-curve law along the graph, moves with horizontal component of the velocity field normalized to $+1$. Notice that $|G_f'(t)| \ge 1$ $\forall t$.

Figure 7.1. A cylindrical helix.

7.4 Example (Cylindrical helix). If $\gamma(t) = (a\cos t, a\sin t, bt)$, $t \in \mathbb{R}$, then $\gamma'(t) = (-a\sin t, a\cos t, b)$, $t \in \mathbb{R}$. We see that the point $\gamma(t)$ moves with constant (scalar) speed along a *helix*, see Figure 7.1.

7.5 Example (Different parametrizations). Different curves may have the same trace, as we have seen for uniform circular motion. As another example, the curves $\gamma_1(t) := (t, 0)$, $\gamma_2(t) := (t^3, 0)$ and $\gamma_3(t) := (t(t^2-1), 0)$, $t \in \mathbb{R}$, are different parametrizations of the abscissa-axis of $\mathbb{R}^2$; of course, the three parametrizations give rise to different motions along the x-axis. Similarly, the curves $\sigma_1(t) = (t^3, t^2)$ and $\sigma_2(t) = (t, (t^2)^{1/3})$, $t \in \mathbb{R}$, are different parametrizations of (a) Figure 7.2. Notice that σ_1 is a C^∞-parametrization, while σ_2 is continuous but not of class C^1.

7.6 Example (Polar curves). Many curves are more conveniently described by a *polar parametrization*: instead of giving the evolution of Cartesian coordinates of $\gamma(t) := (x(t), y(t))$, we give two real functions $\theta(t)$ and $\rho(t)$ that describe respectively, the angle evolution of $\gamma(t)$ measured from the positive part to the abscissa axis and the distance of $\gamma(t)$ from the origin, so that in Cartesian coordinates

$$\gamma(t) = (\rho(t)\cos\theta(t), \rho(t)\sin\theta(t)).$$

If the independent variable t coincides with the angle θ, $\theta(t) = t$, we obtain a *polar curve*

$$\gamma(\theta) = (\rho(\theta)\cos\theta, \rho(\theta)\sin\theta).$$

In the literature there are many *classical curves* that have been studied for their relevance in many questions. Listing them would be incredibly long, but we shall illustrate some of them in Section 7.1.3.

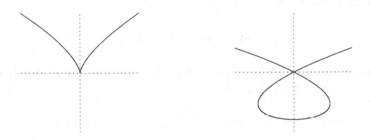

Figure 7.2. (a) $\gamma(t) = (t^3, t^2)$, (b) $\gamma(t) = (t^3 - 4t, t^2 - 4)$.

a. The calculus

Essentially the entire calculus, with the exception of the mean value theorem, can be carried on to curves.

7.7 Definition. *Let* $\gamma \in C^0([a,b];\mathbb{R}^n)$, $\gamma = (\gamma^1, \gamma^2, \ldots, \gamma^n)$. *The integral of* γ *on* $[a,b]$ *is the vector in* $\mathbb{R}^n$

$$\int_a^b \gamma(s)\,ds := \left(\int_a^b \gamma^1(s)\,ds, \int_a^b \gamma^2(s)\,ds, \ \ldots, \ \int_a^b \gamma^n(s)\,ds \right).$$

7.8 Proposition. *If* $\gamma \in C^0([a,b];\mathbb{R}^n)$, *then* $\left| \int_a^b \gamma(s)\,ds \right| \le \int_a^b |\gamma(s)|\,ds$.

Proof. Suppose that $\int_a^b \gamma(s)\,ds \ne 0$, otherwise the claim is trivial. For all $v = (v^1, v^2, \ldots, v^n) \in \mathbb{R}^n$ we have

$$\left(v \, \Big| \int_a^b \gamma(s)\,ds \right) = \sum_{i=1}^n v^i \int_a^b \gamma^i(s)\,ds = \int_a^b v^i \gamma^i(s)\,ds = \int_a^b (v|\gamma(s))\,ds;$$

using Cauchy's inequality we deduce

$$\left| \left(v \, \Big| \int_a^b \gamma(s)\,ds \right) \right| = \left| \int_a^b (v \mid \gamma(s))\,ds \right| \le \int_a^b |(v \mid \gamma(s))|\,ds$$

$$\le |v| \int_a^b |\gamma(s)|\,ds$$

for all $v \in \mathbb{R}^n$. Therefore it suffices to choose $v := \int_a^b \gamma(s)\,ds$ to find the desired result.
□

If $\gamma \in C^1([a,b],\mathbb{R}^n)$ and $n > 1$, the mean value theorem does not hold. Indeed, if $\gamma(t) = (\cos t, \sin t)$, $t \in [0, 2\pi]$, and $s \in [0, 2\pi]$ is such that

$$0 = \gamma(2\pi) - \gamma(0) = 2\pi\gamma'(s),$$

we reach a contradiction, since $|\gamma'(s)| = |(-\sin s, \cos s)| = 1$. However, the fundamental theorem of calculus, when applied to the components yields the following.

7.9 Theorem. *Let* $\gamma \in C^1([a,b];\mathbb{R}^n)$. *Then*

$$\gamma(b) - \gamma(a) = \int_a^b \gamma'(s)\,ds.$$

Finally, we notice that *Taylor's formula* extends to curves simply writing it for each component,

$$\gamma(t) = \gamma(t_0) + \gamma'(t_0)(t - t_0) + \frac{1}{2}\gamma''(t_0)(t - t_0)^2 + \cdots$$

$$+ \frac{1}{k!}\gamma^{(k)}(t_0)(t - t_0)^k + \frac{1}{k!}\int_{T_0}^t (t - s)^k \gamma^{(k+1)}(s)\,ds. \qquad (7.1)$$

Figure 7.3. Some trajectories: from the left, (a) simple curve, (b) simple closed curve, (c), (d), (f) curves that are not simple.

b. Self-intersections

Traces of curves may have *self-intersections*, i.e., in general, curves are not injective. In (b) Figure 7.2 the trace of the curve $\gamma(t) = (t^3 - 4t, t^2 - 4)$ $t \in \mathbb{R}$ self-intersects at the origin. One defines the *multiplicity* of a curve $\gamma \in C^0(I, \mathbb{R}^n)$ at $x \in \mathbb{R}^n$ as the number of t's such that $\gamma(t) = x$,

$$N(\gamma, I, x) := \#\Big\{ t \in I \,\Big|\, \gamma(t) = x \Big\}.$$

Of course, the trace of γ is the set of points with multiplicity at least 1. We shall distinguish two cases.

(i) $\gamma : I \to \mathbb{R}^n$ *is not closed,* i.e., $\gamma(a) \neq \gamma(b)$. In this case we say that γ is *simple* if γ is not injective i.e., all points of its trajectory have multiplicity 1. Notice that, if $I = [a, b]$, then γ is simple if and only if γ is an homeomorphism of $[a, b]$ onto $\gamma([a, b])$, $[a, b]$ being compact, see Corollary 6.19. In contrast, if I is not compact, I and $\gamma(I)$ in general are not homeomorphic. For instance let $I = [0, 2\pi[$ and $\gamma(t) := (\cos t, \sin t)$, $t \in I$ be the uniform circular motion. Then $\gamma(I)$ is the unit circle that is not homeomorphic to I, see Exercise 6.72.

(ii) γ *is a closed curve,* i.e., $I = [a, b]$ and $\gamma(a) = \gamma(b)$. In this case we say that γ is a *simple closed curve* if the restriction of γ to $[a, b[$ is injective, or, equivalently, if all points of the trajectory of γ, but $\gamma(a) = \gamma(b)$ have multiplicity 1.

A (closed) curve γ has self-intersections if it is not a (closed) simple curve.

7.10 ¶. Show that any closed curve $\gamma : [a, b] \to \mathbb{R}^n$ can be seen as a continuous map from the unit circle $S^1 \subset \mathbb{R}^n$. Furthermore show that its trace is homeomorphic to S^1 if γ is simple.

7.11 ¶. Study the curves $(x(t), y(t))$, $x(t) = 2t/(1 + t^2)$, $y(t) = (t^2 - 1)/(1 + t^2)$, $x(t) = t^2/(1 + t^6)$, $y(t) = t^3/(1 + t^6)$.

c. Equivalent parametrizations

Many properties of curves are independent of the choice of the parameter, that is, are invariant under homemorphic changes of the parameter. This is the case for the multiplicity function and, as we shall see later, of the length. For this reason, it is convenient to introduce the following definition

7.12 Definition. *Let I, J be intervals and let $\gamma \in C^0(I, \mathbb{R}^n)$ and $\delta \in C^0(J, \mathbb{R}^n)$. We say that δ is* equivalent *to γ if there is a continuous one-to-one map $h : J \to I$ such that*

$$\delta(s) = \gamma(h(s)) \qquad \forall \, s \in J.$$

In other words δ is equivalent to γ if δ reduces to γ modulo a continuous change of variable in the time axis. Since the inverse $h^{-1} : I \to J$ of a continuous one-to-one map is also continuous, see [GM1], we have that γ is equivalent to δ iff δ is equivalent to γ. Actually one sees that *the relation of equivalence among curves is an equivalence relation.*

Trivially, two equivalent curves have the same trace and the same multiplicity function; the converse is in general false.

7.13 Example. $\gamma(t) = (\cos t, \sin t)$, $t \in [0, 2\pi]$ and $\delta(t) = (\cos t, \sin t)$, $t \in [0, 4\pi]$ have the same trace but are not equivalent since their multiplicity functions are different.

However, we have the following.

7.14 Theorem. *Two simple curves with the same trace are equivalent.*

Proof. Assume for simplicity that the two curves $\gamma \in C^0(I, \mathbb{R}^n)$ and $\delta \in C^0(J, \mathbb{R}^n)$, I and J being intervals, are not closed. Set $h := \gamma^{-1} \circ \delta$ which clearly is a one-to-one and continuous map from J to I. h is then a homeomorphism, see [GM1], and clearly

$$\delta(t) = \gamma \circ \gamma^{-1} \circ \delta(t) = \delta(h(t)) \qquad \forall \, t \in J.$$

$\square$

The notion of equivalence between curves can be made more precise.

7.15 Definition. *Let $\gamma \in C^0(I, \mathbb{R}^n)$ and $\delta \in C^0(J, \mathbb{R}^n)$ be two equivalent curves, and let $h : J \to I$ be a homeomorphism such that $\delta(t) = \gamma(h(t))$ $\forall t \in J$. We say that γ and δ have* the same orientation *if h is monotone-increasing and have* opposite orientation *if h is monotone-decreasing.*

Since every homeomorphism between intervals is either strictly increasing or strictly decreasing, see [GM1], two equivalent curves either have the same orientation or have opposite orientations. In this way, the set of curves can be partitioned into *equivalence classes* and each class decomposes into two disjoint subclasses: equivalent curves with the same orientation and equivalent curves with opposite orientation.

7.1.2 Regular curves and tangent vectors

a. Regular curves

We say that a curve γ of class C^1 is *regular* if $\gamma'(t) \neq 0$ $\forall t$. It is also convenient to reconsider the notion of equivalence in the category of curves of class C^1.

7.16 Definition. *Let I, J be intervals. Two curves $\gamma \in C^1(I, \mathbb{R}^n)$, $\delta \in C^1(J, \mathbb{R}^n)$ of class C^1 are C^1-equivalent if there exists a one-to-one map $h : J \to I$ of class C^1 with $h'(t) \neq 0$ $\forall t \in J$ such that $\gamma(s) = \gamma(h(s))$ $\forall s \in J$.*

Clearly C^1-equivalent curves have the same trace. We can prove that being C^1-equivalent is an equivalence relation between regular curves; actually we shall prove the following result after Proposition 7.37.

7.17 Theorem. *Let γ and δ be two curves of class C^1, and suppose they are regular. Then γ and δ are C^1-equivalent if and only if they are C^0-equivalent.*

Since every function of class C^1 with $h' \neq 0$ $\forall t$ is either strictly increasing or strictly decreasing, since h' cannot change sign, any two C^1-equivalent curves either have the same orientation or have opposite orientation. In this way the set of C^1-curves can be partitioned into *equivalence classes* and each class decomposes into two disjoint subclasses: C^1-equivalent curves with the same orientation and C^1-equivalent curves with opposite orientation.

b. Tangent vectors

Let $\gamma : I \to \mathbb{R}^n$ be a simple, regular curve of class C^1 and let $\Gamma := \gamma(I)$ be its trace. If $x \in \Gamma$, there exists a unique $t \in I$ such that $\gamma(t) = x$.

7.18 Definition. *The space of tangent vectors to the trace Γ at $x \in \Gamma$ is defined as the space of all multiples of $\gamma'(t)$,*

$$\operatorname{Tan}_x \Gamma := \operatorname{Span}\left\{\gamma'(t)\right\} \qquad \gamma(t) = x.$$

The unit tangent vector to γ at $x := \gamma(t)$ is defined by

$$n_\gamma(x) := \frac{\gamma'(t)}{|\gamma'(t)|}, \qquad \gamma(t) = x.$$

Notice that the previous definition makes sense since one proves that $\operatorname{Span}\{\gamma'(t)\}$ where $\gamma(t) = x$, depends only on the trace of γ and on x. In fact, if $\gamma : I \to \mathbb{R}^n$ and $\delta : J \to \mathbb{R}^n$ are two curves with the same trace Γ, then Theorems 7.14 and 7.17 yield that γ and δ are C^1-equivalent, i.e., there exists $h : J \to I$ one-to-one and of class C^1 with $h'(s) \neq 0$ $\forall s \in J$ such that $\delta(s) = \gamma(h(s))$ $\forall s \in I$. Differentiating, we get

$$\delta'(s) = h'(s)\delta'(h(s)),$$

that is, $\delta'(s)$ and $\gamma'(t)$ are multiples of each other when $\delta(s) = \gamma(t) =: x$. Moreover,

$$n_\delta(x) = \frac{\delta'(s)}{|\delta'(s)|} = \operatorname{sgn}(h'(s)) \frac{\gamma'(h(s))}{|\gamma'(h(s))|} = \operatorname{sgn}(h'(s))n_\gamma(x),$$

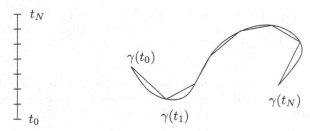

Figure 7.4. A polygonal line inscribed on a curve.

that is, two C^1-equivalent curves with the same orientation have the same unit tangent vector, and two C^1-equivalent curves with opposite orientation have the opposite unit tangent vector.

Remaining in a classic context, it is convenient to also introduce the families of *piecewise-C^1 curves* and *piecewise regular curves*.

7.19 Definition. *A curve* $\gamma : [a, b] \to \mathbb{R}^n$ *is said to be piecewise-C^1 (respectively, regular) if* $\gamma \in C^0(I, \mathbb{R}^n)$ *and there exist finitely many points* $a = t_0 < t_1 < \cdots < t_N = b$ *such that the restrictions* $\gamma_{|[t_i, t_{i-1}]}$ *are of class* C^1 *(respectively, regular) for all* $i = 1, \dots, N$.

We emphasize that in Definition 7.19 γ is required to be continuous everywhere in $[a, b]$, while derivatives are required to exist everywhere except at finite many points where only left- and right-derivatives exist. Notice also that piecewise-C^1 curves are Lipschitz continuous.

7.20 ¶. Let $\gamma : [a, b] \to \mathbb{R}^n$ be a piecewise regular curve. Show that every point in $\gamma([a, b])$ has finite multiplicity. Show a piecewise regular curve that has infinitely many points of multiplicity 2.

7.21 ¶. Show that

$$\gamma(b) - \gamma(a) = \int_a^b \gamma'(s) \, ds$$

if $\gamma : [a, b] \to \mathbb{R}^n$ is piecewise C^1.

c. Length of a curve

Recall that a partition σ of $[a, b]$ is a choice af finitely many points $t_0, \dots, t_N$ with $a = t_0 < t_1 < \cdots < t_N = b$. Denote by $\mathcal{S}$ the family of partitions of $[a, b]$. For each partition $\sigma = \{t_0, t_1, \dots, t_N\} \in \mathcal{S}$ one computes the length of the polygonal line $P(\sigma)$ joining the points $\gamma(t_0), \gamma(t_1), \dots, \gamma(t_N)$ in the listed order, Figure 7.4,

$$P(\sigma) := \sum_{i=1}^{N} |\gamma(t_i) - \gamma(t_{i-1})|.$$

Figure 7.5. The graph of $f(x) = x\sin(1/x)$, $x \in [0,1]$, is not rectifiable.

7.22 Definition. *Let $\gamma \in C^0([a,b]; \mathbb{R}^n)$. The length of γ is defined as*

$$L(\gamma) := \sup\Big\{ P(\sigma) \,\Big|\, \sigma \in \mathcal{S} \Big\}$$

and we say that γ is rectifiable *or γ has* finite total variation *if $L(\gamma) < +\infty$.*

In other words the length of a curve is the supremum of the lengths of all inscribed polygonals. The following is easily seen.

7.23 Proposition. *If γ and δ are equivalent, then $L(\gamma) = L(\delta)$. In particular γ and δ are either both rectifiable or not, and the length of a simple curve depends only on its trace.*

7.24 ¶. Prove Proposition 7.23.

7.25 ¶. Let $\gamma : [a,b] \to \mathbb{R}^n$ be a curve and let $a < c < b$. Show that $L(\gamma) = L(\gamma_{|[a,c]}) + L(\gamma_{[c,b]})$.

7.26 ¶. Show that if $\gamma(t) = (\cos t, \sin t)$, $t \in [0, 2\pi]$, we have $L(\gamma) = 2\pi$, while if $\gamma(t) = (\cos t, \sin t)$, $t \in [0, 4\pi]$, we have $L(\gamma) = 4\pi$.

7.27 Example. Curves $\gamma \in C^0([a,b]; \mathbb{R}^n)$ need not be rectifiable, i.e., of finite length. Indeed the curve graph of f, $\gamma(x) = (x, f(x))$ where

Figure 7.6. A closed curve that is not rectifiable.

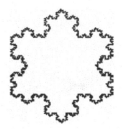

Figure 7.7. An approximation of the von Koch curve.

$$f(x) := \begin{cases} x \sin(1/x) & \text{if } x \in]0,1], \\ 0 & \text{if } x = 0 \end{cases}$$

has infinite length, see Figure 7.5. Indeed, if

$$x_n := \frac{1}{n\pi + \pi/2}, \qquad n \in \mathbb{N},$$

the length of $\gamma_{|[x_{n-1},x_n]}$ is larger than $x_n |\sin 1/x_n| = x_n$, hence for any n

$$L(\gamma) \geq L(\gamma_{|[x_n,1]}) \geq \sum_{k=1}^{n-1} x_k = \sum_{k=1}^{n-1} \frac{1}{k\pi + \pi/2},$$

i.e., $L(\gamma) = \infty$. Notice that γ belongs to $C^0([0,1], \mathbb{R}^n) \cap C^1(]0,1], \mathbb{R}^n)$, but γ' is not bounded in a neighborhood of 0.

7.28 Example (The von Koch curve). Clearly a bounded region of the plane may be enclosed by a curve of arbitrarily large length, think of the coasts of Great Britain or of Figure 7.6. A curve of infinite length enclosing a finite area is the *von Koch curve* that is constructed as follows. Start from an equilateral triangle, replace the middle third of each line segment with the two sides of an equilateral triangle whose third side is the middle third that we want to remove. Then one iterates the procedure indefinitely. One can show that the iterated curves converge uniformly to a curve, called the *von Koch curve*, which

(i)   is a continuous simple curve,
(ii) has infinite length and encloses a finite area,
(iii) is not differentiable at any point.

7.29 ¶. Show that each iteration in the construction of von Koch's curve increases its length by a factor 4/3, and, given any two points on the curve, the length of the arc between the two points is infinity. Finally, show that the surface enclosed by von Koch's curve is 8/5 of the surface of the initial triangle.

7.30 Example (The Peano curve). Continuous nonsimple curves may be quite pathological. Giuseppe Peano (1858–1932) showed in 1890 an example of a continuous curve $\gamma : [0,1] \to [0,1] \times [0,1]$ whose trace is the *entire* square: any such curve is called a *Peano curve*. Following David Hilbert (1862–1943), one such curve may be constructed as follows. Consider the sequence of continuous curves $\gamma_i : [0,1] \to \mathbb{R}^2$ as in Figure 7.8. The curve at step i is obtained by modifying the curve at step $(i-1)$ in an interval of width 2^{-i} and in a corresponding square of side 2^{-i} on the target. The sequence of these curves therefore converges uniformly to a continuous curve, whose trace is the

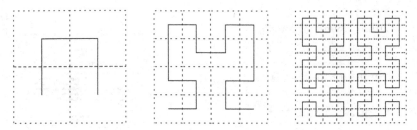

Figure 7.8. Construction of a Peano curve according to Hilbert.

entire square. Of course, γ is not injective, otherwise we would conclude that $[0,1]$ is homeomorphic to the square $[0,1]^2$, compare Proposition 6.69.

Another way of constructing a Peano curve, closer to the original proof of Peano who used ternary representations of reals, is the following. Represent each $x \in [0,1]$ in its dyadic expression, $x = \sum_{0=1}^{\infty} b_i/2^i$, $b_i \in \{0,1\}$, choosing not to have representations ending with period 1. If $x = \sum_{i=0}^{\infty} b_i/2^i \in [0,1]$, set

$$\gamma(x) := \Big(\sum_{i=0}^{\infty} \frac{b_{2i+1}}{2^{i+1}}, \sum_{i=0}^{\infty} \frac{b_{2i}}{2^i} \Big).$$

Using the fact that the alignment "changes" by a small quantity if x varies in a sufficiently small interval, we easily infer that γ is continuous. On the other hand, γ is trivially surjective.

No pathological behavior occurs for curves of class C^1. In particular, there is a formula for computing their length.

7.31 Theorem. *Let $\gamma \in C^1([a,b]; \mathbb{R}^n)$. Then γ is rectifiable and*

$$L(\gamma) = \int_a^b |\gamma'(s)|\, ds.$$

Proof. Let $\sigma \in \mathcal{S}$ be a partition of $[a,b]$, $P(\sigma)$ the length of the polygonal line corresponding to σ. The fundamental theorem of calculus yields

$$\gamma(t_i) - \gamma(t_{i-1}) = \int_{t_{i-1}}^{t_i} \gamma'(s)\, ds,$$

hence

$$|\gamma(t_i) - \gamma(t_{i-1})| \le \int_{t_{i-1}}^{t_i} |\gamma'(s)|\, ds.$$

Summing over i, we conclude

$$P(\sigma) = \sum_{i=1}^{N} |\gamma(t_i) - \gamma(t_{i-1})| \le \int_a^b |\gamma'(s)|\, dx$$

i.e., $L(\gamma) = \sup_\sigma P(\sigma) \le \int_a^b |\gamma'(s)|\, ds < \infty$, for σ arbitrary. This shows that γ is rectifiable.

It remains to show that

$$\int_a^b |\gamma'(s)|\, dx \le L(\gamma) \tag{7.2}$$

or, equivalently, for any $\epsilon > 0$, there is a partition σ_ϵ such that

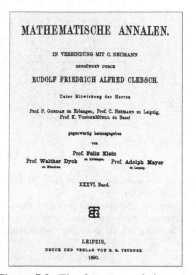

Figure 7.9. The first page of the paper of Giuseppe Peano (1858–1932) appeared in *Matematische Annalen*.

$$\int_a^b |\gamma'(s)|\, ds \leq P(\sigma_\epsilon) + \epsilon.$$

We observe that for every $s \in [t_{i-1}, t_i]$ we have

$$\gamma(t_i) - \gamma(t_{i-1}) = \int_{t_{i-1}}^{t_i} \gamma'(t)\, dt = \int_{t_{i-1}}^{t_i} (\gamma'(t) - \gamma'(s))\, dt + \gamma'(s)(t_i - t_{i-1}),$$

consequently

$$|\gamma'(s)| \leq \frac{1}{t_i - t_{i-1}} |\gamma(t_i) - \gamma(t_{i-1})| + \epsilon \qquad (7.3)$$

provided we choose the partition $\sigma_\epsilon := (t_0, t_1, \ldots, t_N)$ in such a way that

$$|\gamma'(t) - \gamma'(s)| \leq \epsilon \qquad \text{if } s, t \in [t_{i-1}, t_i]$$

(such a choice is possible since $\gamma' : [a, b] \to \mathbb{R}^n$ is uniformly continuous in $[a, b]$ by the Heine–Cantor theorem, Theorem 6.35). The conclusion then follows integrating with respect to s on $[t_{i-1}, t_i]$ and summing over i. $\qquad\square$

Of course Theorem 7.31 also holds for piecewise-C^1 curves: if $\gamma \in C^0[a, b]$, $a = t_0 < t_1 < \cdots < t_N = b$ and $\gamma \in C^1([t_{i-1}, t_i]; \mathbb{R}^n)$ $\forall\, i = 1, \ldots N$, then

$$L(\gamma) = \sum_{i=1}^N \int_{t_{i-1}}^{t_i} |\gamma'(t)|\, dt = \int_a^b |\gamma'(t)|\, dt < +\infty.$$

7.32 Lipschitz curves. Lipschitz curves, i.e., curves $\gamma : [a, b] \to \mathbb{R}^n$ for which there is $L > 0$ such that

$$|\gamma(t) - \gamma(s)| \leq L|t - s| \qquad \forall\, t, s \in [a, b],$$

are rectifiable. In fact, for every partition γ, with $a = t_0 < t_1 < \ldots < t_N = b$ we have

$$P(\sigma) = \sum_{i=1}^{N} |\gamma(t_{i-1}) - \gamma(t_i)| \leq L(b - a).$$

Quite a bit more complicated is the problem of finding an explicit formula for the length of a Lipschitz curve or, more generally, of a rectifiable curve. This was solved with the contributions of Henri Lebesgue (1875–1941), Giuseppe Vitali (1875–1932), Tibor Radó (1895–1965), Hans Rademacher (1892–1969) and Leonida Tonelli (1885–1946) using several results of a more refined theory of integration, known as Lebesgue integration theory.

7.33 ¶ The length formula holds for primitives. Let $\gamma : [a, b] \to \mathbb{R}^n$ be a curve. Suppose there exists a Riemann integrable function $\psi : [a, b] \to \mathbb{R}^n$ such that

$$\gamma(t) = \gamma(a) + \int_a^t \psi(s)\, ds \qquad \forall t \in [a, b].$$

Show that γ is rectifiable and $L(\gamma) = \int_a^b |\psi(t)|\, dt$.

7.34 ¶. Show that two regular curves that are C^1-equivalent have the same length. [*Hint:* Use the formula of integration by substitution.]

7.35 Example (Length of graphs). Let $f \in C^1([a, b], \mathbb{R})$. The graph of f, G_f : $[a, b] \to \mathbb{R}^2$, $G_f(t) = (t, f(t))$, is regular and $G'_f(t) = (1, f'(t))$. Thus the length of G_f is

$$L(G_f) = \int_a^b \sqrt{1 + |f'(x)|^2}\, dx.$$

7.36 Example (Length in polar coordinates). (i) Let $\rho(t) : [a, b] \to \mathbb{R}_+$, $\theta : [a, b] \to \mathbb{R}$ be continuous functions and let $\gamma(t) = (x(t), y(t))$ be the corresponding plane curve in polar coordinates, $\gamma(t) = (\rho(t) \cos \theta(t), \rho(t) \sin \theta(t))$. Since $|\gamma'|^2 = x'^2 + y'^2 = \rho'^2 + \rho^2 \theta'^2$, we infer

$$L(\gamma) = \int_a^b \sqrt{\rho'^2 + \rho^2 \theta'^2}\, dt,$$

in particular, for a *polar curve* $\gamma(t) = (\rho(t) \cos \theta(t), \rho(t) \sin \theta(t))$, we have

$$L(\gamma) = \int_a^b \sqrt{\rho'^2 + \rho^2}\, dt.$$

(ii) Let $\rho(t) : [a, b] \to \mathbb{R}_+$, $\theta : [a, b] \to \mathbb{R}$ and $f : [a, b] \to \mathbb{R}$ be continuous and let $\gamma(t) := (x(t), y(t), z(t))$, $t \in [a, b]$, be the curve in space defined by *cylindrical coordinates* $(\rho(t), \theta(t), f(t))$, i.e., $\gamma(t) := (\rho(t) \cos \theta(t), \rho(t) \sin \theta(t), f(\theta(t)))$. Since $|\gamma'|^2 = \rho'^2 + \rho^2 \theta'^2 + f'^2 \theta'^2$, we infer

$$L(\gamma) = \int_b^b \sqrt{\rho'^2 + \rho^2 \theta'^2 + f'^2 \theta'^2}\, dt.$$

(iii) For a curve in *spherical coordinates* $(\rho(t), \theta(t), \varphi(t))$, that is, for the curve $\gamma(t) = (x(t), y(t), z(t))$, $t \in [a, b]$ where

$$x(t) = \rho(t) \sin \varphi(t) \cos \theta(t), \ \ y(t) = \rho(t) \sin \varphi(t) \sin \theta(t), \ \ z(t) = \rho(t) \cos \varphi(t),$$

the length is

$$L(\gamma) = \int_a^b \sqrt{\rho'^2 + \rho^2 \varphi'^2 + \rho^2 \sin^2 \varphi\, \theta'^2}\, dt.$$

Figure 7.10. Arc length or curvilinear coordinate.

d. Arc length and C^1-equivalence

Let $\gamma \in C^1([a,b];\mathbb{R}^n)$ be a curve of class C^1 and regular, $\gamma'(t) \neq 0$ $\forall t$. The function $s_\gamma : [a,b] \to \mathbb{R}$ that for each $t \in [a,b]$ gives the length of $\gamma_{|[a,t]}$,

$$s_\gamma(t) := L(\gamma_{|[a,t]}) = \int_a^t |\gamma'(s)|\, ds,$$

is called the *arc length* or *curvilinear abscissa* of γ. We have

(i) $s_\gamma(t)$ is continuous, not decreasing and maps $[a,b]$ onto $[0,L]$, L being the length of γ. Moreover s_γ is differentiable at every point and

$$s'_\gamma(t) = |\gamma'(t)| \qquad \forall t \in [a,b],$$

(ii) since γ is regular, $\gamma'(t) \neq 0$ $\forall\, t \in [a,b]$, $s_\gamma(t)$ is in fact strictly increasing; consequently, its inverse $t_\gamma : [0,L] \to [a,b]$ is strictly increasing, too, and by the differentiation theorem of the inverse, see [GM1], t_γ is of class C^1 and

$$t'_\gamma(s) = \frac{1}{|\gamma'(t_\gamma(s))|} \qquad \forall\, s \in [0,L]. \qquad (7.4)$$

With the previous notation, the *reparametrization by arc length* of γ is defined as the curve $\delta_\gamma : [0,L] \to \mathbb{R}^n$ given by

$$\delta_\gamma(s) := \gamma(t_\gamma(s)) \qquad s \in [0,L].$$

Differentiating, we get

$$|\delta'_\gamma(s)| = \left| \frac{d\gamma(t_\gamma(s))}{ds} \right| = |\gamma'(t_\gamma(s))|\, |t'_\gamma(s)| = 1 \qquad \forall s.$$

As a consequence, the arc length reparametrization of a regular curve γ of length L is a curve $\delta : [0,L] \to \mathbb{R}^n$ that is C^1-equivalent to γ, has the same orientation of γ and for which $|\delta'(s)| = 1$ $\forall s \in [0,L]$. It is actually the unique reparametrization with these properties.

7.37 Proposition. *Let $\varphi : [a,b] \to \mathbb{R}^n$ and $\psi : [c,d] \to \mathbb{R}^n$ be two C^1-equivalent curves with the same orientation, $\psi(s) = \varphi(h(s))$ $\forall s \in [c,d]$, for some $h : [c,d] \to [a,b]$ of class C^1 with $h' > 0$, and length L. Then*

$$s_\psi(t) = s_\varphi(h(t)), \;\; \forall t \in [c,d], \quad and \quad t_\psi(s) = t_\varphi(s) \qquad \forall s \in [0,L],$$

hence $\delta_\varphi(s) = \delta_\psi(s)$ $\forall s \in [0,L]$.

Figure 7.11. Maria Agnesi (1718–1799) and a page from the *Editio princeps* of the works of Archimedes of Syracuse (287BC–212BC).

Proof. If $\psi(s) = \varphi(h(s))$ $\forall s \in [c, d]$, $h \in C^1$, $h' > 0$, then for any $t \in [c, d]$

$$s_\psi(t) = \int_c^t |\psi'(\tau)| \, d\tau = \int_c^t |\varphi'(h(\tau))| |h'(\tau)| \, d\tau$$

$$= \int_a^{h(t)} |\varphi'(\tau)| \, d\tau = s_\varphi(h(t)),$$

hence $\delta_\psi := \psi \circ t_\psi = \psi \circ h^{-1} \circ t_\varphi = \varphi \circ h \circ h^{-1} \circ t_\varphi = \varphi \circ t_\varphi = \delta_\varphi$. $\square$

Proof of Theorem 7.17. Assume that $\delta \in C^1([c, d], \mathbb{R}^n)$, $\gamma \in C^1([a, b], \mathbb{R}^n)$ γ regular, $h : [c, d] \to [a, b]$ is continuous and increasing and $\delta(s) = \gamma(h(s))$ $\forall s \in [c, d]$. Then the functions

$$\beta(s) := L(\delta_{|[c,s]}) = \int_c^s |\delta'(s)| \, ds, \quad s \in [c, d],$$

$$\alpha(t) := L(\gamma_{|[a,t]}) = \int_a^t |\gamma'(\tau)| \, d\tau \quad t \in [a, b],$$

are of class C^1 and $\beta(s) = L(\delta_{|[c,s]}) = L(\gamma_{|[a,h(s)]}) = \alpha(h(s))$ $\forall s \in [c, d]$, see Proposition 7.37. Since γ is regular, $\alpha(t)$ is invertible with inverse of class C^1, hence $h(s) = \alpha^{-1}(\beta(s))$ and h is of class C^1. $\square$

7.1.3 Some celebrated curves

Throughout the centuries, mathematicians, artists, scholars of natural sciences and layman have had an interest in plane curves, their variety of forms, and their occurrence in many natural phenomena. As a consequence

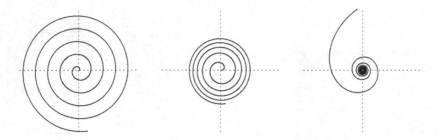

Figure 7.12. (a) Archimedes's spiral, (b) Fermat's spiral, (c) Hyperbolic spiral.

there is a large literature which attempts to classify plane curves according to their properties focusing on their constructive aspects or by simply providing catalogs. In this section we shall present some of these famous curves.

a. Spirals

Spirals are probably among the most known curves, the first and simplest being the *spiral of Archimedes*. This is the curve described by a point that moves with constant velocity along a half-line that rotates with constant angular velocity along its origin. If the origin of the half-line is the origin of a Cartesian plane, we have $\rho = vt, \theta = \omega t$, thus the polar form of Archimedes's spiral is

$$\rho = a\theta, \qquad a := \frac{v}{\omega}.$$

Other spirals are obtained assuming that the motion along the half-line is accelerated, for instance

$$\rho = a\theta^n.$$

All these spirals begin at the origin at $\theta = 0$ and move away from the origin as θ increases.

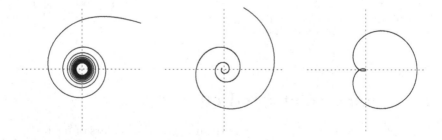

Figure 7.13. (a) Lituus, (b) Logarithmic spiral, (c) Cayley's sextic.

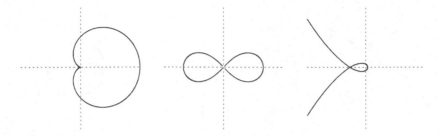

Figure 7.14. (a) Cardioid, (b) Lemniscate, (c) L'Hospital cubic.

ARCHIMEDEAN SPIRALS. These are the curves defined by

$$\rho^m = a\theta, \qquad m \in \mathbb{Q}.$$

Among them, see Figures 7.12 and 7.13, we mention

- *Archimedes's spiral* $\rho = a\theta$,
- *Fermat's spiral* $\rho^2 = a^2\theta$,
- the *hyperbolic or inverse spiral* $\rho = a/\theta$,
- the *lituus* $\rho^2 = a^2/\theta$,
- the *logarithmic* or *equiangular spiral* $\theta = \log_A \rho$, i.e., $\rho = A^\theta, A > 1$.
  It is the *spiralis mirabilis* of Johann Bernoulli (1667–1748). It (actually, its tangent at every point) forms a constant angle with any ray from the origin, and every ray intersects the logarithmic spiral in a sequence of points with distances in a geometric progression. It is probably the spiral that one finds most frequently in nature, expressing growth proportional to the organism, as in shells, pine cones, sunflowers or in galaxies.

SINUSOIDAL SPIRALS. A large variety of curves is described by the *sinusoidal spirals* $\rho^n = a^n \cos(n\theta)$, n rational. For instance,

- *Cayley's sextic* $\rho = 4a\cos^3(\theta/3)$, see Figure 7.13, that we can also write in an implicit form as the set of points (x, y) such that $4(x^2 + y^2 - ax)^3 = 27a^2(x^2 + y^2)^2$,

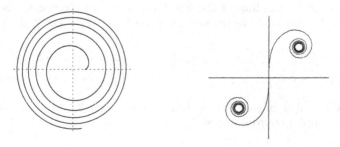

Figure 7.15. (a) Parabolic spiral $a = 1$, $b = 0.7$, (b) Euler's spiral.

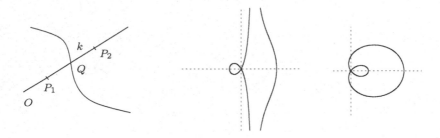

Figure 7.16. (a) The conchoid, (b) The conchoid of Nicomedes $a = 4$, $b = 2$, (c) Limacon of Pascal $a = b = 1$.

- *Cardioid* $\rho = 2a(1 + \cos\theta)$, see Figure 7.14, that we can write implicitly as the set of points (x, y) such that $(x^2 + y^2 - 2ax)^2 = 4a^2(x^2 + y^2)$,
- *Lemniscate of Bernoulli* $\rho^2 = a^2 \cos(2\theta)$, see Figure 7.14, equivalently as the set of points (x, y) such that $(x^2 + y^2)^2 = a^2(x^2 - y^2)$,
- *Cubic of de l'Hospital:* $\rho \cos^3(\theta/3) = a$, see Figure 7.14.

Other well-known spirals are, see Figure 7.15,

PARABOLIC SPIRALS. $(\rho - a)^2 = b^2\theta$,

EULER'S SPIRAL. $\gamma(t) = (x(t), y(t))$ where $x(t) = \pm \int_0^t \frac{\sin t}{\sqrt{t}}\, dt$, $y(t) = \pm \int_0^t \frac{\cos t}{\sqrt{t}}\, dt$, $0 \leq t < \infty$.

b. Conchoids

According to Diadochus Proclus (411–485), Nicomedes (280BC–210BC) studied the problem of the trisection of an angle by means of the conchoids.

Let O be a fixed point, and let ℓ be a line through O intersecting a trajectory C at a point Q. The locus of point P_1 and P_2 on ℓ such that $P_1Q = QP_2 = k = $ const is a *conchoid* of C with respect to O, see Figure 7.16.

CONCHOID OF NICOMEDES. It is the conchoid of a line with respect to a point not on the line. If the line ℓ is $x = b$ and the point 0 is $(0,0)$, then the conchoid has parametric equations $\gamma(t) = (x(t), y(t))$ where

$$\begin{cases} x(t) = b \pm \frac{ab}{\sqrt{t^2 + b^2}}, \\ y(t) = t \pm \frac{at}{\sqrt{t^2 + b^2}}, \end{cases} \quad \text{i.e.,} \quad \begin{cases} x = b + a\cos\theta, \\ y = (b + a\cos\theta)\tan\theta, \end{cases}$$

by the change of variable $t = b \tan\theta$, see Figure 7.16. We can write it also in polar coordinates as

$$\rho(\theta) = a + \frac{b}{\cos\theta},$$

or as the set of points (x, y) such that

$$(x^2 + y^2)(x - b)^2 = a^2 x^2.$$

LIMACON OF PASCAL. (Etienne Pascal (1588–1640), the father of Blaise
Pascal.) It is the conchoid of a circle of radius a with respect to a
point O on the circle. If 0 is the origin and $\rho = 2a \cos \theta$ is the polar
equation of the circle of center $(a, 0)$ through $(0, 0)$, the polar equation
of the limacon is $\rho = 2a \cos \theta + b$, see Figure 7.16. Choosing $b = 2a$
the limacon becomes a cardioid.

CONCHOID OF DÜRER. Let $Q = (q, 0)$ and $R = (0, r)$ be points such that
$q + r = b$. The locus of points P and P', on the straight line through
Q and R, with distance a from Q is Dürer's conchoid (Albrecht Dürer
(1471–1528)), see Figure 7.18. Its Cartesian equation may be found
by eliminating q and r from the equations

$$\begin{cases} b = q + r, \\ a^2 = (x - q)^2 + y^2, \\ y = -\frac{r}{q}x + r. \end{cases}$$

c. Cissoids

Given two curves C_1 and C_2 and a fixed point O, we let Q_1 and Q_2 be the
intersections of a line through 0 with C_1 and C_2, respectively. The locus of
points P on such lines such that $OP = OQ_2 - OQ_1 = Q_2 Q_1$ is the *cissoid*
of C_1 and C_2 with respect to O, see (a) Figure 7.17.

The cissoids of a circle and a tangent line with respect to a fixed point
of the circle that is not opposite to the point of tangency is the *cissoid of
Diocles* introduced by Diocles (240BC–180BC) in his attempts to doubling
the cube, see (b) Figure 7.17. If O is the origin, and the circle has equation $(x - a/2)^2 + y^2 = a^2/4$, the intersections points are $C = a(1, \tan \theta)$,
$B = a \cos \theta(\cos \theta, \sin \theta)$, hence Diocles's cissoid has the Cartesian equation
$y^2(a - x) = x^3$, or, equivalently, polar equation

$$\rho = a \sin \theta \tan \theta.$$

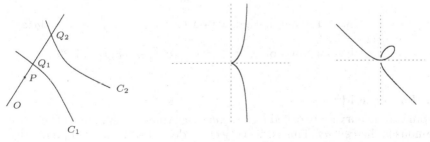

Figure 7.17. (a) The cissoid, (b) Cissoid of Diocles, (c) Folium of Descartes.

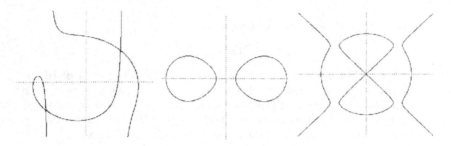

Figure 7.18. (a) Dürer's conchoid, (b) Oval of Cassini, (c) The devil curve.

d. Algebraic curves

These are loci of zeros of polynomials. The degree of the polynomial may be taken as measure of complexity: curves that are zeros of second order polynomials are well classified, see Example 3.69. We list here a few more algebraic curves, see Figure 7.19.

WITCH OF AGNESI. It has an equation $y(x^2+a^2) = a^3$ and it is the trace of the curve $\gamma(t) = (x(t), y(t))$ where $x(t) = at$, $y(t) = a/(1+t^2)$, $t \in \mathbb{R}$.

STROPHOID OF BARROW. It has an equation $x(x-a)^2 = y^2(2a-x)$ and it is the trace of the curve $\gamma(t) = (x(t), y(t))$ where $x(t) = 2a\cos^2 t$, $y(t) = a\tan t(1 - 2\cos^2 t)$, $t \in \mathbb{R}$.

EIGHT CURVE or LEMNISCATE OF GERONO. It has an equation $x^4 = a^2(x^2 - y^2)$ and it is the trace of the curve $\gamma(t) = (x(t), y(t))$ where $x(t) = a\cos t$, $y(t) = a\sin t\cos t$, $t \in \mathbb{R}$.

CURVES OF LISSAJOUS. They are the traces of curves $\gamma(t) = (x(t), y(t))$ where $x(t) = a\sin(\alpha t + d)$, $y(t) = b\sin t$, $t \in \mathbb{R}$ in which each coordinate moves as a simple harmonic motion. One shows that such curves are algebraic closed curves iff α is rational.

FOLIUM OF DESCARTES. It has an equation $x^3 + y^3 = 3axy$ and arises as trace of the curve $\gamma(t) = (x(t), y(t))$ where $x(t) = 3at/(1 + t^3)$, $y(t) = 3at^2/(1 + t^3)$, $t \in \mathbb{R}$, see Figure 7.17.

DEVIL'S CURVE. It has an equation $y^4 - x^4 + ay^2 + bx^2 = 0$, see Figure 7.18.

DOUBLE FOLIUM. It has an equation $(x^2 + y^2)^2 = 4axy^2$, see Figure 7.20.

TRIFOLIUM. It has an equation $(x^2 + y^2)(y^2 + x(x + a)) = 4axy^2$, see Figure 7.20.

OVALS OF CASSINI. They have equation $(x^2 + y^2 + a^2)^2 = b^4 + 4a^2x^2$, see Figure 7.18.

ASTROID. It has an equation $x^{2/3} + y^{2/3} = a^{2/3}$, see Figure 7.20.

e. The cycloid

Nonrational curves are called *transcendental*. Among them one of the most famous is the *cycloid*. This is the trajectory described by a fixed point (the tyre valve) of a circle (a tyre) rolling on a line, see Figure 7.21.

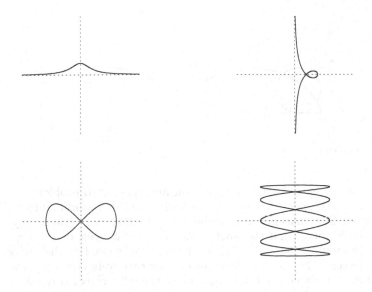

Figure 7.19. Some algebraic curves: from the top-left (a) the witch of Agnesi, (b) the strophoid of Barrow, (c) the lemniscate of Gerono, (d) the Lissajous curve for $n = 5$, $d = \pi/2$.

If the center of the circle is $C = (0, R)$, the radius R, $P = (0, 0)$ and we parametrize the movement with the angle θ that CP makes with the vertical through C, then $P = P(\theta)$, $C = C(\theta)$, the cycloid has period 2π, and we have

$$P(\theta) - C(\theta) = \begin{pmatrix} -R\sin\theta \\ R(1 - \cos\theta) \end{pmatrix}.$$

Since the circle rolls, $C(\theta)$ simply translates parallel to the axis of $R\theta$. We then conclude that the cycloid is the trace of the curve $\gamma : \mathbb{R} \to \mathbb{R}^2$ defined by

$$\gamma(t) = \begin{pmatrix} R(\theta - \sin\theta) \\ R(1 - \cos\theta) \end{pmatrix}.$$

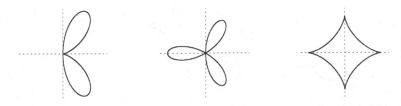

Figure 7.20. From the left: (a) the double folium , (b) the trifolium, (c) the astroid.

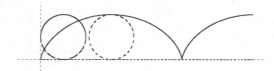

Figure 7.21. The cycloid.

The cycloid solves at least two important and celebrated problems.

As we know, the pendulus is not isochrone, but it is approximately isocronic for small oscillations, see Section 6.3.1 of [GM1]. Christiaan Huygens (1629–1695) found that the isochronal curve is the cycloid.

Johann Bernoulli (1667–1748) showed that the cycloid is the curve of quickest descent, that is, the curve connecting two points on a vertical plane on which a movable point descends under the influence of gravitation in the quickest possible way.

Other curves of the same nature as the cycloids are the *epicycloids* and the *hypercycloids*, see Figure 7.22. These are obtained from a circle that rolls around the inside or the outside of another circle (or another curve).

f. The catenary

Another celebrated transcendental curve is the *catenary*. It describes the form assumed by a perfect flexible inextensible chain of uniform density hanging from two supports, already discussed by Galileo Galilei (1564–1642). Answering a challenge of Jacob Bernoulli (1654–1705), it was proved

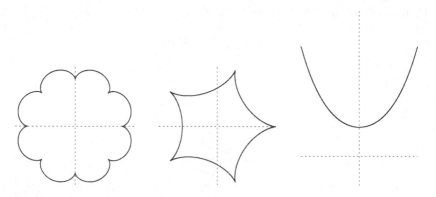

Figure 7.22. (a) The epicycloid $x(t) = 9\cos t - \cos(9t)$, $y(t) = 9\sin t - \sin(9t)$, (b) the ipocycloid $x(t) = 8\cos t + 2\cos(4t)$, $y(t) = 8\sin t - 2\sin(4t)$, (c) a catenary.

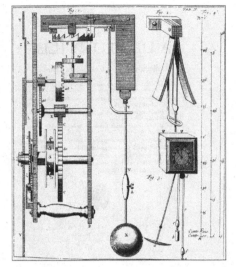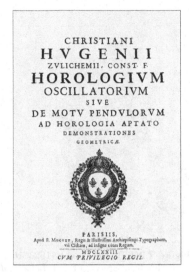

Figure 7.23. The pendulum clock from the *Horologium Oscillatorium* of Christiaan Huygens (1629–1695).

by Gottfried von Leibniz (1646–1716) and Christiaan Huygens (1629–1695) that the equation of the catenary hung at the same height at both sides is

$$y = \frac{a}{2}\left(e^{x/a} + e^{-x/a}\right) = a \cosh \frac{x}{a}.$$

7.2 Curves in Metric Spaces

Of course we may also consider curves in a general metric space X, as continuity is the only requirement. Let us start introducing the notion of *total variation*, a notion essentially due to Camille Jordan (1838–1922).

a. Functions of bounded variation and rectifiable curves

Let X be a metric space and $f : [a, b] \subset \mathbb{R} \to X$ be any map. Denote by $\mathcal{S}$ the family of finite partitions $\sigma = \{t_0, \ldots, t_N\}$ with $a = t_0 < t_1 < \cdots < t_N = b$ of the interval $[a, b]$ and, in correspondence to each partition σ, set $|\sigma| := \max_{i=1,\ldots,N}(|t_i - t_{i-1}|)$ and

$$V_\sigma(f) := \sum_{i=0}^{N-1} d(f(t_i), f(t_{i+1})),$$

that we have denoted by $P(\sigma)$ in the case of curves into $\mathbb{R}^n$.

Figure 7.24. Blaise Pascal (1623–1662) and the frontispiece of his *Lettres de Dettonville* about the *Roulettes*.

7.38 Definition. *The* total variation *of a map* $f : [a, b] \to X$ *is the number (eventually $+\infty$)*

$$V(f, [a, b]) := \sup_{\sigma \in \mathcal{S}} V_\sigma(f).$$

We say that f has bounded total variation *if $V(f, [a, b]) < \infty$. When the curve $f : [a, b] \to X$ is continuous, $V(f, [a, b])$ is called the* length *of f and curves with bounded total variation, that is with finite length, are called* rectifiable.

Either directly or repeating the arguments used in studying the length of curves into $\mathbb{R}^n$, it is easy to show the following.

7.39 Proposition. *We have*

(i) *if $[a, b] \subset [c, d]$, then $V(f, [a, b]) \leq V(f, [c, d])$,*

(ii) *$V(f, [a, b]) \geq d(f(a), f(b))$ and, if f is real-valued and increasing, then $V(f, [a, b]) = f(b) - f(a)$,*

(iii) *every Lipschitz-continuous function $f : [a, b] \to X$ has bounded total variation and $V(f, [a, b]) \leq \operatorname{Lip}(f)(b - a)$,*

(iv) *the total variation is a subadditive set-function, meaning*

$$V(f, [a, b]) \leq V(f, [a, c]) + V(f, [c, b]) \qquad if\ a < c < b;$$

moreover, if f is continuous at c, then $V(f, [a, b]) = V(f, [a, c]) + V(f, [c, b])$,

(v) *$V(f, [a, b]) = \lim_{|\sigma| \to 0^+} V_\sigma(f, [a, b])$.*

Figure 7.25. Frontispieces of two editions of 1532 and of 1606 of *Institutionum geometricorum* by Albrecht Dürer (1471–1528).

7.40 ¶. Let $f : [a,b] \to X \times X$ where X is a metric space. Show that f has bounded variation if and only if the two components of $f = (f_1, f_2)$, $f_{1,2} : [a,b] \to X$ have bounded variation.

We say that two curves $\varphi : [a,b] \to X$ and $\psi : [c,d] \to X$ into X are *equivalent* if there exists a homeomorphism $\sigma : [c,d] \to [a,b]$ such that $\psi(s) = \varphi(h(s)) \; \forall x \in [a,b]$. From the definitions we have the following.

7.41 Proposition. *Two equivalent curves have the same total variation.*

From (iv) and (v) of Proposition 7.39 we also have the following.

7.42 Proposition. *Let $\varphi : [a,b] \to X$ be a rectifiable (continuous) curve. Then the real-valued function $t \to V(\varphi, [a,t])$, $t \in [a,b]$ is continuous and increasing.*

7.43 ¶. Prove the claims in Propositions 7.39, 7.41 and 7.42.

b. Lipschitz and intrinsic reparametrizations

We saw that every regular Euclidean curve may be reparametrized with velocity one.

For curves in an arbitrary metric space we have

7.44 Theorem. *Let $\gamma : [a,b] \to X$ be a simple rectifiable curve on a metric space X of length L. Then there exists a homeomorphism $\sigma : [0,L] \to [a,b]$ such that $\gamma \circ \sigma : [0,L] \to X$ is Lipschitz continuous with Lipschitz constant one.*

Figure 7.26. The sets E_k of the middle third Cantor set.

We call that parametrization of the trace of γ the *intrinsic parametrization* of $\gamma([a, b])$.

Proof. Let $x \in [a, b]$ and $V(x) := V(\gamma, [a, x])$. We have $L = V(\gamma, [a, b])$ and, on account of Proposition 7.42, $V(x)$ is continuous and increasing. Since γ is simple, $V(x)$ is strictly increasing hence a homeomorphism between $[a, b]$ and $[0, L]$. Set $\sigma := V^{-1}$. We then infer for $0 \leq x < y \leq L$

$$d(\gamma(\sigma(y)), \gamma(\sigma(x))) \leq V(\gamma \circ \sigma, [x, y]) = V(\gamma, [\sigma(x), \sigma(y)])$$
$$= V(\sigma(y)) - V(\sigma(x)) = x - y,$$

i.e., $\varphi \circ \sigma$ is Lipschitz continuous with Lipschitz constant one. $\qquad\square$

7.2.1 Real functions with bounded variation

It is worth adding a few more comments about the class of real-valued functions $f : [a, b] \to \mathbb{R}$ with finite bounded variation, denoted by $BV([a, b])$.

7.45 Theorem. *We have*

(i) *$BV([a, b])$ is a linear space and $||f|| := |f(a)| + V(f, [a, b])$ is a norm on it,*
(ii) *$BV([a, b])$ contains the convex cone of increasing functions,*
(iii) *every $f \in BV([a, b])$ is the difference of two increasing functions.*

Proof. We leave to the reader the task of proving (i) and (ii) and we prove (iii). For $f \in BV([a, b])$ and $t \in [a, b]$ set

$$\varphi(t) := V(f, [a, t]), \qquad \psi(t) := \varphi(t) + f(t), \qquad t \in [a, b].$$

For $x, y \in [a, b]$, $x < y$, we have

$$\psi(y) - \psi(x) = [\varphi(y) - \varphi(x)] + [f(y) - f(x)];$$

now the subadditivity of the total variation yields

$$\varphi(y) - \varphi(x) = V(f, [x, y]) \geq |f(y) - f(x)|,$$

in particular

$$\psi(y) - \psi(x) \geq 0.$$

Therefore φ and ψ are both increasing with bounded total variation, and $f(t) = \psi(t) - \varphi(t)$ $\forall t$. $\qquad\square$

A surprising consequence is the following.

7.46 Corollary. *Every function in $BV([a, b])$ has left- and right-limits at every point of $[a, b]$.*

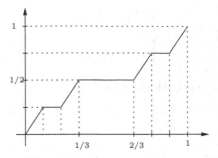

Figure 7.27. An approximate Cantor–Vitali function.

If we reread the proof of (iii) Theorem 7.45, on account of Proposition 7.42 we infer

7.47 Proposition. *Every continuous function $f : [a, b] \to \mathbb{R}$ with finite total variation is the difference of two continuous increasing functions.*

a. The Cantor–Vitali function

The Cantor ternary set is defined, see of [GM2], as $C = \cap_k E_k$ where $E_0 := [0, 1]$, E_1 is obtained from E_0 be removing the open middle third of E_0, and E_{k+1} by removing from each interval of E_k its open middle third.
Define for $k = 0, 1, \ldots$ and $j = 1, \ldots, 2^k$, the base points

$$b_{0,1} = 0 \quad b_{k+1,j} = \begin{cases} \frac{1}{3} b_{k,j} & \text{if } j = 1, \ldots, 2^k \\ \frac{2}{3} + \frac{1}{3} b_{k,j} & \text{if } j = 2^k + 1, \ldots, 2^{k+1}, \end{cases}$$

then the intervals that have been removed from E_{k-1} to get E_k at step k are

$$I_{k-1,j} := b_{k-1,j} + 3^{-k+1}]\frac{1}{3}, \frac{2}{3}[, \qquad j = 1, \ldots, 2^{k-1},$$

and the intervals whose union is E_k are

$$J_{k,j} := b_{k,j} + 3^{-k}[0, 1], \qquad j = 1, \ldots, 2^k.$$

Therefore

$$C = \bigcap_{k=0}^{\infty} \left(\bigcup_{j=1}^{2^k} J_{j,k} \right).$$

Strongly related to Cantor's set is the *Cantor–Vitali function* introduced by Giuseppe Vitali (1875–1932). To define it, we first consider the *approximate Cantor–Vitali functions* $V_k : [0, 1] \to \mathbb{R}$ defined inductively by

$$V_0(x) := x, \qquad V_{k+1}(x) := \begin{cases} \frac{1}{2} V_k(x/3) & \text{if } x \in [0, 1/3], \\ \frac{1}{2} & \text{if } x \in [1/3, 2/3], \\ \frac{1}{2} + \frac{1}{2} V_k(3^{-1}(x - 2/3)) & \text{if } x \in [2/3, 1], \end{cases}$$

see Figure 7.27. One easily checks that for $k = 0, 1, \ldots$

(i) We have $V_k(0) = 0$, $V_k(1) = 1$, $V_k(b_{j,k}) = \frac{j-1}{2^k}$, $V_k(b_{j,k} + 3^{-k}) = \frac{j}{2^k}$ and

$$V_k(x) = \frac{2j - 1}{2^{m+1}} \qquad \text{if } x \in I_{m,j}, \ m = 0, \ldots, k - 1, \ j = 1, \ldots, 2^m.$$

(ii) We have

$$V_k(x) = \left(\frac{3}{2}\right)^k \int_0^x \chi_{E_k}(t) \, dt$$

where χ_{E_k} is the characteristic function of the set E_k that we used to define the Cantor ternary set.

(iii) We have

$$|V_k(x) - V_k(y)| \le |x - y|^\alpha \qquad \forall x, y \in [0, 1],$$

where $\alpha = \log 2 / \log 3$, in particular the V_k's are equi-Hölder. In fact, by symmetry it suffices to prove the claim for $x, y \in [0, 3^{-k}]$ where V_k is linear with slope $(3/2)^k$. For $0 \le x \le y \le 3^{-k}$ we have

$$|V_k(x) - V_k(y)| = \left(\frac{3}{2}\right)^k |x - y| = \left(\frac{3}{2}\right)^k |x - y|^{1-\alpha} |x - y|$$

$$\le \left(\frac{3}{2}\right)^k \left(\frac{1}{3}\right)^{k(1-\alpha)} |x - y|^\alpha = |x - y|^\alpha$$

as $2 \, 3^{-\alpha} = 1$.

(iv) We have

$$|V_{k+1}(x) - V_k(x)| \le 2^{-k+1} \qquad \forall x \in [0, 1].$$

In particular (iv) implies that the sequence $\{V_k\}$ converges uniformly to a function $V(x)$, which is by (iii) Hölder-continuous with exponent $\alpha = \log 2 / \log 3$. The function V is called the *Cantor–Vitali function* and satisfies the following properties

∘ V is not decreasing, hence it has bounded total variation,
∘ in each interval of $[0, 1] \setminus E_k$, $V(x) = V_k(x)$ is constant, in particular V is differentiable outside the Cantor set with $V'(x) = 0 \ \forall x \in [0, 1] \setminus C$,
∘ $V([0, 1]) = [0, 1]$, and V maps $[0, 1] \setminus C$ into the denumerable set

$$D := \left\{ y \in \mathbb{R} \,\Big|\, y = \frac{j}{2^k}, \ j = 0, 1, \ldots, 2^k, \ k \in \mathbb{N} \right\},$$

hence V maps C onto $[0, 1] \setminus D$.

7.48 Homeomorphisms do not preserve fractal dimensions. The function

$$\varphi : [0,1] \to [0,2], \qquad \varphi(x) := x + V(x),$$

is continuous and strictly increasing, hence a homeomorphism between $[0,1]$ and $[0,2]$. In Theorem 8.109 we shall see that the algebraic dimension of $\mathbb{R}^n$ is a topological invariant, that is, $\mathbb{R}^n$ and $\mathbb{R}^m$ are homeomorphic if and only if $n = m$. This is not true in general for the fractal dimension, see Chapter 8 of [GM2]. In fact, φ maps the complement of Cantor's set in $[0,1]$ into the countable union of intervals of total measure 1, $\mathcal{H}^1(\varphi([0,1]) \setminus C) = 1$, hence $\mathcal{H}^1(\varphi(C)) = 1$ and $\dim_{\mathcal{H}}(\varphi(C)) = 1$, while $\dim_{\mathcal{H}}(C) = \log 2 / \log 3$.

7.49 ¶. Let $f : \mathbb{R}^n \to \mathbb{R}^n$ be a Lipschitz-continuous map with Lipschitz inverse. Show that f preserves the fractal dimension, $\dim_{\mathcal{H}}(f(A)) = \dim_{\mathcal{H}} A$. [*Hint:* Recall that $\mathcal{H}^k(f(A)) \leq \mathrm{Lip}(f) \, \mathcal{H}^k(A)$, see Section 8.2.4 of [GM2].]

7.3 Exercises

7.50 ¶. We invite the reader to study some of the curves described in this chapter, try to convince himself that the figures are quite reasonable, and compute the lengths of some of those curves and, when possible, the enclosed areas.

7.51 ¶. Compute the total variation of the following functions $f : [0,2] \to \mathbb{R}$

$$f(x) := x(x^2 - 1),$$

$$f(x) := 3\chi_{[0,1]}(x) + 2\chi_{[1,2]}(x), \qquad \text{where} \qquad \chi_A(x) := \begin{cases} 1 & \text{if } x \in A, \\ 0 & \text{if } x \notin A. \end{cases}$$

7.52 ¶. Let $g(x) = \sqrt{x}$, $x \in [0,1]$, and let $f : [0,1] \to \mathbb{R}$ be given by

$$f(x) = \begin{cases} \frac{1}{n^2} & \text{if } x \in [\frac{1}{n}, \frac{1}{n} + \frac{1}{n^2}], \\ 0 & \text{otherwise.} \end{cases}$$

Show that f, g and $g \circ f$ have bounded total variation.

7.53 ¶. Let $f, g \in BV([0,1])$. Show that $\min(f,g)$, $\max(f,g)$, $|f| \in BV([0,1])$.

7.54 ¶. Show that the Cantor middle third set C is compact and perfect, i.e., $\mathrm{int}(\overline{C}) = \emptyset$.

8. Some Topics from the Topology of $\mathbb{R}^n$

As we have already stated, *topology* is the study of the properties shared by a geometric figure and all its bi-continuous transformations, i.e., the study of *invariants by homeomorphisms*. Its origin dates back to the problem of Königsberg bridges and Euler's theorem about polyhedra, to Riemann's work on the geometric representation of functions, to Betti's work on the notion of multiconnectivity and, most of all, to the work of J. Henri Poincaré (1854–1912). Starting from his research on differential equations in mechanics, Poincaré introduced relevant topological notions and, in particular, the idea of associating to a geometric figure (using a rule that is common to all figures) an *algebraic* object, such as a group, that is a topological invariant for the figure and that one could compute. The *fundamental group* and *homology groups* are two important examples of algebraic objects introduced by Poincaré: this is the beginning of *combinatorial* or *algebraic topology*. With the development of what we call today *general topology* due to, among others, René-Louis Baire (1874–1932), Maurice Fréchet (1878–1973), Frigyes Riesz (1880–1956), Felix Hausdorff (1869–1942), Kazimierz Kuratowski (1896–1980), and the interaction between general and algebraic topology due to L. E. Brouwer (1881–1966), James Alexander (1888–1971), Solomon Lefschetz (1884–1972), Pavel Alexandroff (1896–1982), Pavel Urysohn (1898–1924), Heinz Hopf (1894–1971), L. Agranovich Lyusternik (1899–1981), Lev G. Schnirelmann (1905–1938), Harald Marston Morse (1892–1977), Eduard Čech (1893–1960), the study of topology in a wide sense is consolidated and in fact receives new incentives thanks to the work of Jean Leray (1906–1998), Élie Cartan (1869–1951), Georges de Rham (1903–1990). Clearly, even a short introduction to these topics would deviate us from our course; therefore we shall confine ourselves to illustrating some fundamental notions and basic results related to the topology of $\mathbb{R}^n$, to the notion of *dimension* and, most of all, to the existence of fixed points.

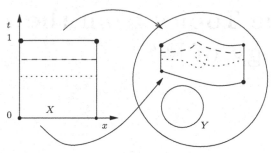

Figure 8.1. A homotopy.

8.1 Homotopy

In this section we shall briefly discuss the different flavors of the notion of *homotopy*. They correspond to the intuitive idea of continuous deformation of one object into another.

8.1.1 Homotopy of maps and sets

a. Homotopy of maps

In the following, the ambient spaces X, Y, Z will be metric spaces.

8.1 Definition. *Two continuous maps $f, g : X \to Y$ are called* homotopic *if there exists a continuous map $H : [0,1] \times X \to Y$ such that $H(0,x) = f(x)$, $H(1,x) = g(x) \ \forall \ x \in X$. In this case we say that H establishes or is* a homotopy *of f to g.*

It is easy to show that the homotopy relation $f \sim g$ on maps from X to Y is an *equivalence relation*, i.e., it is

(i) (REFLEXIVE) $f \sim f$.
(ii) (SYMMETRIC) $f \sim g$ iff $g \sim f$.
(iii) (TRANSITIVE) if $f \sim g$ and $g \sim h$, then $f \sim h$.

Therefore $C^0(X, Y)$ can be partitioned into classes of homotopic functions. It is worth noticing that, since

$$C^0([0,1], C^0(X,Y)) = C^0([0,1] \times X, Y), \tag{8.1}$$

we have the following.

8.2 Proposition. *f and $g \in C^0(X, Y)$ are homotopic if and only if they belong to the same path-connected component of $C^0(X, Y)$ endowed with uniform distance. The subsets of $C^0(X, Y)$ of homotopy equivalent maps are the path-connected components of the metric space $C^0(X, Y)$ with uniform distance.*

Figure 8.2. Frontispieces of the introduction to combinatorial topology by Kurt Reidemeister (1893–1971) and Pavel Alexandroff (1896–1982) in its Italian translation.

8.3 ¶. Let X, Y be metric spaces. Show the equality (8.1), which we understand as an isometry of metric spaces.

8.4 ¶. Let Y be a convex subset of a normed linear space. Then every continuous map $f : X \to Y$ from an arbitrary metric space X is homotopic to a constant. In particular, constant maps are homotopic to each other. [*Hint:* Fix $y_0 \in Y$ and consider the homotopy $H : [0,1] \times X \to Y$ given by $H(t,x) := ty_0 + (1-t)f(x)$.]

8.5 ¶. Let X be a convex set of a normed linear space. Then every continuous map $f : X \to Y$ into an arbitrary metric space is homotopic to a constant function. [*Hint:* Fix $x_0 \in X$ and consider the homotopy $H : [0,1] \times X \to Y$ given by $H(t,x) := f(tx_0 + (1-t)x)$.]

8.6 ¶. Two constant maps are homotopic iff their values can be connected by a path.

8.7 ¶. Let X be a linear normed space. Show that the homotopy classes of maps $f : X \to Y$ correspond to the path-connected components of Y.

According to Exercises 8.4, 8.5 and 8.6, all maps into $\mathbb{R}^n$ or defined on $\mathbb{R}^n$ are homotopic to constant maps. However, this is not always the case for maps from or into $S^n := \{x \mid ||x|| = 1\}$, the unit sphere of $\mathbb{R}^{n+1}$.

8.8 Proposition. *We have*

(i) *Let $f, g : X \to S^n$ be two continuous maps such that $f(x)$ and $g(x)$ are never antipodal, i.e., $g(x) \neq -f(x)$ $\forall x \in X$, then f and g are homotopic; in particular, if $f : X \to S^n$ is not onto, then f is homotopic to a constant.*

Figure 8.3. The figure suggests a homotopy of closed curves, that is a continuous family of closed paths, from a knotted loop to S^1. But, it can be proved that there is no family of homeomorphisms of the ambient space $\mathbb{R}^3$ that, starting from the identity, deforms the initial knotted loop into S^1.

(ii) *Let $B^{n+1} := \{x \in \mathbb{R}^{n+1} \mid |x| \leq 1\}$. A continuous map $f : S^n \to Y$ is homotopic to a constant if and only if f has a continuous extension $F : B^{n+1} \to Y$.*

Proof. (i) Since $f(x)$ and $g(x)$ are never antipodal, the segment $tg(x) + (1-t)f(x)$, $t \in [0,1]$, never goes through the origin; a homotopy of f to g is then

$$H(t,x) := \frac{tg(x) + (1-t)f(x)}{|tg(x) + (1-t)f(x)|}, \qquad (t,x) \in [0,1] \times X.$$

Notice that $y \to \frac{y}{|y|}$ is the radial projection from $\mathbb{R}^{n+1}$ onto the sphere S^n, hence $H(t,x)$ is the radial projection onto the sphere of the segment $tg(x) + (1-t)f(x)$, $t \in [0,1]$. The second part of the claim follows by choosing $y_0 \in S^n \setminus f(X)$ and $g(x) := -y_0$.

(ii) If $F : B^{n+1} \to Y$ is a continuous function such that $F(x) = f(x) \ \forall x \in S^n$, then the map $H(t,x) := F(tx)$, $(t,x) \in [0,1] \times S^n$, is continuous, hence a homotopy of $H(0,x) = F(0)$ to $H(1,x) = f(x)$. Conversely, if $H : [0,1] \times S^n \to Y$ is a homotopy of a constant map $g(x) = p \in Y$ to f, $H(0,x) = p$, $H(1,x) = f(x) \ \forall x \in X$, then the map $F : B^{n+1} \to Y$ defined by

$$F(x) := \begin{cases} H(|x|, x/|x|) & \text{if } x \neq 0, \\ p & \text{if } x = 0 \end{cases}$$

is a continuous extension of f to B^{n+1} with values into Y. □

b. Homotopy classes

Denote by $[X,Y]$ the set of homotopy classes of continuous maps $f : X \to Y$ and by $[f] \in [X,Y]$ the equivalence class of f. The following two propositions collect some elementary facts.

8.9 Proposition. *We have*

(i) (COMPOSITION) *Let $f, f' : X \to Y$, $g, g' : Y \to Z$ be continuous maps. If $f \sim f'$ and $g \sim g'$, then $g \circ f \sim g' \circ f'$.*

(ii) (RESTRICTION) *If $f, g : X \to Y$ are homotopic and $A \subset X$, then $f_{|A}$ is homotopic to $g_{|A}$ as maps from A to Y.*

(iii) (CARTESIAN PRODUCT) $f, g : X \to Y_1 \times Y_2$ *are homotopic if and only if $\pi_i \circ f$ and $\pi_i \circ g$ are homotopic (with values in Y_i) where $i = 1, 2$ and π_i, $i = 1, 2$ denote the projections on the factors.*

A trivial consequence of Proposition 8.9 is that the set $[X, Y]$ is a topological invariant of both X and Y. In a sense $[X, Y]$ gives the number of "different" ways that X can be mapped into Y, hence measures the "topological complexity" of Y relative to that of X.

Let $\varphi : X \to Y$ be a continuous map and let Z be a metric space. Then φ defines a *pull-back* map

$$\varphi^{\#} : [Y, Z] \to [X, Z]$$

defined by $\varphi^{\#}[f] := [f \circ \varphi]$, as Proposition 8.9 yields that the homotopy class of $f \circ \varphi$ depends on the homotopy class of f. Similarly φ induces a *push-forward* map

$$\varphi_{\#} : [Z, X] \to [Z, Y]$$

defined by $\varphi_{\#}[g] := [\varphi \circ g]$.

8.10 Proposition. *We have the following.*

(i) *Let $\varphi, \psi : X \to Y$ be continuous and homotopic, $\varphi \sim \psi$. Then $\varphi^{\#} = \psi^{\#}$ and $\varphi_{\#} = \psi_{\#}$.*

(ii) *Let $\varphi : X \to Y$ and $\eta : Y \to Z$ be continuous. Then*

$$(\eta \circ \varphi)^{\#} = \varphi^{\#} \circ \eta^{\#} \qquad and \qquad (\eta \circ \varphi)_{\#} = \eta_{\#} \circ \varphi_{\#}.$$

c. Homotopy equivalence of sets

8.11 Definition. *Two metric spaces X and Y are said* homotopy equivalent, *or are said to have the same* homotopy type, *if there exist two continuous maps $f : X \to Y$ and $g : Y \to X$ such that $g \circ f \sim \mathrm{Id}_X$ and $f \circ g \sim \mathrm{Id}_Y$.*

If $f : X \to Y$ and $g : Y \to X$ define a homotopy equivalence between X and Y, then for every space Z we infer from Proposition 8.10

$$g^{\#} \circ f^{\#} = \mathrm{Id}_{[Y,Z]}, \qquad f^{\#} \circ g^{\#} = \mathrm{Id}_{[X,Z]}.$$

Similarly

$$g_{\#} \circ f_{\#} = \mathrm{Id}_{[Z,X]}, \qquad f_{\#} \circ g_{\#} = \mathrm{Id}_{[Z,Y]};$$

hence $[Z, X]$ and $[Z, Y]$ are in a one-to-one correspondence.

8.12 Definition. *A space X is called* contractible *if it is homotopy equivalent to a space with only one point, equivalently, if the identity map $i : X \to X$ of X is homotopic to a constant map.*

By definition if X is contractible to $x_0 \in X$, then X is homotopic equivalent to $\{x_0\}$, hence $[Z, X]$ and $[X, Z]$ reduces to a point for any space Z.

Figure 8.4. $\mathbb{R}^n$ is contractible.

8.13 Example. $\mathbb{R}^n$ is contractible. In fact, $H(t,x) := (1-t)x$, $(t,x) \in [0,1] \times \mathbb{R}^n$, contracts $\mathbb{R}^n$ to the origin.

In general, describing the set $[X,Y]$ is a very difficult task even for the simplest case of the homotopy of spheres, $[S^k, S^n]$, $k, n \geq 1$. However, the following may be useful.

8.14 Definition. *Let X be a metric space. We say that $A \subset X$ is a* retract *of X if there exists a continuous map $\rho : X \to A$, called a* retraction, *such that $\rho(x) = x \ \forall x \in A$. Equivalently A is a retract of X if the identity map $\mathrm{Id}_A : A \to A$ extends to a continuous map $r : X \to A$.*
    We say that $A \subset X$ is a deformation retract *of X if A is a retract of X and the identity map $\mathrm{Id}_X \to X$ is homotopic to a retraction of X to A.*

Let $A \subset X$ be a deformation retract of X and denote by $i_A : A \to X$ the inclusion map. Since $\mathrm{Id}_X : X \to X$ is homotopic to the retraction map $r : X \to A$, we have

$$r \circ i_A = \mathrm{Id}_A, \qquad i_A \circ r = r \sim \mathrm{Id}_X,$$

hence A and X are homotopic equivalent. By the above, for every space Z we have $[A, Z] = [X, Z]$ and $[Z, A] = [Z, X]$ as sets, thus reducing the computation of $[Z, A]$ and of $[X, Z]$ respectively, to the smaller sets $[Z, X]$ and $[A, Z]$.
    The following observation is useful.

8.15 Proposition. *Let $A \subset X$ be a subset of a metric space X. Then A is a deformation retract of X if and only if A is a retract of X and $\mathrm{Id}_X : X \to X$ is homotopic to a continuous map $g : X \to A$.*

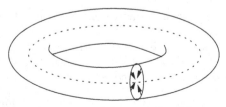

Figure 8.5. S^1 is a deformation retract of the torus $T \subset \mathbb{R}^3$.

Figure 8.6. S^n is a deformation retract of $B^n \setminus \{0\}$.

Proof. It is enough to prove sufficiency. Let $r : X \to A$ be a retraction and let $h : [0,1] \times X \to X$ be a homotopy of Id_X to g, $h(0,x) = x$, $h(1,x) = g(x)$ $\forall x \in X$. Then the map

$$k(t,x) = \begin{cases} r(h(2t,x)) & \text{if } 0 \le t \le \frac{1}{2}, \\ h(2-2t,x) & \text{if } \frac{1}{2} \le t \le 1 \end{cases}$$

is continuous since $h(1,x) = r(h(1,x))$ $\forall x$ and shows that Id_X is homotopic to $r : X \to A$. $\square$

8.16 ¶. Show that every point of a space X is a retract of X.

8.17 ¶. Show that $\{0,1\} \subset \mathbb{R}$ is not a retract of $\mathbb{R}$.

8.18 ¶. Show that a retract $A \subset X$ of a space X is a closed set.

8.19 ¶. The possibility of retracting X onto A is related to the possibility of extending continuous maps on A to continuous maps on X. Show

Proposition. *$A \subset X$ is a retract of X if and only if for any topological space Z any continuous map $f : A \to Z$ extends to a continuous map $F : X \to Z$.*

8.20 ¶. Show that S^n is a deformation retract of $B^{n+1} \setminus \{0\}$, see Figure 8.6.

8.21 ¶. With reference to Figure 8.8, show that $M \setminus \partial M$ is not a retract of M, but M and $M \setminus \partial M$ are homotopy equivalent since they have a deformation retract in common.

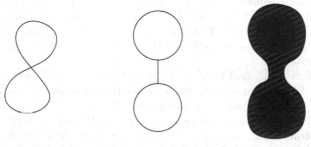

Figure 8.7. The first two figures are homotopy equivalent since they are both deformation retracts of the third figure.

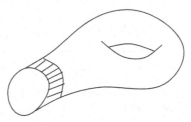

Figure 8.8. $M \setminus \partial M$ is not a retract of M, but M and $M \setminus \partial M$ are homotopy equivalent.

d. Relative homotopy

Intuitively, see Figure 8.1, the maps $H_t : X \to Y$, $t \in [0,1]$ defined by $H_t(x) := H(t,x)$, are a continuous family of continuous maps that deform f to g.

In particular, it is important to note that, in considering homotopy of maps, the target space is relevant and must be kept fixed in the discussion. As we shall see in the sequel, maps with values in Y that are nonhomotopic may become homotopic when seen as maps with values in $Z \supset Y$.

Also, it is worth considering homotopies of a suitable restricted type. For instance, when working with paths with fixed endpoints, it is better to consider homotopies such that for each t all curves $x \to H_t(x) := H(t,x)$, $x \in [0,1]$, have the same fixed endpoints for all $t \in [0,1]$. Similarly, when working with closed curves, it is worthwhile considering homotopies $H(t,x)$ such that every curve $x \to H_t(x) := H(t,x)$ is closed for all $t \in [0,1]$.

8.22 Definition. *Let $\mathcal{C} \subset C^0(X,Y)$. We say that $f, g \in \mathcal{C}$ are homotopic relative to $\mathcal{C}$ if there exists a continuous map $H[0,1] \times X \to Y$ such that $H(0,x) = f(x)$, $H(1,x) = g(x)$ and the curves $x \to H_t(x) := H(t,x)$ belong to $\mathcal{C}$ for all $t \in [0,1]$.*

It is easy to check that the relative homotopy is an equivalence relation. The set of relative homotopy classes with respect to $\mathcal{C} \subset C^0(X,Y)$ is denoted by $[X,Y]_{\mathcal{C}}$.

Some choices of the subset $\mathcal{C} \subset C^0([X,Y])$ are relevant.

(i) Let $Z \subset Y$ and $\mathcal{C} := \{f \in C^0([0,1],Y) \mid f(X) \subset Z\}$. In this case a homotopy relative to $\mathcal{C}$ is a homotopy of maps with values in Z.

(ii) Let $X = [0,1]$, $a,b \in Y$ and $\mathcal{C} := \{f \in C^0(X,Y) \mid f(0) = a,\ f(1) = b\}$. Then a homotopy relative to $\mathcal{C}$ is called a *homotopy with fixed endpoints*.

(iii) Let $X = [0,1]$, and let $\mathcal{C} := \{f \in C^0([0,1],Y) \mid f(0) = f(1)\}$ be the class of closed curves, or *loops*, in Y. In this case two curves homotopic relative to $\mathcal{C}$ are said *loop-homotopic*.

Recall that a closed curve $\gamma : [0,1] \to X$ can be reparametrized as a continuous map $\delta : S^1 \to X$ from the unit circle $S^1 \subset \mathbb{C}$. Now let $\gamma_1, \gamma_2 : [0,1] \to X$ be two loops and let $\delta_1, \delta_2 : S^1 \to X$ be two corresponding reparametrizations on S^1. Then, recalling that homotopies are simply paths in the space of continuous maps, it is trivial to show that γ_1 and γ_2 are loop-homotopic if and only if δ_1 and δ_2 are homotopic as maps from S^1 into X. Therefore $[[0,1], X]_C = [S^1, X]$.

Finally, notice that the intuitive idea of continuous deformation has several subtle aspects, see Figure 8.3.

8.1.2 Homotopy of loops

a. The fundamental group with base point

Let X be a metric space and let $x_0 \in X$. It is convenient to consider loops $\gamma : [0,1] \to X$ with $\gamma(0) = \gamma(1) = x_0$. We call them *loops with base point x_0*. Also, one can introduce a restricted form of homotopy between loops with base point x_0 by considering loop-homotopies $H(t,x)$ such that $x \to H(t,x)$ has base point x_0 for every t. We denote the corresponding homotopy equivalence relation and homotopy classes repectively, by $\sim_{x_0}$ and $[\ ]_{x_0}$. Finally,

$$\pi_1(X, \{x_0\})$$

denotes the set of loop-homotopy with base point x_0 classes of loops with base point x_0.

8.23 ¶. Show that $\pi_1(X, x_0)$ reduces to a point if X is contractible and $x_0 \in X$. [*Hint:* Show that $\pi_1(X, x_0) \subset [S^1, X]$.]

b. The group structure on $\pi_1(X, x_0)$

Given two loops $\varphi, \psi : [0,1] \to X$ with base point x_0, we may consider the *junction* of φ and ψ denoted by $\varphi * \psi$ as the loop with base point r_0 defined by

$$\varphi * \psi(t) := \begin{cases} \varphi(2t) & \text{if } t \in [0, 1/2], \\ \psi(2t - 1) & \text{if } t \in [1/2, 1]. \end{cases}$$

Since the homotopies with fixed endpoints can be joined, too, we have $\varphi_1 * \psi_1 \sim \varphi * \psi$ if $\varphi_1, \varphi_2, \psi_1, \psi_2$ are loops with base points x_0 such that $\varphi_1 \sim_{x_0} \varphi$ and $\psi_1 \sim_{x_0} \psi$. Thus the junction induces an operation on $\pi_1(X, \{x_0\})$,

$$* : \pi_1(X, \{x_0\}) \times \pi_1(X, \{x_0\}) \to \pi_1(X, \{x_0\})$$

still denoted by $*$, defined by $[\varphi]_{x_0} * [\psi]_{x_0} := [\varphi * \psi]_{x_0}$. It is easy to see the following.

8.24 Proposition. *The map $*$ has the following properties.*

(i) (ASSOCIATIVITY) *Let $f, g, h : [0,1] \to X$ be three loops with base point x_0. Then $([f]_{x_0} * [g]_{x_0}) * [h]_{x_0} = [f]_{x_0} * ([g]_{x_0} * [h]_{x_0})$.*

(ii) (RIGHT AND LEFT IDENTITIES) *Let $f : [0,1] \to X$ be a loop with base point x_0 and let $e_{x_0} : [0,1] \to X$ be the constant map, $e_{x_0}(t) := x_0$. Then $[e_{x_0}]_{x_0} * [f]_{x_0} = [f]_{x_0} * [e_{x_0}]_{x_0} = [f]_{x_0}$.*
(iii) (INVERSE) *Let $e_{x_0} : [0,1] \to X$ be the constant map $e_{x_0}(t) := x_0$ and, for a loop $f : [0,1] \to X$ with base point x_0, let $\overline{f} : [0,1] \to X$ be the map $\overline{f}(t) := f(1-t)$. Then $[f]_{x_0} * [\overline{f}]_{x_0} = [\overline{f}]_{x_0} * [f]_{x_0} = [e_{x_0}]_{x_0}$.*

In this way the junction of loops defines a natural *group structure* on $\pi_1(X, \{x_0\})$, where $[f]_{x_0}^{-1} = [\overline{f}]_{x_0}$.

8.25 Definition. *Let X be a space and $x_0 \in X$. The set $\pi_1(X, \{x_0\})$ of homotopy classes of loops with base point x_0 has a natural group structure induced by the junction operation of loops. We then call $\pi_1(X, \{x_0\})$ the* fundamental group *of X, or the* first homotopy group *of X, with base point x_0.*

c. Changing base point

By definition $\pi_1(X, x_0)$ depends on the base point x_0. However, if $x_0, x_1 \in X$, suppose that there exists a path $\alpha : [0,1] \to X$ from x_0 to x_1 and let $\overline{\alpha} : [0,1] \to X$, $\overline{\alpha}(t) := \alpha(1-t)$, be the *reverse* path from x_1 to x_0. For every loop γ with base point x_0, the curve $\alpha * \gamma * \overline{\alpha}$ is a loop with base point x_1. Since evidently $\alpha * \gamma_1 * \overline{\alpha} \sim \alpha * \gamma_2 * \overline{\alpha}$ if $\gamma_1 \sim \gamma_2$, α defines a map $\alpha_* : \pi_1(X, x_0) \to \pi_1(X, x_1)$ by

$$\alpha_*([\gamma]_{x_0}) := [\alpha * \gamma * \overline{\alpha}]_{x_1}, \tag{8.2}$$

where we have denoted by $[\ ]_{x_0}$ and $[\ ]_{x_1}$ respectively, the homotopy classes of curves with base point x_0 and x_1. It is trivial to see that α_* is a group isomorphism, thus concluding the following.

8.26 Proposition. *$\pi_1(X, x_0)$ and $\pi_1(X, x_1)$ are isomorphic as groups for all $x_0, x_1 \in X$ if X is path-connected.*

Thus, for a path-connected space X, all groups $\pi_1(X, x_0)$, $x_0 \in X$ are the same group up to an isomorphism. We call it the *fundamental group* or the *first homotopy group* of X, and we denote it by $\pi_1(X)$. However, the map α_* defined by (8.2) depends explictly on α.

For convenience, let $h_\alpha := \alpha_*$. Examples show that in general $h_\alpha \neq h_\beta$ if α and β have the same endpoints, but we have

$$h_\beta^{-1} h_\alpha([\gamma]_{x_1}) = h_\beta^{-1}([\alpha\gamma\overline{\alpha}]_{x_0}) = [\overline{\beta}\alpha\gamma\overline{\alpha}\beta]_{x_1} = [\overline{\beta}\alpha]_{x_1} * [\gamma]_{x_1} * [\overline{\beta}\alpha]_{x_1}^{-1}.$$

This implies that

(i) $h_\alpha = h_\beta$ if α and β are homotopic with the same endpoints,
(ii) h_α is always the same map, independently from α, if $\pi_1(X, x_1)$ is a commutative group.

Figure 8.9. Camille Jordan (1838–1922) and the frontispiece of the Japanese translation of the *Lecture Notes of Elementary Topology and Geometry* by J. M. Singer and J. A. Thorpe.

Thus, attaching a path to x_0 to any curve $\gamma : S^1 \to X$, we can construct a loop with base point x_0 and, at the homotopy level, this construction is actually a map $h : [S^1, X] \to \pi_1(X, x_0)$. It is clear that h is one-to-one, since its inverse is just the inclusion of $\pi_1(X, x_0)$ into $[S^1, X]$.

8.27 Proposition. *Let X be path-connected. If $\pi_1(X)$ is commutative, then the map $h : [S^1, X] \to \pi_1(X)$ described above is bijective.*

8.28 Definition. *We say that a space X is simply connected if X is path-connected and $\pi_1(X, x_0)$ reduces to a point for some $x_0 \in X$ (equivalently for any $x_0 \in X$ by Proposition 8.26).*

8.29 ¶. Show that X is simply connected if X is path-connected and contractible.

d. Invariance properties of the fundamental group

Let us now look at the action of continuous maps on the fundamental group. Let X, Y be metric spaces and let $x_0 \in X$. To any continuous map $f : X \to Y$ one associates a map

$$f_\# : \pi_1(X, x_0) \to \pi_1(Y, f(x_0))$$

defined by $f_\#([\gamma]_{x_0}) := [f \circ \gamma]_{f(x_0)}$. It is easy to see that the above definition makes sense, and that actually $f_\#$ is a group homomorphism.

8.30 Proposition. *We have the following.*

(i) *Let $f : X \to Y$ and $g : Y \to Z$ be two continuous maps. Then $(g \circ f)_\# = g_\# \circ f_\#$.*

(ii) *If* Id $: X \to X$ *is the identity map and* $x_0 \in X$, *then* Id$_\#$ *is the identity map on* $\pi_1(X, \{x_0\})$.

(iii) *Suppose Y is path-connected, and let $F : [0,1] \times X \to Y$ be a homotopy of two maps f and g from X into Y. Then the curve $\alpha(t) := F(t, x_0)$, $t \in [0,1]$, joins $f(x_0)$ to $g(x_0)$ and $g_\# = \alpha_* \circ f_\#$.*

Proof. (i) and (ii) are trivial. To prove (iii), it is enough to show that $f \circ \gamma$ and $\overline{\alpha} \circ g \circ \gamma \circ \alpha$ are homotopic for every loop γ with base point x_0. A suitable homotopy is given by the map $H(t, x) : [0,1] \to X \to Y$ defined by

$$H(t, x) := \begin{cases} \overline{\alpha}(2x) & \text{if } x \le \frac{1-t}{2}, \\ F\left(t, \gamma\left(\frac{4x+2t-2}{3t+1}\right)\right) & \text{if } \frac{1-t}{2} \le x \le \frac{t+3}{4}, \\ \alpha(4x - 3) & \text{if } x \ge \frac{t+3}{4}. \end{cases}$$

$\square$

Of course, Proposition 8.30 (i) and (ii) imply that a homeomorphism $h : X \to Y$ induces an isomorphism between $\pi_1(X, x_0)$ and $\pi_1(Y, h(x_0))$. Therefore, on account of Proposition 8.26, *the fundamental group is a topological invariant of path-connected spaces.* Actually, from (iii) Proposition 8.30 we infer the following.

8.31 Theorem. *Let X, Y be two path-connected homotopy equivalent spaces. Then $\pi_1(X)$ and $\pi_1(Y)$ are isomorphic.*

Proof. Let $f : X \to Y$, $g : Y \to X$ be continuous such that $g \circ f \sim \text{Id}_X$ and $f \circ g \sim \text{Id}_Y$ and let $x_0 \in X$. Then we have two induced maps

$$f_\# : \pi_1(X, x_0) \to \pi_1(Y, f(x_0)),$$
$$g_\# : \pi_1(Y, f(x_0)) \to \pi_1(X, g(f(x_0))).$$

Let $H : [0,1] \times X \to X$ be the homotopy of Id$_X$ to $g \circ f$ and let $K : [0,1] : Y \to Y$ be the homotopy of Id$_Y$ to $f \circ g$. If $\alpha_1(t) := H(t, x_0)$, $\alpha_2(t) := K(t, f(x_0))$, then by Proposition 8.30 we infer

$$g_\# \circ f_\# = (g \circ f)\# = (\alpha_1) * \circ (\text{Id}_X)\# = (\alpha_1)_*,$$
$$f_\# \circ g_\# = (f \circ g)\# = (\alpha_2) * (\text{Id}_Y)\# = (\alpha_2)_*.$$

Since $(\alpha_1)_*$ and $(\alpha_2)_*$ are isomorphisms, $f_\#$ is injective and surjective. $\square$

8.1.3 Covering spaces

a. Covering spaces

A useful tool to compute, at least in some cases, the fundamental group, is the notion of *covering space*.

8.32 Definition. *A covering of Y is a continuous map $p : X \to Y$ from a topological space X, called the total space, onto Y such that for all $x \in Y$ there exists an open set $U \subset Y$ containing x such that $p^{-1}(U) = \cup_\alpha V_\alpha$, where V_α are pairwise disjoint open sets and $p_{|V_\alpha}$ is a homeomorphism between V_α and U. Each V_α is called a* slice *of $p^{-1}(U)$.*

Figure 8.10.

8.33 Example. Let Y be any space. Consider the disjoint union of k-copies of Y, that we can write as a Cartesian product $X := Y \times \{1, 2, \ldots, k\}$. Then the projection map $p : X \to Y$, $p((y, i)) = y$, is a covering of X.

8.34 Example. Let S^1 be the unit circle of $\mathbb{C}$. Then the circular motion $p : \mathbb{R} \to S^1$, $p(\theta) = e^{i \, 2\pi\theta}$ is a covering of S^1.

8.35 Example. Let $X \subset \mathbb{R}^3$ be the trace of the regular helix $\gamma(t) = (\cos t, \sin t, t)$. Then $p : X \to S^1$ where $p : \mathbb{R}^3 \to \mathbb{R}^2$, $p(x, y, z) := (x, y)$, is the orthogonal projection on $\mathbb{R}^2$, is another covering of S^1.

8.36 ¶. Let $p : X \to Y$ be a covering of Y. Suppose that Y is connected and that for some point $y_0 \in Y$ the set $p^{-1}(y_0)$ is finite and contains k points. Show that $p^{-1}(y)$ contains k points for all $y \in Y$. In this case, we say that $p : X \to Y$ is a k-*fold covering* of Y.

8.37 ¶. Show that $p : \mathbb{R}_+ \to S^1$, $p(t) := e^{i \, t}$, is not a covering of S^1.

8.38 ¶. Show that, if $p : \widetilde{X} \to X$ and $q : \widetilde{Y} \to Y$ are coverings respectively, of X and Y, then $p \times q : \widetilde{X} \times \widetilde{Y} \to X \times Y$, $p \times q(x, y) := (p(x), q(y))$, is a covering of $X \times Y$. In particular, if $p : \mathbb{R} \to S^1$ is defined by $p(t) := e^{i \, 2\pi t}$, then the map $p \times p : \mathbb{R} \times \mathbb{R} \to S^1 \times S^1$ is a covering of the *torus* $S^1 \times S^1$. Figure 8.10 shows the covering map for the standard torus of $\mathbb{R}^3$ that is homeomorphic to the torus $S^1 \times S^1 \subset \mathbb{R}^4$.

8.39 ¶. Think of S^1 as a subset of $\mathbb{C}$. Show that the map $p : S^1 \to S^1$, $p(z) = z^2$, is a two-fold covering of S^1. More generally, show that the map $S^1 \to S^1$ defined by $p(t) := z^n$ is a $|n|$-covering of S^1 if $n \in \mathbb{Z} \setminus \{0\}$.

8.40 ¶. Show that the map $p : \mathbb{R}_+ \times \mathbb{R} \to \mathbb{R}_+ \times S^1$ defined by $p(s, \theta) = (s, e^{i\theta})$ is a covering of $S^1 \times \mathbb{R}_+$.

8.41 ¶. Show that the map $p : \mathbb{R}_+ \times \mathbb{R} \to \mathbb{R}^2 \setminus \{0\}$ defined by $p(\rho, \theta) := \rho e^{i\theta}$ is a covering of $\mathbb{R}^2 \setminus \{0\}$.

b. Lifting of curves

In connection with coverings the notion of (continuous) *lift* is crucial.

8.42 Definition. *Let $p : X \to Y$ be a covering of Y and let $f : Z \to Y$ be a continuous map. A continuous map $\widetilde{f} : Z \to X$ such that $p \circ \widetilde{f} = f$ is called a* lift *of f on X.*

8.43 Example. Let $p : \mathbb{R} \to S^1$ be the covering of S^1 given by $p(t) = e^{i\,t}$. A lift of $f : [0, 1] \to S^1$ is a continuous map $h : [0, 1] \to \mathbb{R}$ such that $f(t) = e^{i\,h(t)}$. Looking at t as a time variable, $h(t)$ is the angular evolution of $f(t)$ as $f(t)$ moves on S^1.

8.44 Example. Not every function can be lifted. For instance, consider the covering $p : \mathbb{R} \to S^1$, $p(t) = e^{i\,2\pi t}$. Then the identity map on S^1 cannot be lifted to a continuous map $h : S^1 \to \mathbb{R}$. In fact, parametrizing maps from S^1 as closed curves parametrized on $[0, 2\pi]$, h would be periodic. On the other hand, if h was a lift of $z = e^{i\,t}$, we would have $e^{it} = e^{i\,h(t)}$, which implies that $h(t) = t+$const, a contradiction.

However, curves can be lifted to curves that are not necessarily closed. Let X be a metric space. We say that X is *locally path-connected* if every point $x \in X$ has an open path-connected neighborhood U.

8.45 Proposition. *Let $p : X \to Y$ be a covering of Y and let $x_0 \in X$. Suppose that X and Y are path-connected and locally path-connected. Then*

(i) *each curve $\beta : [0, 1] \to Y$ with $\beta(0) = p(x_0)$ has a unique continuous lift $\alpha : [0, 1] \to X$ such that $p \circ \alpha = \beta$ and $\alpha(0) = x_0$,*
(ii) *for every continuous map $k : [0, 1] \times [0, 1] \to Y$ with $k(0, 0) = p(x_0)$, there is a unique continuous lift $h : [0, 1] \times [0, 1] \to X$ such that $h(0, 0) = x_0$ and $p(h(t, s)) = k(t, s)$ for all $(t, s) \in [0, 1] \times [0, 1]$.*

Proof. Step 1. Uniqueness in (i). Suppose that for the two curves α_1, α_2 we have $p(\alpha_1(t)) = p(\alpha_2(t))$ $\forall t \in [0, 1]$ and $\alpha_1(0) = \alpha_2(0)$. The set $E := \{t \,|\, \alpha_1(t) = \alpha_2(t)\}$ is closed in $[0, 1]$; since p is a local homeomorphism, it is easily seen that E is also open in $[0, 1]$. Therefore $E = [0, 1]$.

Step 2. Existence in (i). We consider the subset

$$E := \Big\{ t \in [0, 1] \,\Big|\, \exists \text{ a continuous curve } \alpha_t : [0, t] \to X$$

$$\text{such that } \alpha(0) = x_0 \text{ and } p(\alpha_t(\theta)) = \beta(\theta) \; \forall \theta \in [0, t] \Big\}$$

and shall prove that E is open and closed in $[0, 1]$ consequently, $E = [0, 1]$ as it is not empty.

Let $\tau \in E$ and let U be an open neighborhood of $\alpha_\tau(\tau)$ for which $p_{|U}$ is a homeomorphism. For σ sufficiently small, $\sigma < \sigma_0$, the curve $s \to \gamma(s) := (p_{|U})^{-1}(\beta(s))$, $s \in [\tau, \tau + \sigma]$, is continuous, $\gamma(\tau) = \alpha_\tau(\tau)$ and $p(\alpha(s)) = \beta(s)$, $\forall s \in [\tau, \tau + \sigma]$. Therefore for the curve $\alpha_\sigma : [0, \sigma] \to X$ defined by

$$\alpha_\sigma(s) := \begin{cases} \alpha_\tau(s) & \text{if } 0 \le s \le \tau, \\ \gamma(s) & \text{if } \tau < s \le \tau + \sigma, \end{cases}$$

we have $\alpha_\sigma(t) = \beta(t)$ for all $t \in [0, \sigma]$, i.e., $\tau + \sigma_0 \in E$ for some $\sigma_0 > 0$ if $\tau \in E$, or, in other words, E is open in $[0, 1]$.

We now prove that E is closed by showing that $T := \sup E \in E$. Let $\{t_n\} \subset E$ be a nondecreasing sequence that converges to T and for every n, let $\alpha_n : [0, t_n] \to X$ be such that $p(\alpha_n(t)) = \beta(t)$ $\forall t \in [0, t_n]$. Because of the uniqueness $\alpha_r(t) = \alpha_s(t)$ for all $t \in [0, r]$ if $s < r$, consequently a continuous curve $\alpha : [0, T[\to X$ is defined so that $p(\alpha(t)) = \beta(t)$ $\forall t \in [0, T[$. It remains to show that we can extend continuously α at T. Let V be an open neighborhood of $\beta(T)$ such that $p^{-1}(V) = \cup_j U_j$ where U_α are pairwise disjoint open sets that are homeomorphic to V. Then $\beta(t) \in V$ for $\bar{t} < t \le T$. Since $a([\bar{t}, T[)$ is connected and the U_α's are pairwise disjoint, we infer that $\alpha(t)$ must belong to a unique U_α, say U_1, for $\bar{t} < t < T$. It suffices now to extend by continuity α by setting $\alpha(T) := (p_{|U_1})^{-1}(\beta(T))$.

Step 3. *(ii)* Uniqueness follows from (i). In fact, if $P \in [0,1] \times [0,1]$, γ_P is the segment joining $(0,0)$ to P and $h_1, h_2 : [0,1] \to [0,1] \to X$ are such that $p \circ h_1 = p \circ h_2$ in $[0,1] \times [0,1]$ with $h_1(0,0) = h_2(0,0)$, from (i) we infer $h_1(P) = h_2(P)$, as $h_1|_{\gamma_P} = h_2|_{\gamma_P}$.

Let us prove existence. Again by (i) there is a curve $\alpha(t)$ with $\alpha(0) = x_0$ and $p(\alpha(t)) = k(0,t)$ for all t, and, for each t, a curve $s \to h(s,t)$ such that $p(h(s,t)) = k(s,t)$ with $h(0,t) = \alpha(t)$. Of course $k(0,0) = \alpha(0) = x_0$ and it remains to show that $h : [0,1] \times [0,1] \to X$ is continuous. Set $R_s := [0,s[\times[0,1]$ and $R_0 := \{0\} \times [0,1]$. Suppose h is not continuous and let $(\bar{s}, \bar{t})$ be a point in the closure of the points of discontinuity of h. Let U be an open and connected neighborhood of $h(\bar{t}, \bar{s})$ such that $p_{|U}$ is a homeomorphism. By lifting $k_{|p(U)}$ we find a rectangle $R \subset \mathbb{R}^2$ that has $(\bar{s}, \bar{t})$ as an interior point and a continuous function $w : R \to U$ with $w(\bar{t}, \bar{s}) = h(\bar{t}, \bar{s})$ such that $p(w(t,s)) = k(t,s) = p(h(t,s))$ for all $(t,s) \in R_{\bar{s}}$. Since w and h are continuous in $R_{\bar{s}}$, they agree in $R_s \cap R$. On the other hand, both $h(t,s)$ and $w(t,s)$ lift the same function $k(t,s)$, thus by (i) they agree, hence $h(t,s) = w(t,s)$ is continuous in a neighborhood of $(\bar{t}, \bar{s})$: a contradiction. $\square$

8.46 Proposition. *Let X and Y be path-connected and locally path-connected metric spaces and let $f : X \to Y$ be a covering of Y. Let $\alpha, \beta : [0,1] \to Y$ be two curves with $\alpha(0) = \beta(0)$ and $\alpha(1) = \beta(1)$ that are homotopic with fixed endpoints and let $a, b : [0,1] \to X$ be their continuous lifts that start at the same point $a(0) = b(0)$. Then $a(1) = b(1)$, and a and b are homotopic with fixed endpoints.*

Proof. From (i) Proposition 8.45 we know that α, β can be lifted uniquely to two curves $a, b : [0,1] \to X$ with $a(0) = b(0) = a_0$, $p(a_0) = \alpha(0)$. Let $k : [0,1] \times [0,1] \to Y$ be a homotopy between α and β, i.e., $k(0,t) = \alpha(t)$, $k(1,t) = \beta(t)$, $k(s,0) = \alpha(0) = \beta(0)$, $k(s,1) = \alpha(1) = \beta(1)$. By (ii) Proposition 8.45 we can lift k to h, so that $p(h(s,t)) = k(s,t)$ and $k(0,0) = a(0) = b(0)$. Then h is a homotopy between a and b and in particular $a(1) = h(0,1) = b(1)$. $\square$

8.47 Theorem. *Let X and Y be path-connected and locally path-connected metric spaces and let $p : X \to Y$ be a covering of Y. If Y is simply connected, then $p : X \to Y$ is a homeomorphism.*

Proof. Suppose there are $x_1, x_2 \in X$ with $p(x_1) = p(x_2)$. Since X is connected, there is a curve $a : [0,1] \to X$ with $a(0) = x_1$ and $a(1) = x_2$. Let $b : [0,1] \to X$ be the constant curve $b(t) = x_1$. The image curves $\alpha(t) := p(a(t))$ and $\beta(t) := p(b(t))$ are closed curves, hence homotopic, Y being simply connected. Proposition 8.46 then yields $x_2 = a(1) = b(1) = x_1$. $\square$

8.48 Theorem. *Let X and Y be path-connected and locally path-connected, and let $p : X \to Y$ be a covering of Y. Suppose that Z is path-connected and simply connected. Then any continuous map $f : Z \to Y$ has a lift $\widetilde{f} : Z \to X$. More precisely, given $z_0 \in Z$ and $x_0 \in X$, such that $p(x_0) = f(z_0)$, there exists a unique continuous map $f : Z \to X$ such that $\widetilde{f}(z_0) = x_0$ and $p \circ \widetilde{f} = f$.*

Proof. Let $z \in Z$ and let $\gamma : [0,1] \to Z$ be a curve joining z_0 to z. Then the curve $\alpha(t) := f(\gamma(t))$, $t \in [0,1]$, in Y has a lift to a curve $a : [0,1] \to X$ with $a(0) = x_0$, see (i) Proposition 8.45, and Proposition 8.46 shows that $a(1)$ depends on $\alpha(1) = f(z)$

and does not depend on the particular curve γ. Thus we define $\widetilde{f}(z) := a(1)$, and by definition $f(z_0) = x_0$ and $p \circ \widetilde{f} = f$. We leave to the reader to check that $\widetilde{f}$ is continuous.

$\square$

c. Universal coverings and homotopy

8.49 Definition. *Let Y be a path-connected and locally path-connected metric space. A covering $p : X \to Y$ is said to be a* universal covering *of Y if X is path-connected, locally path-connected and simply connected.*

From Theorems 8.47 and 8.48 we immediately infer

8.50 Theorem (Universal property). *Let X, Y, Z be path-connected and locally path-connected metric spaces. Let $p : X \to Y$, $q : Z \to Y$ be two coverings of Y and suppose Z simply connected. Then q has a lift $\widetilde{q} : Z \to X$ which is also a covering of X. Moreover $\widetilde{q}$ is a homeomorphism if X is simply connected, too.*

The relevance of the universal covering space in computing the homotopy appears from the following.

8.51 Theorem. *Let X and Y be path-connected and locally path-connected metric spaces and let $p : X \to Y$ be the universal covering of Y. Then $\forall y_0 \in Y$ $\pi_1(Y, y_0)$ and $p^{-1}(y_0) \subset X$ are one-to-one.*

Proof. Fix $q \in p^{-1}(x_0)$. For any curve α in Y with base point x_0, denote by $a : [0, 1] \to X$ its lift with $a(0) = q$. Clearly a is a curve in X which ends at $a(1) \in p^{-1}(x_0)$. Moreover, if β is loop-homotopic to α in Y, then necessarily $a(1) = b(1)$, so the map $\alpha \to a(1)$ is actually a map

$$\varphi_q : \pi_1(Y, x_0) \to p^{-1}(x_0).$$

Of course φ_q is surjective since any curve in X with endpoints in $p^{-1}(x_0)$ projects onto a closed loop in Y with base point x_0. Moreover, if $\varphi_q([\gamma]) = \varphi_q([\delta])$, then the lifts c and d that start at the same point end at at the same point; consequently c and d are homotopic, as X is simply connected. Projecting the homotopy between c and d onto Y yields $[\gamma] = [\delta]$.

$\square$

d. A global invertibility result

Existence of a universal covering $p : X \to Y$ of a space Y can be proved in the setting of topological spaces. Observe that if X and Y are path-connected and locally path-connected, and if $p : X \to Y$ is a universal covering of Y, then Y is *locally simply connected*, i.e., such that $\forall y \in Y$ there exists an open set $V \subset Y$ containing y such that every loop in V with base point at x is homotopic (in Y) to the constant loop x. It can be proved in the context of topological spaces that *any path-connected, locally path-connected and locally simply-connected Y has a universal covering $p : X \to Y$.* We do not deal with such a general problem and confine ourselves to discussing whether a given continuous map $f : X \to Y$ is a covering of Y.

Let X, Y be metric spaces. A continuous map $f : X \to Y$ is a *local homeomorphism* if every $x \in X$ has an open neighborhood U such that $f_{|U}$ is a homeomorphism onto its image. We say that f is a *proper* map if $f^{-1}(K)$ is compact in X for every compact $K \subset Y$.

Clearly a homeomorphism from X onto its image $f(X) \subset Y$ is a local homeomorphism and a proper map. Also, if $p : X \to Y$ is a covering of Y then p is a local homeomorphism. We have

8.52 Theorem. *Let X be path-connected and locally path-connected and let $f : X \to Y$ be a local homeomorphism and a proper map. Then X and $f(X)$ are open, path-connected and locally path-connected and $f : X \to f(X)$ is a covering of $f(X)$.*

Before proving Theorem 8.52, let us introduce the *Banach indicatrix* of $f : X \to \mathbb{R}^n$ as the map

$$N_f : Y \to \mathbb{N} \cup \{\infty\}, \qquad N_f(y) := \#\Big\{x \in X \,|\, f(x) = y\Big\}.$$

Evidently $f(X) = \{y \,|\, N_f(y) \geq 1\}$ and f is injective iff $N_f(y) \leq 1 \; \forall y$.

8.53 Lemma. *Let $f : X \to Y$ be a local homeomorphism and a proper map. Then N_f is bounded and locally constant on $f(X)$.*

Proof. Since f is a local homeomorphism, the set $f^{-1}(y) = \{x \in X \,|\, f(x) = y\}$ is discrete and in fact $f^{-1}(y)$ is finite, since f is proper. Let $N_f(\overline{y}) = k$ and $f^{-1}(\overline{y}) = \{\overline{x}_1, \ldots, \overline{x}_k\}$. Since f is a local homeomorphism, we can find open disjoint neighborhoods U_1 of $\overline{x}_1$, ..., U_k of $\overline{x}_k$ and an open neighborhood V of $\overline{y}$ such that $f_{|U_j} : U_j \to V$ are homeomorphisms. In particular, for every $y \in V$ there is a unique $x_j \in U_j$ such that $f(x_j) = y$. It follows that $N_f(y) \geq k \; \forall y \in V$. We now show that for every $\overline{y}$ there exists a neighborhood W of $\overline{y}$, $W \subset V$, such that $N_f(y) \leq k$ holds for all $y \in W$. Suppose, in fact, that for a $\overline{y}$ there is no neighborhood W such that $N(y) \leq k$ for $y \in W$, then there is a sequence $\{y_i\} \subset W$, $y_i \to \overline{y}$ with $N(y_i) > k$, and points $\xi_i \notin U_1 \cup \cdots \cup U_k$ with $f(\xi_i) = y_i$. The set $f^{-1}(\{y_i\} \cup \{\overline{y}\})$ is compact since f is proper, thus possibly passing to a subsequence $\{\xi_i\}$ converges to a point $\overline{\xi}$ and necessarily $\overline{\xi} \notin U_1 \cup \cdots \cup U_k$; passing to the limit we also find $f(\overline{\xi}) = \overline{y}$: a contradiction since $\overline{\xi}$ is different from $\overline{x}_1, \ldots, \overline{x}_k$. $\quad\square$

Proof of Theorem 8.52. From Lemma 8.53 we know that, for every $y \in Y$, $f^{-1}(y)$ contains finitely many points $\{x_1, x_2, \ldots, x_N\}$ where N is locally constant. If U_i, $i = 1, \ldots, N$, $U_i \ni x_i$ and $V \ni y$ are open and homeomorphic sets, we then set

$$V = \cap_{i=1}^N V_i \cap \Big\{y \in Y \,\Big|\, N_f(y) = N\Big\}, \qquad W_i := (f_{|U_i})^{-1}(V).$$

Clearly V is open and $f^{-1}(V)$ is a finite sum of disjoint open sets that are homeomorphic to V. $\quad\square$

As a consequence of Theorem 8.47 we then infer the following useful *global invertibility theorem*.

8.54 Theorem. *Let X be path-connected and locally path-connected, and let $f : X \to Y$ be a local homeomorphism that is proper. If $f(X)$ is simply connected, then f is injective, hence a homeomorphism between X and $f(X)$.*

Proof. $f : X \to f(X)$ is a covering by Theorem 8.52. Theorem 8.47 then yields that f is one-to-one, hence a homeomorphism of X onto $f(X)$.   □

8.1.4 A few examples

a. The fundamental group of S^1

The map $p : \mathbb{R} \to S^1$, $p(t) = e^{i\,2\pi t}$ is a universal covering of S^1. Therefore for any $x_0 \in S^1$, $p^{-1}(x_0) = \mathbb{Z}$ as sets. Therefore, see Theorem 8.51, one can construct an injective and surjective map

$$\varphi_{x_0} : \pi_1(S^1, x_0) \to p^{-1}(x_0) = \mathbb{Z}$$

that maps $[\alpha]$ to the end value $a(1) \in \mathbb{Z}$ of the lift a of α with $a(0) = 0$. We have

8.55 Lemma. $\varphi_{x_0} : \pi_1(S^1, x_0) \to \mathbb{Z}$ *is a group isomorphism.*

Proof. Let α, β be two loops in S^1 with base point x_0 and a, b the liftings with $a(0) = b(0) = 0$. If $n := \varphi_0([a])$ and $m = \varphi_0([\beta])$, we define $c : [0, 1] \to \mathbb{R}$ by

$$c(s) = \begin{cases} a(2s) & s \in [0, 1/2], \\ n + b(2s - 1) & s \in [1/2, 1]. \end{cases}$$

It is not difficult to check that c is the lift of $\alpha * \beta$ with $c(0) = 0$ so that

$$\varphi_0([\alpha] * [\beta]) = \varphi_0([\alpha * \beta]) = c(1) = n + m = \varphi_0([\alpha]) + \varphi_0([\beta]).$$

□

Since φ_{x_0} is a group isomorphism and $\mathbb{Z}$ is commutative, $\pi_1(S^1, x_0)$ is commutative, and there is an injective and bijective map $h : [S^1, S^1] \to \pi_1(X, x_0)$, see Proposition 8.27. The composition map

$$\deg : C^0(S^1, S^1) \to \mathbb{Z}, \qquad \deg(\gamma) := \varphi_{x_0}(h([\gamma]))$$

is called the *degree* on S^1, and by construction we have the following.

8.56 Theorem. *Two maps* $f, g : S^1 \to S^1$ *have the same degree if and only if they are homotopic.*

Later we shall see that we can recover the degree mapping more directly.

8.57 ¶. Show that the fundamental group of $\mathbb{R}^2 \setminus \{0\}$ is $\mathbb{Z}$.

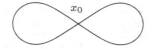

Figure 8.11. A figure eight.

b. The fundamental group of the figure eight

The *figure eight* is the union of two circles A and B with a point x_0 in common.

If a is a loop based at x_0 that goes clockwise once around A, and a^{-1} is the loop that goes counterclockwise once around A, and similarly for b, b^{-1}, then the cycle $aba^{-1}b^{-1}$ is a loop that cannot be unknotted in $A \cup B$ while $aa^{-1}bb^{-1}$ can.

More precisely, one shows that the fundamental group of the figure eight is the *noncommutative free group on the generators a and b*. Indeed, this can be proved using the following special form of the so-called Seifert–Van Kampen theorem.

8.58 Theorem. *Suppose $X = U \cup V$, where U, V are open path-connected sets and $U \cap V$ is path-connected and simply connected. Then for any $x_0 \in U \cap V$, $\pi_1(X, x_0)$ is the free product of $\pi_1(U, x_0)$ and $\pi_1(V, x_0)$.*

8.59 ¶. Show that the fundamental group of $\mathbb{R}^2 \setminus \{x_0, x_1\}$ is isomorphic to the fundamental group of the figure eight.

8.60 ¶. Show that $\pi_1(X \times Y, (x_0, y_0))$ is isomorphic to $\pi_1(X, x_0) \times \pi_1(Y, y_0)$, in particular the fundamental group of the torus $S^1 \times S^1$ is $\mathbb{Z} \times \mathbb{Z}$.

8.61 ¶. Let $X = A_1 \cup A_2 \cup \cdots \cup A_n$ where each A_i is homeomorphic to S^1, and $A_i \cap A_j = \{x_0\}$ if $i \neq j$. Show that $\pi_1(X, x_0)$ is the free group on n generators $\alpha_1, \ldots, \alpha_n$ where α_i is represented by a path that goes around A_i once.

8.62 ¶. Let X be the space obtained by removing n points of $\mathbb{R}^2$. Show that $\pi_1(X, x_0)$ is a free group on n generators $\alpha_1, \ldots, \alpha_n$, where α_i is represented by a closed path which goes around the ith hole once.

c. The fundamental group of S^n, $n \geq 2$

The following result is also a consequence of Theorem 8.58.

8.63 Theorem. *Let $X = U \cup V$ where U and V are simply connected open sets of X and $U \cap V$ is path-connected. Then X is simply connected, i.e., $\pi_1(X, x_0) = 0$.*

As a consequence we have the following.

8.64 Proposition. *The sphere $S^n \subset \mathbb{R}^{n+1}$ is simply connected, i.e., $\pi_1(S^n, x_0) = 0$ if $n \geq 2$.*

Proof. Let p_S and p_N be respectively, the south pole and the north pole of the sphere. The stereographic projection from the south (north) pole establishes a homeomorphism between $S^n \setminus \{p_S\}$ (respectively, $S^n \setminus \{p_N\}$) and $\mathbb{R}^n$. Thus $\pi_1(S^n \setminus \{p_S\}, x_0) = \pi_1(S^n \setminus \{p_N\}, x_0) = 0 \ \forall x_0 \neq p_s, p_N$. By Theorem 8.63 it suffices to show that $S^n \setminus \{p_S, p_N\}$ is path-connected. For that we notice that the stereographic projection is a homeomorphism between $S^n \setminus \{p_S, p_N\}$ and $\mathbb{R}^n \setminus \{0\}$ which in turn is path-connected if $n \geq 2$. $\qquad \square$

Since the fundamental groups of $\mathbb{R}^{n+1} \setminus \{0\}$ and S^n are isomorphic, see Theorem 8.31, equivalently we can state

8.65 Proposition. $\mathbb{R}^n \setminus \{0\}$ *is simply connected if $n > 2$.*

8.66 ¶. Show that $\mathbb{R}^n$, $n > 2$, and $\mathbb{R}^2$ are not homeomorphic.

8.1.5 Brouwer's degree

a. The degree of maps $S^1 \to S^1$

A more analytic presentation of the mapping degree for maps $S^1 \to S^1$ is the following. Think of S^1 as the unit circle in the complex plane, so that the rotations of S^1 write as complex multiplication, and represent loops in S^1 as maps $f : S^1 \to S^1$ or by 2π-periodic functions $\theta \to f(e^{i\theta})$, $f : S^1 \to S^1$.

8.67 Lemma. *Let $f : S^1 \to S^1$ be continuous. There exists a unique continuous function $h : \mathbb{R} \to \mathbb{R}$ such that*

$$\begin{cases} f(e^{i\theta}) = f(1)e^{ih(\theta)} & \forall \theta \in \mathbb{R}, \\ h(0) = 0. \end{cases} \tag{8.3}$$

Proof. Consider the covering $p : \mathbb{R} \to S^1$ of S^1 given by $p(t) := e^{it}$. The loop $g(z) := f(z)/f(1)$ has base point $1 \in S^1$. Then by the lifting argument, Proposition 8.45, there exists a lift $h : \mathbb{R} \to \mathbb{R}$ such that (8.3) holds. The uniqueness follows directly from (8.3). In fact, if h_1, h_2 verify (8.3), then $h_1(\theta) - h_2(\theta) = k(\theta)2\pi$ where $k(\theta) \in \mathbb{Z}$. As h_1 and h_2 are continuous, $k(\theta)$ is constant, hence $k(\theta) = k(0) = 0$. $\qquad \square$

Let $f : S^1 \to S^1$ be continuous and let $h : \mathbb{R} \to \mathbb{R}$ be as in (8.3). Of course, for every θ we have

$$h(\theta + 2\pi) - h(\theta) = 2k(\theta)\pi$$

for some integer $k(\theta) \in \mathbb{Z}$. Since h is continuous, k is continuous, hence constant. Observe that $k = h(2\pi) - h(0) = h(2\pi)$ and k is independent of the initial point $f(1)$. In particular, $f : S^1 \to S^1$ and $f/f(1) : S^1 \to S^1$ have the same degree.

8.68 Definition. *Let $f : S^1 \to S^1$ and let h be as in (8.3). There is a unique integer $d \in \mathbb{Z}$ such that*

$$h(\theta + 2\pi) - h(\theta) = d\,2\pi \qquad \forall \theta \in \mathbb{R}.$$

The number d is called the winding number, *or* degree, *of the map $f : S^1 \to S^1$, and it is denoted by $\deg(f)$.*

8.69 Theorem. *Two continuous maps $f_0, f_1 : S^1 \to S^1$ have the same degree if and only if they are homotopic.*

Proof. Let $f : S^1 \to S^1$. We have already observed that $f(z)$ and $f(z)/f(1)$ have the same degree. On the other hand, $f(z)$ and $f(z)/f(1)$ are also trivially homotopic. To prove the theorem it is therefore enough to consider maps f_0, f_1 with the same base point, say $f(1) = 1$.

(i) Assume f_0, f_1 are homotopic with base point $1 \in S^1$. By the lifting argument, the liftings h_0, h_1 of f_0, f_1 characterized by (8.3) have $h_1(2\pi) = h_2(2\pi)$, hence

$$\deg(f_1) = h_1(2\pi) - h_1(0) = h_1(2\pi) = h_2(2\pi) = h_2(2\pi) - h_2(0) = \deg(f_2).$$

Conversely, let $f : S^1 \to S^1$ be of degree d and let h be given by (8.3). Then the map $k : [0,1] \times S^1 \to S^1$ defined by

$$k(t, \theta) := \exp\left(t h(\theta) + d\,(1-t)\theta\right)$$

establishes a homotopy of f to the map $\varphi : S^1 \to S^1$, $\varphi(z) = z^d$. Therefore, if f_0 and f_1 have the same degree d and base point $1 \in S^1$, then they are both homotopic to the same map $\varphi(z) = z^d$. $\qquad\square$

Finally we observe that $\deg(z^d) = d \; \forall d \in \mathbb{Z}$ and that, if f and g have the same base point, $\deg(g * f) = \deg(g) + \deg(f)$.

b. An integral formula for the degree

Let $f : S^1 \to S^1$ and let $h : \mathbb{R} \to \mathbb{R}$ be as in (8.3). Clearly, thinking of θ as a time variable, $h(\theta)$ is the angle evolution of the point $f(e^{i\theta})$ on the circle. The degree of f corresponds to the total angle evolution, that is to the number of revolutions that $f(z)$ does as z goes around S^1 once counterclockwise, counting the revolutions positively if $f(z)$ goes counterclockwise and negatively if $f(z)$ goes clockwise.

Suppose $f : [0, 2\pi] \to S^1$ is a loop of class C^1, that is $\theta \to f(e^{i\theta})$ is of class C^1, and let $h : \mathbb{R} \to \mathbb{R}$ be as in (8.3). Differentiating (8.3) we get

$$i e^{i\theta} f'(e^{i\theta}) = i f(1) e^{i h(\theta)} h'(\theta) = i f(e^{i\theta}) h'(\theta)$$

and taking the modulus $|h'(\theta)| = |f'(e^{i\theta})|$. Therefore, h' is the angular velocity of $f(z)$ times ± 1 depending on the direction of motion of $f(z)$ when z moves as $e^{i\theta}$ on the unit circle. In coordinates, writing $f := f_1 + i f_2$, we have $f' = f_1' + i f_2'$, hence

$$h'(\theta) = e^{i\theta} \frac{f'(e^{i\theta})}{f(e^{i\theta})} = i e^{i\theta}\left(- f_1'(e^{i\theta}) f_2(e^{i\theta}) + f_2'(e^{i\theta}) f_1(e^{i\theta}) \right).$$

We conclude using the fundamental theorem of calculus

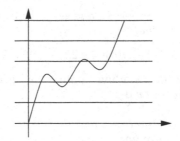

Figure 8.12. Counting the degree.

8.70 Proposition (Integral formula for the degree). *Let $f : S^1 \to S^1$ be of class C^1. Then the lift h of f in (8.3) is given by*

$$h(t) = \int_0^t h'(\theta)\, d\theta = \int_0^t ie^{i\theta}\Big(- f_2(e^{i\theta})f_1'(e^{i\theta}) + f_1(e^{i\theta})f_2'(e^{i\theta}) \Big)\, d\theta. \quad (8.4)$$

In particular

$$
\begin{aligned}
\deg(f) &= \frac{1}{2\pi} \int_0^{2\pi} h'(\theta)\, d\theta \\
&= \frac{1}{2\pi} \int_0^{2\pi} ie^{i\theta}\Big(- f_2(e^{i\theta})f_1'(e^{i\theta}) + f_1(e^{i\theta})f_2'(e^{i\theta}) \Big)\, d\theta.
\end{aligned} \quad (8.5)
$$

One can define the lifting and degree of smooth maps by (8.5), showing the homotopy invariance in the context of regular maps, and then extending the theory to continuous functions by an approximation procedure.

c. Degree and inverse image

The degree of $f : S^1 \to S^1$ is strongly related to the number of roots of the equation $f(x) = y$ counted with a suitable sign.

8.71 Proposition. *Let $f : S^1 \to S^1$ be a continuous map with degree $d \in \mathbb{Z}$. For every $y \in S^1$, there exist at least $|d|$ points $x_1, x_2, \ldots, x_d$ in S^1 such that $f(x_i) = y$, $i = 1, \ldots, d$. Furthermore, if $f : S^1 \to S^1$ goes around S^1 never turning back, i.e., if $f(e^{i\theta}) = e^{ih(\theta)}$ where $h : [0, 2\pi] \to \mathbb{R}$ is strictly monotone, then the equation $f(x) = y$, $y \in S^1$, has exactly $|d|$ solutions.*

Proof. Let $h : \mathbb{R} \to \mathbb{R}$ be as in (8.3) so that $h(2\pi) = 2\pi d$, and let $s \in [0, 2\pi[$ be such that $e^{is} = y$. For convenience suppose $d > 0$.

The intermediate value theorem yields d distinct points $\theta_1, \theta_2, \ldots, \theta_d$ in $[0, 2\pi[$ such that $h(\theta_1) = s$, $h(\theta_2) = s + 2\pi$, $\ldots$, $h(\theta_d) = s + 2(d - 1)\pi$, hence at least d distinct points $x_1, x_2, \ldots, x_d$ such that $f(x_j) = f(e^{i\theta_j}) = e^{ih(\theta_j)} = e^{is} = y$, see Figure 8.12. They are of course exactly d points $x_1, x_2, \ldots, x_d$ if h is strictly monotone. $\quad\square$

With the previous notation, suppose $f : S^1 \to S^1$ is of class C^1, let h be as in (8.3) and let $y \in S^1$ and $s \in [0, 2\pi[$ be such that $y = e^{is}$. Assume that y is chosen so that the equation $f(x) = y$ has a finite number of solutions and set

$$N_+(f, y) := \#\left\{\theta \in S^1 \,\middle|\, h(\theta) = s \pmod{2\pi}, \ h'(\theta) > 0\right\},$$

$$N_-(f, y) := \#\left\{\theta \in S^1 \,\middle|\, h(\theta) = s \pmod{2\pi}, \ h'(\theta) < 0\right\}.$$

Then one sees that

$$\deg(f) = N_+(f, y) - N_-(f, y), \tag{8.6}$$

see Figure 8.12.

8.72 The fundamental theorem of algebra. Using the degree theory we can easily prove that *every complex polynomial*

$$P(z) := z^m + a_1 z^{m-1} + \cdots + a_{m-1} z + a_0$$

has at least a complex root. Set $S_\rho^1 := \{z \,|\, |z| = \rho\}$. For ρ sufficiently large, P maps S_ρ^1 in $\mathbb{R}^2 \setminus \{0\}$. Also $\deg(P/|P|) = m$. In fact, by considering the homotopy

$$P_t(z) := z^m + t(a_1 z^{m-1} + \cdots + a_0) \qquad t \in [0, 1],$$

of $P(z)$ to z^m, we have

$$|P_t(z)| \geq |z|^m \left(1 - t\left(\frac{|a_1|}{|z|} + \cdots + \frac{|a_0|}{|z|^m}\right)\right) \qquad \forall |z| \neq 0.$$

Thus $|P_t(z)| > 0 \ \forall \ t \in [0, 1]$ provided $|z|$ is large enough, consequently $P_t(z)/|P_t(z)|$, $t \in [0, 1]$, $z \in S^1$, establishes a homotopy of $P/|P|$ to z^m from S_ρ^1 into S^1, and we conclude that

$$\deg(P/|P|) = \deg(z^m) = m.$$

8.73 ¶. Show that $f : S^1 \to S^1$ has at least $d - 1$ fixed points if $\deg f = d$. [*Hint:* See Figure 8.12.]

d. The homological definition of degree for maps $S^1 \to S^1$

Let $f : S^1 \to \Sigma^1$ be a continuous map, where for convenience we have denoted the target space S^1 by Σ^1. We fix in S^1 and Σ^1 two orientations, for instance the counterclockwise orientation, and we divide S^1 in small arcs whose images by f do not contain antipodal points (this is possible since $f : S^1 \to \Sigma^1$ is uniformly continuous) and let $z_1, \ldots, z_n, z_{n+1} = z_1 \in S^1$ the points of such subdivision indexed according to the chosen orientation in S^1. For each $i = 1 \ldots, n$ we denote by α_i the minimal arc connecting $f(z_i)$ with $f(z_{i+1})$. We give it the positive sign if $f(z_i)$ precedes $f(z_{i+1})$ with respect to the chosen orientation of Σ^1, negative otherwise. Finally, for $\zeta \in \Sigma^1$, $\zeta \neq f(z_i) \ \forall \ i$, we denote by $p(\zeta)$ and $n(\zeta)$ the number of arcs α_i respectively, positive and negative that contain ζ. Then

$$p(\zeta) - n(\zeta) = \deg(f) \in \mathbb{Z}$$

as we can see looking at the lift of f.

Figure 8.13. Frontispiece of lecture notes by Louis Nirenberg and a page from a paper by L. E. Brouwer (1881–1966).

8.2 Some Results on the Topology of $\mathbb{R}^n$

Though the presentation of these topics would require more space and advanced techniques, and in any case, it leads us away from the main path, we think that it is worthwhile to present here some results that are relevant in the sequel. However, we shall confine ourselves to illustrate the ideas and refer to the literature for complete proofs and more details. In the next two paragraphs we collect a few relevant results on the topology of maps into S^n that we freely use in the rest of this section.

8.2.1 Brouwer's theorem

a. Brouwer's degree

A topological degree, called *Brouwer's degree*, can be defined for continuous maps $f : S^n \to S^n$, $n \geq 2$, either by extending the homological type arguments in the case $n = 1$ or, more generally, in terms of *homology groups* or, analytically, in terms of a sum with sign of the numbers of inverse images of a point, either pointwise or in the mean. Intuitively, one counts *how many times* the target S^n is covered algebraically by the source S^n via the map f.[1]

---

[1] Both approaches require the development of more advanced and relevant techniques; we refer the reader e.g., to

- J. Dugundji, *Topology*, Allyn and Bacon, 1966,
- L. Nirenberg, *Topics in Nonlinear Analysis*, AMS-CIMS, New York, 2001,

In this way we end up with a map

$$\deg : C^0(S^n, S^n) \to \mathbb{Z}$$

such that

(i) $\deg(\mathrm{Id}) = 1$,
(ii) $\deg(f) = 0$ if f is constant,
(iii) $\deg(f) = (-1)^{n+1}$ if $f(x) = -x$,

and we have the following.

8.74 Theorem (Brouwer). *Let $f_0, f_1 : S^n \to S^n$ be continuous and homotopic. Then $\deg(f_0) = \deg(f_1)$.*

Indeed the degree completely characterizes the homotopy classes of continuous maps from S^n into S^n. In fact, we have the following.

8.75 Theorem (Hopf). *Two continuous maps of S^n into itself are homotopic if and only if they have the same degree. Moreover, for each $d \in \mathbb{Z}$ there is a map $f : S^n \to S^n$ with $\deg(f) = d$.*

A map $f : S^n \to S^n \subset \mathbb{R}^{n+1}$ is called *antipodal* if $f(-x) = -f(x)$ $\forall x \in S^n$. For instance, $\mathrm{Id} : S^n \to S^n$ and $-\mathrm{Id} : S^n \to S^n$ are antipodal.

8.76 Theorem (Borsuk antipodal theorem). *Let $f : S^n \to S^n$ be a continuous antipodal map. Then $\deg(f)$ is odd; in particular f is not homotopic to a constant map.*

b. Extension of maps into S^n

The following two extension theorems for maps into S^n are also crucial. We refer the reader e.g., to J. Dugundji, *Topology*, Allyn and Bacon, 1966,

8.77 Theorem. *We have the following.*

(i) *Let $A \subset S^n$ be a closed set. Every continuous map $f : A \to S^n$ extends to a continuous map $F : S^n \to S^n$.*
(ii) *Let $A \subset S^{n+1}$ be closed and $f : A \to S^n$ be continuous. Pick a point $x_i \in U_i$ in every bounded connected component U_i of $A^c := S^{n+1} \setminus A$. Then there is a continuous extension $F : S^{n+1} \setminus \cup_i \{p_i\} \to S^n$.*

8.78 Theorem (Borsuk). *Let $A \subset \mathbb{R}^k$, $k \geq 1$, be closed and let $f : A \to S^n$ be continuous. Then f can be extended to a continuous map $F : \mathbb{R}^k \to S^n$ if and only if f is homotopic to a constant map.*

Observing that A^c has a unique unbounded connected component if A is a compact subset of $\mathbb{R}^n$, and using the stereographic projection, it is not difficult to infer from Theorem 8.77 (i), (ii).

---

o one of the several books on degree theory.

8.79 Theorem. *We have the following.*

(i) *Let $A \subset \mathbb{R}^n$ be compact. Then any continuous function from A into S^n can be extended to a continuous $F : \mathbb{R}^n \to S^n$.*

(ii) *Let $A \subset \mathbb{R}^{n+1}$ be compact and $f : A \to S^n$ be continuous. Pick a point $p_i \in U_i$ in every connected component U_i of A^c. Then f can be extended to a continuous map $F : \mathbb{R}^{n+1} \setminus \cup_i \{p_i\} \to S^n$.*

As a consequence of the Hopf theorem and Proposition 8.8 we immediately infer the following.

8.80 Proposition. *A function $f : S^n \to S^n$ has a continuous extension $F : \mathrm{cl}\,(B^{n+1}) \to S^n$ if and only if $\deg(f) = 0$.*

8.81 Corollary. *Let $f : S^n \to \mathbb{R}^{n+1} \setminus \{0\}$ be a continuous map. Then there exists a continuous extension $F : \mathrm{cl}\,(B^{n+1}) \to \mathbb{R}^{n+1} \setminus \{0\}$ of f if and only if $\deg(f/|f|) = 0$.*

c. Brouwer's fixed point theorem

Since the identity from S^n into S^n has degree one, and the constant maps have degree zero, from the homotopic invariance of the degree we conclude the following.

8.82 Theorem (Brouwer). *The identity map* $\mathrm{Id} : S^n \to S^n$ *is not homotopic to a constant map.*

In other words, we cannot peel an orange without piercing the peel.

Brouwer's theorem, whose content is quite intuitive, at least in dimension $n = 2$, has several interesting and surprising consequences. In fact, we have the following.

8.83 Theorem. *The following claims are equivalent*

(i) (BROUWER'S THEOREM) *The identity map* $\mathrm{Id} : S^n \to S^n$ *is not homotopic to a constant map.*

(ii) *There is no continuous map $F : B \to S^n$, $B = \mathrm{cl}\,(B^{n+1})$, such that $F(x) = x \; \forall x \in S^n$, that is, S^n is not a retract of B.*

(iii) (BROWER'S FIXED POINT THEOREM, I) *Every continuous map $f : B \to B$, $B := \mathrm{cl}\,(B^{n+1})$, has a fixed point, i.e., there is at least one $x \in B$ such that $f(x) = x$.*

Proof. (i) $\Rightarrow$ (ii) If $F : B \to S^n$ is a continuous function with $F(x) = x \; \forall x \in S^n$, then $H(t, x) := F(tx)$, $(t, x) \in [0, 1] \times S^n$, is a homotopy of the identity to $F(0)$. A contradiction.

(ii) $\Rightarrow$ (iii) Suppose that there is a continuous $F : B \to B$ such that $F(x) \neq x$ for all $x \in B$. Then, and we leave this to the reader, the map $G : B \to S^n$ that maps $x \in B$ into the unique point of S^n on the half-line from $f(x)$ to x would be a continuous map from B in S^n with $G(x) = x \; \forall x \in S^n$, contradicting (ii).

(iii) $\Rightarrow$ (i) Suppose that there is a homotopy $H : [0,1] \times S^n \to S^n$ between the identity and a constant map, $H(1,x) = x$, $H(0,x) = p \in S^n$. Then the function $F : B \to S^n$ defined by

$$F(x) := \begin{cases} H\left(|x|, \frac{x}{|x|}\right) & \text{if } x \neq 0, \\ p & \text{if } x = 0, \end{cases}$$

would be a continuous extension of the identity on S^n to B, hence $-F(x) : B \to B$ would have no fixed point. $\qquad\square$

8.84 ¶. Let $U \subset \mathbb{R}^{n+1}$ be a bounded open set. Prove that there exists no continuous retraction $r : \overline{U} \to \partial U$ with $r(x) = x$ on ∂U. [*Hint:* Let $0 \in U$, $B(0,k) \supset \overline{U}$ and consider the continuous map $f : \overline{B(0,k)} \to \overline{B(0,k)}$ defined by

$$f(x) := \begin{cases} k \dfrac{r(x)}{|r(x)|} & \text{if } x \in \overline{U}, \\ k \dfrac{x}{|x|} & \text{if } x \in \overline{B(0,k)} \setminus U.] \end{cases}$$

d. Fixed points and solvability of equations in $\mathbb{R}^{n+1}$

Going through the proof of Theorem 8.83, we can deduce a number of results concerning the solvability of equations of the type $F(x) = 0$.

Let $f : S^n \to \mathbb{R}^{n+1} \setminus \{0\}$ be a continuous map. Since f never vanishes, the map $f/|f|$ continuously maps S^n into S^n. We call *degree of f with respect to the origin* the number

$$\deg(f, 0) := \deg\left(\frac{f}{|f|}\right).$$

8.85 Proposition. *Let* $f : S^n \to \mathbb{R}^{n+1} \setminus \{0\}$ *be a continuous map with* $\deg(f, 0) \neq 0$. *Then every extension* $F : B \to \mathbb{R}^n$ *of* f, $B := \mathrm{cl}\,(B^{n+1})$, *has a zero in* B^{n+1}.

Proof. Suppose this is not true. Then there exists a continuous extension $F : B \to \mathbb{R}^{n+1} \setminus \{0\}$ of f. Hence $F(x)/|F(x)|$ is a continuous map from B into S^n. According to Proposition 8.80, $F(x)/|F(x)| = f(x)/|f(x)|$ has degree zero, a contradiction. $\qquad\square$

Let us illustrate a few situations in which Proposition 8.85 applies.

8.86 Proposition. *Let* $F : \mathrm{cl}\,(B^{n+1}) \to \mathbb{R}^{n+1}$ *be a continuous map such that* $F(x)$ *never points opposite to x for all* $x \in S^n$. *Then* $F(x) = 0$ *has a solution.*

Proof. Let $f := F_{|S^n} : S^n \to \mathbb{R}^{n+1}$. Since, by assumption $F(x) + \lambda x \neq 0 \;\forall\, \lambda \geq 0$, $\forall x \in S^n$, f has no zeros and therefore

$$h(t, x) := tf(x) + (1 - t)x, \quad t \in [0,1], x \in S^n,$$

never vanishes. Hence $h(t,x)/|h(t,x)|$ is a homotopy of $f/|f| : S^n \to S^n$ to the identity map $\mathrm{Id} : S^n \to S^n$. It follows that $\deg(f, 0) = \deg(f/|f|) = 1$. We conclude, on account of Proposition 8.85 that F, being an extension of f, has at least one zero in B^{n+1}. $\qquad\square$

8.87 Theorem (Brouwer's fixed point, II). *Let $F : B \to \mathbb{R}^n$, $B :=$ cl(B^n), be a continuous map with $F(\partial B) \subset B$. Then F has a fixed point.*

Proof. Set $\phi(x) := x - F(x)$, $x \in B$, and suppose $\phi(x) \neq 0 \ \forall x$, otherwise we are through. In this case ϕ never points opposite for each $x \in \partial B$. Indeed, if $x - F(x) + \lambda x = 0$ for some $\lambda \geq 0$ and $x \in \partial B$, then $F(x) = (1+\lambda)x$. Now $\lambda > 0$ is impossible since $|F(x)| \leq 1$, and, if $\lambda = 0$, then $F(x) = x$ on ∂B which we have ruled out. Thus, $F(x) - x = 0$ has a solution inside B. $\square$

It is worth noticing that Brouwer's theorem still holds if we replace cl(B^n) with any set which is homeomorphic to the closed ball of $\mathbb{R}^n$. Moreover, it also holds in the following form.

8.88 Theorem (Brouwer fixed point theorem, III). *Every continuous map $f : K \to K$ from a convex compact set K into itself has a fixed point.*

Proof. According to Dugundji's theorem, Theorem 6.42, f has a continuous extension $F : \mathbb{R}^n \to \mathbb{R}^n$, whose image is contained in K, K being convex and closed. If B is a ball containing K, then $F(B) \subset B$ and by Brouwer's fixed point theorem, Theorem 8.87, F has a fixed point $x \in B$, i.e., $F(x) = x$, and, since $F(x) \in K$, we conclude that $x \in K$. $\square$

e. Fixed points and vector fields

Every $(n + 1)$-dimensional *vector field* in a domain $A \subset \mathbb{R}^{n+1}$ may be regarded as a map $\varphi : A \subset \mathbb{R}^{n+1} \to \mathbb{R}^{n+1}$, once we fix the coordinates. If φ is continuous and nonzero, the degree of φ with respect to the origin is called the *characteristic* of the vector field. The Brower's degree properties and Proposition 8.8 then read in terms of vector fields as follows.

8.89 Proposition. *We have the following.*

(i) (BROUWER) *Let φ be a nonvanishing vector field in cl(B^{n+1}). Then $\varphi_{|S^n}$ has characteristic zero.*

(ii) *The outward normal to B^{n+1} at $x \in S^n = \partial B^{n+1}$ is x. Therefore the outward normal field to S^n, $x \to x/|x|$, $x \in \mathbb{R}^{n+1} \setminus \{0\}$, has characteristic one.*

(iii) *The inward normal at $x \in S^n$ is $-x$. Therefore the inward normal field to S^n, $x \to -x/|x|$, $x \in \mathbb{R}^{n+1} \setminus \{0\}$, has characteristic $(-1)^{n+1}$.*

(iv) *Let φ and ψ be two continuous nonvanishing vector fields on S^n that are never opposite on S^n. Then φ and ψ have the same characteristic.*

Let us draw some consequences.

8.90 Proposition. *Each nonvanishing vector field on $\varphi :$ cl$(B^{n+1}) \to \mathbb{R}^{n+1}$ must contain at least an inward normal and an outward normal vector.*

Proof. In fact, $\varphi_{|S^n}$ has characteristic zero by (i) Proposition 8.89. Since $\varphi_{|S^n}$ and the field of outward (inward) normals have different characteristics, we infer from Proposition 8.89 (iv) that $\varphi_{|S^n}$ must contain an inward (outward) normal. $\square$

8.91 Theorem (Poincaré–Brouwer). *Every continuous nonvanishing vector field on an even-dimensional sphere S^{2n} must contain at least one normal vector. In particular, there can be no continuous nonvanishing tangential vector fields to S^n.*

Proof. By (ii), (iii) Proposition 8.89, the inward and outward normal vector fields in S^{2n} have different characteristics. Since any unitary vector field must have characteristics differing from one of these two fields, the result follows from (iv) Proposition 8.89.   □

8.92 Proposition. *Let $f : S^{2n} \to S^{2n}$ be a continuous map. Then either f has a fixed point $x = f(x)$, or there is an $x \in S^{2n}$ such that $f(x) = -x$.*

Proof. Suppose $f : S^{2n} \to S^{2n}$ has no fixed point. Then the vector field $g : S^{2n} \to S^{2n}$ given by $g(x) := \frac{f(x)-x}{|f(x)-x|}$, $x \in S^{2n}$, is continuous and of modulus one. Thus it contains a normal vector, i.e., $f(x) - x = \lambda x$ for some $x \in S^n$ and $\lambda \in \mathbb{R}$. Since $|f(x)| = |x| = 1$ we infer $1 = |f(x)| = |\lambda + 1||x| = |\lambda + 1|$, i.e., either $\lambda = 0$ or $\lambda = -2$. We cannot have $\lambda = 0$ since otherwise $f(x) - x = 0$ and x would be a fixed point. Thus necessarily $\lambda = -2$, i.e., $f(x) = -x$.   □

8.93 ¶. Let $\phi : \mathbb{R}^n \to \mathbb{R}^n$ be a continuous map that is *coercive*, that is

$$\frac{(\phi(x) \mid x)}{|x|} \to \infty \qquad \text{uniformly as } |x| \to \infty.$$

Show that ϕ is onto $\mathbb{R}^n$. [*Hint:* Show that for every $x, y \in \mathbb{R}^n$ $\phi(x) - y$ never points opposite to x for $|x| = R$, R large.]

8.94 ¶. Let $\phi : \mathbb{R}^n \to \mathbb{R}^n$ be a continuous map such that

$$\limsup_{|x| \to +\infty} \frac{|\phi(x)|}{|x|} < 1.$$

Show that ϕ has a fixed point.

8.95 ¶. Let us state another equivalent form of Brouwer's fixed point theorem.

Theorem (Miranda) *Let $f \cdot Q := \{x \in \mathbb{R}^n \mid |x_i| \leq 1, \ i = 1, \ldots, n\} \to \mathbb{R}^n$ be a continuous map such that for $i = 1, \ldots, n$ we have*

$$f_i(x_1, \ldots, x_{i-1}, -1, x_{i+1}, \ldots, x_n) \geq 0,$$
$$f_i(x_1, \ldots, x_{i-1}, 1, x_{i+1}, \ldots, x_n) \leq 0.$$

Then there is at least one $x \in Q$ such that $f(x) = 0$.

Show the equivalence between the above theorem and Brouwer's fixed point theorem. [*Hint:* To prove the theorem, first assume that strict inequalities hold. In this case show that for a suitable choice of $\epsilon_1, \ldots, \epsilon_n \in \mathbb{R}$ the transformation

$$x'_i = x_i + \epsilon_i f_i(x), \qquad i = 1, \ldots, n,$$

maps Q into itself, and use Brouwer's theorem. In the general case, apply the above to $f(x) - \delta x$ and let δ tend to 0.

Conversely, if F maps Q into itself, consider the maps $f_i(x) = F_i(x) - x_i$, $i = 1, \ldots, n$.]

8.96 ¶. Show that there is a nonvanishing tangent vector field on an odd-dimensional sphere S^{2n-1}. [*Hint:* Think $S^{2n-1} \subset \mathbb{R}^{2n}$. Then the field

$$x = (x_1, x_2, \ldots, x_{2n}) \to (-x_{n+1}, -x_{n+2}, \ldots, x_{2n}, x_1, x_2, \ldots, x_n)$$

defines a map from S^{2n-1} into itself that has no fixed point.]

Figure 8.14. Karol Borsuk (1905–1982) and a page from one of his papers.

8.2.2 Borsuk's theorem

Also Borsuk's theorem, Theorem 8.76, has interesting equivalent formulations and consequences.

8.97 Theorem. *The following statements hold and are are equivalent.*

(i) (BORSUK–ULAM) *There is no continuous antipodal map $f : S^n \to S^{n-1}$.*

(ii) *Each continuous $f : S^n \to \mathbb{R}^n$ sends at least one pair of antipodal points to the same point.*

(iii) (LYUSTERNIK–SCHNIRELMANN) *In each family of $n+1$ closed subsets covering S^n at least one set must contain a pair of antipodal points.*

Proof. Borsuk's theorem $\Rightarrow$ (i) If $f : S^n \to S^{n-1}$ is a continuous antipodal map, and if we regard S^{n-1} as the equator of S^n, $S^{n-1} \subset S^n$, f would give us a nonsurjective map $f : S^n \to S^n$, hence homotopic to a constant. On the other hand f has odd degree by Borsuk's theorem, a contradiction.

(i) $\Rightarrow$ (ii) Suppose that there is a continuous $g : S^n \to \mathbb{R}^n$ such that $g(x) \neq g(-x)$. Then the map $f : S^n \to S^{n-1}$ defined by

$$f(x) := \frac{g(-x) - g(x)}{|g(-x) - g(x)|}$$

would yield a continuous antipodal map.

(ii) $\Rightarrow$ (iii) Let $F_1, \ldots, F_{n+1}$ be $n + 1$ closed sets covering S^n and let $\alpha : S^n \to S^n$ be the map $\alpha(x) = -x$. Suppose that $\alpha(F_i) \cap F_i = \emptyset$ for all $i = 1, \ldots, n$. Then we can find continuous functions $g_i : S^n \to [0, 1]$ such that $g_i^{-1}(0) = F_i$ and $g_i^{-1}(1) = \alpha(F_i)$. Next we define $g : S^n \to \mathbb{R}^n$ as $g(x) = (g_1(x), \ldots, g_n(x))$. By the assumption there is $x_0 \in S^n$ such that $g_i(x_0) = g_i(-x_0)\ \forall\ i$, thus $x_0 \notin \overset{n}{\underset{i=1}{\cup}} F_i$ and $x_0 \notin \overset{n}{\underset{i=1}{\cup}} \alpha(F_i)$, consequently $x_0 \in F_{n+1} \cap \alpha(F_{n+1})$, a contradiction.

(iii) $\Rightarrow$ (i) Let $f : S^n \to S^{n-1}$ be a continuous map. We decompose S^{n-1} into $(n+1)$ closed sets $A_1, \ldots, A_{n+1}$ each of which has diameter less than two; this is possible by projecting the boundary of an n-simplex enclosing the origin and S^{n-1}. Defining $F_i := f^{-1}(A_i)$, $i = 1, \ldots, n+1$, according to the assumption there is an x_0 and a k such that $x_0 \in F_k \cap \alpha(F_k)$. But then $f(x_0)$ and $f(-x_0)$ belong to F_k and so f cannot be antipodal. $\qquad\square$

8.98 Theorem. $\mathbb{R}^n$ *is not homeomorphic to* $\mathbb{R}^m$ *if* $n \neq m$.

Proof. Suppose $n > m$ and let $h : \mathbb{R}^n \to \mathbb{R}^m$ be a continuous map. Since $n - 1 \geq m$, from (ii) Theorem 8.97 we conclude that $h_{|S^{n-1}} : S^{n-1} \to \mathbb{R}^m \subset \mathbb{R}^{n-1}$ must send two antipodal points into the same point, so that h cannot be injective. $\qquad\square$

8.99 Remark. As a curiosity, (ii) of Theorem 8.97 yields that at every instant there are two antipodal points in the earth with the same temperature and atmospheric pressure.

8.100 ¶. Show that every continuous map $f : S^n \to S^n$ such that $f(x) \neq f(-x)\ \forall\ x$ is surjective.

8.2.3 Separation theorems

8.101 Definition. *We say that a set* $A \subset \mathbb{R}^{n+1}$ *separates* $\mathbb{R}^{n+1}$ *if its complement* $A^c := \mathbb{R}^{n+1} \setminus A$ *is not connected.*

8.102 Theorem. *Let* $A \subset \mathbb{R}^{n+1}$ *be compact. Then*

(i) *each connected component of* $\mathbb{R}^{n+1} \setminus A$ *is a path-connected open set,*

(ii) A^c *has exactly one unbounded connected component,*

(iii) *the boundary of each connected component of* A^c *is contained in* A,

(iv) *if* A *separates* $\mathbb{R}^{n+1}$, *but no proper subset does so, then the boundary of each connected component of* A^c *is exactly* A.

Proof. (i) follows e.g., from Corollary 6.68, since connected components of A^c are open sets.

(ii) Let $\overline{B}$ be a closed ball such that $\overline{B} \supset A$. Then B^c is open, connected and $\overline{B}^c \subset A^c$. Thus $\overline{B}^c$ is contained in a unique connected component of A^c.

(iii) Let U be any connected component of A^c and $x \in \partial U$. We claim that x does not belong to any connected component of A^c, consequently $x \notin A^c$. In fact, $x \notin U$, and, if x was in some component V, there would exist $B(x, \epsilon) \subset V$. $B(x, \epsilon)$ would then also intersect U, thus $U \cap V \neq \emptyset$: a contradiction.

(iv) Let U be any connected component of A^c. Since A separates $\mathbb{R}^{n+1}$, there is another connected component V of A^c and, because $V \subset \mathbb{R}^{n+1} \setminus \overline{U}$, necessarily $\mathbb{R}^{n+1} \setminus \overline{U} \neq \emptyset$. Consequently $\mathbb{R}^{n+1} \setminus \partial U$ splits as

$$\mathbb{R}^{n+1} \setminus \partial U = U \cup (\mathbb{R}^{n+1} \setminus \overline{U})$$

which are disjoint and nonempty, so ∂U separates $\mathbb{R}^{n+1}$. Since by (iii) $\partial U \subset A$ and is closed, it follows from the hypotheses on A that $\partial U = A$. $\qquad\square$

8.103 Theorem (Borsuk's separation theorem). *Let $A \subset \mathbb{R}^{n+1}$ be compact. Then A separates $\mathbb{R}^{n+1}$ if and only if there exists a continuous map $f : A \to S^n$ that is not homotopic to a constant.*

Proof. Define the map $\beta_{p|A}$ as

$$\beta_p : \mathbb{R}^{n+1} \setminus \{p\} \to S^n, \qquad x \to \frac{x - p}{|x - p|}.$$

Assume that A separates $\mathbb{R}^{n+1}$. Then $\mathbb{R}^{n+1} \setminus A$ has at least one bounded component U. Choosing any $p \in U$ we shall show that $\beta_{p|A}$ cannot be extended to a continuous function on the closed set $A \cup U$, consequently on $\mathbb{R}^{n+1}$; hence $\beta_{p|A}$ is not homotopic to a constant map by Proposition 8.8. In fact, if $F : A \cup U \to S^n$ were a continuous extension of $\beta_{p|A}$ we choose $R > 0$ such that $B(p, R) \supset A \cup U$ and define $g : \overline{B(p, R)} \to \partial B(p, R)$ as

$$g(x) := \begin{cases} p + R\dfrac{x - p}{|x - p|} & \text{if } x \in \overline{B(p, R)} \setminus U, \\[2mm] p + RF(x) & \text{if } x \in \overline{U}. \end{cases}$$

Then g would be continuous in $\overline{B(p, R)}$ and $g = \text{Id}$ on $\partial B(p, R)$: this contradicts Brouwer's theorem.

Conversely, suppose that A does not separate A^c. Then A^c has exactly one connected component, which is necessarily unbounded. By Theorem 8.79, f extends to $F : \mathbb{R}^{n+1} \to S^n$. Therefore F and consequently $f = F_{|A}$ are homotopic to a constant map. $\qquad\square$

In particular, Borsuk's separation theorem tells us that the separation property is invariant by homeomorphisms.

8.104 Corollary. *Let A be a compact set in $\mathbb{R}^n$ and let $h : A \to \mathbb{R}^n$ be a homeomorphism onto its image. Then A separates $\mathbb{R}^n$ if and only if $h(A)$ separates $\mathbb{R}^n$.*

As a consequence we have the following.

8.105 Theorem (Jordan's separation theorem). *A homeomorphic image of S^n in $\mathbb{R}^{n+1}$ separates $\mathbb{R}^{n+1}$, and no proper closed subset of S^n does so. In particular $h(S^n)$ is the complete boundary of each connected component of $\mathbb{R}^{n+1} \setminus h(S^n)$.*

It is instead much more difficult to prove the following general *Jordan's theorem*.

8.106 Theorem (Jordan). *Let $h : S^n \to \mathbb{R}^{n+1}$ be a homeomorphism between S^n and its image. Then $\mathbb{R}^{n+1} \setminus h(S^n)$ has exactly two connected components, each having $h(S^n)$ as its boundary.*

Jordan's theorem in the case $n = 1$ is also known as the *Jordan curve theorem*. We also have

8.107 Theorem (Jordan–Borsuk). *Let K be a compact subset of $\mathbb{R}^{n+1}$ such that $\mathbb{R}^n \setminus K$ has k connected components, and let h be a homeomorphism of K into its image on $\mathbb{R}^{n+1}$. Then $\mathbb{R}^{n+1} \setminus h(K)$ has k connected components.*

Particularly relevant is the following theorem that follows from Borsuk's separation theorem, Theorem 8.103.

8.108 Theorem (Brouwer's invariance domain theorem). *Let U be an open set of $\mathbb{R}^{n+1}$ and let $h : U \subset \mathbb{R}^{n+1} \to \mathbb{R}^{n+1}$ be a homeomorphism between U and its image. Then $h(U)$ is an open set in $\mathbb{R}^{n+1}$.*

Proof. Let $y \in h(U)$. We shall show that there is an open set $W \subset \mathbb{R}^{n+1}$ such that $y \in W \subset h(U)$. Set $x = h^{-1}(y)$, and $B := B(x, \epsilon)$ so that $\overline{B} \subset U$. Then

(i) $\mathbb{R}^{n+1} \setminus h(\overline{B})$ is connected by Corollary 8.104 since $\overline{B}$ is homeomorphic to $\overline{B(0, 1)}$ and $\overline{B(0, 1)}$ does not separate $\mathbb{R}^{n+1}$,

(ii) $h(\overline{B}, \partial B) = h(\overline{B}) \setminus h(\partial B)$ is connected since it is homeomorphic to $B(x, \epsilon)$.

By writing

$$\mathbb{R}^{n+1} \setminus h(\partial B) = \left(\mathbb{R}^n \setminus h(\overline{B}) \right) \cup h(\overline{B} \setminus \partial B)$$

we see that $\mathbb{R}^n \setminus h(\partial B)$ is the union of two nonempty, disjoint connected sets, that are necessarily the connected components of $\mathbb{R}^n \setminus h(\partial B)$; since $h(\partial B)$ is compact, they are also open in $\mathbb{R}^{n+1}$. Thus we can take $W := h(\overline{B} \setminus \partial B)$. □

A trivial consequence of the domain invariance theorem is that if A is *any* subset of $\mathbb{R}^{n+1}$ and $h : A \to \mathbb{R}^{n+1}$ is a homeomorphism between A and its image $h(A)$, then h maps the interior of A onto the interior of $h(A)$ and the boundary of A onto the boundary of $h(A)$.

Using Theorem 8.108 we can also prove

8.109 Theorem. $\mathbb{R}^n$ *and* $\mathbb{R}^m$ *are not homeomorphic if* $n \neq m$.

Proof. Suppose $m > n$. If $\mathbb{R}^n$ were homeomorphic to $\mathbb{R}^m$, then the image of $\mathbb{R}^n$ into $\mathbb{R}^m$ under such homeomorphism would be open in $\mathbb{R}^m$. However, the image is not open under the map $(x_1, \ldots, x_n) \to (x_1, \ldots, x_n, 0, \ldots, 0)$. □

8.3 Exercises

8.110 ¶ Euler's formula. Prove Euler's formula for convex polyhedra in $\mathbb{R}^3$: $V - E + F = 2$, where $V := \#$ vertices, $E := \#$ edges, $F := \#$ faces, see Theorem 6.60 of [GM1]. [*Hint:* By taking out a face, deform the polyhedral surface into a plane polyhedral surface for which $V - E + F$ decreases by one. Thus it suffices to show that for the plane polyhedral surface we have $V - E + F = 1$. Triangularize the face, noticing that this does not change $V - E + F$; eliminate from the exterior the triangles, this does not change $V - E + F$ again, reducing in this way to a single triangle for which $V - E + F = 3 - 3 + 1 = 1$.]

8.111 ¶. Prove

Proposition. *Let Δ be an open set of $\mathbb{C}$. Δ is simply connected if and only if Δ is path-connected and $\mathbb{C} \setminus \Delta$ has no compact connected components.*

[*Hint:* Use Jordan's theorem to show that Δ^c has a bounded connected component if Δ is not simply connected. To prove the converse, use that $\mathbb{R}^2 \setminus \{x_0\}$ is not simply connected.]

8.112 ¶. Prove

Theorem (Perron–Frobenius). *Let $A = [a_{ij}]$ be an $n \times n$ matrix with $a_{ij} \geq 0$ $\forall i, j$. Then A has an eigenvector x with nonnegative coordinates corresponding to a nonnegative eigenvalue.*

[*Hint:* If $Ax = 0$ for some $x \in D := \{x \in \mathbb{R}^n \mid x^i \geq 0 \ \forall i, \ \sum_{i=1}^n x^i = 1\}$ we have finished the proof. Otherwise $f(x) := Ax/(\sum_i (Ax)^i)$ has a fixed point in D.]

8.113 ¶. Prove

Theorem (Rouché). *Let $B = B(0, R)$ be a ball in $\mathbb{R}^n$ with center at the origin. Let $f, g \in C^0(\overline{B})$ with $|g(x)| < |f(x)|$ on ∂B. Then $\deg(f, 0) = \deg(f + g, 0)$.*

Continuity in Infinite-Dimensional Spaces

Vito Volterra (1860–1940), David Hilbert (1862–1943) and Stefan Banach (1892–1945).

9. Spaces of Continuous Functions, Banach Spaces and Abstract Equations

The combination of the structure of a vector space with the structure of a metric space naturally produces the structure of a *normed space* and a *Banach space*, i.e., of a *complete* linear normed space. The abstract definition of a linear normed space first appears around 1920 in the works of Stefan Banach (1892–1945), Hans Hahn (1879–1934) and Norbert Wiener (1894–1964). In fact, it is in these years that the Polish school around Banach discovered the principles and laid the foundation of what we now call *linear functional analysis*. Here we shall restrain ourselves to introducing some definitions and illustrating some basic facts in Sections 9.1 and 9.4.

Important examples of Banach spaces are provided by spaces of continuous functions that play a relevant role in several problems. In Section 9.3 we shall discuss the *completeness* of these spaces, some *compactness criteria* for subsets of them, in particular the *Ascoli–Arzelà* theorem, and finally the *density* of subspaces of smoother functions in the class of continuous functions, as the *Stone–Weierstrass* theorem.

Finally, Section 9.5 is dedicated to establishing some principles that ensure the existence of solutions of functional equations in a general context. We shall discuss the *fixed point theorems of Banach* and of *Caccioppoli–Schauder*, the *Leray–Schauder principle* and the *method of super- and subsolutions*. Later, in Chapter 11 we shall discuss some applications of these principles.

9.1 Linear Normed Spaces

9.1.1 Definitions and basic facts

9.1 Definition. *Let X be a linear space over $\mathbb{K} = \mathbb{R}$ or $\mathbb{C}$. A norm on X is a function $|| \; || : X \to \mathbb{R}_+$ satisfying the following properties*

(i) $||x|| \in \mathbb{R} \; \forall x \in X$,
(ii) $||x|| \geq 0$ *and* $||x|| = 0$ *if and only if* $x = 0$,
(iii) $||\lambda x|| = |\lambda| \, ||x|| \; \forall \, x \in X, \; \forall \, \lambda \in \mathbb{K}$,
(iv) $||x + y|| \leq ||x|| + ||y|| \; \forall \, x, y \in X$.

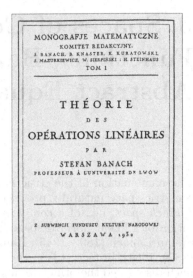

Figure 9.1. Stefan Banach (1892–1945) and the frontispiece of the *Théorie des operations linéaires.*

If $\| \; \|$ is a norm on X, we say that $(X, \| \; \|)$ is a linear normed space *or simply that X is a* normed space *with norm $\| \; \|$.*

Let X be a linear space. A norm $\| \; \|$ on X induces a natural distance on X, defined by

$$d(x, y) := \|x - y\| \qquad \forall x, y \in X,$$

which is invariant by translations, i.e., $d(x+z, y+z) = d(x, y) \; \forall x, y, z \in X$. Therefore, topological notions such as open sets, closed sets, compact sets, convergence of sequences, etc., and metric notions, such as completeness and Cauchy sequences, see Chapter 5, are well defined in a linear normed space. For instance, if X is a normed space with norm $\| \; \|$, we say that $\{x_n\} \subset X$ *converges to* $x \in X$ if $\|x_n - x\| \to 0$ as $n \to \infty$. Notice also that the norm $\| \; \| : X \to \mathbb{R}$ is a continuous function and actually a Lipschitz-continuous function,

$$\Big| \|x\| - \|y\| \Big| \leq \|x - y\|,$$

see Example 5.25.

9.2 Definition. *A real (complex) normed space $(X, \| \; \|)$ that is complete with respect to the distance $d(x, y) := \|x - y\|$ is called a real (complex) Banach space.*

9.3 Remark. By Hausdorff's theorem, see Chapter 5, every normed linear space X can be completed into a metric space, that is, X is homeomorphic to a dense subset of a complete metric space. Indeed, the completed metric space and the homeomorphism inherit the linear structure, as one easily sees. Thus *every normed space X is isomorphic to a dense subset of a Banach space.*

9.4 Example. With the notation above:
   (i) $\mathbb{R}$ with the Euclidean norm $|x|$ is a Banach space. In fact, $|x|$ is a norm on $\mathbb{R}$, and Cauchy sequences converge in norm, compare Theorem 2.35 of [GM2].
   (ii) $\mathbb{R}^n$, $n \geq 1$, is a normed space with the Euclidean norm

$$||x|| := \Big(\sum_{i=1}^{n} |x^i|^2 \Big)^{1/2}, \qquad x = (x^1, x^2, \dots, x^n),$$

   see Example 3.2. It is also a Banach space, see Section 5.3.
   (iii) Similarly, $\mathbb{C}^n$ is a Banach space with the norm $||z|| := (\sum_{i=1}^{n} |z^i|^2)^{1/2}$, $z = (z^1, z^2, \dots, z^n)$.

9.5 ¶ Convex sets. In a linear space, we may consider convex subsets and convex functions.

Definition. *$E \subset X$ is* convex *if $\lambda x + (1 - \lambda)y \in E$ for all $x, y \in E$ and for all $\lambda \in [0, 1]$. $f : X \to \mathbb{R}$ is called* convex *if $f(\lambda x + (1 - \lambda)y) \leq \lambda f(x) + (1 - \lambda)f(y)$ for all $x, y \in X$ and all $\lambda \in [0, 1]$.*

Show that the balls $B(x_0, r) := \{x \in X \mid ||x - x_0|| < r\}$ of a normed space X are convex.

a. Norms induced by inner and Hermitian products

Let X be a real (complex) linear space with an inner (Hermitian) product $(x|y)$. Then $||x|| := \sqrt{(x|x)}$ is a norm on X, see Propositions 3.7 and 3.16. But in general, norms on linear vector spaces are not induced by inner or Hermitian products.

9.6 Proposition. *Let $||\ ||$ be a norm on a real (respectively, complex) normed linear space X. A necessary and sufficient condition for the existence of an inner (Hermitian) product $(\ |\ )$ such that $||x|| = (x|x)\ \forall x \in X$ is that the* parallelogram law *holds,*

$$||x + y||^2 + ||x - y||^2 = 2(||x||^2 + ||y||^2) \qquad \forall x, y \in X.$$

9.7 ¶. Show Proposition 9.6. [*Hint:* First show that if $||x||^2 = (x|x)$, then the parallelogram law holds. Conversely, in the real case set

$$(x \mid y) := \frac{1}{4}(||x + y||^2 - ||x - y||^2),$$

and show that it is an inner product, while in the complex case, set

$$(x|y) := \frac{1}{4}(||x + y||^2 - ||x - y||^2) + i\frac{1}{4}(||x + iy||^2 - ||x - iy||^2),$$

and show that $(x|y)$ is a Hermitian product.

9.8 ¶. For $p \geq 1$, $||x||_p := \Big(\sum_{i=1}^{n} |x^i|^p \Big)^{1/p}$, $x = (x^1, x^2, \dots, x^n)$, is a norm in $\mathbb{R}^n$, cf. Exercise 5.13. Show that it is induced by an inner product if and only if $p = 2$.

b. Equivalent norms

9.9 Definition. *Two norms $|| \ ||_1$ and $|| \ ||_2$ on a linear vector space X are said to be* equivalent *if there exist two constants $0 < m < M$ such that*

$$m \, ||x||_1 \le ||x||_2 \le M \, ||x||_1 \qquad \forall x \in X. \tag{9.1}$$

If $|| \ ||_1$ and $|| \ ||_2$ are equivalent, then trivially the normed spaces $(X, || \ ||_1)$ and $(X, || \ ||_2)$ have the same convergent sequences (to the same limits) and the same Cauchy sequences. Therefore $(X, || \ ||_1)$ *is a Banach space if and only if* $(X, || \ ||_2)$ *is a Banach space.* Since the induced distances are translation invariant, we have the following.

9.10 Proposition. *Let $|| \ ||_1$ and $|| \ ||_2$ be two norms on a linear vector space X. The following statements are equivalent*

(i) *$|| \ ||_1$ and $|| \ ||_2$ are equivalent norms,*
(ii) *the relative induced distances are topologically equivalent,*
(iii) *for any $\{x_n\} \subset X$, $||x_n||_1 \to 0$ if and only if $||x_n||_2 \to 0$.*

Proof. Obviously (i) $\Rightarrow$ (ii) $\Rightarrow$ (iii). Let us prove that (iii) $\Rightarrow$ (i). (iii) implies that the identity map $i : (X, || \ ||_1) \to (X, || \ ||_2)$ is continuous at 0. Therefore there exists $\delta > 0$ such that $||z||_2 \le 1$ if $||z||_1 \le \delta$. For $x \in \mathbb{R}$, $x \ne 0$, if $z := (\delta/||x||_1))x$ we have $||z||_1 = \delta$ hence $||z||_2 \le 1$, i.e., $||x||_2 \le \frac{1}{\delta} ||x||_1$. Exchanging the role of $|| \ ||_1$ and $|| \ ||_2$ and repeating the argument, we also get the inequality $||x||_1 \le \frac{1}{\delta_1} ||x||_2 \ \forall x \in X$ for some $\delta_1 > 0$, hence (i) is proved. $\qquad \square$

9.11 ¶. Let X and Y be two Banach spaces. Show that their Cartesian product, called the *direct sum*, is a Banach space with the norm $||(x, y)||_{1, X \times Y} := ||x||_X + ||y||_Y$. Show that

$$||(x, y)||_{p, X \times Y} := \sqrt[p]{||x||_X^p + ||y||_Y^p}, \ p \ge 1,$$

$$||(x, y)||_{\infty, X \times Y} := \max(||x||_X, ||y||_Y),$$

are equivalent norms.

c. Series in normed spaces

In a linear vector space X, finite sums of elements of X are elements of X. Therefore, given a sequence $\{x_n\}$ in X, we can consider the series $\sum_{n=0}^{\infty} x_n$, i.e., the sequence of partial sums $\left\{ \sum_{k=0}^{n} x_k \right\}$. If, moreover, X is a normed space, we can inquire about the convergence of series in X.

9.12 Definition. *Let X be a normed vector space with norm $|| \ ||$. A series $\sum_{n=0}^{\infty} x_n$, $x_n \in X$, is said to be* convergent *in X if the sequence of its partial sums, $s_n := \sum_{k=0}^{n} x_k$ converges in X, i.e., there exists $s \in X$ such that $||s_n - s|| \to 0$. In this case we write*

$$s = \sum_{k=0}^{\infty} x_k \qquad in \ X$$

instead of $||s_n - s|| \to 0$ and s is said to be the sum *of the series.*

9.13 Remark. Writing $s = \sum_{k=0}^{\infty} x_k$ might make one forget that the sum of the series s is a limit. In dubious cases, for instance if more than one convergence is involved, it is worth specifying in which normed space $(X, \|\ \|_X)$, equivalently with respect to which norm $\|\ \|_X$, the limit has been computed by writing

$$s = \sum_{k=0}^{\infty} x_k \qquad \text{in the norm } X,$$

or

$$s = \sum_{k=0}^{\infty} x_k \qquad \text{in } X,$$

or, even better, writing $\left\| s - \sum_{k=0}^{n} x_k \right\|_X \to 0$.

9.14 Definition. *Let X be a normed space with norm $\|\ \|$. We say that the series $\sum_{n=0}^{\infty} x_n$, $\{x_n\} \subset X$, is absolutely convergent if the series of the norms $\sum_{n=0}^{\infty} \|x_n\|$ converges in $\mathbb{R}$.*

We have seen, compare Proposition 2.39 of [GM2], that every absolutely convergent series in $\mathbb{R}$ is convergent. In general, we have the following.

9.15 Proposition. *Let X be a normed space with norm $\|\ \|$. Then all the absolutely convergent series of elements of X converge in X if and only if X is a Banach space. Moreover, if $\sum_{n=0}^{\infty} x_n$ is convergent, then*

$$\left\| \sum_{k=p}^{\infty} x_k \right\| \le \sum_{k=p}^{\infty} \|x_k\|.$$

Proof. Let X be a Banach space, and let $\sum_{k=0}^{\infty} x_k$ be absolutely convergent. The sequence of partial sums of $\sum_{k=0}^{\infty} \|x_k\|$ is a Cauchy sequence in $\mathbb{R}$, hence $\sum_{k=p}^{q} \|x_k\| \to 0$ as $p, q \to \infty$. From the triangle inequality we infer that

$$\left\| \sum_{k=p}^{q} x_k \right\| \le \sum_{k=p}^{q} \|x_k\|$$

hence $\left\| \sum_{k=p}^{q} x_k \right\| \to 0$ as $p, q \to \infty$, i.e., the sequence of partial sums of $\sum_{k=0}^{\infty} x_k$ is a Cauchy sequence in X. Consequently, it converges in norm in X, since X is a Banach space.

Conversely, let $\{x_k\} \subset X$ be a Cauchy sequence. By induction select n_1 such that $\|x_n - x_{n_1}\| < 1$ if $n \ge n_1$, then $n_2 > n_1$ such that $\|x_n - x_{n_2}\| < 1/2$ if $n \ge n_2$ and so on. Then $\{x_{n_k}\}$ is a subsequence of $\{x_k\}$ such that

$$\|x_{n_{k+1}} - x_{n_k}\| \le 2^{-k} \qquad \forall k,$$

and consequently the series $\sum_{k=1}^{\infty} (x_{n_{k+1}} - x_{n_k})$ is absolutely convergent, hence convergent to a point $y \in X$ by assumption, i.e.,

$$\left\| \sum_{k=1}^{p} (x_{n_{k+1}} - x_{n_k}) - y \right\| \to 0, \qquad \text{as } p \to +\infty.$$

Since this simply amounts to $||x_{n_p} - x|| \to 0$, $x := y + x_{n_1}$, $\{x_{n_k}\}$ converges to x, and, as $\{x_n\}$ is a Cauchy sequence, we conclude that in fact the entire sequence $\{x_n\}$ converges to x. Finally, the estimate follows from the triangle inequality

$$\left|\left|\sum_{k=p}^{q} x_k\right|\right| \le \sum_{k=p}^{q} ||x_k||$$

as $q \to \infty$, since we are able to pass to the limit as $\sum_{n=0}^{\infty} x_k$ converges. $\qquad\square$

9.16 ¶ Commutativity. Let X be a Banach space and let $\{x_n\} \subset X$ be such that $\sum_n x_n$ is absolutely convergent. Then $\sum_n x_{\sigma(n)}$, for $\{x_{\sigma(n)}\}$ a rearrangement of $\{x_n\}$, is also absolutely convergent, and $\sum_n x_n = \sum_n x_{\sigma(n)}$.

9.17 ¶ Associativity. Let X be a Banach space and let $\{x_n\} \subset X$. Let $\{I_n\}$ be a sequence of nonempty subsets of $\mathbb{N}$ with $I_n \cap I_m = \emptyset$ if $n \neq m$ and $\cup_n I_n = \mathbb{N}$. If $\sum_n x_n$ is absolutely convergent, then

$$\sum_{n=1}^{\infty} \left(\sum_{k \in I_n} x_k \right)$$

is absolutely convergent and $\sum_{k=0}^{\infty} x_k = \sum_{n=1}^{\infty} \left(\sum_{k \in I_n} x_k \right)$.

d. Finite-dimensional normed linear spaces

In a finite-dimensional vector space, there is only one topology induced by a norm: the Euclidean topology. In fact, if $\mathbb{K} = \mathbb{R}$ or $\mathbb{K} = \mathbb{C}$, we have the following.

9.18 Theorem. *In $\mathbb{K}^n$ any two norms are equivalent.*

Proof. It suffices to prove that any norm ρ on $\mathbb{K}^n$ is equivalent to the Euclidean norm $|\ |$. Let $(e_1, e_2, \ldots, e_n)$ be the standard basis of $\mathbb{K}^n$. If $x = (x^1, x^2, \ldots, x^n)$ and $y = (y^1, y^2, \ldots, y^n)$, we have

$$\rho(x - y) = \rho\left(\sum_{i=1}^{n}(x^i - y^i)e_i\right) \le \sum_{i=1}^{n} |x^i - y^i| \rho(e_i)$$

$$\le \left(\sum_{i=1}^{n} \rho(e_i)^2\right)^{1/2} |x - y|,$$

hence $\rho : \mathbb{K}^n \to \mathbb{R}_+$ is continuous. Since the unit ball $B := \{x \in \mathbb{K}^n \,|\, |x| = 1\}$ of $\mathbb{K}^n$ is compact, we infer that ρ attains a maximum value M and a minimum value m on B. Since the minimum value is attained, we infer that $m > 0$, otherwise ρ would be zero at some point of B. Therefore $0 < m \le \rho(x) \le M$ on B, and, on account of the 1-homogeneity of the norm,

$$m|x| \le \rho(x) \le M|x| \qquad \forall x \in \mathbb{K}^n,$$

i.e., $||\ ||$ is equivalent to the Euclidean norm. $\qquad\square$

9.19 Corollary. *Every finite-dimensional normed space X is a Banach space. In particular, any finite-dimensional subspace of X is closed and $K \subset X$ is compact is and only if K is closed and bounded.*

Proof. Let ρ be a norm on X and let $\mathcal{E} : \mathbb{K}^n \to X$ be a coordinate map on X. Since $\mathcal{E}$ is linear and nonsingular, $\rho \circ \mathcal{E}$ is a norm on $\mathbb{K}^n$ and $\mathcal{E}$ is trivially an isometry between the two normed spaces $(\mathbb{K}^n, \rho \circ \mathcal{E})$ and (X, ρ). Since $\rho \circ \mathcal{E}$ is equivalent to the Euclidean norm, $(\mathbb{K}^n, \rho \circ \mathcal{E})$ is a Banach space and therefore (X, ρ) is a Banach space, too. The second claim is obvious. □

9.20 ¶. Let X be a normed space of dimension n. Then any system of coordinates $\mathcal{E} : X \to \mathbb{K}^n$ is a linear continuous map between $\mathbb{K}^n$ with the Euclidean metric and the normed space X.

A key ingredient in the proof of Theorem 9.18 is the fact that the closed unit ball in $\mathbb{R}^n$ is compact. This property is characteristic of finite-dimensional spaces.

9.21 Theorem (Riesz). *The closed unit ball of a normed linear space X is compact if and only if X is finite dimensional.*

For the proof we need the following lemma, due to Frigyes Riesz (1880–1956), which in this context plays the role of the orthogonal projection theorem in spaces with inner or Hermitian products, see Theorem 3.27 and Chapter 10.

9.22 Lemma. *Let Y be a closed linear subspace of a normed space $(X, ||\ ||)$. Then there exists $\overline{x} \in X$ such that $||\overline{x}|| = 1$ and $||\overline{x} - x|| \geq 1/2$ $\forall\, x \in Y$.*

Proof. Take $x_0 \in X \setminus Y$ and define $d := \inf\{||y - x_0||\,|\,y \in Y\}$. We have $d > 0$, otherwise we could find $\{y_n\} \subset Y$ with $y_n \to x_0$ and $x_0 \in Y$ since Y is closed. Take $y_0 \in Y$ with $||y_0 - x_0|| \leq 2d$ and set $\overline{x} = \frac{x_0 - y_0}{||x_0 - y_0||}$. Clearly $||\overline{x}|| = 1$ and $y_0 + y||x_0 - y_0|| \in Y$ if $y \in Y$, hence

$$||y - \overline{x}|| = \left\|y - \frac{x_0 - y_0}{||x_0 - y_0||}\right\| = \frac{\Big\|y||x_0 - y_0|| - x_0 + y_0\Big\|}{||x_0 - y_0||} \geq \frac{d}{2d} = \frac{1}{2}$$

□

Proof of Theorem 9.21. Let $B := \{x \in X\,|\,||x|| \leq 1\}$. If X has dimension n, and $\mathcal{E} : \mathbb{K}^n \to X$ is a system of coordinates, then $\mathcal{E}$ is an isomorphism, hence a homeomorphism. Since B is bounded and closed, $\mathcal{E}^{-1}(B)$ is also bounded and closed, hence compact in $\mathbb{K}^n$, see Corollary 9.19. Therefore $B = \mathcal{E}(\mathcal{E}^{-1}(B))$ is compact in X.

We now prove that B is not compact if X has infinite dimension. Take x_1 with $||x_1|| = 1$. By Lemma 9.22, we find x_2 with $||x_2|| = 1$ at distance at least $1/2$ from the subspace Span$\{x_1\}$, in particular $||x_1 - x_2|| \geq \frac{1}{2}$. Again by Lemma 9.22, we find x_3 with $||x_3|| = 1$ at distance at least $1/2$ from Span$\{x_1, x_2\}$, in particular $||x_3 - x_1|| \geq \frac{1}{2}$ and $||x_3 - x_2|| \geq \frac{1}{2}$. Iterating this procedure we construct a sequence $\{x_n\}$ of points in the unit sphere such that $||x_i - x_j|| \geq \frac{1}{2}$ $\forall i, j,\ i \neq j$. Therefore $\{x_n\}$ has no convergent subsequence, hence the unit sphere is not compact. □

9.23 Remark. We emphasize that, in any infinite-dimensional normed space we have constructed a sequence of unit vectors, a subsequence of which is never Cauchy.

9.1.2 A few examples

In Sections 9.2 and 9.4 we shall discuss respectively, the relevant Banach spaces of linear continuous operators and of bounded continuous functions. Here we begin with a few examples.

a. The space ℓ_p, $1 \leq p < \infty$

Let $(Y, || \ ||_Y)$ be a normed space and $p \in \mathbb{R}$, $p \geq 1$. For a sequence $\xi = \{\xi_i\} \subset Y$ we define

$$||\xi||_{\ell_p(Y)} := \left(\sum_{i=1}^{\infty} ||\xi_i||_Y^p \right)^{1/p}.$$

Then the space of sequences

$$\ell_p(Y) := \left\{ \xi = \{\xi_i\} \ \Big| \ ||\xi||_{\ell_p(Y)} < +\infty \right\}$$

is a linear space with norm $||\xi||_{\ell_p(Y)}$. Moreover, we have the following.

9.24 Proposition. $\ell_p(Y)$ is a Banach space if Y is a Banach space.

Proof. Let $\{\xi_k\}$, $\xi_k := \{\xi_i^{(k)}\}$, be a Cauchy sequence in $\ell_p(Y)$. Since for any i

$$||\xi_n^i - \xi_m^i||_Y \leq ||\xi_n - \xi_m||_{\ell_p(Y)}, \tag{9.2}$$

the sequence $\{\xi_i^{(k)}\}_k$ is a Cauchy sequence in Y, hence it has a limit $\xi_i \in Y$,

$$||\xi_i^{(k)} - \xi_i||_Y \to 0 \qquad \text{as } k \to \infty.$$

We then set $\xi := \{\xi_i\}$ and prove that $\{\xi_k\}$ converges to ξ in $\ell_p(Y)$. Fix $\epsilon > 0$, then for all $n, m \geq n_0(\epsilon)$ we have

$$||\xi_n - \xi_m||_{\ell_p(Y)} < \epsilon$$

hence, for all $r \in \mathbb{N}$

$$\sum_{i=1}^{r} ||\xi_i^{(n)} - \xi_i^{(m)}||_Y^p < \epsilon^p$$

and, since $x \to ||z - x||_Y$ is continuous in Y, as $m \to \infty$,

$$\sum_{i=1}^{r} ||\xi_i^{(n)} - \xi_i||_Y^p \leq \epsilon^p$$

for $n \geq n_0(\epsilon)$ and all r. Letting $r \to \infty$, we find $||\xi_n - \xi||_{\ell_p(Y)} \leq \epsilon$ for $n \geq n_0$, i.e., $\xi_n \to \xi$ in $\ell_p(Y)$. Finally, the triangle inequality shows that $\xi \in \ell_p(Y)$. $\qquad\square$

b. A normed space that is not Banach

The map

$$f \; \rightarrow \; ||f||_p = ||f||_{p,]a,b[} := \left(\int_a^b |f(t)|^p \, dt \right)^{1/p}, \qquad p \geq 1,$$

defines a norm on the space of continuous functions $C^0([a, b], \mathbb{R})$. Indeed, if Y is a linear normed space with norm $|| \;\; ||_Y$,

$$||f||_p = ||f||_{L^p(]a,b[,Y)} := \int_a^b ||f(t)||_Y^p \, dt, \qquad p \geq 1,$$

defines a norm on the space of continuous functions with values in Y. In fact, $t \rightarrow ||f(t)||_Y^p$ is a continuous real-valued map, hence Riemann integrable, thus $||f||_p$ is well defined. Clearly $||f||_p = 0$ if and only if $f(t) = 0 \; \forall t$ and $||f||_p$ is positively homogeneous of degree 1. It remains to prove the triangle inequality for $f \rightarrow ||f||_p$, called the *Minkowski inequality*,

$$||f + g||_p \leq ||f||_p + ||g||_p \qquad \forall f, g \in C^0([a, b], Y).$$

The claim is trivial if one of the two functions is zero. Otherwise, we use the convexity of $\phi : Y \rightarrow \mathbb{R}$ where $\phi(y) := ||y||_Y^p$, i.e.,

$$\phi(\lambda x + (1 - \lambda)y) \leq \lambda \phi(x) + (1 - \lambda)\phi(y), \qquad \forall x, y \in Y, \; \forall \lambda \in [0, 1], \quad (9.3)$$

with $x = f(t)/||f||_p$, $y = g(t)/||g||_p$ and $\lambda = ||f||_p/(||f||_p + ||g||_p)$, and we integrate to get

$$\frac{1}{||f||_p + ||g||_p} \int_a^b ||f(t) + g(t)||_Y^p \, dt \leq 1,$$

that is Minkowski's inequality.

It turns out that $C^0([a, b], \mathbb{R})$ *normed with* $|| \;\; ||_p$ *is not complete*, see Example 9.25. Its completion is denoted by $L^p(]a, b[)$. Its characterization is one of the outcomes of the *Lebesgue theory of integration*.

9.25 Example. Define, see Figure 9.2,

$$f_n(t) = \begin{cases} 0 & \text{if } -1 \leq t \leq 0, \\ nt & \text{if } 0 < t \leq 1/n, \\ 1 & \text{if } 1/n \leq t \leq 1 \end{cases} \qquad \text{and} \qquad f(t) = \begin{cases} 0 & \text{if } -1 \leq t \leq 0, \\ 1 & \text{if } 0 < t \leq 1. \end{cases}$$

The sequence $\{f_n\}$ converges to f in norm $|| \;\; ||_p$ as

$$||f_n - f||_p^p = \int_0^{1/n} (1 - nt)^p \, dt \leq \frac{1}{n} \rightarrow 0.$$

If $g \in C^0([-1, 1])$ is the limit of $\{f_n\}$, then $||f - g||_p = 0$, consequently $g = f = 0$ on $[-1, 0]$ and $g = f = 1$ on $]0, 1]$, a contradiction, since g is continuous.

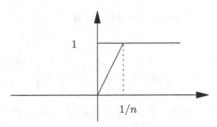

Figure 9.2. Pointwise approximation of the Heaviside function.

c. Spaces of bounded functions

Let A be any set and Y be a normed space with norm $|| \; ||_Y$. The *uniform norm* of a function $f : A \to Y$ is defined by the number (possibly infinity)

$$||f||_{\mathcal{B}(A,Y)} := \sup_{x \in A} ||f(x)||_Y.$$

$||f||_{\mathcal{B}(A,Y)}$ defines a norm on the space of *bounded functions* $f : A \to Y$

$$\mathcal{B}(A,Y) := \Big\{ f : A \to Y \; \Big| \; ||f||_{\mathcal{B}(A,Y)} < +\infty \Big\}$$

which then becomes a normed space. The norm $||f||_{\mathcal{B}(A,Y)}$ on $\mathcal{B}(A,Y)$ is also denoted by $||f||_{\infty,A}$ or even by $||f||_\infty$ when no confusion can arise. The topology induced on $\mathcal{B}(A,Y)$ by the uniform norm is called the *topology of uniform convergence*, see Example 5.19. In particular, we say that a sequence $\{f_n\} \subset \mathcal{B}(A,Y)$ *converges uniformly* in A to $f \in \mathcal{B}(A,Y)$ and we write

$$f_n(x) \to f(x) \qquad \text{uniformly in } A,$$

if

$$||f_n - f||_{\mathcal{B}(A,Y)} \to 0.$$

9.26 Proposition. *If Y is a Banach space, then $\mathcal{B}(A,Y)$ is a Banach space.*

Proof. Let $\{f_n\} \subset \mathcal{B}(A,Y)$ be a Cauchy sequence with respect to $|| \; ||_\infty$. For any $\epsilon > 0$ there is a n_0 such that $||f_n - f_m||_\infty \leq \epsilon$ for all $n, m \geq n_0$. Therefore, for all $x \in A$ and $n, m \geq n_0$

$$||f_n(x) - f_m(x)||_Y \leq \epsilon. \tag{9.4}$$

Consequently, for all $x \in A$, $\{f_n(x)\}$ is a Cauchy sequence in Y hence it converges to an element $f(x) \in Y$. Letting $m \to \infty$ in (9.4), we find

$$||f_n(x) - f(x)||_Y \leq \epsilon \qquad \forall \, n \geq n_0 \text{ and } \forall x \in A,$$

i.e., $||f - f_n||_\infty \leq \epsilon$ for $n \geq n_0$, hence $||f||_\infty \leq ||f_n||_\infty + \epsilon$, i.e., $f \in \mathcal{B}(A,Y)$ and $f_n \to f$ uniformly in $\mathcal{B}(A,Y)$ since ϵ is arbitrary. $\qquad \square$

9.27 ¶. Let Y be finite dimensional and let $(e_1, e_2, \ldots, e_n)$ be a basis of Y. We can write f as $f(x) = f_1(x)e_1 + \cdots + f_n(x)e_n$. Thus $f \in \mathcal{B}(A,Y)$ if and only if all the components of f are bounded real functions.

d. The space $\ell_\infty(Y)$

A special case occurs when $A = \mathbb{N}$. In this case $\mathcal{B}(A, Y)$ is the space of bounded sequences of Y, that we better denote by

$$\ell_\infty(Y) := \mathcal{B}(\mathbb{N}, Y).$$

Therefore, by Proposition 9.26, $\ell_\infty(Y)$ is a Banach space with the uniform norm

$$||\xi||_{\ell_\infty(Y)} := ||\xi||_{\mathcal{B}(\mathbb{N},Y)} = \sup_i ||\xi_i||_Y,$$

if Y is complete.

9.28 ¶. Show that for $1 \leq p < q < \infty$ we have

  (i) $\ell_p(\mathbb{R}) \subset \ell_q(\mathbb{R}) \subset \ell_\infty(\mathbb{R})$,
  (ii) $\ell_p(\mathbb{R})$ is a proper subspace of $\ell_q(\mathbb{R})$,
  (iii) the identity map $\mathrm{Id} : \ell_p(\mathbb{R}) \to \ell_q(\mathbb{R})$ is continuous,
  (iv) $\ell_1(\mathbb{R})$ is a dense subset of $\ell_q(\mathbb{R})$ with respect to the convergence in $\ell_q(\mathbb{R})$.

9.29 ¶. Show that, if $p, q \geq 1$ and $1/p + 1/q = 1$, then

$$\sum_{i=1}^\infty |\xi_i \eta_i| \leq \frac{||\xi||_{\ell_p(\mathbb{R})}^p}{p} + \frac{||\eta||_{\ell_q(\mathbb{R})}^q}{q} \tag{9.5}$$

for all $\{\xi_n\} \in \ell_p(\mathbb{R})$ and $\{\eta_n\} \in \ell_q(\mathbb{R})$. Moreover, show that

$$||\xi||_{\ell_p(\mathbb{R})} = \sup\left\{ \Big| \sum_{n=1}^\infty \xi_n \eta_n \Big| \,\Big|\, ||\eta||_{\ell_q(\mathbb{R})} \leq 1 \right\}. \tag{9.6}$$

[*Hint:* For proving (9.5) use the Young inequality $ab \leq a^p/p + b^q/q$. Using (9.5), show that $\geq$ holds in (9.6). By a suitable choice $b = b(a)$ and again using Young's inequality, finally show equality in (9.6).]

9.2 Spaces of Bounded and Continuous Functions

In this section we discuss some basic properties of the space of continuous and bounded functions from a metric space into a Banach space.

9.2.1 Uniform convergence

a. Uniform convergence

Let X be a metric space and let Y be a normed space with norm $|| \; ||_Y$. Then, as we have seen in Proposition 9.26, the space $\mathcal{B}(X, Y)$ of bounded functions from X into Y is a normed space with uniform norm, and $\mathcal{B}(X,Y)$ is a Banach space provided Y is a Banach space.

We denote by $C_b(X,Y)$ the subspace of $\mathcal{B}(X,Y)$ of bounded and continuous functions from X into Y,

$$C_b(X,Y) := C^0(X,Y) \cap \mathcal{B}(X,Y).$$

Observe that, by the Weierstrass theorem $C_b(X,Y) = C(X,Y)$ if X is compact, and that, trivially, $C_b(X,Y)$ is a normed space with uniform norm.

9.30 Proposition. *$C_b(X,Y)$ is a closed subspace of $\mathcal{B}(X,Y)$.*

Proof. Let $\{f_n\} \subset C_b(X,Y)$ be such that $f_n \to f$ uniformly. For any $\epsilon > 0$, we choose $n_0 = n_0(\epsilon)$ such that $||f - f_{n_0}||_{\infty,X} < \epsilon$. It follows that

$$||f||_{\infty,X} \leq ||f - f_{n_0}||_{\infty,X} + ||f_{n_0}||_{\infty,X} < +\infty,$$

i.e., $f \in \mathcal{B}(X,Y)$. Moreover, since f_{n_0} is continuous, for a fixed $x_0 \in X$ there exists $\delta > 0$ such that $||f_{n_0}(x) - f_{n_0}(x_0)||_Y < \epsilon$ whenever $x \in X$ and $d_X(x,x_0) < \delta$. Thus, for $d(x,x_0) < \delta$, we deduce that

$$||f(x) - f(x_0)||_Y \leq ||f(x) - f_{n_0(x)}||_Y + ||f_{n_0}(x) - f_{n_0}(x_0)||_Y$$
$$+ ||f_{n_0}(x_0) - f(x_0)||_Y \leq 3\epsilon$$

i.e., f is continuous at x_0. In conclusion, $f \in C^0(X,Y) \cap \mathcal{B}(X,Y)$. □

Immediate consequences are the following corollaries.

9.31 Corollary. *The uniform limit of a sequence of continuous functions is continuous.*

9.32 Corollary. *Let X be a metric space and let Y be a Banach space. Then $C_b(X,Y)$ with uniform norm is a Banach space.*

9.33 ¶. Show that the space $C^1([a,b],\mathbb{R})$ of real functions $f : [a,b] \to \mathbb{R}$, which are of class C^1, is a Banach space with the norm

$$||f||_{C^1} := \sup_{x \in [0,1]} |f(x)| + \sup_{x \in [0,1]} |f'(x)|.$$

[*Hint:* If $\{f_k\}$ is a Cauchy sequence in $C^1([a,b])$, show that $f_k \to f$, $f_k' \to g$, uniformly. Then passing to the limit in

$$f_k(x) - f_k(a) = \int_a^x f_k'(t)\,dt,$$

show that f is differentiable and $f'(x) = g(x)\ \forall x$.]

9.34 ¶. Let X be a metric space and let Y be a complete metric space. Show that the space of bounded and continuous functions from X into Y, endowed with the metric

$$d_\infty(f,g) := \sup_{x \in X} d_Y(f(x),g(x)),$$

is a complete metric space.

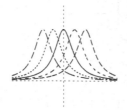

Figure 9.3. Consider a wave shaped function, e.g., $f(x) = 1/(1+x^2)$, and its translates $f_n(x) := 1/(1 + (x + n)^2)$. Then $||f_n||_\infty = 1$, while $f_n(x) \to 0$ for all $x \in \mathbb{R}$.

b. Pointwise and uniform convergence

Let A be a set and let Y be a normed space normed by $|\ |_Y$. We say that $\{f_n\}$, $f_n : A \to Y$, *converges pointwise to* $f : A \to Y$ *in* A if

$$|f_n(x) \to f(x)|_Y \to 0 \qquad \forall x \in A,$$

while we say that $\{f_n\}$ *converges uniformly to* f in A if

$$||f_n - f||_{\infty, A} \to 0.$$

Since for all $x \in A$ $||f_n(x) - f(x)||_Y \le ||f_n - f||_{\infty, X}$, uniform convergence trivially implies pointwise convergence while the converse is generally false. For instance, a sequence of continuous functions may converge pointwise to a discontinuous function, and in this case, the convergence cannot be uniform, as shown by the sequence $f_n(x) := x^n$, $x \in [0, 1[$, that converges to the function f which vanishes for all $x \in [0, 1[$, while $f(1) = 1$. Of course, a sequence of continuous functions may also converge pointwise and not uniformly to a continuous function, compare Figure 9.3.

More explicitly, $f_n \to f$ pointwise in A if

$$\forall x \in A, \forall \epsilon > 0 \ \exists \, \overline{n} = \overline{n}(x, \epsilon) \text{ such that } |f_n(x) - f(x)|_Y < \epsilon \text{ for all } n \ge \overline{n},$$

while, $f_n \to f$ uniformly in A if

$$\forall \, \epsilon > 0 \ \exists \, \overline{n} = \overline{\epsilon} \text{ such that } |f_n(x) - f(x)|_Y < \epsilon \text{ for all } n \ge \overline{n} \text{ and all } x \in A.$$

Therefore, we have pointwise convergence or uniform convergence according to whether the index $\overline{n}$ depends on or is independent of the point x.

c. A convergence diagram

For series of functions $f_n : A \to Y$, we shall write

$$f(x) = \sum_{n=1}^{\infty} f_n(x) \qquad \forall x \in A$$

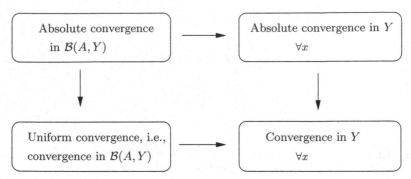

Figure 9.4. The relationships among the different notions of convergence for series of functions.

if the partial sums converge pointwise in A, and

$$f(x) = \sum_{n=1}^{\infty} f_n(x) \qquad \text{uniformly in } A$$

if the partial sums converge uniformly. Simply writing $\sum_{n=1}^{\infty} f_n(x) = f(x)$ is, in fact, ambiguous.

Summarizing, we introduced four different types of convergence for series of functions from a set A into a normed space Y. More precisely, if $\{f_n\} \subset \mathcal{B}(A, Y)$ and $f \in \mathcal{B}(A, Y)$, we say that

(i) $\sum_{n=0}^{\infty} f_n(x)$ *converges pointwise* to f if for all $x \in A$ $\sum_{n=0}^{\infty} f_n(x) = f(x)$ in Y, i.e., for all $x \in A$, $||\sum_{n=0}^{p} f_n(x) - f(x)||_Y \to 0$ as $p \to \infty$,

(ii) $\sum_{n=0}^{\infty} f_n(x)$ *converges absolutely in Y for all* $x \in A$ i.e., for any fixed $x \in A$, the series of nonnegative real numbers $\sum_{n=0}^{\infty} ||f_n(x)||_Y$ converges,

(iii) $\sum_{n=0}^{\infty} f_n(x)$ *converges uniformly in A* to f if $\sum_{n=0}^{\infty} f_n = f$ in $\mathcal{B}(A, Y)$, i.e., $||\sum_{n=0}^{p} f_n - f||_{\mathcal{B}(A,Y)} \to 0$ as $p \to \infty$,

(iv) $\sum_{n=0}^{\infty} f_n(x)$ *converges absolutely in $\mathcal{B}(A, Y)$* if the series of nonnegative real numbers $\sum_{n=0}^{\infty} ||f_n||_{\mathcal{B}(A,Y)}$ converges.

Clearly (iv) implies (ii), and (iii) implies (i). Moreover, (iv) implies (iii) and (ii) implies (i) if Y is a Banach space; the other implications are false, see Example 9.35 below.

9.35 Example. Consider functions $f : \mathbb{R}_+ \to \mathbb{R}$. Choosing $f_n(x) := (-1)^n/n$, we see that $\sum_{n=1}^{\infty} f_n(x)$ converges pointwise and uniformly, but not absolutely in $\mathbb{R}$ or in $\mathcal{B}(\mathbb{R}, \mathbb{R})$.

Let $f(x) := \sin x/x$, $x > 0$, and, for any $n \in \mathbb{N}$,

$$f_n(x) := \begin{cases} \dfrac{\sin x}{x} & \text{if } n < x \leq n+1, \\ 0 & \text{otherwise.} \end{cases}$$

Since $c_1/n \leq ||f_n||_\infty \leq c_2/n$, $\sum_n f_n$ does not converge absolutely in $\mathcal{B}(\mathbb{R}_+, \mathbb{R})$. But $f(x) = \sum_{n=0}^{\infty} f_n(x)$ converges pointwise $\forall x \in \mathbb{R}_+$ and also absolutely in $\mathbb{R}$ for all $x \in \mathbb{R}_+$. Finally,

$$f(x) - \sum_{n=0}^{p} f_n(x) = \begin{cases} \dfrac{\sin x}{x} & \text{if } x > p \\ 0 & \text{otherwise,} \end{cases}$$

hence

$$\left\| f - \sum_{n=0}^{p} f_n \right\|_\infty \le \frac{c_2}{p} \to 0 \qquad \text{as } p \to \infty,$$

therefore $\sum_n f_n$ converges uniformly, that is in $\mathcal{B}(\mathbb{R}_+, \mathbb{R})$.

Here the convergence is uniform in $\mathcal{B}(\mathbb{R}_+, \mathbb{R})$ but not absolute in $\mathcal{B}(\mathbb{R}_+, \mathbb{R})$, because the functions f_n take their maxima at different points and the maximum of the sum is much smaller than the sum of the maxima.

9.36 Theorem (Dini). *Let X be a compact metric space and let $\{f_n\}$ be a monotonic sequence of functions $f_n : X \to \mathbb{R}$ that converges pointwise to a continuous function f. Then f_n converges uniformly to f.*

9.37 ¶. Show Dini's theorem. [*Hint:* Assuming that f_n converges by decreasing to 0, for all $\epsilon > 0$ and for all $x \in X$ there exists a neighborhood V_x of x such that $|f_n(x)| < \epsilon$ $\forall x \in V_x$ for all n larger than some $n(x)$. Then use the compactness of X. Alternatively, use the uniform continuity theorem, Theorem 6.35.]

9.38 ¶. Show a sequence $\{f_n\}$ that converges pointwise to zero and does not converge uniformly in any interval of $\mathbb{R}$. [*Hint:* Choose an ordering of the rationals $\{r_n\}$ and consider the sequence $f_n(x) := \sum_{n=0}^{\infty} \varphi(x - r_n)$ where φ is the function in Figure 9.3.]

9.39 ¶. Let E be a dense subset of a metric space X, let Y be a Banach space and let $\{f_n\}$, $f_n : X \to Y$ be a sequence of continuous functions that converges uniformly in E. Show that $\{f_n\}$ converges uniformly in X.

9.40 ¶ Hausdorff distance. Let Z be a metric space with metric d bounded in $Z \times Z$. In the space $\mathcal{C}$ of bounded closed sets of Z, we define the *Hausdorff distance* by

$$\rho(E, F) := \sup \left\{ \sup_{x \in E} d(x, F), \sup_{x \in F} d(x, E) \right\}.$$

Show that ρ is a distance on $\mathcal{C}$. Now suppose that X is a compact metric space and Y is a normed space. Show that $\{f_n\}$ converges uniformly to f if and only if the graphs in $X \times Y$ of the f_n's converge to the graph of f with respect to the Hausdorff distance.

d. Uniform convergence on compact subsets

9.41 Example. We have seen in Theorem 7.14 of [GM2] that a power series with radius of convergence $\rho > 0$ converges totally, hence uniformly, on every disk of radius $r < \rho$. This does not mean that it converges uniformly in the open disk $\{|z| < \rho\}$. For instance, the geometric series $\sum_{n=0}^{\infty} x^n$ has radius of convergence 1 and if $|x| < 1$

$$\frac{1}{1-x} - \sum_{n=0}^{p} x^n = \frac{x^{p+1}}{1-x},$$

consequently for all p,

$$\sup_{x \in]-1,1[} \left| \frac{1}{1-x} - \sum_{n=0}^{p} x^n \right| = +\infty.$$

Figure 9.5. Giulio Ascoli (1843–1896) and the first page of *Sulle serie di funzioni* by Cesare Arzelà (1847–1912).

9.42 Definition. *Let $A \subset \mathbb{R}^n$ and let Y be a normed space. We say that a sequence of functions $\{f_n\}$, $f_n : A \to Y$, converges uniformly on compact subsets of A to $f : A \to Y$ if for every compact subset $K \subset A$ we have $\|f_n - f\|_{\infty, K} \to 0$ as $n \to \infty$.*

9.43 ¶. Let $f_n, f : \overline{\Omega} \to \mathbb{R}$ be continuous functions defined on the closure of an open set Ω of $\mathbb{R}^n$. Show that $\{f_n\}$ converges uniformly to f on $\overline{\Omega}$ if and only if $\{f_n\}$ converges uniformly to f in Ω.

9.44 ¶. Let $\{f_n\}$, $f_n : A \subset \mathbb{R}^n \to Y$, be a sequence of continuous functions that converges uniformly on compact subsets of A to $f : A \to Y$. Show that f is continuous.

9.2.2 A compactness theorem

At the end of the nineteenth century, especially in the works of Vito Volterra (1860–1940), Giulio Ascoli (1843–1896), Cesare Arzelà (1847–1912) there appears the idea of considering functions $\mathcal{F}$ whose values depend on the values of a function, the so-called *funzioni di linee*, functions of lines; one of the main motivations came from the calculus of variations. This eventually led to the notion of abstract spaces of Maurice Fréchet (1878–1973). In this context, a particularly relevant result is the compactness criterion now known as the Ascoli–Arzelà theorem.

a. Equicontinuous functions

Let X be a metric space and Y a normed space.

9.45 Definition. *We say that a subset* $\mathcal{F}$ *of* $\mathcal{B}(X,Y)$ *is* equibounded, *or* uniformly bounded, *by some constant* $M > 0$ *if we have* $||f(x)||_Y \leq M$ $\forall x \in X, \forall f \in \mathcal{F}$.

We say that the family of functions $\mathcal{F}$ *is* equicontinuous *if for all* $\epsilon > 0$ *there is* $\delta > 0$ *such that*

$$||f(x) - f(y)||_Y < \epsilon \qquad \forall x, y \in X \text{ with } d_X(x,y) < \delta, \text{ and } \forall f \in \mathcal{F}.$$

9.46 Definition (Hölder-continuous functions). *Let* X, Y *be metric spaces. We say that a function* $f : X \rightarrow Y$ *is* Hölder-continuous *with exponent* α, $0 < \alpha \leq 1$, *if there is a constant* M *such that*

$$d_Y(f(x), f(y)) \leq M d_X(x,y)^\alpha,$$

and we denote by $C^{0,\alpha}(X,Y)$ *the space of these functions.*

Clearly the space $C^{0,1}(X,Y)$ is the space $\text{Lip}(X,Y)$ of *Lipschitz-continuous* functions from X into Y.

On $C^{0,\alpha}(X,Y) \cap \mathcal{B}(X,Y)$, $0 < \alpha \leq 1$, we introduce the norm

$$||f||_{C^{0,\alpha}} = \sup_{x \in X} ||f(x)||_Y + \sup_{x,y \in X, x \neq y} \frac{||f(x) - f(y)||_Y}{||x - y||_X^\alpha}, \qquad (9.7)$$

and it is easy to show the following.

9.47 Proposition. $C^{0,\alpha}(X,Y) \cap \mathcal{B}(X,Y)$ *endowed with the norm* (9.7) *is a Banach space if* Y *is a Banach space.*

Bounded subsets with respect to the norm (9.7), of $C^{0,\alpha}(X,Y) \cap \mathcal{B}(X,Y)$ provide examples of equicontinuous families.

See the exercises at the end of this chapter for more on Hölder-continuous functions.

b. The Ascoli–Arzelà theorem

9.48 Theorem (Ascoli–Arzelà). *Every sequence of functions* $\{f_n\}$ *in* $C^0([a,b])$ *which is equibounded and equicontinuous has a subsequence that is uniformly convergent.*

More generally, we have the following.

9.49 Theorem. *Let* X *be a compact metric space. A subset* $\mathcal{F}$ *of* $C^0(X, \mathbb{R})$ *is relatively compact if and only if* $\mathcal{F}$ *is equibounded and equicontinuous.*

Proof. We recall that a subset of a metric space is *relatively compact* or *precompact* if and only if it is *totally bounded*, see Theorem 6.8.

If $\mathcal{F}$ is relatively compact, then $\mathcal{F}$ is totally bounded and, in particular, equibounded in $C^0(X, \mathbb{R})$. For any $\epsilon > 0$, let $f_1, \ldots, f_{n_\epsilon} \in \Phi$ be an ϵ-net for $\mathcal{F}$, i.e., $\forall\, f \in \mathcal{F}$ $||f - f_i|| < \epsilon$ for some f_i. Since the f_i's are uniformly continuous, there is a δ_ϵ such that

$$d_X(x,y) < \delta_\epsilon \text{ implies } |f_i(x) - f_i(y)| < \epsilon, \qquad i = 1, 2, \ldots, n_\epsilon.$$

Figure 9.6. Two pages respectively from *Le curve limiti di una varietà data di curve* by Giulio Ascoli (1843–1896) and from *Sulle funzioni di linee* by Cesare Arzelà (1847–1912).

Given $f \in \mathcal{F}$ we choose i_0, $1 \le i_0 \le n_\epsilon$ in such a way that $\|f - f_{i_0}\| < \epsilon$. Then

$$|f(x) - f(y)| \le |f(x) - f_{i_0}(x)| + |f_{i_0}(x) - f_{i_0}(y)| + |f_{i_0}(y) - f(y)|$$
$$\le 2\|f - f_{i_0}\| + |f_{i_0}(x) - f_{i_0}(y)| \le 3\epsilon,$$

for $d_X(x, y) < \delta_\epsilon$, hence $\mathcal{F}$ is an equicontinuous family.

Conversely, suppose $\mathcal{F}$ is equibounded and equicontinuous. Again by Theorem 6.8, it suffices to show that $\mathcal{F}$ is totally bounded. Let $\epsilon > 0$. From the compactness of X and the equicontinuity of $\mathcal{F}$, we infer that there exists a finite family of open balls $B(x_i, r_i)$ that cover X and such that

$$|f(y) - f(x_i)| < \epsilon \qquad \forall y \in B(x_i, r_i), \qquad \forall f \in \mathcal{F}.$$

Since the set $K := \{f(x_i) \mid 1 \le i \le n, \ f \in \mathcal{F}\}$ is bounded, we find $y_1, y_2, \ldots, y_m \in \mathbb{R}$ such that $K \subset \cup_{i=i}^m B(y_i, \epsilon)$. The set $\mathcal{F}$ is covered by the finite union of the sets

$$\mathcal{F}_\pi := \Big\{ f \in \mathcal{F} \ \Big| \ f(x_i) \in B(y_{\pi(i)}, \epsilon) < \epsilon, \ i = 1, \ldots, n \Big\},$$

with π varying among the bijective maps $\pi : \{1, \ldots, n\} \to \{1, \ldots, n\}$. Therefore, it suffices to show that diam $\mathcal{F}_\pi \le 4\epsilon$. Since for $f_1, f_2 \in \mathcal{F}_\pi$ and $x \in B(x_i, r_i)$ we have

$$|f_1(x) - f_2(x)| \le |f_1(x) - f_1(x_i)| + |f_1(x_i) - y_{\pi(i)}|$$
$$+ |y_{\pi(i)} - f_2(x_i)| + |f_2(x_i) - f_2(x)| \le 4\epsilon,$$

the proof is concluded.   □

9.50 ¶. Notice that the sequence $\{f_n\}$ of wave shaped functions in Figure 9.3,

$$f_n(x) := \frac{1}{1 + (x + n)^2}, \qquad x \in \mathbb{R}, \ n \in \mathbb{N},$$

is equicontinuous and equibounded, but not relatively compact.

9.51 ¶. Theorem 9.49 can be formulated in slightly more general forms that are proved to hold with the same proof of Theorem 9.49.

Theorem. *Let X be a compact metric space, and let Y be a Banach space. A subset $\mathcal{F} \subset C(X, Y)$ is relatively compact if and only if $\mathcal{F}$ is equicontinuous and, for every X, the set $\mathcal{F}_x$ of all values $f(x)$ of $f \in \mathcal{F}$ is relatively compact in Y.*

Theorem. *Let X and Y be metric spaces. Suppose X is compact. A sequence $\{f_n\} \subset C(X, Y)$ converges uniformly if and only if $\{f_n\}$ is equicontinuous and there exists a compact set $K \subset Y$ such that $\{f_n(x)\}$ is contained in a δ-neighborhood of K for n sufficiently large.*

9.52 ¶. Show the following.

Proposition. *Let X, Y be two metric spaces and let $\Omega \subset X$ be compact. Then the subsets of $C^{0,\alpha}(\Omega, Y)$ that are bounded in the $\| \; \|_{C^{0,\alpha}}$ norm are relatively compact in $C^0(\Omega, Y)$.*

9.3 Approximation Theorems

In this section we deal with the following questions: Can we approximate a continuous function uniformly, and with given precision, by a polynomial? Under which conditions are classes of smooth functions dense with respect to the uniform convergence in the class of continuous functions?

9.3.1 Weierstrass and Bernstein theorems

a. Weierstrass's approximation theorem

In 1885 Karl Weierstrass (1815–1897) proved the following.

9.53 Theorem (Weierstrass, I). *Every continuous function in a closed bounded interval $[a, b]$ is the uniform limit of a sequence of polynomials.*

In particular, for every n there exists a polynomial $Q_n(x)$ (of degree $d = d(n)$ sufficiently large) such that $|f(x) - Q_n(x)| \leq 2^{-n} \; \forall \, x \in [a, b]$. If we set

$$P_1(x) := Q_1(x), \qquad P_n(x) := Q_n(x) - Q_{n-1}(x), \qquad n > 1,$$

we therefore conclude that *every continuous function $f(x)$ can be written in a closed and bounded interval as the (infinite) sum of polynomials,*

$$f(x) = \sum_{n=0}^{\infty} P_n(x) \qquad \text{uniformly in } [a, b].$$

We recall that, in general, a continuous function is not the sum of a power series, since the sum of a power series is at least a function of class C^∞, compare [GM2].

Many proofs of Weierstrass's theorem are nowadays available; in this section we shall illustrate some of them. This will allow us to discuss a number of facts that are individually relevant.

A first proof of Theorem 9.53. We first observe, following Henri Lebesgue (1875–1941), that in order to approximate uniformly in $[a, b]$ any continuous function, it suffices to approximate the function $|x|$, $x \in [-1, 1]$.

In fact, any continuous function in $[a, b]$ can be approximated, uniformly in $[a, b]$, by continuous and piecewise linear functions. Thus it suffices to approximate continuous and piecewise linear functions. Let $f(x)$ be one of such functions. Then there exist points $x_0 = a < x_1 < x_2 < \cdots < x_r < x_{r+1} = b$ such that $f'(x)$ takes a constant value d_k in each interval $]x_k, x_{k+1}[$. Then, in $[a, b]$ we have

$$f(x) = f(a) + \sum_{k=0}^{r}(d_k - d_{k-1})\varphi_{x_k}(x), \qquad d_{-1} = 0,$$

where

$$\varphi_c(x) := \max(x - c, 0) = \frac{1}{2}((x - c) + |x - c|).$$

If we are able to approximate $|x - x_k|$, $x \in [a, b]$, uniformly by polynomials $\{Q_{k,n}\}$, then the polynomials

$$P_n(x) := f(a) + \sum_{k=0}^{r}(d_k - d_{k-1})\frac{1}{2}((x - x_k) + Q_{k,n}(x))$$

approximate $f(x)$ uniformly in $[a, b]$. By a linear change of variable, it then suffices to approximate $|x|$ uniformly in $[-1, 1]$.

This can be done in several ways. For instance, noticing that if $x \in [-1, 1]$, then $1 - |x|$ solves the equation in z

$$z = \frac{1}{2}[z^2 + (1 - x^2)],$$

one considers the discrete process

$$\begin{cases} z_{n+1}(x) = \frac{1}{2}[z_n^2(x) + (1 - x^2)] & n \geq 0, \\ z_0(x) = 1. \end{cases}$$

It is then easily seen that the polynomials $z_n(x)$ satisfy
   (i) $z_n(x) \geq 0$ in $[-1, 1]$,
   (ii) $z_n(x) \geq z_{n+1}(x)$,
   (iii) $z_n(x)$ converges pointwise to $1 - |x|$ if $x \in [-1, 1]$.
Since $1 - |x|$ is continuous, Dini's theorem, Theorem 9.36, yields that the polynomials $z_n(x)$ converge uniformly to $1 - |x|$ in $[-1, 1]$.

Alternatively one shows, using the binomial series, that

$$\sqrt{1 - x} = \sum_{n=0}^{\infty} c_n x^n, \qquad c_n = \binom{1/2}{n}(-1)^n,$$

in $]-1, 1[$. Then one proves that the series converges absolutely in $C^0([-1, 1])$, hence uniformly in $[-1, 1]$. In fact, we observe that $c_n := \binom{1/2}{n}(-1)^n$ is negative for $n \geq 1$ hence,

 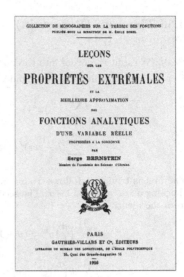

Figure 9.7. The first page of Weierstrass's paper on approximation by polynomials and the *Leçons sur les propriétés extrémales* by Sergei Bernstein (1880–1968).

$$\sum_{n=0}^{p} |c_n| = 2 - \sum_{n=0}^{p} c_n = 2 - \lim_{x \to 1^-} \sum_{n=0}^{p} c_n x^n$$

$$\leq 2 - \lim_{x \to 1^-} \sum_{n=0}^{\infty} c_n x^n = 2 - \lim_{x \to 1^-} \sqrt{1-x} = 2.$$

Replacing $1 - x$ with x^2, it follows that

$$|x| = \sum_{n=0}^{\infty} c_n (1 - x^2)^n \qquad \text{uniformly in } [-1, 1].$$

$\square$

9.54 ¶. Add details to the previous proof.

b. Bernstein's polynomials

Another proof of Theorem 9.53, grounded in probablistic ideas, see Exercise 9.57, and giving explicit formulas for the approximating polynomials, is due to Sergei Bernstein (1880–1968).

It is enough to consider functions defined in $[0, 1]$ instead of in a generic interval $[a, b]$.

9.55 Definition. *Let $f \in C^1([0, 1])$. Bernstein polynomials of f are*

$$B_n(x) := \sum_{k=0}^{n} f\left(\frac{k}{n}\right) \binom{n}{k} x^k (1 - x)^{n-k}, \qquad n \geq 0.$$

9.56 Theorem (Bernstein). *Bernstein's polynomials of f converge uniformly in $[0,1]$ to f.*

Proof. We split the proof into three steps.

Step 1. The following identities hold

$$\sum_{k=0}^{n} \binom{n}{k} x^k (1-x)^{n-k} = 1, \tag{9.8}$$

$$\sum_{k=0}^{n} (k-nx)^2 \binom{n}{k} x^k (1-x)^{n-k} = nx(1-x). \tag{9.9}$$

The first is trivial: it follows from the binomial formula $(a+b)^n = \sum_0^n \binom{n}{k} a^k b^{n-k}$ by choosing $a = x$ and $b = 1-x$. The second needs some computation. Fix $n \geq 1$. Starting from the identities

$$\sum_{k=0}^{n} \binom{n}{k} y^k = (y+1)^n,$$

$$\sum_{k=0}^{n} k \binom{n}{k} y^k = yD\Big(\sum_{k=0}^{n} \binom{n}{k} y^k \Big) = ny(y+1)^{n-1},$$

$$\sum_{k=0}^{n} k^2 \binom{n}{k} y^k = ny(ny+1)(y+1)^{n-2},$$

we replace y by $x/(1-x)$ and multiply each of the equalities by $(1-x)^n$. It follows that

$$\sum_{k=0}^{n} \binom{n}{k} x^k (1-x)^{n-k} = 1,$$

$$\sum_{k=0}^{n} k \binom{n}{k} x^k (1-x)^{n-k} = nx,$$

$$\sum_{k=0}^{n} k^2 \binom{n}{k} x^k (1-x)^{n-k} = nx(nx+1-x).$$

Multiplying each of the previous identities respectively, by $n^2 x^2$, $-2nx$ and -1, and summing, we infer (9.9).

Step 2. As $x(1-x) \leq \frac{1}{4}$, (9.9) yields

$$\sum_{k=0}^{n} \frac{(k-nx)^2}{n^2} \binom{n}{k} x^k (1-x)^{n-k} \leq \frac{1}{4n}. \tag{9.10}$$

Fix $\delta > 0$ and $x \in [0,1]$, and denote by $\Delta_n(x)$ the set of k in $\{0, 1, \dots, n\}$ such that

$$\left| \frac{k}{n} - x \right| \geq \delta;$$

(9.10) then yields

$$\sum_{k \in \Delta_n(x)} \binom{n}{k} x^k (1-x)^k \leq \frac{1}{4n\delta^2}, \tag{9.11}$$

that is, for n large, the terms that mostly contribute to the sum in (9.8) are the ones with index k such that

$$\left| \frac{k}{n} - x \right| < \delta.$$

Step 3. Set $M := \sup_{x \in [0,1]} |f(x)|$ and, given $\epsilon > 0$, let δ be such that $|f(x) - f(y)| < \epsilon$ for $|x - y| < \delta$. Then we have

$$B_n(x) - f(x) = \sum_{k=0}^{n} \left[f\left(\frac{k}{n}\right) - f(x) \right] \binom{n}{k} x^k (1-x)^{n-k}$$

$$= \sum_{k \in \Delta_n(x)} \left[f\left(\frac{k}{n}\right) - f(x) \right] \binom{n}{k} x^k (1-x)^{n-k}$$

$$+ \sum_{k \in \Gamma_n(x)} \left[f\left(\frac{k}{n}\right) - f(x) \right] \binom{n}{k} x^k (1-x)^{n-k}$$

$$=: \Sigma_\Delta + \Sigma_\Gamma$$

where

$$\Gamma_n := \left\{ 0, \ldots, n \right\} \setminus \Delta_n = \left\{ k \in \{0, \ldots, n\} \;\Big|\; \left| \frac{k}{n} - x \right| < \delta \right\}.$$

For $k \in \Gamma_n(x)$, i.e., $|k/n - x| < \delta$, we have $|f(k/n) - f(x)| < \epsilon$, hence

$$|\Sigma_\Gamma| \le \epsilon \sum_{k=0}^{n} \binom{n}{k} x^k (1-x)^{n-k} = \epsilon;$$

on the other hand, if $|k/n - x| \ge \delta$, (9.11) yields

$$|\Sigma_\Delta| \le 2M \frac{1}{4n\delta^2} = \frac{M}{2n\delta^2}.$$

Therefore, we conclude for n large enough so that $M/(2n\delta^2) \le \epsilon$,

$$|B_n(x) - f(x)| < 2\epsilon \qquad \text{uniformly in } [0,1].$$

$\square$

9.57 ¶. The previous proof has the following probabilistic formulation. Let $0 \le p \le 1$ and let $X_n(p)$ be a random variable with binomial distribution

$$P(\{X_n(p) = r/n\}) = \binom{n}{r} p^r (1-p)^{n-r}.$$

If $f : [0,1] \to \mathbb{R}$ is a function, the expectation of $f(X_n(t))$ is given by

$$\mathbf{E}\left[f(X_n(t)) \right] = \sum_{r=0}^{\infty} f\left(\frac{r}{n}\right) \binom{r}{n} t^r (1-t)^{n-r}$$

and one shows in the theory of probability that $\mathbf{E}\left[f(X_n(t)) \right]$ converge uniformly to f.

c. Weierstrass's approximation theorem for periodic functions

We denote by C_T^0 the class of continuous periodic functions in $\mathbb{R}$ with period $T > 0$.

9.58 Theorem (Weierstrass, II). *Every function $f \in C_T^0$ is the uniform limit of trigonometric polynomials with period T.*

In Section 9.3 we shall give a direct proof of this theorem and in Section 11.5 we shall give another proof of it: the Féjer theorem. It is worth noticing that, in general, a continuous function is neither the uniform nor the pointwise sum of a trigonometric series.

Here we shall prove that the claims in Theorems 9.58 and 9.53 are equivalent. By a linear change of variable we may assume that $T = 2\pi$. First let us prove the following.

9.59 Lemma. *Let $f \in C^0([-\pi, \pi])$ be even. Then for any $\epsilon > 0$ there is an even trigonometric polynomial*

$$T(x) := \sum_{k=0}^{n} a_k \cos kx$$

such that $|f(x) - T(x)| < \epsilon \ \forall \ x \in [-\pi, \pi]$.

Proof. We apply Theorem 9.53 to the continuous function $g(y) := f(\arccos y)$, $y \in [-1, 1]$, to obtain

$$\left| f(\arccos y) - \sum_{k=0}^{n} c_k y^k \right| < \epsilon \qquad \text{in } [-1, 1],$$

hence

$$\left| f(x) - \sum_{k=0}^{n} c_k \cos^k x \right| < \epsilon \qquad \text{in } [0, \pi].$$

To conclude, it suffices to notice that $\sum_{k=0}^{n} c_k \cos^k x$ is an even polynomial.   □

Proof of Theorem 9.58. Let $f \in C^0_{2\pi}(\mathbb{R})$. We consider the two even functions in $[-\pi, \pi]$

$$f(x) + f(-x), \qquad (f(x) - f(-x)) \sin x.$$

Then Lemma 9.59 yields for any $\epsilon > 0$

$$f(x) + f(-x) = T_1(x) + \alpha_1(x), \qquad (f(x) - f(-x)) \sin x = T_2(x) + \alpha_2(x)$$

$T_1(x)$ and $T_2(x)$ being two even trigonometric polynomials, and for the remainders α_1 and α_2 one has

$$|\alpha_1(x)|, \ |\alpha_2(x)| \le \epsilon \qquad \text{in } [-\pi, \pi].$$

Multiplying the first equation by $\sin^2 x$ and the second by $\sin x$ and summing we find

$$f(x) \sin^2 x = T_3(x) + \alpha_3(x), \tag{9.12}$$

for $T_3(x)$ a trigonometric polynomial and $||\alpha_3||_{\infty, [-\pi, \pi]} \le 2\epsilon$. The same argument applies to $f(x - \pi/2)$, yielding

$$f\left(x - \frac{\pi}{2}\right) \sin^2 x = T_4(x) + \alpha_4(x)$$

where T_4 is a trigonometric polynomial and $||\alpha_4||_{\infty, [-\pi, \pi]} \le 2\epsilon$. By changing the variable x in $x + \frac{\pi}{2}$, we then infer

$$f(x) \cos^2 x = T_5(x) + \alpha_5(x) \tag{9.13}$$

where $T_5(x) := T_4(x + \frac{\pi}{2})$ and $||\alpha_5||_{\infty, [-\pi, \pi]} \le 2\epsilon$. Summing (9.12) and (9.13) we finally conclude the proof.   □

9.60 Remark. Actually, the two Weiertrass theorems are equivalent. We have already proved Theorem 9.58 using Theorem 9.53. We now outline how to deduce the first Weierstrass theorem, Theorem 9.53, from Theorem 9.58, leaving the details to the reader. Given $f \in C^0([-\pi, \pi])$, the function

$$g(x) := f(x) + \frac{f(-\pi) - f(\pi)}{2\pi} x$$

satisfies $g(\pi) = g(-\pi)$, hence g can be extended to a continuous periodic map of period 2π. According to Theorem 9.58, for any $\epsilon > 0$ we find a trigonometric polynomial

$$T_\epsilon(x) := a_0 + \sum_{k=1}^{n(\epsilon)} (a_k^\epsilon \cos kx + b_k^\epsilon \sin kx)$$

with $|g(x) - T_\epsilon(x)| \leq \epsilon$ for all $x \in [-\pi, \pi]$. Next, we approximate $\sin kx$ and $\cos kx$ by polynomials (e.g., by Taylor polynomials), concluding that there is a polynomial $Q_\epsilon(x)$ with $|T_\epsilon(x) - Q_\epsilon(x)| < \epsilon \; \forall x \in [-\pi, \pi]$, hence $|g(x) - Q_\epsilon(x)| < 2\epsilon$ in $[-1, 1]$.

9.3.2 Convolutions and Dirac approximations

We now introduce a procedure that allows us to find smooth approximations of functions.

a. Convolution product

Here we confine ourselves to considering only continuous functions defined on the entire line. The choice of the entire line as a domain is not a restriction, since every continuous function on an interval $[a, b]$ can be extended to a continuous function in $\mathbb{R}$ and, actually for any $\delta > 0$, to a continuous function that vanishes outside $[a - \delta, b + \delta]$.

9.61 Example (Integral means). Let $f : \mathbb{R} \to \mathbb{R}$ be continuous. For any $\delta > 0$ consider the *mean function* of f

$$f_\delta(x) := \frac{1}{2\delta} \int_{x-\delta}^{x+\delta} f(\xi) \, d\xi, \qquad x \in \mathbb{R}. \tag{9.14}$$

Simple consequences of the fundamental theorem of calculus are

 (i) $f_\delta(x)$ is Lipschitz continuous,
 (ii) $f_\delta(x) \to f(x)$ pointwise,
while from the estimate

$$|f_\delta(x) - f(x)| \leq \sup_{|y-x|<\delta} |f(y) - f(x)|$$

and Theorem 6.35

 (iii) $f_\delta(x) \to f(x)$ uniformly on every bounded interval of $\mathbb{R}$.

The above allows us, of course, to uniformly approximate continuous functions with Lipschitz-continuous functions on every bounded interval.

9.62 Definition. *Let $f, g : \mathbb{R} \to \mathbb{R}$ be two Riemann integrable functions. Suppose that $g(x - t)f(t)$ is summable in $\mathbb{R}$ for any $x \in \mathbb{R}$. Then the function*

$$g * f(x) := \int_{\mathbb{R}} g(x - y)f(y)\, dy, \qquad x \in \mathbb{R},$$

called the convolution product *of f and g, is well defined.*

Clearly the map $(f, g) \to g * f$

(i) is a bilinear operator,
(ii) $g * f = -f * g$ since

$$g * f(x) = \int_{\mathbb{R}} g(x - y)f(y)\, dy = -\int_{\mathbb{R}} g(z)f(x - z)\, dz = -f * g(x),$$

(iii) if f and g are summable in $[a, b]$ and f vanishes outside the interval $[a, b]$, then $g * f$ is well defined in $\mathbb{R}$ and

$$|g * f(x)| \leq \int_a^b |g(x - y)|\, |f(y)|\, dy \leq \|g\|_{\infty, [x-b, x-a]} \int_a^b |f(y)|\, dy. \tag{9.15}$$

9.63 Example. The function f_δ in (9.14) is the convolution product of f and

$$g(x) = \begin{cases} 1 & \text{se } |x| \leq \delta, \\ 0 & \text{se } |x| > \delta. \end{cases}$$

9.64 Example. If $g(t) = \sum_{k=0}^n c_k t^k$ is a polynomial of degree n, then for any f that vanishes outside an interval,

$$g * f(x) = \sum_{k=0}^n c_k \sum_{j=0}^k \binom{k}{j}(-1)^{n-j}\left(\int_{\mathbb{R}} y^{n-j} f(y)\, dy\right) x^j$$

is again a polynomial of degree n.

9.65 Example. If $f = 0$ outside $[-\pi, \pi]$, then

$$e^{ikt} * f(x) = \int_{-\pi}^{\pi} f(y)e^{ik(x-y)}\, dy = 2\pi c_k e^{ikx}$$

i.e., $e^{ikt} * f$ is the kth harmonic component of the periodic extension of f, compare Section 11.5.

9.66 Theorem. *Let $g \in C^k(\mathbb{R})$, and let f be Riemann summable. Suppose that either f or g vanishes outside a bounded interval $[a, b]$. Then $g * f \in C^k(\mathbb{R})$ and $D_k(g * f)(x) = (D_k g) * f(x)\ \forall x \in \mathbb{R}$.*

Proof. We prove the claim when $f = 0$ outside $[a, b]$, the other case $g = 0$ outside $[a, b]$ is similar. By (9.15) we then have

$$|g * f(x)| \leq ||g||_{\infty, [x-b, x-a]} ||f||_1 \qquad ||f||_1 := \int_a^b |f(y)| \, dy,$$

hence

$$||g * f||_{\infty, [c, d]} \leq ||g||_{\infty, [c-b, d-a]} ||f||_1.$$

(i) We now prove that $g * f \in C^0(\mathbb{R})$ if $g \in C^0(\mathbb{R})$. In fact,

$$g * f(x + h) - g * f(x) = \int (g(x - y + h) - g(x - y)) f(y) \, dy = G * f(x),$$

where $G(x) := g(x + h) - g(x)$. Therefore, using (9.15), we get

$$|g * f(x + h) - g * f(x)| \leq ||g(x + h) - g(x)||_{\infty, [x-b, x-a]} ||f||_1 \to 0 \qquad (9.16)$$

as $h \to 0$ since $||g(t + h) - g(t)||_{\infty, [x-b, x-a]} \to 0$, because of the uniform continuity of g on compact sets.

(ii) Similarly, we prove that $g * f \in C^1(\mathbb{R})$ if $f \in C^1(\mathbb{R})$. We have

$$\frac{g * f(x + h) - g * f(x)}{h} - \int g'(x - y) f(y) \, dy$$

$$= \int_{\mathbb{R}} \left(\frac{g(x - y + h) - g(x - y)}{h} - g'(x - y) \right) f(y) \, dy$$

$$= H * f(x),$$

where $H(x) := \frac{g(x+h) - g(x)}{h} - g'(x)$. Again, by (9.15),

$$\left| \frac{g * f(x + h) - g * f(x)}{h} - \int g'(x - y) f(y) \, dy \right|$$

$$\leq \left\| \frac{g(t + h) - g(t)}{h} - g'(t) \right\|_{\infty, [x-b, x-a]} ||f||_1.$$

Since

$$\left| \frac{g(x + h) - g(x)}{h} - g'(x) \right| = \left| \frac{1}{h} \int_x^{x+h} (g'(y) - g'(x)) \right|$$

$$\leq \frac{1}{h} \int_x^{x+h} |g(y) - g(x)| \, dy$$

$$\leq \sup_{|y-x| < |h|} |g'(y) - g'(x)| \to 0 \qquad \text{as } h \to 0$$

because of the uniform continuity of g' on compact sets, we then conclude that $g * f$ is differentiable at x and that $(g * f)'(x) = g' * f(x) \; \forall x \in \mathbb{R}$. Finally $(g * f)' = g' * f$ is continuous by (i).

(iii) The general case is then proved by induction. $\square$

9.67 Remark. Let f and g be summable and let one of them vanish outside a bounded interval. If f instead of g is of class $C^k(\mathbb{R})$, then, recalling that $g * f = -f * g$, we infer from Theorem 9.66 that $g * f \in C^k(\mathbb{R})$ and $D_k(g * f)(x) = g * (D_k f)(x)$. Therefore if both f and g are of class $C^k(\mathbb{R})$, then

$$D_k(g * f)(x) = (D_k g) * f(x) = g * (D_k f)(x)$$

and, in general, $g * f$ is as smooth as the smoother of f and g.

b. Mollifiers

9.68 Definition. *A function* $k(x) \in C^\infty(\mathbb{R})$ *such that*

$$k(x) = k(-x), \quad k(x) \geq 0, \quad k(x) = 0 \text{ for } |x| \geq 1, \quad \int_{\mathbb{R}} k(x)\, dx = 1$$

is called a smoothing kernel.

9.69 ¶. The function

$$\varphi(x) := \begin{cases} \exp\left(-\dfrac{1}{1-x^2}\right) & \text{if } |x| < 1, \\ 0 & \text{if } |x| \geq 1 \end{cases}$$

is $C^\infty(\mathbb{R})$, nonnegative, even and with finite integral. Hence the map $k(x) := \frac{1}{A}\varphi(x)$, where $A := \int_{\mathbb{R}} \varphi(x)\, dx$, is a smoothing kernel.

Given a smoothing kernel $k(x)$, we can generate the family

$$k_\epsilon(x) := \epsilon^{-1} k\left(\frac{x}{\epsilon}\right) \qquad \epsilon > 0.$$

Trivially, $k_\epsilon(-x) = k_\epsilon(x)$ and

$$k_\epsilon \in C_c^\infty(\mathbb{R}), \quad k_\epsilon(x) > 0, \quad k_\epsilon(x) = 0 \text{ per } |x| \geq \epsilon, \quad \int_{\mathbb{R}} k_\epsilon(x)\, dx = 1.$$

Also $k_\epsilon(x) = 0$ for $|x| \geq \epsilon$ and $||k_\epsilon||_\infty = ||k||_\infty/\epsilon$.

9.70 Definition. *Given a smoothing kernel k, the* mollifiers *or* smoothing operators S_ϵ *are defined by*

$$S_\epsilon f(x) := k_\epsilon * f(x) = \int_{\mathbb{R}} k_\epsilon(x - y) f(y)\, dy.$$

We have

$$S_\epsilon f(x) = k_\epsilon * f(x) = \int_{x-\epsilon}^{x+\epsilon} k_\epsilon(x - y) f(y)\, dy$$

$$= \frac{1}{\epsilon} \int_{x-\epsilon}^{x+\epsilon} k\left(\frac{x - y}{\epsilon}\right) f(y)\, dy = \int_{-1}^{1} k(z)\, f(x - \epsilon z)\, dz.$$

Since the functions k_ϵ are of class C^∞, the functions $S_\epsilon f(x)$, $x \in \mathbb{R}$, are of class C^∞ by Theorem 9.66. Moreover, as shown by the next theorem, they converge to f in norms that are as strong as the differentiability of f; for instance, they converge uniformly or in norm C^1 if f is continuous or if f is of class C^1, respectively.

9.71 Proposition. *Let* $f \in C^0(\mathbb{R})$. *Then*

(i) $S_\epsilon f \in C^\infty(\mathbb{R})$, $\forall \epsilon > 0$;

(ii) *If $f = 0$ in $[a, b]$, then $S_\epsilon f(x) = 0$ in $[a + \epsilon, b - \epsilon]$;*

(iii) $(S_\epsilon f)'(x) = \frac{1}{\epsilon} \int_\mathbb{R} k'\left(\frac{x-y}{\epsilon}\right) f(y)\, dy$;

(iv) $S_\epsilon f \to f$ *as $\epsilon \to 0$ uniformly in any bounded interval $[a, b]$.*

Moreover, if $f \in C^1(\mathbb{R})$, then $(S_\epsilon f)'(x) = (S_\epsilon f')(x)\ \forall x \in \mathbb{R}$ and $S_\epsilon f'(x) \to f'(x)$ uniformly on any bounded interval $[a, b]$.

Proof. (i), (iii) follow from Theorem 9.66, and (ii) follows from the definition.

(iv) If $f \in C^0(\mathbb{R})$ and $x \in \mathbb{R}$ we have

$$|f(x) - S_\epsilon f(x)| = \left| \int K_\epsilon(x - y)[f(y) - f(x)]\, dy \right| = |k_\epsilon * (f - f(x))(x)|$$

$$\leq \sup_{|y-x|<\epsilon} |f(y) - f(x)| \int_\mathbb{R} k_\epsilon(y)\, dy = \sup_{|y-x|<\epsilon} |f(y) - f(x)|.$$

Since f is uniformly continuous on bounded intervals in $\mathbb{R}$, $\sup_{|y-x|<\epsilon} |f(y)-f(x)| \to 0$, consequently $|S_\epsilon(x) - f(x)| \to 0$ as $\epsilon \to 0$ uniformly on compact sets of $\mathbb{R}$.

If $f \in C^1(\mathbb{R})$, we have already proved in Theorem 9.66 that $S_\epsilon f$ is of class C^1 and that $(S_\epsilon f)'(x) = S_\epsilon f'(x)$. Applying (iv) to $S_\epsilon f$ and $S_\epsilon f'$ we then reach the claim. □

c. Approximation of the Dirac mass

The family $\{k_\epsilon\}$ is often referred to as an *approximation of the Dirac delta*. In applications, the Dirac δ is often "defined" as a function vanishing at every point but zero and with the property that

$$\int_{-\infty}^{+\infty} \delta(x)\, dx = 1;$$

sometimes it is "defined", with respect to convolution, as if it would operate like

$$\int_{-\infty}^{+\infty} \delta(x - \xi) f(\xi)\, d\xi = f(x).$$

Of course, no such function exists in the classical sense; but it can be thought of as a *linear operator* from $C_c^0(\mathbb{R})$ into $\mathbb{R}$

$$\delta_{x_0} : C_c^0(\mathbb{R}) \to \mathbb{R}, \qquad \delta_{x_0} f := f(x_0).$$

We shall avoid dealing directly with δ, as the correct context for doing this is the *theory of distributions*, and we set

9.72 Definition. *A sequence of* nonnegative *functions $D_n : \mathbb{R} \to \mathbb{R}$ with the properties that for any interval $[a, b]$ and for any $\rho > 0$ we have*

$$\int_{B(0,\rho)} D_n(x)\, dx \to 1, \qquad \int_{[a,b]\setminus B(0,\rho)} D_n(x)\, dx \to 0, \quad as\ n \to \infty, \quad (9.17)$$

is called an approximation of the δ.

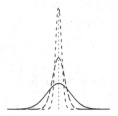

Figure 9.8. Approximations of the Dirac δ.

9.73 ¶. Let $\{D_n\}$ be an approximation of δ and let f be a continuous function in $[a, b]$. Show that

$$\lim_{n \to \infty} \int_a^b D_n(x - y) f(y) \, dx = f(x) \qquad \forall x \in]a, b[.$$

It is easy to prove the following.

9.74 Theorem. *Let $\{D_n\}$ be an approximation of δ. Suppose that each D_n is continuous in $\mathbb{R}$ and let f be a continuous function in $[a, b]$. Then the functions*

$$f_n(x) := \int_a^b D_n(x - y) f(y) \, dy, \qquad x \in [a, b],$$

converge uniformly to f in every interval $[c, d]$ strictly contained in $[a, b]$.

Theorem 9.74 uses, in an essential way, the fact that the approximations of δ are nonnegative. For instance, the result in Theorem 9.74 does not hold for the sequence of the Dirichlet kernels, since the Fourier series of f does not converge to f if f is merely continuous, compare Section 11.5.

9.75 ¶. Prove Theorem 9.74.

9.76 ¶. Consider in $[-1, 1]$ the sequence of functions

$$D_n(x) := c_n (1 - x^2)^n \qquad \text{where} \qquad c_n := \frac{1}{\int_{-1}^1 (1 - t^2)^n \, dt}.$$

Show that for every $\rho \in]0, 1[$

$$\lim_{n \to \infty} \frac{\int_\rho^1 (1 - t^2)^n \, dt}{\int_0^1 (1 - t^2)^n \, dt} = 0.$$

Infer that $\{D_n\}$ is an approximation of δ, hence the functions

$$f_n(x) := c_n \int_0^1 [1 - (t - x)^2]^n f(t) \, dt$$

converge uniformly to f on compact sets of $] - 1, 1[$.

Finally, observing that the functions $f_n(x)$ are actually polynomials of degree not greater than $2n$, called *Stieltjes polynomials*, deduce from the above Weierstrass's theorem, Theorem 9.53.

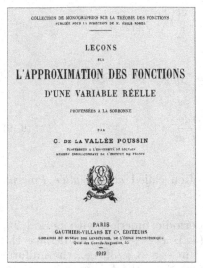
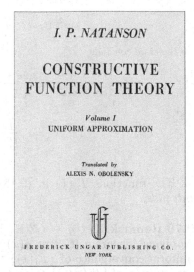

Figure 9.9. Frontispieces of *L'approximation des fonctions* by Charles de la Vallée–Poussin (1866–1962) and of J. P. Natanson *Constructive Function Theory*, New York, 1964.

Consider the functions

$$D_n(t) := c_n \cos^{2n}\left(\frac{t}{2}\right), \qquad t \in [-\pi, \pi]$$

where, see 2.66 of [GM2],

$$c_n := \frac{1}{\int_{-\pi}^{\pi} \cos^{2n}\left(\frac{t}{2}\right) dt} = \frac{1}{2\pi}\frac{(2n)!!}{(2n-1)!!}.$$

As proved below, we have the following.

9.77 Lemma. *The sequence $\{D_n\}$ is an approximation of δ.*

Hence, as a consequence of Theorem 9.74, we can state the following.

9.78 Theorem (de la Vallée Poussin). *Let $f \in C^0([-\pi, \pi])$. The functions*

$$T_n(x) := \frac{(2n)!!}{(2n-1)!!}\frac{1}{2\pi}\int_{-\pi}^{\pi} f(t)\cos^{2n}\left(\frac{t-x}{2}\right) dt \qquad (9.18)$$

converge uniformly to f in every interval $[a, b]$ with $-\pi < a < b < \pi$.

Proof of Lemma 9.77. (i) Since $\cos t$ is decreasing in $[0, \pi/2]$, we have

$$\int_{\rho}^{\pi/2} \cos^{2n}(t)\, dt \leq \left(\frac{\pi}{2} - \rho\right)\cos^{2n}(\rho) \leq \frac{\pi}{2}\cos^{2n}(\rho);$$

on the other hand, since $\cos t$ is concave in $[0, \pi/2]$, we have $\cos t \geq 1 - 2t/\pi$, hence

$$\int_0^{\pi/2} \cos^{2n}(t)\, dt \geq \int_0^{\pi/2} \left(1 - \frac{2t}{\pi}\right)^{2n} = \frac{\pi}{2(2n+1)};$$

we therefore conclude

$$\frac{\int_\rho^{\pi/2} \cos^{2n}(t)\, dt}{\int_0^{\pi/2} \cos^{2n}(t)\, dt} \leq \frac{2(2n+1)}{\pi}\,\frac{\pi}{2}\cos^{2n}(\rho) \;\to\; 0 \qquad \text{as } n \to \infty$$

and

$$\lim_{n\to\infty} \frac{\int_0^\rho \cos^{2n}(t)\, dt}{\int_0^{\pi/2} \cos^{2n}(t)\, dt} = 1. \tag{9.19}$$

$\square$

The functions $T_n(x)$ in (9.18) are often called *de la Vallée Poussin integrals*.

9.79 Remark. Let $g \in C^0(\mathbb{R})$ be a periodic function with period 2π. Applying Theorem 9.78 to $g(x) := f(3x)$, $x \in [-\pi, \pi]$, we deduce the uniform convergence of $\{T_n(x)\}$ to $g(x)$ in $[-\pi/3, \pi/3]$, i.e., the uniform convergence of $\{T_n(x/3)\}$ to $g(x)$ in $[-\pi, \pi]$. Since the T_n's are trigonometric polynomials of degree at most $2n$, we may deduce at once the second Weierstrass theorem from Theorem 9.78.

9.3.3 The Stone–Weierstrass theorem

Weierstrass's theorems can be generalized to and seen as consequences of the following theorem proved in 1937 by Marshall Stone (1903–1989) and known as the *Stone–Weierstrass theorem*.

Let X be a compact metric space and let $C(X) = C^0(X, \mathbb{R})$ be the Banach space of continuous functions with uniform norm. An *algebra of functions* $\mathcal{A}$ is a real (complex) linear space of functions $f : X \to \mathbb{R}$ (respectively, $f : X \to \mathbb{C}$) such that $fg \in \mathcal{A}$ if f and $g \in \mathcal{A}$. We say that $\mathcal{A}$ *distinguishes between the points of* X if for any two distinct points x and y in X there is a function f in $\mathcal{A}$ such that $f(x) \neq f(y)$. We say that $\mathcal{A}$ *contains the constants* if the constant functions belong to $\mathcal{A}$.

9.80 Theorem (Stone–Weierstrass). *Let X be a compact metric space and let $\mathcal{A}$ be an algebra of continuous real-valued functions, $\mathcal{A} \subset C^0(X, \mathbb{R})$. If $\mathcal{A}$ contains constants and if it also distinguishes between the points of X, then $\mathcal{A}$ is dense in the Banach space $C^0(X, \mathbb{R})$.*

Let $\mathcal{A}$ be an algebra of bounded and continuous functions. As we have seen, the function $|y|$ can be approximated uniformly in $[0, 1]$ by polynomials. Consequently, if $f \in \mathcal{A}$, by considering instead of f the function $h := f/\|f\|_\infty$, we can approximate $x \to |f(x)|$ uniformly by the functions $P_n(f(x))$ where $\{P_n\}$ is a sequence of polynomials. Since the $P_n(f(x))$'s belong to $\mathcal{A}$, as $\mathcal{A}$ is an algebra of functions and $f \in \mathcal{A}$, we conclude that $|f|$ belongs to the uniform closure of $\mathcal{A}$, and also

$$\max{(f,g)} := \frac{1}{2}(f + g + |f - g|), \qquad \min(f,g) := \frac{1}{2}(f + g - |f - g|)$$

are in the uniform closure of $\mathcal{A}$, if both f and g are in the uniform closure of $\mathcal{A}$.

A linear space of functions R with the property that $\max{(f,g)}$ and $\min{(f,g)}$ are in R if f and $g \in R$ is called a *linear lattice*: the above can be then restated as *the closure of $\mathcal{A}$ is a linear lattice*. To prove that $\mathcal{A}$ is dense in $C^0(X,\mathbb{R})$, it therefore suffices to prove the following.

9.81 Theorem. *Let X be a compact metric space. A linear lattice $R \subset C^0(X,\mathbb{R})$ is dense, provided it contains the constants and distinguishes between the points in X.*

Proof. First we show that, for any $f \in C^0(X,\mathbb{R})$ and any couple of distinct points $x, y \in X$, we can find a function $\psi_{x,y} \in R$ such that

$$\psi_{x,y}(x) = f(x) \qquad \psi_{x,y}(y) = f(y).$$

In fact by hypothesis, we can choose $w \in R$ such that $w(x) \neq w(y)$; then the function

$$\psi_{x,y}(t) := \frac{(f(x) - f(y))\, w(t) - (f(x)w(y) - f(y)w(x))}{w(x) - w(y)}$$

has the required property.

Given $f \in C^0(X,\mathbb{R})$, $\epsilon > 0$ and $y \in X$, for every $x \in X$ we find a ball $B(x, r_x)$ such that $\psi_{x,y}(t) > f(t) - \epsilon \ \forall t \in B(x, r_x)$. Since X is compact, we can cover it by a finite number of these balls $\{B_{x_i}\}$ and we set $\varphi_y := \max \psi_{x_i,y}$. Then $\varphi_y(y) = f(y)$ and $\varphi_y \in R$ since R is a lattice. We now let y vary, and for any y we find $B(y, r_y)$ such that $\varphi_y(t) \leq f(t) + \epsilon \ \forall t \in B(y, r_y)$. Again covering X by a finite number of these balls $\{B(y_i, r_{y_i})\}$, and setting $\varphi := \max_i \varphi_{y_i}$, we conclude $\varphi \subset R$ and $|\varphi(t) - f(t)| < \epsilon$ $\forall t \in X$, i.e., the claim. $\qquad\square$

Of course real polynomials in $[a, b]$ form an algebra of continuous functions that contains constants and distinguishes between the points of $[0, 1]$. Thus the Stone–Weierstrass theorem implies the first Weierstrass theorem and even more, we have the following.

9.82 Corollary. *Every real-valued continuous function on a compact set $K \subset \mathbb{R}^n$ is the uniform limit in K of a sequence of polynomials in n variables.*

Theorem 9.80 does *not* extend to algebra of complex-valued functions. In fact, in the theory of functions of complex variables one shows that the uniform limits of polynomials are actually *analytic functions* and the map $z \to \overline{z}$, which is continuous, is not analytic. However, we have the following.

9.83 Theorem. *Let $\mathcal{A} \subset C^0(X,\mathbb{C})$ be an algebra of continuous complex-valued functions defined on a compact metric space X. Suppose that $\mathcal{A}$ distinguishes between the points in X, contains all constant functions and contains the conjugate $\overline{f}$ of f if $f \in \mathcal{A}$. Then $\mathcal{A}$ is dense in $C^0(X,\mathbb{C})$.*

Figure 9.10. René-Louis Baire (1874–1932) and the frontispiece of his *Leçons sur les functions discontinues.*

Proof. Denote by $\mathcal{A}_0$ the subalgebra of $\mathcal{A}$ of real-valued functions. Of course $\Re f = \frac{1}{2}(f + \bar{f})$ and $\Im f = \frac{1}{2i}(f - \bar{f})$ belong to $\mathcal{A}_0$ if f and $g \in \mathcal{A}$. Since $f(x) \neq f(y)$ implies that $\Re f(x) \neq \Re f(y)$ or $\Im f(x) \neq \Im f(y)$, $\mathcal{A}_0$ also distinguishes between the points of X and, trivially, contains the real constants. It follows that $\mathcal{A}_0$ is dense in $C^0(X, \mathbb{R})$ and consequently, $\mathcal{A}$ is dense in $C(X, \mathbb{C})$. $\square$

The real-valued trigonometric polynomials

$$a_0 + \sum_{k=1}^{n}(a_k \cos kx + b_k \sin kx) \tag{9.20}$$

form an algebra that distinguishes between the points of $[0, 2\pi[$ and contains the constants. Thus, trigonometric polynomials are dense among continuous real-valued periodic functions of period 2π.

More generally, from Theorem 9.83 we infer the following.

9.84 Theorem. *All continuous complex-valued functions defined on the unit sphere $\{z \in \mathbb{C} \mid |z| = 1\}$ are uniform limits of complex-valued trigonometric polynomials*

$$\sum_{k=-n}^{n} c_k e^{ik\theta}.$$

9.3.4 The Yosida regularization

a. Baire's approximation theorem

The next theorem relates semicontinuous functions to continuous functions.

9.85 Theorem (Baire). *Let X be a metric space and let $f : X \to \mathbb{R}$ be a function that is bounded from above and upper semicontinuous. Then there is a decreasing sequence of continuous, actually Lipschitz continuous, functions $\{f_n\}$ such that $f_n(x) \to f(x)$ for all $x \in X$.*

Proof. Consider the so-called *Yosida regularization* of f

$$f_n(x) := \sup_{y \in X} \{f(y) - nd(y, x)\}.$$

Obviously $f(x) \le f_n(x) \le \sup f$, $f_n(x) \ge f_{n+1}(x)$. We shall now show that each f_n is Lipschitz continuous with Lipschitz constant less than n. Let $x, y \in X$ and assume that $f_n(x) \ge f_n(y)$. For all $\eta > 0$ there is $x' \in X$ such that

$$f_n(x) < f(x') - nd(x, x') + \eta$$

hence

$$0 \le f_n(x) - f_n(y) \le f(x') - nd(x, x') + \eta - (f(x') - nd(y, x'))$$
$$= n(d(y, x') - d(x, x')) + \eta \le nd(x, y) + \eta$$

thus

$$|f_n(x) - f_n(y)| \le nd(x, y),$$

since η is arbitrary.

Let us show that $f_n(x_0) \downarrow f(x_0)$. Denote by M the $\sup_{x \in X} f(x)$. Since $f(x_0) \ge \limsup_{x \to x_0} f(x)$, for any $A > f(x_0)$ there is a spherical neighborhood $B(x_0, \delta)$ of x_0 such that $f(x) < A \ \forall x \in B(x_0, \delta)$, hence

$$f(x) - nd(x_0, x) \le \begin{cases} A & \text{if } d(x, x_0) < \delta, \\ M - n\delta & \text{if } d(x, x_0) \ge \delta. \end{cases}$$

Then $f(x) - nd(x_0, x) \le A \ \forall x \in X$, provided n is sufficiently large, $n \ge \frac{M - f(x_0)}{\delta}$, hence

$$f(x_0) \le f_n(x_0) = \sup_X (f(x) - nd(x_0, x)) \le A.$$

Since $A > f(x_0)$ is arbitrary, we conclude $f(x_0) = \lim_{n \to \infty} f_n(x_0)$.   □

Suppose that $X = \mathbb{R}^n$. An immediate consequence of Dini's theorem, Theorem 9.36, and of Baire's theorem, is the following.

9.86 Theorem. *Let $f : \mathbb{R}^n \to \mathbb{R}$ be a function that is bounded from above and upper semicontinuous. Then there exists a sequence of Lipschitz-continuous functions that converges uniformly on compact sets to f.*

b. Approximation in metric spaces

Yosida regularization also turns out to be useful to approximate uniformly continuous functions from a metric (or normed) space into $\mathbb{R}$ by Lipschitz-continuous functions.

Let X be a normed space with norm $\| \ \|$.

9.87 Proposition. *The class of uniformly continuous functions $f : X \to \mathbb{R}$ is closed with respect to the uniform convergence.*

Recall that the *modulus of continuity* of $f : X \to \mathbb{R}$ is defined for all $t \in [0, +\infty[$ by

$$\omega_f(t) := \sup\Big\{ |f(x) - f(y)| \,\Big|\, x, y \in X, \ \|x - y\| \leq t \Big\}. \tag{9.21}$$

Clearly f is uniformly continuous on X if and only if $\omega_f(t) \to 0$ as $t \to 0$.

9.88 ¶. Prove Proposition 9.87.

Lipschitz-continuous functions from X to $\mathbb{R}$ are of course uniformly continuous, therefore uniform limits of Lipschitz-continuous functions are uniformly continuous too, on account of Proposition 9.87. We shall now prove the converse, compare Example 9.61.

9.89 Theorem. *Every uniformly continuous function $f : X \to \mathbb{R}$ is the uniform limit of a sequence of Lipschitz-continuous functions.*

In order to prove Theorem 9.89, we introduce the function $\delta_f(s)$ that measures the uniform distance of f from the class $\text{Lip}_s(X)$ of Lipschitz-continuous functions $g : X \to \mathbb{R}$ with Lipschitz constants not greater than s

$$\delta_f(s) := \inf\Big\{ \|f - g\|_\infty \,\Big|\, g \in \text{Lip}_s(X) \Big\}.$$

9.90 ¶. Show that $s \to \delta_f(s)$ is nonincreasing and that f is the uniform limit of a sequence of Lipschitz-continuous functions if and only if $\delta_f(s) \to 0$ as $s \to \infty$.

Then we introduce the *Yosida regularization* of $f : X \to \mathbb{R}$ by

$$f_s(x) := \inf_{y \in X} \Big\{ f(y) + s \, \|x - y\| \Big\}.$$

9.91 ¶. Show that

(i) f_s is Lipschitz-continuous with Lipschitz constant s,
(ii) $f_s(x) \leq f_t(x) \ \forall x$ if $s < t$, and, actually
(iii) f_s is the largest s-Lipschitz-continuous function among functions less than or equal to f.

9.92 Proposition. *Let $f : X \to \mathbb{R}$ be a uniformly continuous function. Then*

$$\delta_f(s) = \frac{1}{2} \sup_{t \geq 0} \Big\{ \omega_f(t) - st \Big\}. \tag{9.22}$$

Moreover, the minimum distance of f from $\mathrm{Lip}\,_s(X)$ is obtained at $g_s(x) := f_s(x) + \delta_f(s)$, i.e.,

$$\delta_f(s) = ||f - g_s||_\infty.$$

Proof. Let $g \in \mathrm{Lip}\,_s(X)$. Then

$$|f(x) - f(y)| \leq 2\,||f - g||_\infty + s||x - y||.$$

For x, y such that $||x - y|| \leq t$, by taking the infimum with respect to g, we infer

$$|f(x) - f(y)| \leq 2\,\delta_f(s) + st$$

and, taking the supremum in x and y with $||x - y|| \leq t$, we get $\omega_f(t) \leq 2\delta_f(s) + st$, hence

$$\frac{1}{2} \sup_{t \geq 0} \Big\{ \omega_f(t) - st \Big\} \leq \delta_f(s). \tag{9.23}$$

Let us prove that the inequality (9.23) is actually an equality and the second part of the claim. For $x, y \in X$ we have

$$f(x) - f(y) \leq \omega_f(||x - y||) = \Big(\omega_f(||x - y||) - s||x - y|| \Big) + s||x - y||$$

$$\leq \sup_{t \geq 0} \Big\{ \omega_f(t) - st \Big\} + s||x - y||.$$

By taking the supremum in y we get

$$0 \leq f(x) - f_s(x) \leq \sup_{t \geq 0} \Big\{ \omega_f(t) - st \Big\} < 2\delta_f(s) \qquad \forall x \in X$$

hence, by (9.23) we infer

$$||f - f_s||_\infty \leq \sup_{t \geq 0} \Big\{ \omega_f(t) - st \Big\} \leq 2\delta_f(s). \tag{9.24}$$

Therefore, for $g_s(x) := f_s(x) + \delta_f(x)$ we have

$$||f - g_s||_\infty \leq \delta_f(s),$$

and, since $g_s \in \mathrm{Lip}\,_s(X)$, we conclude $||f - g_s||_\infty = \delta_f(s)$. Moreover, by (9.24)

$$\delta_f(s) = ||f - g_s||_\infty = ||f - f_s||_\infty - \delta_f(s) \leq \sup_{t \geq 0} \Big\{ \omega_f(t) - st \Big\} - \delta_f(s) \leq \delta_f(s)$$

i.e.,

$$\frac{1}{2} \sup_{t \geq 0} \Big\{ \omega_f(t) - st \Big\} = \delta_f(s).$$

$\square$

9.93 ¶. Show that if $f^s := -(-f)_s$, then $f^s \in \mathrm{Lip}\,_s(X)$ and

$$\delta_f(s) = ||f - g^s||_\infty, \qquad g^s(x) := f^s(x) - \delta_f(s).$$

Proof of Theorem 9.89. It is enough to prove that $\inf_{s>0} \delta_f(s) = 0$. First, we notice that $\omega_f(t)$ is nondecreasing and subadditive. In fact, if x, y are such that $||x-y|| \leq a+b$ and we write

$$z := \frac{b}{a+b}x + \frac{a}{a+b}y,$$

then $|z - x| \leq a$ and $|z - y| \leq b$; consequently,

$$|f(x) - f(z)| \leq \omega_f(a) \qquad \text{and} \qquad |f(y) - f(z)| \leq \omega_f(b)$$

that yield at once

$$\omega_f(a + b) \leq \omega_f(a) + \omega_f(b).$$

Next, we observe that for any $\epsilon > 0$ and $t = m\epsilon + \sigma$, $m \in \mathbb{N}$, $\sigma < \epsilon$, we have $m < t/\epsilon$ and

$$\omega_f(t) = \omega_f(m\epsilon + \sigma) \leq m\omega_f(\epsilon) + \omega_f(\sigma) \leq \omega_f(\epsilon)\frac{t}{\epsilon} + \omega_f(\epsilon).$$

Therefore

$$\delta_f\left(\frac{\omega_f(\epsilon)}{\epsilon}\right) = \frac{1}{2}\sup_{t \geq 0}\left\{\omega_f(t) - \frac{\omega_f(\epsilon)}{\epsilon}t\right\}$$

$$\leq \frac{1}{2}\sup_{t \geq 0}\left\{\frac{\omega_f(\epsilon)}{\epsilon}t + \omega_f(\epsilon) - \frac{\omega_f(\epsilon)}{\epsilon}t\right\} = \frac{1}{2}\omega_f(\epsilon).$$

From the last inequality we easily infer $\inf_{s>0} \delta_f(s) = 0$. $\qquad\qquad\square$

9.4 Linear Operators

9.4.1 Basic facts

In finite-dimensional spaces, linear maps are continuous, but this is no more true in infinite-dimensional normed spaces, see Example 9.96.

9.94 ¶. Show that

Proposition. *Let X and Y be normed spaces. Suppose that X is finite dimensional. Then every linear map $L : X \to Y$ is continuous.*

The following proposition characterizes linear maps between two Banach spaces that are continuous.

9.95 Proposition. *Let X and Y be normed spaces and let $L : X \to Y$ be a linear map. Then the following conditions are equivalent*

- (i) *L is continuous in X,*
- (ii) *L is continuous at 0,*
- (iii) *L is bounded on the unit ball, i.e., there exists $K \geq 0$ such that $||L(x)||_Y \leq K$ $\forall x$ with $||x||_X \leq 1$,*
- (iv) *there exists a constant K such that $||L(x)||_Y \leq K||x||_X$ $\forall x \in X$,*
- (v) *L is Lipschitz continuous.*

Proof. If L is continuous, then trivially, (ii) holds. If (ii) holds, then there exists $\delta > 0$ such that $||L(x)||_Y \leq 1$ provided $||x||_X < \delta$. This yields $||L(x)||_Y \leq 1/\delta$ if $||x||_X \leq 1$, since L is linear, i.e., (iii). Assuming (iii) and the linearity of L, we infer that for all $x \in X$, $x \neq 0$,

$$\frac{||L(x)||_Y}{||x||_X} = \left|\left|L\left(\frac{x}{||x||_X}\right)\right|\right|_Y \leq K,$$

i.e., (iv). (iv) in turn implies (v) since

$$||L(x) - L(y)||_Y = ||L(x - y)||_Y \leq K||x - y||_X \qquad \forall x, y \in X,$$

and trivially, (v) implies (i).   □

9.96 Example. Let X be a normed space and let $\{e_n\} \subset X$ be a countable system of independent vectors with $||e_n|| = 1$, and let $Y \subset X$ be the subspace of finite linear combinations of $\{e_n\}$. Consider the operator $L : Y \to \mathbb{R}$ defined on $\{e_n\}$ by $L(e_n) := n$ $\forall n$ and linearly extended to Y. Evidently L is linear and not bounded.

Linear maps between Banach spaces are often called *linear operators*.

a. Continuous linear forms and hyperplanes

Consider a linear map $L : X \to \mathbb{K}$ defined on a linear normed space X, often called also a *linear form*. If L is not identically zero, we can find $\overline{x}$ such that $\overline{x} \notin \ker L$ and we can decompose every $x \in X$ as

$$x = \frac{L(x)}{L(\overline{x})}\overline{x} + \left(x - \frac{L(x)}{L(\overline{x})}\overline{x}\right);$$

in other words

$$X = \operatorname{Span}\{\overline{x}\} \oplus \ker L.$$

However it may happen that $\ker L$ is dense in X.

9.97 Proposition. *Let $L : X \to \mathbb{R}$ be a linear map defined on a normed space X. Then $\ker L$ is closed if and only if L is continuous.*

Proof. Trivially, $\ker L := L^{-1}(0)$ is closed if L is continuous. Conversely, if $\ker L = X$, then L is constant, hence continuous. Otherwise we can choose $\overline{x}$ such that $L(\overline{x}) = 1$. Since $\ker L$ is closed, also $H := \overline{x} + \ker L$ is closed; since $0 \notin H$, we can then find a ball $B(0, r)$ such that $B(0, r) \cap H = \emptyset$. We now prove that L is continuous showing that $|L(x)| < 1 \; \forall x \in B(0, r)$.

In fact, if $|L(x)| \geq 1$ for some $x \in B(0, r)$, then

$$\left|\left|\frac{x}{L(x)}\right|\right| = \frac{1}{|L(x)|}||x|| < r \qquad \text{while} \qquad L\left(\frac{x}{L(x)}\right) = 1.$$

Since $H = \{x \mid L(x) = 1\}$, we conclude that $x/L(x) \in H \cap B(0, r)$, a contradiction.   □

9.98 Corollary. *If $L : X \to \mathbb{R}$ is a linear map on a normed space X, then $\ker L$ is either closed or dense in X.*

In fact, the closure of $\ker L$ is a linear subspace that may agree either with $\ker L$ or X.

b. The space of linear continuous maps

For any linear map $L : X \to Y$ between two normed spaces with norms $||\ ||_X$ and $||\ ||_Y$, we define

$$||L||_{\mathcal{L}(X,Y)} := \sup_{||x||_X \leq 1} ||L(x)||_Y, \tag{9.25}$$

i.e.,

$$||L||_{\mathcal{L}(X,Y)} = ||L||_{\infty,B}, \qquad B = \Big\{ x \in X \ \Big| \ ||x||_X = 1 \Big\},$$

or, equivalently,

$$||L||_{\mathcal{L}(X,Y)} = \inf \Big\{ K \in \mathbb{R} \ \Big| \ ||L(x)||_Y \leq K ||x||_X \Big\}$$

so that

$$||L(x)||_Y \leq ||L||_{\mathcal{L}(X,Y)} \, ||x||_X \qquad \forall x \in X.$$

One can shorten this to

$$||Lx|| \leq ||L|| \, ||x||$$

once the norms used to evaluate x, Lx and L are understood.

From Proposition 9.95 it follows that $L : X \to Y$ *is continuous if and only if* $||L||_{\mathcal{L}(X,Y)} < +\infty$. For this reason, linear continuous maps from X to Y are also called *bounded operators*.

It is now easily seen that $||L||_{\mathcal{L}(X,Y)}$ is a norm on the linear space $\mathcal{L}(X,Y)$ of linear continuous maps from X into Y.

9.99 Theorem. *Let X be a normed space and let Y be a Banach space. Then $\mathcal{L}(X,Y)$ is a Banach space.*

Proof. Let $\{L_n\}$ be a Cauchy sequence in $\mathcal{L}(X,Y)$. For any $\epsilon > 0$ there is $n_0(\epsilon)$ such that $||L_n - L_m|| < \epsilon$ for $n, m \geq n_0$. In particular, $||L_n(x) - L_m(x)||_Y \leq \epsilon ||x||_X$ for all $x \in X$, i.e., for every $x \in X$ $\{L_n(x)\}$ is a Cauchy sequence in Y hence converges to an element $L(x) \in Y$, $L_n(x) \to L(x)$ as $n \to \infty$. Letting n to infinity in

$$L_n(x+y) = L_n(x) + L_n(y), \qquad L_n(\lambda x) = \lambda L_n(x),$$

we see that L is linear. Letting $m \to \infty$ in $||L_n(x) - L_m(x)||_Y \leq \epsilon$ valid for $||x||_X \leq 1$, $n \geq n_0(\epsilon)$, we also find $||L_n(x) - L(x)||_Y \leq \epsilon$ for $||x||_X \leq 1$ and $n \geq n_0(\epsilon)$. This implies $||L|| \leq ||L_n|| + \epsilon$ and $||L_n - L|| \leq \epsilon$ for $n \geq n_0$, which in turn yields $L_n \to L$ in $\mathcal{L}(X,Y)$. $\square$

c. Norms on matrices

For any n, let $\mathbb{K} = \mathbb{R}$ ($\mathbb{K} = \mathbb{C}$) and consider $\mathbb{R}^n$ (respectively $\mathbb{C}^n$) as an Euclidean (Hermitian) space endowed with the standard Euclidean (Hermitian) product and let $|\ |$ be the associated norm. Let $L : \mathbb{K}^n \to \mathbb{K}^m$ be linear, let $\mathbf{L} \in M_{m,n}(\mathbb{K})$ be the associated matrix, $L(x) =: \mathbf{L}x$, and let $\mu_1, \mu_2, \ldots, \mu_n$ be the singular values of L, that is the eigenvalues of the matrix $(\mathbf{L}^T \mathbf{L})^{1/2}$ ordered in increasing order. Then

$$||L||^2 = \sup_{|x|=1} |L(x)|^2 = \sup_{|x|=1} (L^*L(x)|x) = \mu_n^2.$$

Now define the ℓ^2- norm of $L \in \mathcal{L}(\mathbb{K}^n, \mathbb{K}^m)$, by

$$||L||_2 := \left(\sum_{i,j} (\mathbf{L}_j^i)^2 \right)^{1/2}.$$

Of course, $|| \ ||_2$ and $|| \ ||$ are equivalent norms in $\mathcal{L}(\mathbb{K}^n, \mathbb{K}^m)$, since $\mathcal{L}(\mathbb{K}^n, \mathbb{K}^m)$ is finite dimensional. More precisely we compute

$$||L||_2^2 = \operatorname{tr}(\mathbf{L}^T\mathbf{L}) = \operatorname{tr}(L^*L) = \mu_1^2 + \ldots \mu_n^2,$$

and therefore,

9.100 Proposition. *Let $L \in M_{m,n}(\mathbb{K})$. Then $||L||$ is the maximum of the singular values of L and $||L|| \leq ||L||_2 \leq \sqrt{n} \, ||L||$. Moreover, $||L|| = ||L||_2$ if and only if* Rank $L = 1$.

Proof. Let $\mu_1, \mu_2, \ldots, \mu_n$ be the singular values of L ordered in nondecreasing order. By the above, $||L|| = \mu_n \leq ||L||_2 \leq \sqrt{n}||L||$ while equality $||L|| = ||L||_2$ is equivalent to $\mu_1 = \cdots = \mu_{n-1} = 0$, and this happens if and only if Rank $L = 1$. $\square$

9.101 ¶. Let $T : \ell_2 \to \mathbb{R}$ be a bounded operator and for $i = 1, \ldots,$ let $e_i = \{\delta_{in}\}_n$. Then

$$||T||^2 = \sum_{j=1}^{\infty} |T'(e_i)|^2.$$

d. Pointwise and uniform convergence for operators

In $\mathcal{L}(X, Y)$ we may define two notions of convergence.

9.102 Definition. *Let $\{L_n\} \subset \mathcal{L}(X, Y)$.*

(i) *We say that $\{L_n\}$ converges pointwise to L if $\forall x \in X$ we have*

$$||L_n(x) - L(x)||_Y \to 0,$$

(ii) *we say that $\{L_n\}$ converges to L in norm or uniformly, if*

$$||L_n - L||_{\mathcal{L}(X,Y)} = \sup_{||x||_X \leq 1} ||L_n(x) - L(x)||_Y \to 0 \qquad \text{as } n \to \infty.$$

Trivially, $L_n \to L$ pointwise if $L_n \to L$ uniformly. But the converse is in general false and holds true if X is finite dimensional.

9.103 Example. Recall that a sequence $\{x_n\}$ is in $\ell_2(\mathbb{R})$ if and only if $\sum_{k=1}^{\infty} x_k^2 < +\infty$. For any $n \in \mathbb{N}$ let $e^{(n)} := \{\delta_{kn}\}_k$. Of course, $||e^{(n)}||_2 = 1 \ \forall n$.

Consider the sequence of linear forms $\{L_n\}$ on $\ell_2(\mathbb{R})$ defined by $L_n(\{x_k\}) = x_n$. For any $x \in \ell_2(\mathbb{R})$ we have $\sum_{k=1}^{\infty} x_k^2 < +\infty$, hence $L_n(x) = x_n \to 0$ as $n \to \infty$, i.e., $L_n \to 0$ pointwise. On the other hand,

$$||L_n - 0||_{\mathcal{L}(\ell_2, \mathbb{R})} = ||L_n||_{\mathcal{L}(\ell_2, \mathbb{R})} = \sup_{||x||_2 \leq 1} |L_n(x)| \geq L_n(e^{(n)}) = 1$$

and $\{L_n\}$ does not converge uniformly to 0.

e. The algebra End (X)

Let X, Y and Z be linear normed spaces and let $f : X \to Y$ and $g : Y \to Z$ be linear continuous operators. The composition $g \circ f : X \to Z$ is again a linear continuous operator from X to Z, and for every $x \in X$ we have

$$||(g \circ f)(x)||_Z \leq ||g|| \, ||f(x)||_Y \leq ||g|| \, ||f|| \, ||x||_X,$$

hence

$$||g \circ f|| \leq ||g|| \, ||f||. \tag{9.26}$$

9.104 Example. In general $||g \circ f|| < ||g|| \, ||f||$. For instance, if $X = \mathbb{R}^2$ and f and g are the orthogonal projections on the axes, we have $||f|| = ||g|| = 1$ and $f \circ g = g \circ f = 0$, hence $||g \circ f|| = ||f \circ g|| = 0$.

9.105 Example. Let $T : \mathbb{R}^2 \to \mathbb{R}^2$ be defined by $T(x) = \mathbf{T}x$ where $\mathbf{T} := \begin{pmatrix} 1 & 0 \\ 0 & \epsilon \end{pmatrix}$, $0 < \epsilon << 1$. Then $T^{-1}(x) = \mathbf{T}^{-1}x$ where $\mathbf{T}^{-1} = \begin{pmatrix} 1 & 0 \\ 0 & 1/\epsilon \end{pmatrix}$. We then compute $||T|| = 1$, $||T^{-1}|| = 1/\epsilon$ and $||T|| \, ||T^{-1}|| >> 1 = ||\operatorname{Id}|| = ||T^{-1} \circ T||$.

Let X be a Banach space with norm $|| \; ||_X$, denote the Banach space $\mathcal{L}(X, X)$ by $\operatorname{End}(X)$ and the norm on $\operatorname{End}(X)$ by $|| \; ||$. The product of composition defines in $\operatorname{End}(X)$ a structure of *algebra* in which the product satisfies the inequality (9.26): this is expressed by saying that $\operatorname{End}(X)$ *is a Banach algebra*. Clearly, if $L \in \operatorname{End}(X)$ and $L^n = L \circ L \circ \cdots \circ L$, then by (9.26) we have

$$||L^n|| \leq ||L||^n. \tag{9.27}$$

Again, in general, we may have a strict inequality.

9.106 Proposition. *Let X be a Banach space and $L \in \operatorname{End}(X)$. If $||L|| < 1$, then $\operatorname{Id} - L$ is invertible,*

$$(\operatorname{Id} - L)^{-1} = \sum_{n=0}^{\infty} L^n \qquad in \; \operatorname{End}(X)$$

and $||(\operatorname{Id} - L)^{-1}|| \leq \frac{1}{1-||L||}$. In particular, for any $y \in X$ the equation $x - Lx = y$ has a unique solution, $x = \sum_{n=0}^{\infty} L^n y$ and $||x|| \leq \frac{1}{1-||L||} ||y||$.

Proof. The series $\sum_{n=0}^{\infty} L^n$ is absolutely convergent, since

$$\sum_{n=0}^{\infty} ||L^n|| \leq \sum_{n=0}^{\infty} ||L||^n = \frac{1}{1 - ||L||},$$

hence convergent. In particular, $S := \sum_{n=0}^{\infty} L^n \in \operatorname{End}(X)$ and $||S|| \leq \frac{1}{1-||L||}$. Finally

$$(\operatorname{Id} - L) \sum_{k=0}^{n} L^k = \operatorname{Id} - L^{n+1} \to \operatorname{Id} \qquad in \; \operatorname{End}(X)$$

since $||L^{n+1}|| \leq ||L||^{n+1} \to 0$. $\qquad\square$

f. The exponential of an operator

Again by (9.27) we get, similarly to Proposition 9.106, the following.

9.107 Proposition. *Let X be a Banach space and $L \in \text{End}(X)$.*

(i) *Let $f(z) := \sum_{n=0}^{\infty} a_n z^n$ be a power series with radius of convergence $\rho > 0$. If $\|L\| < \rho$, then the series $\sum_{n=0}^{\infty} a_n L^n$ converges in $\text{End}(X)$ and defines a linear continuous operator*

$$f(L) := \sum_{n=0}^{\infty} a_n L^n \in \text{End}(X).$$

(ii) *The series $\sum_{k=0}^{\infty} \frac{1}{k!} L^k$ converges in $\text{End}(X)$ and defines the linear continuous operator*

$$e^L = \exp(L) := \sum_{k=0}^{\infty} \frac{1}{k!} L^k \in \text{End}(X).$$

9.108 ¶. Show the following.

Proposition. *Let X be a Banach space and let $A, B \in \text{End}(X)$. Then we have*

(i) $\left(\text{Id} + \frac{A}{n}\right)^n \to e^A$ *in* $\text{End}(X)$,

(ii) $\left\|e^A\right\| \leq e^{\|A\|}$,

(iii) *If A and B commute, i.e., $AB = BA$, then*

$$e^A B = B e^A, \qquad e^{A+B} = e^A e^B,$$

(iv) *if A has an inverse, then $(e^A)^{-1} = e^{-A}$,*

(v) *if X is finite-dimensional, $X = \mathbb{R}^n$, we have*

$$e^{PAP^{-1}} = P e^A P^{-1}, \qquad \text{if } P \text{ has an inverse,}$$
$$\det e^A = e^{\text{tr}\, A}, \qquad \text{if } A \text{ is symmetric.}$$

9.4.2 Fundamental theorems

In this subsection, we briefly illustrate four of the most important theorems about the structure of linear continuous operators on normed spaces. The first three, the *principle of uniform boundedness*, the *open mapping theorem* and the *closed graph theorem* are a consequence of Baire's category theorem, see Chapter 5, and are due to Stefan Banach (1892–1945); the fourth one, known as *Hahn–Banach* theorem, was proved independently by Hans Hahn (1879–1934) in 1926 and by Banach in 1929.

a. The principle of uniform boundedness

The following important theorem is known as the Banach–Steinhaus theorem and also as the *principle of uniform boundedness*.

9.109 Theorem (Banach–Steinhaus). *Let X be a Banach space and Y be a normed linear space. Let $\{T_\alpha\}$ be a family of bounded linear operators from X to Y indexed on an arbitrary set $\mathcal{A}$ (possibly nondenumerable). If*

$$\sup_{\alpha \in \mathcal{A}} ||T_\alpha x||_Y \leq C(x) < +\infty \qquad \forall x \in X,$$

then there exists a constant C such that

$$\sup_{\alpha \in \mathcal{A}} ||T_\alpha||_{\mathcal{L}(X,Y)} \leq C < +\infty.$$

Proof. Set

$$X_n := \Big\{ x \in X \, \Big| \, ||T_\alpha x||_Y \leq n \ \forall \alpha \in \mathcal{A} \Big\}.$$

X_n is closed and by hypothesis $\cup_n X_n = X$. By Baire's category theorem, it follows that there exists $n_0 \in \mathbb{N}$, $x_0 \in X$ and $r_0 > 0$ such that $B(x_0, r_0) \subset X_{n_0}$, that is

$$||T_\alpha(x_0 + r_0 z)||_Y \leq n_0 \qquad \forall \alpha \in \mathcal{A},$$

hence $\forall z \in B(0, 1)$,

$$||T_\alpha(z)||_Y \leq \frac{1}{r_0} ||T_\alpha(r_0 z + x_0 - x_0)||_Y$$

$$\leq \frac{1}{r_0}(n_0 + ||T_\alpha(x_0)||_Y) \leq \frac{n_0 + C(x_0)}{r_0} =: C.$$

$\square$

The following corollaries are trivial consequences.

9.110 Corollary. *Let $\{T_k\}$ be a sequence of bounded linear operators from a Banach space X into a normed space Y. Suppose that for each $x \in X$ $\lim_{k\to\infty} T_k x =: Tx$ exists in Y. Then the limit operator T is also a bounded linear operator from X to Y and we have*

$$||T||_{\mathcal{L}(X,Y)} \leq \liminf_{k\to\infty} ||T_k||_{\mathcal{L}(X,Y)}.$$

9.111 Corollary. *Let $1 \leq p < +\infty$. Any linear operator from $\ell_p(\mathbb{R})$ into a normed linear space Y is continuous.*

Proof. In fact, the linear operators $\{L_n\}$ defined by

$$L_n(\xi) := L((\xi_1, \xi_2, \ldots, \xi_n, 0, 0, \ldots)) \qquad \text{if } \xi = (\xi_1, \xi_2, \ldots)$$

are clearly continuous and $L_n(\xi) \to L(\xi)$ in ℓ_p. Therefore L is continuous by the Banach–Steinhaus theorem. $\square$

The following theorem, again due to Banach, is also a consequence of Baire's category theorem.

9.112 Theorem. *Given a sequence of bounded linear operators $\{T_k\}$ from a Banach space X into a normed linear space Y, the set*

$$\left\{ x \in X \;\middle|\; \liminf_{k \to \infty} \|T_k x\|_Y < +\infty \right\}$$

either coincides with X or is a set of the first category of X.

This in turn implies the following.

9.113 Corollary (Principle of condensation of singularities). *For $p = 1, 2, \ldots,$ let $\{T_{p,q}\}$, $q = 1, 2, \ldots,$ be a sequence of bounded linear operators from a Banach space into a normed space Y_p. Assume that for each p there exists $x_p \in X$ such that $\limsup_{q \to \infty} \|T_{p,q} x_p\|_{\mathcal{L}(X, Y_p)} = \infty$. Then the set*

$$\left\{ x \in X \;\middle|\; \limsup_{q \to \infty} \|T_{p,q}\|_{\mathcal{L}(X, Y_p)} = +\infty \text{ for all } p = 1, 2, 3, \ldots \right\}$$

is of second category.

The above principle gives a general method of finding functions with many singularities. For instance one can find in this way a continuous function $x(t)$ of period 2π such that the partial sum of its Fourier expansion

$$S_n f(t) := \frac{a_0}{2} + \sum_{k=1}^{n} (a_k \cos kt + b_k \sin kt)$$

satisfies the condition

$$\limsup_{n \to \infty} |S_n f(t)| = \infty$$

in a set $P \subset [0, 2\pi]$ which has the power of the continuum.[1]

b. The open mapping theorem

9.114 Theorem (Banach's open mapping theorem). *Let X and Y be Banach spaces and let T be a surjective bounded linear operator from X into Y. Then T is open, i.e., it maps open sets of X onto open sets of Y.*

Proof. We divide the proof into two steps.

Step 1. First we prove that there is a $\delta > 0$ such that

$$\overline{TB_X(0,1)} \supset B_Y(0, 2\delta).$$

Set $X_n := n\overline{TB_X(0,1)}$. All X_n are closed and, since T is surjective, $\cup_{n=1}^{\infty} X_n = Y$. By Baire's category theorem, see Theorem 5.118, it follows that for some n, X_n has a nonvoid interior. By homogeneity, $\overline{T(B_X(0,1))}$ has a nonvoid interior, too, i.e., there exists $y_0 \in Y$ and $\delta > 0$ such that $B_Y(y_0, 4\delta) \subset \overline{T(B_X(0,1))}$. By symmetry $-y_0 \in \overline{TB_X(0,1)}$, and, as $\overline{TB_X(0,1)}$ is convex, $B_Y(0, 2\delta) \subset \overline{TB_X(0,1)}$.

Step 2. We shall now prove that

$$TB_X(0,1) \supset B_Y(0, \delta),$$

that is the claim. Observe that by Step 1 and homogeneity

$$\overline{TB_X(0,r)} \supset B_Y(0, 2\delta r) \qquad \forall r > 0. \tag{9.28}$$

---

[1] For proofs we refer the interested reader to e.g., K. Yosida, *Functional Analysis*, Springer–Verlag, Berlin, 1964.

We want to prove that the equation $Tx = y$ has a solution $x \in B_X(0,1)$ for any $y \in B_Y(0, \delta)$. Let $y \in Y$ be such that $||y||_Y < \delta$. By (9.28) there exists $x_1 \in X$ such that $||x_1||_X < 1/2$ and $||Tx_1 - y||_Y < \delta/2$. Similarly, considering the equation $Tx = y - Tx_1$, one can find $x_2 \in X$ such that $||x_2||_X < 1/4$ and $||y - Tx_1 - Tx_2||_Y < \delta/4$. By induction, we then construct points $x_n \in X$ such that $||x_n||_X < 2^{-n}$ and $||y - \sum_{k=1}^n Tx_k||_Y < \delta/2^k$. Therefore the series $\sum_{k=1}^\infty x_k$ is absolutely convergent in X with sum less than 1, hence it converges to some $x \in X$ with $||x||_X < 1$, and $||y - Tx||_Y = 0$. $\square$

9.115 ¶. Show the converse of the open mapping theorem: if $T : X \to Y$ is an open, bounded linear operator between Banach spaces, then T is surjective.

A trivial consequence of Theorem 9.114 is the following.

9.116 Corollary (Banach's continuous inverse theorem). *Let X, Y be Banach spaces and let $T : X \to Y$ be a surjective and one-to-one bounded linear operator. Then T^{-1} is a bounded operator.*

9.117 Remark. Let X and Y be Banach spaces and let $T : X \to Y$ be a linear continuous operator. Often one says that the equation $Tx = y$ is *well posed* if for any $y \in Y$ it has a unique solution $x \in X$ which depends continuously on y. Corollary 9.116 says that the equation $Tx = y$ is well-posed if X and Y are Banach spaces and $Tx = y$ is uniquely solvable $\forall y \in Y$.

c. The closed graph theorem

Let X, Y be two Banach spaces. Then $X \times Y$ endowed with the norm

$$||(x,y)||_{X \times Y} := ||x||_X + ||y||_Y$$

is also a Banach space.

9.118 Theorem (Banach's closed graph theorem). *Let X and Y be Banach spaces and let $T : X \to Y$ be a linear operator. Then $T : X \to Y$ is bounded if and only if its graph*

$$G_T := \Big\{ (x,y) \in X \times Y \,\Big|\, y = Tx \Big\}$$

is closed in $X \times Y$.

Proof. If T is continuous, then trivially G_T is closed. Conversely, G_T is a closed linear subspace of $X \times Y$, hence G_T is a Banach space with the induced norm of $X \times Y$. The linear map $\pi : G_T \to X$, $\pi((x, Tx)) := x$, is a bounded linear operator that is one-to-one and onto; hence, by the Banach continuous inverse theorem, the inverse map of π, $\pi^{-1} :$ $X \to G_T$, $x \to (x, Tx)$, is a bounded linear operator, i.e., $||x||_X + ||Tx||_Y \leq C ||x||_X$ for some constant C. T is therefore bounded. $\square$

Figure 9.11. Hans Hahn (1879–1934) and Hugo Steinhaus (1887–1972).

d. The Hahn–Banach theorem

The Hahn–Banach theorem is one of the most important results in linear functional analysis. Basically, it allows one to extend to the whole space a bounded linear operator defined on a subspace in a controlled way. In particular, it enables us to show that the dual space, i.e., the space of linear bounded forms on X, is rich.

9.119 Theorem (Hahn–Banach, analytical form). *Let X be a real normed space and let $p : X \to \mathbb{R}$ be a sublinear functional, that is, satisfying*

$$p(x + y) \le p(x) + p(y), \qquad p(\lambda x) = \lambda p(x) \qquad \forall \lambda > 0, \ \forall x, y \in X.$$

Let Y be a linear subspace of X and let $f : Y \to \mathbb{R}$ be a linear functional such that $f(x) \le p(x) \ \forall x \in Y$. Then f can be extended to a linear functional $F : X \to \mathbb{R}$ satisfying

$$F(x) = f(x) \ \forall x \in Y, \qquad F(x) \le p(x) \ \forall x \in X.$$

Proof. Denote by $\mathcal{K}$ the set of all pairs (Y_α, g_α) where Y_α is a linear subspace of X such that $Y_\alpha \supset Y$ and g_α is a linear functional on Y_α satisfying

$$g_\alpha(x) = f(x) \ \forall x \in X, \qquad g_\alpha(x) \le p(x) \ \forall x \in Y_\alpha.$$

We define an order in $\mathcal{K}$ by $(Y_a, g_a) \le (Y_\beta, g_\beta)$ if $Y_\alpha \subset Y_\beta$ and $g_\alpha = g_\beta$ on Y_α. Then $\mathcal{K}$ becomes a partially ordered set. Every totally ordered subset $\{(Y_\alpha, g_\alpha)\}$ clearly has an upperbound (Y', g') given by $Y' = \cup_\beta Y_\beta$, $g' = g_\beta$ on Y_β. Hence, by Zorn's lemma, see e.g., Section 3.3 of [GM2], there is a maximal element (Y_0, g_0). If we show that $Y_0 = X$, then the proof is complete with $F = g_0$.

We shall assume that $Y_0 \ne X$ and derive a contradiction. Let $y_1 \notin Y_0$ and consider

$$Y_1 := \mathrm{Span}\Big(Y_0 \cup \{y_1\}\Big) = \Big\{x = y + \lambda y_1 \ \Big| \ y \in Y_0, \ \lambda \in \mathbb{R}\Big\},$$

notice that $y \in Y_0$ and $\lambda \in \mathbb{R}$ are uniquely determined by x, otherwise we get $y_1 \in Y_0$. Define $g_1 : Y_1 \to \mathbb{R}$ by $g_1(y + \lambda y_1) := g_0(y) + \lambda c$. If we can choose c in such a way that

$$g_1(y + \lambda y_1) = g_0(y) + \lambda c \le p(y + \lambda y_1)$$

for all $\lambda \in \mathbb{R}$, $y \in Y_0$, then $(Y_1, g_1) \in \mathcal{K}$ and $(Y_0, g_0) \le (Y_1, g_1)$, $Y_1 \ne Y_0$. This contradicts the maximality of (Y_0, g_0).

To choose c, we notice that for $x, y \in Y_0$

$$g_0(y) - g_0(x) = g_0(y - x) \leq p(y - x) \leq p(y + y_1) + p(-y_1 - x).$$

Hence

$$-p(-y_1 - x) - g_0(x) \leq p(y + y_1) - g_0(y).$$

This implies that

$$A := \sup_{x \in Y_0} \left\{ -p(-y_1 - x) - g_0(x) \right\} \leq \inf_{y \in Y_0} \left\{ p(y + y_1) - g_0(y) \right\} =: B.$$

Thus we can choose c such that $A \leq c \leq B$. Then

$$c \leq p(y + y_1) - g_0(y) \qquad \forall y \in Y_0,$$
$$p(-y_1 - y) - g_0(y) \leq c \qquad \forall y \in Y_0.$$

Multiplying the first inequality by λ, $\lambda > 0$, and the second by λ, $\lambda < 0$, and replacing y with y/λ we conclude that for all $\lambda \neq 0$ and trivially for $\lambda = 0$

$$\lambda c \leq p(y + y_1) - g_0(y).$$

$\square$

9.120 Theorem (Hahn–Banach). *Let X be a normed linear space of $\mathbb{K} = \mathbb{R}$ or $\mathbb{K} = \mathbb{C}$ and let Y be a linear subspace of X. Then for every $f \in \mathcal{L}(Y, \mathbb{K})$ there exists $F \in \mathcal{L}(X, \mathbb{K})$ such that*

$$F(x) = f(x) \; \forall x \in Y, \qquad ||F||_{\mathcal{L}(X,\mathbb{K})} = ||f||_{\mathcal{L}(Y,\mathbb{K})}.$$

Proof. If X is a real normed space, then the assertion follows from Theorem 9.119 with $p(x) = ||f||_{\mathcal{L}(Y,\mathbb{R})}||x||_X$. To prove that $||F(x)||_{\mathcal{L}(X,\mathbb{R})} \leq ||f||_{\mathcal{L}(Y,\mathbb{R})}$, notice that $F(x) = \theta |F(x)|$, $\theta := \pm 1$, then

$$|F(x)| = \theta F(x) = F(\theta x) \leq p(\theta x) = ||f||_{\mathcal{L}(Y,\mathbb{R})}||\theta x||_X = ||f||_{\mathcal{L}(Y,\mathbb{R})}||x||_X.$$

This shows $||F||_{\mathcal{L}(X,\mathbb{R})} \leq ||f||_{\mathcal{L}(Y,\mathbb{R})}$. The opposite inequality is obvious.

Suppose now that X and Y are complex normed spaces. Consider the real-valued map

$$h(x) := \Re f(x), \qquad x \in Y.$$

h is a $\mathbb{R}$-linear bounded form on Y considered as a real normed space since

$$|h(x)| \leq |f(x)| \leq ||f||_{\mathcal{L}(Y,\mathbb{C})} ||x||_X \qquad \forall x \in Y,$$

thus the first part of the proof yields a $\mathbb{R}$-linear bounded map $H : X \to \mathbb{R}$, such that $H(x) = h(x) \; \forall x \in Y$ and $|H(x)| \leq ||f||_{\mathcal{L}(Y,\mathbb{R})}||x||_X \; \forall x \in X$.

Now define

$$F(x) := H(x) - iH(ix) \qquad \forall x \in X,$$

hence $H(x) = \Re F(x)$. It is easily seen that $F : X \to \mathbb{C}$ is a $\mathbb{C}$-linear map and extends f. It remains to show that

$$|F(x)| \leq ||f||_{\mathcal{L}(Y,\mathbb{C})}||x||_X \qquad \forall x \in X.$$

For $x \in X$, we can write $F(x) = re^{i\beta}$ with $r \geq 0$. Hence

$$|F(x)| = r = \Re(e^{-i\beta} F(x)) = \Re F(e^{-i\beta} x) = H(e^{-i\beta} x)$$
$$\leq ||f||_{\mathcal{L}(Y,\mathbb{R})}||e^{-i\beta x}||_X \leq ||f||_{\mathcal{L}(Y,\mathbb{C})}||x||_X.$$

$\square$

Simple consequences are the following corollaries.

9.121 Corollary. *Let X be a normed space and let $\bar{x} \in X$. Then there exists $F \in \mathcal{L}(X, \mathbb{R})$ such that*

$$F(x) = ||\bar{x}||_X, \qquad ||F||_{\mathcal{L}(X, \mathbb{R})} = 1.$$

9.122 Corollary. *Let X be a normed space. Then for all $x \in X$*

$$||x||_X = \sup \Big\{ F(x) \,\Big|\, F \in \mathcal{L}(X, \mathbb{R}), ||F||_{\mathcal{L}(X < \mathbb{R})} \leq 1 \Big\}$$

9.123 Corollary. *Let Y be a linear subspace of a normed linear space X. If Y is not dense in X, then there exists $F \in \mathcal{L}(X, \mathbb{R})$ $F \neq 0$, such that $F(y) = 0 \; \forall y \in Y$.*

9.124 ¶. Prove Corollaries 9.121, 9.122 and 9.123.

We can give a geometric formulation to the Hahn–Banch theorem that is very useful. For the sake of simplicity from now on we shall assume that X is a real normed space, even though the following results hold also for complex normed spaces.

A *closed affine hyperplane* in X is a set of the form

$$H := \Big\{ x \in X \,\Big|\, F(x) = \alpha \Big\}$$

where $F \in \mathcal{L}(X, \mathbb{R})$ and $\alpha \in \mathbb{R}$. It defines the two half-spaces

$$H_- := \Big\{ x \in X \,|\, F(x) < \alpha \Big\}, \qquad H_+ := \Big\{ x \in X \,|\, F(x) \geq \alpha \Big\}.$$

We say that H separates the sets A and B if

$$A \subset H_- \qquad \text{and} \qquad B \subset H_+.$$

9.125 Lemma (Gauge function). *Let $C \subset X$ be an open convex subset of the real normed space X and let $0 \in C$. Define*

$$p(x) := \inf \Big\{ \alpha > 0 \,\Big|\, \frac{x}{\alpha} \in C \Big\}.$$

Then

  (i) *p is sublinear,*
  (ii) *$\exists M$ such that $0 \leq p(x) \leq M \, ||x||_X$,*
(iii) *$C := \Big\{ x \in X \,\Big|\, p(x) < 1 \Big\}$.*

Proof. If $B(0, r) \subset X$, we clearly have $p(x) \leq \frac{1}{r}\|x\|_X \; \forall x \in X$, that is (ii). Let us prove (iii). Suppose $x \in C$. Since C is open, $(1 + \epsilon)x \in C$, if ϵ is small. Hence $p(x) \leq \frac{1}{1+\epsilon} < 1$. Conversely, if $p(x) < 1$, there is α, $0 < \alpha < 1$, such that $\alpha^{-1}x \in C$, hence $x = \alpha(\alpha^{-1}x) + (1 - \alpha)0 \in C$. Finally, let us prove (i). Trivially $p(\lambda x) = \lambda p(x)$ for $\lambda > 0$. For all $x, y \in X$ and $\epsilon > 0$ we know that

$$\frac{x}{p(x) + \epsilon}, \quad \frac{y}{p(y) + \epsilon} \in C,$$

consequently,

$$\frac{tx}{p(x) + \epsilon} + \frac{(1 - t)y}{p(y) + \epsilon} \in C \qquad \forall t \in [0, 1].$$

In particular, for

$$t := \frac{p(x) + \epsilon}{p(x) + p(y) + 2\epsilon}$$

we obtain

$$\frac{x + y}{p(x) + p(y) + 2\epsilon} \in C.$$

This yields $p(x + y) \leq p(x) + p(y) + 2\epsilon$ and the claim follows, since ϵ is arbitrary. $\square$

9.126 Proposition. *Let $C \subset X$ be an open convex subset of the real normed space X and let $\overline{x} \in X$, $\overline{x} \notin C$. Then there exists $f \in \mathcal{L}(X, \mathbb{R})$ such that $f(x) < f(\overline{x}) \; \forall x \in C$. In particular, C and $\overline{x}$ are separated by the closed affine hyperplane $\{x \mid f(x) = f(\overline{x})\}$.*

Proof. By translation we can assume $0 \in C$ and introduce the gauge function $p(x)$ by Lemma 9.125. If $Y := \mathrm{Span}\{\overline{x}\}$ and $g : \mathrm{Span}\{\overline{x}\} \to \mathbb{R}$ is the linear map $g(t\overline{x}) := t$, it is clear that $g(x) \leq p(x) \; \forall x \in \mathrm{Span}\{\overline{x}\}$. By Theorem 9.119, there exists a linear extension f of g such that $f(x) \leq p(x) \; \forall x \in X$. In particular, we have $f(\overline{x}) = 1$ and f is bounded because of (ii) of Lemma 9.125. On the other hand, $f(x) < 1 \; \forall x \in C$ by (iii) of Lemma 9.125. $\square$

9.127 Theorem (Hahn–Banach thereom, geometrical form). *Let A and B be two nonempty disjoint convex sets of a real normed space X. Suppose A is open. Then A and B can be separated by a closed affine hyperplane.*

Proof. Set $C := A - B = \{x - y \mid x \in A, \; y \in B\}$. Trivially C is convex and open as $C := \cup_{y \in B}(A - y)$; moreover, $0 \notin C$ since $A \cap B = \emptyset$. By Proposition 9.126 there exists $f \in \mathcal{L}(X, \mathbb{R})$ such that $f(z) < 0 \; \forall z \in C$, i.e., $f(x) < f(y) \; \forall x \in A \; \forall y \in B$. If we choose α such that

$$\sup_{x \in A} f(x) \leq \alpha \leq \inf_{y \in B} f(y),$$

the affine hyperplane $\{f(x) = \alpha\}$ separates A and B. $\square$

9.5 Some General Principles for Solving Abstract Equations

In this final section we establish some fundamental principles concerning the solvability of abstract equations

$$Au = f$$

where $A : X \to Y$ is a continuous function also called a *continuous nonlinear operator* between Banach spaces. These principles are fully appreciated for instance when dealing with the theory of ordinary or partial differential equations; however in Chapter 11 we shall illustrate some of their applications.

9.5.1 The Banach fixed point theorem

Many problems take the form of finding a fixed point for a suitable transformation. For instance, if A maps X into X where X is a vector space, the equation $Au = 0$ is equivalent to $Au + u = u$, i.e., to finding a fixed point for the operator $A +$ Id. The *contraction mapping theorem*, proved by Stefan Banach (1892–1945) in 1922, an elementary version of which we saw in Theorem 8.48 in [GM2], is surely one of the simplest results that ensures the existence of a fixed point and also gives a procedure to determine it. The method has its origins in the *method of successive approximations* of Émile Picard (1856–1941) and may be regarded as an abstract formulation of it. Let $\{x_n\}$ be defined by

$$x_n = F(x_{n-1}).$$

If $\{x_n\}$ converges to $\overline{x}$ and F is continuous, then $\overline{x}$ is a fixed point of F, $F(\overline{x}) = \overline{x}$.

a. The fixed point theorem

Let X be a metric space. A map $T : X \to X$ is said to be *k-contractive* if $d(T(x), T(y)) \le kd(x, y)\ \forall x, y \in X$, or, in other words, if $T : X \to X$ is Lipschitz continuous with Lipschitz constant

$$\text{Lip}\,(T) = \sup\left\{ \frac{d(Tx, Ty)}{d(x, y)} \,\Big|\, x, y, \in X \right\}$$

less than or equal to k. If $0 \le k < 1$, T is often said simply a *contraction* or a *contractive mapping*. A point $x \in X$ for which $Tx = x$ is called a *fixed point* for T. The *contraction principle* states that contractions have a unique fixed point.

9.128 Theorem (The fixed point theorem of Banach). *Let X be a complete metric space and let $T : X \to X$ be k-contractive with $0 \le k < 1$. Then T has a unique fixed point. Moreover, given $x_0 \in X$, the sequence $\{x_n\}$ defined recursively by $x_{n+1} = T(x_n)$ converges with an exponential rate to the fixed-point, and the following estimates hold*

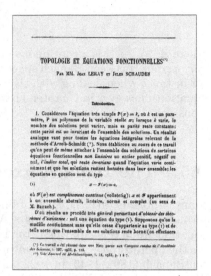

Figure 9.12. The frontispiece of *Leçons sur quelques équations fonctionnelles* by Émile Picard (1856–1941) and the first page of a celebrated paper by Jean Leray (1906–1998) and Juliusz Schauder (1899–1943) appeared in Journal de Mathématiques in 1933.

$$\begin{cases} d(x_{n+1}, x) \leq kd(x_n, x), \\ d(x_n, x) \leq \dfrac{k^n}{1-k}d(x_1, x_0), \\ d(x_{n+1}, x) \leq \dfrac{k}{1-k}d(x_{n+1}, x_n). \end{cases}$$

Proof. The proof is as in Theorem 8.48 of [GM2]. First we prove *uniqueness*. If x, y are two fixed points, from $d(x, y) = d(Tx, Ty) \leq kd(x, y)$, $0 \leq k < 1$ we infer $d(x, y) = 0$, i.e., $x = y$.

Then we prove existence. Choose any $x_0 \in X$ and let $x_{n+1} := T(x_n)$, $n \geq 0$. We have
$$d(x_{n+1}, x_n) \leq kd(x_n, x_{n-1}) \leq k^n d(x_1, x_0) = k^n d(T(x_0), x_0),$$
hence for $p > n$
$$d(x_p, x_n) \leq \sum_{j=n}^{p-1} d(x_{j+1}, x_j) \leq \sum_{j=n}^{p-1} k^j d(x_1, x_0) \leq \frac{k^n}{1-k}d(x_1, x_0).$$

Therefore $d(x_p, x_n) \to 0$ as $n, p \to \infty$, i.e., $\{x_n\}$ is a Cauchy sequence, hence it has a limit $x \in X$ and x is a fixed point as it is easily seen passing to the limit in $x_{n+1} = T(x_n)$. Finally, we leave to the reader the proof of the convergence estimates. $\square$

Notice that the second estimate in Theorem 9.128 allows us to evaluate the number of iterations that are sufficient to reach a desired accuracy; the second estimate allows us to evaluate the accuracy of x_n as an approximate value of x in terms of $d(x_{n+1}, x_n)$.

9.129 ¶. Show that $T : X \to X$ has a unique fixed point if its mth iterate $T^m = T \circ T \circ \cdots \circ T$ is a k-contractive mapping with $0 \leq k < 1$. [*Hint:* x and Tx are both fixed points of T^m.]

9.130 ¶. Let $X := C^0([a, b])$ and let

$$Tf(t) := \int_a^t f(s)\, ds, \qquad a \le t \le b.$$

Show that

$$T^m f(t) = \frac{1}{(m-1)!} \int_a^t (t-s)^{m-1} f(s)\, ds \qquad a \le t \le b,$$

is a contractive map if m is sufficiently large.

9.131 Proposition. *Let X be a Banach space and $T : X \to X$ a k-contractive map with $0 \le k < 1$. Then $\mathrm{Id} - T$ is a bijection from X into itself, i.e., for every $y \in X$ the equation $x - Tx = y$ has a unique solution, moreover*

$$\mathrm{Lip}\,(\mathrm{Id} - T)^{-1} \le \frac{1}{1-k}. \tag{9.29}$$

Proof. For any $y \in X$ the equation $x - Tx = y$ is equivalent to $x = y + Tx =: F(x)$. Since F is k-contractive and $k < 1$, the fixed point theorem shows that $x - Tx = y$ has a unique solution for any given $y \in X$, i.e., $\mathrm{Id} - T$ is bijective. Finally, if $x - Tx = y$, then

$$||x|| \le ||x - Tx|| + ||Tx|| \le ||y|| + k\,||x||$$

i.e., $||x|| \le \frac{1}{1-k}||y||$. □

9.132 ¶. Let X be a Banach space and $T : X \to X$ a Lipschitz-continuous map. Show that the equation

$$Tx + \mu x = y$$

is solvable for any y, provided $|\mu|$ is sufficiently large.

9.133 ¶. Let X be a Banach space and $\mathcal{B} : X \times X \to \mathbb{R}$ a bilinear continuous form such that

$$|\mathcal{B}(x, y)| \le B\,||x||\,||y|| \qquad \forall x, y \in X.$$

Show that the equation

$$x = y + \mathcal{B}(x, x)$$

has at least a solution if $||y|| < 1/(2B)$ and, in this case, show that there is a unique solution satisfying $||x|| \le 1/(2B)$. [*Hint:* First look at the simplest case of $X = \mathbb{R}$ and $\mathcal{B}(x, x) := B\,x^2$.]

b. The continuity method

The solvability of a linear equation $L_1 x = y$ can be reduced to the solvability of a simpler equation $L_0 x = y$ by means of the following.

9.134 Theorem (The continuity method). *Let X be a Banach space, Y a normed space and L_0, L_1 two linear continuous functions from X to Y. For $t \in [0, 1]$ consider the family of linear continuous functions $L_t : X \to Y$ given by*

$$L_t := (1 - t)L_0 + tL_1$$

and suppose that there exists a constant C such that the following a priori estimates hold

Figure 9.13. George Birkhoff (1884–1944) and a page from a paper by Birkhoff and Kellogg in *Transactions*, 1922.

$$||x||_X \leq C||L_t x||_Y \qquad \forall x \in X, \ \forall t \in [0,1]. \tag{9.30}$$

Then, of course the functions L_t, $t \in [0,1]$, are injective; moreover, L_1 is surjective if and only if L_0 is surjective.

Proof. Injectivity follows from the linearity and (9.30). Suppose now that L_s is surjective for some s. Then $L_s : X \to Y$ is invertible and by (9.30) $||(L_s)^{-1}|| \leq C$. We shall now prove that the equation $L_t x = y$ can be solved for any $y \in Y$ provided t is closed to s. For this we notice that $L_t x = y$ is equivalent to

$$L_s x = y + (L_s - L_t)x = y + (t-s)L_0 x - (t-s)L_1 x$$

which, in turn, is equivalent to

$$x = L_s^{-1} y + (t-s)L_s^{-1}(L_0 - L_1)x =: Tx$$

since $L_s : X \to Y$ has an inverse. Then we observe that $||Tx - Tz||_Y \leq C|t-s|(||L_0|| + ||L_1||)$, consequently T is a contractive map if

$$|t-s| \leq \delta := \frac{1}{C(||L_0|| + ||L_1||)},$$

and we conclude that L_t is surjective for all t with $|t-s| < \delta$. Since δ is independent of s, starting from a surjective map L_0 we successively find that L_t with $t \in [0,\delta]$, $[0,2\delta]$, ... is surjective. We therefore prove that L_1 is surjective in a finite number of steps. $\square$

9.135 Remark. Notice that the proof of Theorem 9.134 says that, assuming (9.30), the subset of $[0,1]$

$$S := \left\{ s \in [0,1] \ \middle| \ L_s : X \to Y \text{ is surjective} \right\}$$

is open and closed in $[0,1]$. Therefore $S = [0,1]$ provided $S \neq \emptyset$.

9.5.2 The Caccioppoli–Schauder fixed point theorem

Compared to the fixed point theorem of Banach, the fixed point theorem of Caccioppoli and Schauder is more sophisticated: it extends the finite-dimensional fixed point theorem of Brouwer to infinite-dimensional spaces.

9.136 Theorem (The fixed point theorem of Brouwer). *Let K be a nonempty, compact and convex set of $\mathbb{R}^n$ and let f be a continuous map mapping K into itself. Then f has at least a fixed point in K.*

The generalization to infinite dimensions and to the abstract setting is due to Juliusz Schauder (1899–1943) at the beginning of the Twenties of the past century, however in specific situations it also appears in some of the works of George Birkhoff (1884–1944) and Oliver Kellogg (1878–1957) of 1922 and of Renato Caccioppoli (1904–1959) (independently from Juliusz Schauder) of the 1930's, in connection with the study of partial differential equations.

Brouwer's theorem relies strongly on the continuity of the map f and in particular, on the property that those maps have of transforming bounded sets of a finite-dimensional linear space into relatively compact sets. As we have seen in Theorem 9.21, such a property is not valid anymore in infinite dimensions, thus we need to restrict ourselves to continuous maps that transform bounded sets into relatively compact sets. In fact, the following example shows that a fixed-point theorem such as Brouwer's cannot hold for continuous functions from the unit ball of an infinite-dimensional space into itself.

9.137 Example. Consider the map $f : \ell_2 \to \ell_2$ given by

$$f\Big(\{x_n\}\Big) = \Big\{\sqrt{1 - ||x||_{\ell_2}^2}, x_1, x_2, \dots\Big\}.$$

Clearly f maps the unit ball of ℓ_2 in itself, is continuous and has no fixed point.

a. Compact maps

9.138 Definition. *Let X and Y be normed spaces. The (non)linear operator $A : X \to Y$ is called* compact *if*

(i) *A is continuous,*
(ii) *A maps bounded sets of X into relatively compact subsets of Y, equivalently for any bounded sequence $\{x_k\} \subset X$ we can extract a subsequence $\{x_{n_k}\}$ such that $\{Ax_{n_k}\}$ is convergent.*

9.139 Example. Consider the integral operator $A : C^0([a, b]) \to C^0([a, b])$ that maps $u \in C^0([a, b])$ into $Au(x) \in C^0([a, b])$ defined by

$$Au(x) := \int_a^b F(x, y, u(y)) \, dy$$

where $F(x, y, u)$ is a continuous real-valued function in $\mathbb{R}^3$. For $r > 0$ set $Q_r :=$ $\{(x, y, u) \in \mathbb{R}^3 \mid x, y \in [a, b], \, |u| \le r\}$ and

$$M_r := \left\{ u \in C^0([a, b]) \,\middle|\, ||u||_\infty \le r \right\}.$$

Proposition. $A : M_r \to C^0([a, b])$ *is a compact operator.*

Proof. (i) First we prove that $A : M_r \to C^0([a, b])$ is continuous. Fix $\epsilon > 0$ and observe that, F being uniformly continuous in Q_r, there exists $\delta > 0$ such that

$$|F(x, y, u) - F(x, y, v)| < \epsilon$$

if $(x, y, u), (x, y, v) \in B_r$ and $|u - v| < \delta$. Consequently, we have

$$|F(x, y, u(y)) - F(x, y, v(y))| < \epsilon$$

for $u, v \in M_r$ with $||u - v||_{\infty, [a, b]} < \delta$, hence

$$||Au - Av||_{\infty, [a, b]} = \sup_{x \in [a, b]} \left| \int_a^b [F(x, y, u(y)) - F(x, y, v(y))] \, dy \right| \le \epsilon(b - a). \quad (9.31)$$

(ii) It remains to show that A maps bounded sets into relatively compact sets. To do that, it suffices to show that $A(M_r)$ is relatively compact in $C^0([a, b])$. We now check that $A(M_r) \subset C^0([a, b])$ is a set of equibounded and equicontinuous functions. Then the Ascoli–Arzelà theorem, Theorem 9.48, yields the required property. In fact, the equiboundedness of functions in $A(M_r)$ follows from

$$||Au||_\infty \le (b - a) \sup_{(x, y, z) \in Q_r} |F(x, y, z)|,$$

while the equicontinuity of functions in $A(M_r)$ is just (9.31).   □

Compact operators arise as limits of maps with *finite rank* as shown by the following theorem.

9.140 Theorem. *Let X and Y be Banach spaces and $M \subset X$ a nonempty bounded set. We have*

(i) *If $\{A_n\}$, $A_n : M \to Y$, is a sequence of compact operators that converges to $A : M \to Y$ in $\mathcal{B}(A, Y)$, i.e., $||A_n - A||_{\mathcal{B}(A, Y)} \to 0$ as $n \to \infty$, then A si compact.*

(ii) *Suppose $A : M \to Y$ is compact. Then there exists a sequence $\{A_n\}$ of continuous operators $A_n : X \to Y$ such that $||A_n - A||_{\infty, M} \to 0$ as $n \to \infty$ and each A_n has range in a finite-dimensional subspace of Y as well as in the convex envelope of $A(M)$.*

Proof. (i) Fix $\epsilon > 0$ and choose n so that $||A_n - A||_{\infty, M} < \epsilon$. Since $A_n(M)$ is relatively compact, we can cover $\overline{A_n(M)}$ with a finite number of balls $\overline{A_n(M)} \subset \cup_{i=1}^I B(x_i, \epsilon)$, $i = 1, \ldots, I$. Therefore $\overline{A(M)} \subset \cup_{i=1}^I B(x_i, 2\epsilon)$, i.e., $\overline{A(M)}$ is totally bounded, hence $A(M)$ has compact closure, compare Theorem 6.8.

(ii) Since $A(M)$ is relatively compact, for each n there is a $\frac{1}{n}$-net, i.e., elements $y_j \in A(M)$, $j = 1, \ldots, J_n$ such that $A(M) \subset \cup_{j=1}^{J_n} B(y_j, 1/n)$, or, equivalently,

$$\min_j ||Ax - y_j|| \le \frac{1}{n} \qquad \forall x \in M.$$

Figure 9.14. Renato Caccioppoli (1904–1959) and Carlo Miranda (1912–1982).

Define the so-called *Schauder operators*

$$A_n x := \frac{\sum_{j=1}^{J_n} a_j(x) y_j}{\sum_{j=1}^{J_n} a_j(x)} \qquad x \in M,$$

where, for $x \in M$ and $j = 1, \ldots, J_n$,

$$a_j(x) := \max \left\{ \frac{2}{n} - ||Ax - y_j||, 0 \right\}.$$

It is easily seen that the functions $a_j : M \to \mathbb{R}$ are continuous and do not vanish simultaneously; moreover

$$||A_n x - Ax||_Y = \frac{||\sum a_j(x)(y_j - Ax)||_Y}{\sum a_j(x)} \leq \frac{\sum a_j(x)||y_j - Ax||_Y}{\sum a_j(x)} \leq \frac{1}{n};$$

the claim then easily follows. $\qquad \square$

b. The Caccioppoli–Schauder theorem

9.141 Theorem (Caccioppoli–Schauder). *Let $M \subset X$ be a closed, bounded, convex nonempty subset of a Banach space X. Every compact operator $A : M \to M$ from M into itself has at least a fixed point.*

Proof. Let $u_0 \in M$. Replacing u with $u - u_0$ we may assume that $0 \in M$. From (ii) of Theorem 9.140 there are finite-dimensional subspaces $X_n \subset X$ and continuous operators $A_n : M \to X_n$ such that $||Au - A_n u|| \leq \frac{1}{n}$ and $A_n(M) \subset co(A(M))$. The subset $M_n := X_n \cap M$ is bounded, closed, convex and nonempty (since $0 \in M_n$) and $A_n(M_n) \subset M_n$. Brouwer's theorem then yields a fixed point for $A_n : M_n \to M_n$, i.e.,

$$u_n \in M_n, \qquad A_n u_n = u_n,$$

hence, as the sequence $\{u_n\}$ is bounded,

$$||Au_n - u_n|| = ||Au_n - A_n u_n|| \leq \frac{1}{n}||u_n|| \to 0.$$

Since A is compact, passing to a subsequence still denoted by $\{u_n\}$, we deduce that $\{Au_n\}$ converges to an element $v \in X$. On the other hand $v \in M$, since M is closed, and

$$||u_n - v|| \leq ||v - Au_n|| + ||Au_n - u_n|| \to 0 \qquad \text{as } n \to \infty;$$

thus $u_n \to v$ and from $Au_n = u_n \; \forall n$ we infer $Av = v$ taking into account the continuity of A. $\qquad \square$

c. The Leray–Schauder principle

A consequence of the Caccioppoli–Schauder theorem is the following principle, which is very useful in applications, proved by Jean Leray (1906–1998) and Juliusz Schauder (1899–1943) in 1934 in the more general context of the *degree theory* and often referred to as to the *fixed point theorem of Helmut Schaefer (1925–)* .

9.142 Theorem (Schaefer). *Let X be a Banach space and $A : X \to X$ a compact operator. Suppose that the following a priori estimate holds: there is a positive number $r > 0$ such that, if $u \in X$ solves*

$$u = tAu \qquad \text{for some } 0 \le t \le 1,$$

then

$$||u|| < r.$$

Then the equation

$$v = Av \qquad v \in X$$

has at least a solution.

Proof. Let $M := \{u \in X \mid ||u|| \le r\}$ and consider the composition B of A with the retraction on the ball, i.e.,

$$Bu := \begin{cases} Au & \text{if } ||Au|| \le r, \\ \dfrac{rAu}{||Au||} & \text{if } ||Au|| \ge r. \end{cases}$$

B maps M to M, is continuous and maps bounded sets in relatively compact sets, since A is compact. Therefore the Caccioppoli–Schauder theorem yields a fixed point $\overline{u} \in M$ for B, $B\overline{u} = \overline{u}$. Now, if $||A\overline{u}|| \le r$, $\overline{u}$ is also a fixed point for A; otherwise $||A\overline{u}|| > r$ and

$$\overline{u} = B\overline{u} = \frac{r}{||A\overline{u}||} A\overline{u} = tA\overline{u}, \qquad t := \frac{r}{||A\overline{u}||} \le 1,$$

hence $||\overline{u}|| < r$: it follows that also $||B\overline{u}|| < r$, i.e., $\overline{u} = B\overline{u} = A\overline{u}$ and $\overline{u}$ is again a fixed point for A.   □

Theorems 9.134 and 9.142 may be regarded as special cases of a sort of general principle: *a priori estimates on the possible solution yield existence of a solution.*

9.5.3 The method of super- and sub-solutions

In this section we state an abstract formulation of the following principle that reminds us of the intermediate value theorem: *to find a solution, it often suffices to find a subsolution and a supersolution.*

Figure 9.15. Juliusz Schauder (1899–1943) and Jean Leray (1906–1998).

a. Ordered Banach spaces

9.143 Definition. *An* order cone *in a Banach space X is a subset X_+ such that*

(i) X_+ *is closed, convex nonempty and $X_+ \neq \{0\}$,*
(ii) *if $u \in X_+$ and $\lambda \geq 0$, then $\lambda u \in X_+$,*
(iii) *if $u \in X_+$ and $-u \in X_+$, then $u = 0$.*

An order cone $X_+ \subset X$ defines a total order in X

$$u \leq v \qquad \text{if and only if} \qquad v - u \in X_+,$$

and we say that X is an *ordered Banach space* (by X_+). In this case *intervals* in X are well defined

$$[u, w] := \{v \in X \mid u \leq v \leq w\}.$$

9.144 Definition. *An order cone X_+ is called* normal *if there is a number $c > 0$ such that $\|u\| \leq c\|v\|$ whenever $0 \leq u \leq v$.*

9.145 Example. In $\mathbb{R}^n$ with the Euclidean norm the set

$$X_+ = \mathbb{R}^n_+ := \Big\{(x_1, \ldots, x_n) \,\Big|\, x_i \geq 0 \;\forall i\Big\}$$

is a normal order cone: $0 \leq x \leq y$ implies $|x| \leq |y|$.

9.146 Example. In the Banach space $C^0([a, b])$ with the uniform norm

$$C^0_+([a, b]) := \Big\{u \in C^0([a, b]) \,\Big|\, u(x) \geq 0 \;\forall x \in [a, b]\Big\}$$

is a normal order cone.

9.147 ¶. Let u, v, w, u_n, v_n be elements of an order cone X_+ of a Banach space X. Show that

(i) $u \leq v$ and $v \leq w$ imply $u \leq w$,
(ii) $u \leq v$ and $v \leq u$ imply $u = v$,
(iii) if $u \leq v$, then $u + w \leq v + w$ and $\lambda u \leq \lambda v \;\forall \lambda \geq 0, \;\forall w \in X$,
(iv) if $u_n \leq v_n$, $u_n \to u$ and $v_n \to v$ as $n \to \infty$, then $u \leq v$,
(v) if X_+ is normal, then $u \leq v \leq w$ imply

$$\|v - u\| \leq c\|w - u\| \quad \text{and} \quad \|w - v\| \leq c\|w - u\|.$$

b. Fixed points via sub- and super-solutions

9.148 Theorem. *Let X be a Banach space ordered by a normal order cone, let $u_0, v_0 \in X$ and let $A : [u_0, v_0] \subset X \to X$ be a (possibly nonlinear) compact operator. Suppose that A is monotone increasing, i.e., $Au \leq Av$ whenever $u \leq v$ and that*

(i) *u_0 is a* subsolution *for the equation $Au = u$, i.e., $u_0 \leq Au_0$,*
(ii) *v_0 is a* supersolution *for the equation $Au = u$, i.e., $Av_0 \leq v_0$.*

Then the two iterative processes

$$u_{n+1} = Au_n \qquad and \qquad v_{n+1} = Av_n \qquad \forall n \geq 0$$

started respectively, from u_0 and v_0 converge respectively to solutions u_- and u_+ of the equation $u = Au$. Moreover

$$u_0 \leq u_1 \leq \cdots \leq u_n \leq \cdots \leq u_- \leq u_+ \leq \cdots \leq v_n \leq \cdots \leq v_0.$$

Proof. By induction
$$u_0 \leq \cdots \leq u_n \leq v_n \leq \cdots \leq v_0,$$
since A is monotone. From (v) of Exercise 9.147

$$||v_0 - u_n|| \leq C \, ||v_0 - u_0|| \qquad \forall n,$$

i.e., $\{u_n\}$ is bounded. As A is compact, there exists $u_- \in X$ such that for a subsequence $\{u_{k_n}\}$ of $\{u_n\}$ we have $Au_{k_n} \to u_-$ as $n \to \infty$. Finally $u_- = Au_-$, since A is continuous. One operates similarly with $\{v_n\}$. $\qquad\square$

9.149 Remark. Notice that the conclusion of Theorem 9.148 still holds if we require that A be monotone on the sequences $\{u_n\}$ and $\{v_n\}$ defined by $u_{n+1} = Au_n$ and $v_{n+1} = Av_n$ started respectively, at u_0 and v_0 instead of being monotone in $[u_0, v_0]$.

9.6 Exercises

9.150 ¶. Show that in a normed space $(X, || \; ||)$ the norm $|| \; || : X \to \mathbb{R}_+$ is a Lipschitz-continuous function with best Lipschitz constant one, i.e.,

$$\Big| ||x|| - ||y|| \Big| \leq ||x - y||.$$

9.151 ¶. Show that the set $E_t := \{x \in X \mid f(x) \leq t\}$ is convex for all t if X is a normed space and $f : X \to \mathbb{R}$ is convex.

9.152 ¶. Let X be a normed space with $|| \; ||_X$. Show that $x \to ||x||^p$, $p \geq 1$, is a convex function.

9.153 ¶. Convexity can replace the triangle inequality. Prove the following claim.

 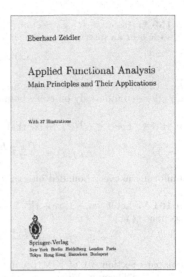

Figure 9.16. Frontispieces of two volumes on functional analysis.

Proposition. *Let X be a linear space and let $f : X \to \mathbb{R}_+$ be a function such that*
 (i) *$f(x) \geq 0$, $f(x) = 0$ iff $x = 0$,*
 (ii) *f is positively homogeneous of degree one: $f(\lambda x) = |\lambda| f(x)$ $\forall x \in X$, $\forall \lambda \geq 0$,*
 (iii) *the set $\{x \mid f(x) \leq 1\}$ is convex.*
Then $f(x)$ is a norm on X.

9.154 ¶. Prove the following variant of Lemma 9.22.

Lemma (Riesz). *Let Y be a closed proper linear subspace of a normed space X. Then, for every $\epsilon > 0$, there exists a point $z \in X$ with $||z|| = 1$ such that $||z - y|| > 1 - \epsilon$ for all $y \in Y$.*

9.155 ¶. Show that $BV([a, b])$ is a Banach space with the norm

$$||f||_{BV} := \sup_{x \in [a,b]} |f(x)| + V_a^b(f).$$

[*Hint:* Compare Chapter 7 for the involved definitions.]

9.156 ¶. Show that in $C^0([a, b])$ the norms $|| \ ||_\infty$ e $|| \ ||_{L^p}$ are not equivalent.

9.157 ¶. Show that in $C^1([0, 1])$

$$|x(0)| + \int_0^1 |x'(t)| \, dt$$

defines a norm, and that the convergence in this norm implies uniform convergence.

9.158 ¶. Denote by c_o the linear space of infinitesimal real sequences $\{x_n\}$ and by c_{oo} the linear subspace of c_o of sequences with only finite many nonzero elements. Show that c_o is closed in ℓ_∞ while c_{oo} is not closed.

9.159 ¶. Recall, see e.g., Section 2.2 of [GM1], that the oscillation of a function $f :$ $\mathbb{R} \to \mathbb{R}$ over an interval around x and radius δ is defined as

$$\omega_{f,\delta}(x) := \sup_{|y-x|<\delta} |f(y) - f(x)|$$

and that $f : \mathbb{R} \to \mathbb{R}$ is continous at x if and only if $\omega_{f,\delta}(x) \to 0$ as $\delta \to 0$. Show that $\omega_{f,\delta}(x) \to 0$ uniformly on every bounded interval of $\mathbb{R}$. [*Hint:* Use Theorem 6.35.]

9.160 ¶. Let $f \in C^1(\mathbb{R})$. Show that

$$\Delta_h f(x) := \frac{f(x+h) - f(x)}{h} \to f'(x) \qquad \text{as } h \to 0$$

uniformly in every bounded interval of $\mathbb{R}$. [*Hint:* Use Theorem 6.35.]

9.161 ¶. Let $f :]x_0 - 1, x_0 + 1[\subset \mathbb{R} \to \mathbb{R}$ be differentiable at x_0. Show that the blow-up sequence $\{f_n\}$,

$$f_n(z) := \frac{f(x_0 + z/n) - f(x_0)}{\frac{1}{n}} \to f'(x_0)z,$$

compare Section 3.1 of [GM1], converges uniformly on every bounded interval of $\mathbb{R}$. [*Hint:* Use Theorem 6.35.]

9.162 ¶. Compute, if it exists,

$$\lim_{n \to \infty} \int_2^4 f_n(x)\, dx, \qquad f_n(x) := \frac{n}{x}(e^{x/n} - 1).$$

9.163 ¶. Discuss the uniform convergence of the sequences of real functions in $[0, 1]$

$$f_n(x) := (-1)^n n(x+1)x^n, \qquad \frac{1}{n}(2 + \sin(nx))e^{1-\cos(nx)}x.$$

9.164 ¶. Discuss the uniform convergence of the following real series

$$\sum_{n=1}^{\infty} \sqrt{1 - \cos\left(\frac{(n+2)x}{n^4}\right)}, \qquad \sum_{n=1}^{\infty}(e^x - e^{-x})\arctan\frac{x+1}{n},$$

$$\sum_{n=0}^{\infty} \frac{n}{n^3 x^2 + 1}, \qquad \sum_{n=1}^{\infty}\left(\frac{\arctan(nx)}{\pi} + \frac{\pi}{n}\right)^n,$$

$$\sum_{n=1}^{\infty} x^n \log\left(1 + \frac{x}{n^2}\right), \qquad \sum_{n=2}^{\infty} \log^n\left(x + \frac{1}{nx}\right).$$

9.165 ¶. Show that $\{u \in C^0([0, 1]) \mid \int_0^1 u(x)\, dx = 0\}$ is a linear subspace of $C^0([0, 1])$ that is not closed.

9.166 ¶. Show that $\{u \in C^0([0, 1]) \mid u(0) = 1\}$ is closed, convex and dense in $C^0([0, 1])$.

9.167 ¶. Show that $\{u \in C^0(\mathbb{R}) \mid \lim_{x \to \pm\infty} u(x) = 0\}$ is a closed subspace of $C_b^0(\mathbb{R})$.

9.168 ¶. Show that the subspace $C_c^0(\mathbb{R})$ of $C_b^0(\mathbb{R})$ of functions with compact support is not closed.

9.169 ¶. Let X be a compact metric space and $\mathcal{F} \subset C^0(X)$. Show that $\mathcal{F}$ is equicontinuous if

(i) the functions in $\mathcal{F}$ are equi-Lipschitz, i.e., $\exists\, M$ such that

$$|f(x) - f(y)| \leq M\, d(x, y) \qquad \forall x, y \in X, \ \forall f \in \mathcal{F},$$

(ii) the functions in $\mathcal{F}$ are equi-Hölder, i.e., $\exists\, M$ and α, $0 < \alpha \leq 1$, such that

$$|f(x) - f(y)| \leq M\, d(x, y)^{\alpha} \qquad \forall x, y \in X, \ \forall f \in \mathcal{F}.$$

9.170 ¶. Let $\mathcal{F} \subset C^0([a, b])$. Show that any of the following conditions implies equicontinuity of the family $\mathcal{F}$.

(i) the functions in $\mathcal{F}$ are of class C^1 and there exists $M > 0$ such that

$$|f'(x)| \leq M \qquad \forall\, x \in [a, b], \ \forall\, f \in \mathcal{F}$$

(ii) the functions in $\mathcal{F}$ are of class C^1 and there exists $M > 0$ and $p > 1$ such that

$$\int_a^b |f'|^p\, dx \leq k \qquad \forall\, f \in \mathcal{F}.$$

9.171 ¶. Let $\mathcal{F} \subset C^0([a, b])$ be a family of equicontinuous functions. Show that any of the following conditions implies equiboundedness of the functions in $\mathcal{F}$.

(i) $\exists\, C$, $\exists\, x_0 \in [a, b]$ such that $|f(x_0)| \leq C \ \forall\, f \in \mathcal{F}$,
(ii) $\exists\, C$ such that $\forall\, f \in \Phi \ \exists\, x \in [a, b]$ with $|f(x)| \leq C$,
(iii) $\exists\, C$ such that $\int_a^b |f(t)|\, dt \leq C$.

9.172 ¶. Let Q be a set and let X be a metric space. Prove that a subset $\mathcal{B}$ of the space of bounded functions from Q in X with the uniform norm is relatively compact if and only if, for any $\epsilon > 0$, there exists a finite partition $Q_1, \ldots, Q_{n_\epsilon}$ of Q such that the *total variation* of every $u \in \mathcal{B}$ in every Q_i is not greater than ϵ.

9.173 ¶. Show that a subset $K \subset \ell_p$, $1 \leq p < \infty$, is compact if and only if

(i) $\sup_{\{x_n\} \in K} \sum_{n=1}^{\infty} |x_n|^p < \infty$,
(ii) $\forall \epsilon > 0 \ \exists\, n_\epsilon$ such that $\sum_{n=n_\epsilon}^{\infty} |x_n|^p \leq \epsilon$ for all $\{x_n\} \in K$.

9.174 ¶. Let X be a complete metric space with the property that the bounded and closed subsets of $C^0(X)$ are compact. Show that X consists of a finite number of points.

9.175 ¶ Hölder-continuous functions. Let Ω be a bounded open subset of $\mathbb{R}^n$. According to Definition 9.46 the space of *Hölder-continuous functions* with exponent α in Ω, $C^{0,\alpha}(\Omega)$, $0 < \alpha \leq 1$ (also called Lipschitz continuous if $\alpha = 1$) is defined as the linear space of continuous functions in Ω such that

$$||u||_{0,\alpha,\Omega} := \sup_{\Omega} |u| + [u]_{0,\alpha,\Omega} < +\infty$$

where

$$[u]_{0,\alpha,\Omega} := \sup_{\substack{x,y \in \Omega \\ x \neq y}} \frac{u(x) - u(y)}{|x - y|^{\alpha}}.$$

One also defines $C_{loc}^{0,\alpha}(\Omega)$ as the space of functions that belong to $C^{0,\alpha}(\Lambda)$ for all relatively compact open subsets Λ, $\Lambda \subset\subset \Omega$. Show that $C^{0,\alpha}(\Omega)$ is a Banach space with the norm $||\ ||_{0,\alpha,\Omega}$.

9.176 ¶. Show that the space $C^k([a,b])$ is a Banach space with norm

$$||u||_{C^k} := \sum_{h=0}^{k} ||D^h u||_{\infty,[a,b]}.$$

Define $C^{k,\alpha}([a,b])$ as the linear space of functions in $C^k([a,b])$ with Hölder-continuous k-derivative with exponent α such that

$$||u||_{k,\alpha,[a,b]} := ||u||_{C^k([a,b])} + [D^k u]_{0,\alpha,[a,b]} < +\infty.$$

Show that $C^{k,\alpha}([a,b])$ is a Banach space with norm $||\ ||_{k,\alpha,[a,b]}$.

9.177 ¶. Show that the immersion of $C^{0,\alpha}([a,b])$ into $C^{0,\beta}([a,b])$ is compact if $0 \leq \alpha < \beta \leq 1$. More generally, show that the immersion of $C^{h,\alpha}([a,b])$ into $C^{k,\beta}([a,b])$ is compact if $k + \beta < h + \alpha$.

9.178 ¶. Let $\Omega \subset \mathbb{R}^2$ be defined by

$$\Omega := \left\{(x,y) \in \mathbb{R}^2 \,\middle|\, y < |x|^{1/2},\ x^2 + y^2 < 1\right\}.$$

By considering the function

$$u(x,y) := \begin{cases} (\operatorname{sgn} x)y^\beta & \text{if } y > 0, \\ 0 & \text{if } y \leq 0 \end{cases}$$

where $1 < \beta < 2$, show that $u \in C^1(\Omega)$, but $u \notin C^{0,\alpha}(\Omega)$ if $\beta/2 < \alpha \leq 1$.

9.179 ¶. Prove the following

Proposition. *Let Ω be a bounded open set in $\mathbb{R}^n$ satisfying one of the following conditions*

(i) *Ω is convex,*
(ii) *Ω is star-shaped,*
(iii) *$\partial\Omega$ is locally the graph of a Lipschitz-continuous function.*
Then $C^{h,\alpha}(\Omega) \subset C^{k,\beta}(\Omega)$ and the immersion is compact if $k + \beta < h + \alpha$.

[*Hint:* Show that in all cases there exists a constant M and an integer n such that $\forall x, y \in \Omega$ there are at most n points $z_1, z_2, \ldots, z_n$ with $z_1 = x$ and $z_n = y$ such that $\sum_{i=1}^{n-1} |z_i - z_{i+1}| \leq M |x - y|$. Use Lagrange mean value theorem.]

9.180 ¶. Show that the space of Lipschitz-continuous functions in $[a,b]$ is dense in $C^0([a,b])$. [*Hint:* Use the mean value theorem.]

9.181 ¶. Show that the space of Lipschitz-continuous functions in $[a,b]$ with Lipschitz constant less than k agrees with the closure in $C^0([a,b])$ of the functions of class C^1 with $\sup_x |f'(x)| \leq k$.

9.182 ¶. Let $\lambda > 0$. Show that

$$\left\{u \in C^0([0,+\infty[) \,\middle|\, \sup_{[0,+\infty[} |u|e^{-\lambda x} < +\infty\right\}$$

is complete with respect to the metric $d(f,g) := \sup_x\{|f(x) - g(x)|\,e^{-\lambda x}\}$.

9.183 ¶. Let $f : [0, 1] \to [0, 1]$ be a diffeomorphism with $f'(x) > 0 \; \forall x \in [0, 1]$. Show that there exists a sequence of polynomials $P_n(x)$, which are diffeomorphisms from $[0, 1]$ into $[0, 1]$, that converges uniformly to f in $[0, 1]$. [*Hint:* Use Weierstrass's approximation theorem.]

9.184 ¶. Define for $\mathbf{A}[a_j^i] \in M_{n,n}(\mathbb{K})$, $\mathbb{K} = \mathbb{R}$ or $\mathbb{K} = \mathbb{C}$,

$$||\mathbf{A}|| := \sup\Big\{ \frac{|\mathbf{A}x|}{|x|} \mid x \neq 0 \Big\}.$$

Show that

   (i) $|\mathbf{A}x| \leq ||\mathbf{A}|| \, |x| \; \forall x \in X$,
   (ii) $||\mathbf{A}|| = \sup\{(\mathbf{A}x|y) \mid |x| = |y| = 1\}$,
   (iii) $||\mathbf{A}||^2 \leq \sum_{i,j=1}^{n} (a_j^i)^2 \leq n ||\mathbf{A}||^2$,
   (iv) $||\mathbf{A}^*|| = ||\mathbf{A}||$,
   (v) $||\mathbf{AB}|| \leq ||\mathbf{A}|| \, ||\mathbf{B}||$.

9.185 ¶. Let $\mathbf{A} \in M_{n,n}(\mathbb{C})$ and $A(x) := \mathbf{A}x$. Show that

   (i) if $||z|| = |z|_\infty := \max(|z^1|, \ldots, |z^n|)$, then

$$||\mathbf{A}|| = \sup_{||z|| \leq 1} ||\mathbf{A}(z)|| = \max_i \Big(\sum_{j=1}^{n} |\mathbf{A}_j^i| \Big),$$

   (ii) if $||z|| = |z|_1 := \sum_{i=1}^{n} |z^i|$, then

$$||\mathbf{A}|| = \sup_{||z|| le 1} ||\mathbf{A}(z)|| = \max_j \Big(\sum_{i=1}^{n} |\mathbf{A}_j^i| \Big).$$

9.186 ¶. Let $\mathbf{A}, \mathbf{B} \in M_{2,2}(\mathbb{R})$ be given by

$$\mathbf{A} = \begin{pmatrix} 0 & 1 \\ 0 & 0 \end{pmatrix}, \qquad \mathbf{B} = \begin{pmatrix} 0 & 0 \\ 1 & 0 \end{pmatrix}.$$

Then $\mathbf{AB} \neq \mathbf{BA}$. Compute $\exp(\mathbf{A})$, $\exp(\mathbf{B})$, $\exp(\mathbf{A})\exp(\mathbf{B})$, $\exp(\mathbf{B})\exp(\mathbf{A})$ and $\exp(\mathbf{A} + \mathbf{B})$.

9.187 ¶. Define

$$\mathcal{N}(n) := \{N \in \text{End}\,(\mathbb{C}^n) \mid N \text{ is normal}\},$$
$$\mathcal{U}(n) := \{N \in \text{End}\,(\mathbb{C}^n) \mid N \text{ is unitary}\},$$
$$\mathcal{H}(n) := \{N \in \text{End}\,(\mathbb{C}^n) \mid N \text{ is self-adjoint}\},$$
$$\mathcal{H}_+(n) := \{N \in \text{End}\,(\mathbb{C}^n) \mid N \text{ is self-adjoint and positive}\}.$$

Show

   (i) if $N \in \mathcal{N}(n)$ has spectral resolution $N = \sum_{j=1}^{n} \lambda_j P_j$, then $\exp(N) \in \mathcal{N}(n)$ and has the spectral resolution $\exp(N) = \sum_{j=1}^{n} e^{\lambda_j} P_j$,
   (ii) $\exp$ is one-to-one from $\mathcal{H}(n)$ onto $\mathcal{H}_+(n)$,
   (iii) the operator $H \to \exp(iH)$ is one-to-one from $\mathcal{H}_n$ onto $\mathcal{U}(n)$.

9.188 ¶. Let $L \in \text{End}(\mathbb{C}^n)$. Then $\text{Id} - L$ is invertible if and only if 1 is not an eigenvalue of L. If L is normal, then $L = \sum_{j=1}^n \lambda_j P_j$, and we have

$$(\text{Id} - L)^{-1} = \sum_{j=1}^n \frac{1}{1 - \lambda_j} P_j.$$

If $||L|| < 1$, then all eigenvalues have modulus smaller than one and

$$(\text{Id} - L)^{-1} = \sum_{n=0}^\infty \sum_{j=1}^n \lambda_j^n P_j = \sum_{n=0}^\infty L^n.$$

9.189 ¶. Let $T, T^{-1} \in \text{End}(X)$. Show that $S \in \text{End}(X)$ and $||S - T|| \leq 1/||T||^{-1}$, then S^{-1} exists, is a bounded operator and

$$||S^{-1} - T^{-1}|| \leq \frac{||T^{-1}||}{1 - ||S - T||\,||T^{-1}||}.$$

9.190 ¶. Let X and Y be Banach spaces. We denote by $\text{Isom}(X, Y)$ the subspace of all *continuous isomorphisms from X into Y*, that is the subset of $\mathcal{L}(X, Y)$ of linear continous operators $L : X \to Y$ with continuous inverse. Prove the following.

Theorem. *We have*

(i) $\text{Isom}(X, Y)$ *is an open set of $\mathcal{L}(X, Y)$.*

(ii) *The map $f \to f^{-1}$ from $\text{Isom}(X, Y)$ into itself is continuous.*

[*Hint:* In the case of finite-dimensional spaces, it suffices to observe that the determinant is a continuous function.]

9.191 ¶. Show that, if f is linear and preserves the distances, then $f \in \text{Isom}(X, Y)$.

9.192 ¶. Show that the linear map $D : C^1([0, 1]) \subset C^0([0, 1]) \to C^0([0, 1])$ that maps f to f' is *not* continuous with respect to the uniform convergence. Show that also the map from C^0 into C^0 with domain C^1

$$f \in C^1([0, 1]) \subset C^0([0, 1]) \to f'(1/2) \in \mathbb{R}$$

is not continuous. In particular, notice that linear subspaces of a normed space are not necessarily closed.

9.193 ¶. Fix $a = \{a_n\} \in \ell_\infty$ and consider the linear operator $L : \ell_1 \to \ell_1$, $(Lx)_n = a_n x_n$. Show that

(i) $||L|| = ||a||_{\ell_\infty}$,

(ii) L is injective iff $a_n \neq 0 \ \forall n$,

(iii) L is surjective and L^{-1} è continuous if and only if $\inf |a_n| > 0$.

9.194 ¶. Show that the equation $2u = \cos u + 1$ has a unique solution in $C^0([0, 1])$.

10. Hilbert Spaces, Dirichlet's Principle and Linear Compact Operators

In a normed space, we can measure the length of a vector but not the angle formed by two vectors. This is instead possible in a Hilbert space, i.e., a Banach space whose norm is induced by an inner (or Hermitian) product. The inner (Hermitian) product allows us to measure the length of a vector, the distance between two vectors and the angle formed by them.

The abstract theory of Hilbert spaces originated from the theory of integral equations of Vito Volterra (1860–1940) and Ivar Fredholm (1866–1927), successively developed by David Hilbert (1862–1943) and J. Henri Poincaré (1854–1912) and reformulated, mainly by Erhard Schmidt (1876–1959), as a theory of linear equations with infinitely many unknowns. The axiomatic presentation, based on the notion of inner product, appeared around the 1930's and is due to John von Neumann (1903–1957) in connection with the developments of quantum mechanics.

In this chapter, we shall illustrate the *geometry of Hilbert spaces*. In Section 10.2 we discuss the *orthogonality principles*, in particular the *projection theorem* and the *abstract Dirichlet principle*. Then, in Section 10.4 we shall discuss the *spectrum of compact operators* partially generalizing to infinite dimensions the theory of finite-dimensional eigenvalues, see Chapter 4.

10.1 Hilbert Spaces

A Hilbert space is a real (complex) Banach space whose norm is induced by an inner (Hermitian) product.

10.1.1 Basic facts

a. Definitions and examples

10.1 Definition. *A real (complex) linear space, endowed with an inner or scalar (respectively Hermitian) product (|) is called a* pre-Hilbert space.

Figure 10.1. David Hilbert (1862–1943) and the *Theorie der linearen Integralgleichungen*.

We have discussed algebraic properties of the inner and Hermitian products in Chapter 3. We recall, in particular, that in a pre-Hilbert space H the function

$$||x|| := \sqrt{(x \mid x)}, \qquad x \in H, \tag{10.1}$$

defines a norm on H for which the *Cauchy–Schwarz inequality*,

$$|(x|y)| \le ||x|| \, ||y|| \qquad \forall x, y \in H,$$

holds. Moreover, *Carnot's theorem*

$$||x + y||^2 = ||x||^2 + ||y||^2 + 2\Re(x|y) \qquad \forall x, y \in H$$

and the *parallelogram law*

$$||x + y||^2 + ||x - y||^2 = 2 \left(||x||^2 + ||y||^2 \right) \qquad \forall x, y \in H$$

hold. In Chapter 3 we also discussed the geometry of real and complex pre-Hilbert spaces of finite dimension. Here we add some considerations that are relevant for spaces of infinite dimension.

A pre-Hilbert space H is naturally a normed space and has a natural topology induced by the inner product. In particular, if $\{x_n\} \subset H$ and $x \in H$, then $x_n \to x$ means that $||x_n - x|| = (x_n - x|x_n - x)^{1/2} \to 0$ as $n \to \infty$. As for any normed vector space, the norm is continuous. We also have the following.

10.2 Proposition. *The inner (or Hermitian) product in a pre-Hilbert space H is continuous on $H \times H$, i.e., if $x_n \to x$ and $y_n \to y$ in H, then $(x_n|y_n) \to (x|y)$. In particular, if $(x|y) = 0$ for all y in a dense subset Y of H, we have $x = 0$.*

Proof. In fact

$$|(x_n|y_n) - (x|y)| = |(x_n - x|y_n) + (x|y_n - y)|$$
$$\leq ||x_n - x|| \, ||y_n|| + ||x|| \, ||y_n - y||;$$

the claim then follows since the sequence $||y_n||$ is bounded, since it is convergent.

If Y is a dense subset of H, we find for any $x \in H$ a sequence $\{y_n\} \subset Y$ such that $y_n \to x$. Taking the limit in $(x \mid y_n) = 0$, we get $(x \mid x) = 0$. □

10.3 ¶ Differentiability of the inner product. Let $u :]a, b[\to H$ be a map from an interval of $\mathbb{R}$ into a pre-Hilbert space H. We can extend the notion of derivative in this context. We say that u is differentiable at $t_0 \in]a, b[$ if the limit

$$u'(t_0) := \lim_{t \to 0} \frac{u(t) - u(t_0)}{t - t_0} \in H$$

exists. Check that

Proposition. *If $u, v :]a, b[\to H$ are differentiable in $]a, b[$, so is $t \to (u(t) \mid v(t))$ and*

$$\frac{d}{dt}(u(t) \mid v(t)) = (u'(t)|v(t)) + (u(t)|v'(t)) \qquad \forall t \in]a, b[.$$

10.4 Definition. *A pre-Hilbert space H that is complete with respect to the induced norm, $||x|| := \sqrt{(x|x)}$, is called a* Hilbert *space.*

10.5 ¶. Every pre-Hilbert space H, being a metric space, can be completed. Show that its completion $\tilde{H}$ is a Hilbert space with an inner product that agrees with the original one when restricted to H.

Exercise 10.5 and Theorem 9.21 yield at once the following.

10.6 Proposition. *Every finite-dimensional pre-Hilbert space is complete, hence a Hilbert space. In particular, any finite-dimensional subspace of a pre-Hilbert space is complete, hence closed. The closed unitary ball of a Hilbert space H is compact if and only if H is finite dimensional.*

10.7 Example. The space of square integrable real sequences

$$\ell_2 = \ell_2(\mathbb{R}) := \left\{ x = \{x_n\} \, \Big| \, x_n \in \mathbb{R}, \, \sum_{i=1}^{\infty} |x_i|^2 < \infty \right\}$$

is a Hilbert space with inner product $(x \mid y) := \sum_{i=1}^{\infty} x_i y_i$, compare Section 9.1.2. Similarly, the space of square integrable complex sequences

$$\ell_2(\mathbb{C}) := \left\{ x = \{x_n\} \, \Big| \, x_n \in \mathbb{C}, \, \sum_{i=1}^{\infty} |x_i|^2 < \infty \right\}$$

is a Hilbert space with the Hermitian product $(x \mid y) := \sum_{i=1}^{\infty} x_i \overline{y}_i$.

10.8 Example. In $C^0([a, b])$

$$(f|g) := \int_a^b f(x)g(x)\, dx$$

defines an inner product with induced norm $||f||_2 := \left(\int_a^b |f(t)|^2\, dt \right)^{1/2}$. As we have seen in Section 9.1.2, $C^0([a, b])$ is *not* complete with respect to this norm. Similarly

$$(f \mid g) := \int_0^1 f(x)\overline{g}(x)\, dx,$$

defines in $C^0([0, 1], \mathbb{C})$ a pre-Hilbert structure for which $C^0([0, 1], \mathbb{C})$ is not complete.

b. Orthogonality

Two vectors x and y of a pre-Hilbert space are said to be *orthogonal*, and we write $x \perp y$, if $(x|y) = 0$. The *Pythagorean theorem* holds for pairwise orthogonal vectors $x_1, x_2, \ldots, x_n$

$$\sum_{i=1}^n ||x_i||^2 = \left|\left| \sum_{i=1}^n x_i \right|\right|^2.$$

Actually, if H is a real pre-Hilbert space, $x \perp y$ if and only if $||x + y||^2 = ||x||^2 + ||y||^2$.

A denumerable set of vectors $\{e_n\}$ is called *orthonormal* if $(e_h|e_k) = \delta_{hk}$ $\forall h, k$. Of course, orthonormal vectors are linearly independent.

10.9 Example. Here are a few examples.

(i) In ℓ_2, the sequence $e_1 = (1, 0, \ldots)$, $e_2 = (0, 1, \ldots)$, $\ldots$, is orthonormal. Notice that it is not a linear basis in the algebraic sense.

(ii) In $C^0([a, b], \mathbb{R})$ with the L^2-inner product

$$(f|g)_{L^2} := \int_a^b f(x)g(x)\, dx$$

the *triginometric system*

$$\frac{1}{b - a}, \quad \cos\left(n\frac{2\pi x}{b - a}\right), \quad \sin\left(n\frac{2\pi x}{b - a}\right), \qquad n = 1, 2, \ldots$$

is orthonormal, compare Lemma 5.45 of [GM2].

(iii) In $C^0([a, b], \mathbb{C})$ with the Hermitian L^2-product $(f|g)_{L^2} := \int_a^b f(x)\overline{g}(x)\, dx$, the *trigonometric system*

$$\frac{1}{b - a}\exp\left(i\frac{2k\pi x}{b - a}\right), \qquad k \in \mathbb{Z},$$

forms again an *orthonormal system*.

10.1.2 Separable Hilbert spaces and basis

a. Complete systems and basis

Let H be a pre-Hilbert space. We recall that a set E of vectors in H are said to be linearly independent if any finite choice of vectors in E are linearly independent. A set $E \subset H$ of linearly independent vectors such that any vector in H is a finite linear combination of vectors in E is called an *algebraic basis* of H.

We say that a system of vectors $\{e_\alpha\}_{\alpha \in \mathcal{A}}$ in a pre-Hilbert space H is *complete* if the smallest closed linear subspace that contains them is H, or equivalently, if all finite linear combinations of the $\{e_\alpha\}$ are dense in H. Operatively, $\{e_\alpha\}_{\alpha \in \mathcal{A}} \subset H$ is complete if for every $x \in H$, there exists a sequence $\{x_n\}$ of *finite linear combinations of the e_α's,*

$$x_n = \sum_{\alpha_1, \dots, \alpha_k \in \mathcal{A}} p_{(n)}^{\alpha_i} e_{\alpha_i}$$

that converges to x. [1]

10.10 Definition. *A complete denumerable system $\{e_n\}$ of a pre-Hilbert space H of linearly independent vectors is called a* basis *of H.*

b. Separable Hilbert spaces

A metric space X is said to be *separable* if there exists a denumerable and dense family in X. Suppose now that H is a separable pre-Hilbert space, and $\{x_n\}$ is a denumerable dense subset of H; then necessarily $\{x_n\}$ is a complete system in H. Therefore, if we inductively eliminate from the family $\{x_n\}$ all elements that are linearly dependent on the preceding ones, we construct an at most denumerable basis of vectors $\{y_n\}$ of H. Even more, applying the iterative process of Gram–Schmidt, see Chapter 3, to the basis $\{y_n\}$, we produce an at most denumerable orthonormal basis of H, thus concluding that every separable pre-Hilbert space has an at most denumerable orthonormal basis. The converse holds, too. If $\{e_n\}$ is an at most denumerable complete system in H and, for all n, V_n is the family of the linear combinations of $e_1, e_2, \dots, e_n$ with rational coefficients (or, in the complex case, with coefficients with rational real and imaginary parts), then $\cup_n V_n$ is dense in H. We therefore can state the following.

10.11 Theorem. *A pre-Hilbert space H is separable if and only if it has an at most denumerable orthonormal basis.*

---

[1] Notice that a basis, in the sense just defined, need not be a basis in the algebraic sense. In fact, though every element in H is the limit of finite linear combinations of elements of $\{e_\alpha\}$, it need not be a finite linear combination of elements of $\{e_\alpha\}$. Actually, it is a theorem that any algebraic basis of an infinite-dimensional Banach space has a nondenumerable cardinality.

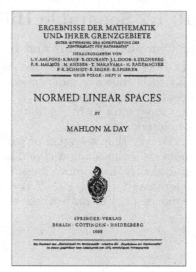

Figure 10.2. Frontispieces of *Geometria nello Spazio Hilbertiano* by Giuseppe Vitali (1875–1932) and a volume on normed spaces.

10.12 Example. The following is an example of a nonseparable pre-Hilbert space: the space of all real functions f that are nonzero in at most a denumerable set of points $\{t_i\}$ (varying with f) and moreover satisfy $\sum_i f(t_i)^2 < \infty$ with inner product $(x \mid y) = \sum x(t)y(t)$, the sum being restricted to points where $x(t)y(t) \neq 0$.

10.13 Remark. Using Zorn's lemma, one can show that every Hilbert space has an orthonormal basis (nondenumerable if the space is nonseparable); also there exist nonseparable pre-Hilbert spaces with no orthonormal basis.

Let H be a separable Hilbert space, let $\{e_n\}$ be an orthonormal basis on H and let $P_n : H \to H$ be the orthogonal projection on the finite-dimensional subspace $H_n := \operatorname{Span}\{e_1, e_2, \ldots, e_n\}$. If $L : H \to Y$ is a linear operator from H into a linear normed space Y, set $L_n(x) := L \circ P_n(x) \; \forall x \in H$. Since the L_n's are obviously continuous, H_n being finite dimensional, and $\|L_n(x) - L(x)\|_Y \to 0 \; \forall x \in H$, we infer from the Banach–Steinhaus theorem the following.

10.14 Proposition. *Any linear map $L : H \to Y$ from a separable Hilbert space into a normed space Y is bounded.*

Therefore linear unbounded operators on a separable pre-Hilbert space $L : D \to Y$ are necessarily defined only on a dense subset $D \subsetneq H$ of a separable Hilbert space. There exist instead noncontinuous linear operators from a nonseparable Hilbert space into $\mathbb{R}$.

10.15 Example. Let X be the Banach space c_0 of infinitesimal real sequences, cf. Exercise 9.158, and let $f : X \to \mathbb{R}$ be defined by $f((\alpha_1, \alpha_2, \ldots)) := \alpha_1$. Then $\ker f =$

$\{(\alpha_n) \in c_0 \,|\, \alpha_1 = 0\}$ is closed. To get an example of a dense hyperplane, let $\{e^n\}$ be the element of c_0 such that $e_k^n = \delta_{k,n}$ and let x^0 be the element of c_0 given by $x_n^0 = 1/n$, so that $\{x^0, e^1, e^2, \dots\}$ is a linearly independent set in c_0. Denote by $\mathcal{B}$ a Hamel basis (i.e., an algebraic basis) in c_0 which contains $\{x^0, e^1, e^2, \dots\}$, and set

$$\mathcal{B} = \left\{ x^0, e^1, e^2, \dots \right\} \cup \left\{ b^i \,\middle|\, i \in I \right\}$$

where $b^i \neq x^0, e^n$ for any i and n. Define

$$f : c_0 \to \mathbb{R}, \qquad f\left(\alpha_0 x^0 + \sum_{n=1}^{\infty} \alpha_n e^n + \sum_{i \in I} \alpha_i b^i\right) = \alpha_0.$$

Since $e^n \in \ker f \; \forall n \geq 1$, $\ker f$ is dense in c_0 but clearly $\ker f \neq c_0$.

10.16 ¶. Formulate similar examples in the Hilbert space of Example 10.12.

c. Fourier series and ℓ_2

We shall now show that there exist essentially only two separable Hilbert spaces: $\ell_2(\mathbb{R})$ and $\ell_2(\mathbb{C})$.

As we have seen, if H is a finite-dimensional pre-Hilbert space, and $(e_1, e_2, \dots, e_n)$ is an orthonormal basis of H, we have

$$x = \sum_{j=1}^{n} (x|e_j)\, e_j, \qquad ||x||^2 = \sum_{j=1}^{n} |(x|e_j)|^2.$$

We now extend these formulas to separable Hilbert spaces.

Let H be a separable pre-Hilbert space and let $\{e_n\}$ be an orthonormal set of H. For $x \in H$, the *Fourier coefficients of x with respect to* $\{e_n\}$ are defined as the sequence $\{(x|e_j)\}_j$, and the *Fourier series of x* as the series

$$\sum_{j=1}^{\infty} (x|e_j) e_j,$$

whose partial n-sum is the orthogonal projection $P_n(x)$ of x into the finite-dimensional space $V_n := \mathrm{Span}\,\{e_1, e_2, \dots, e_n\}$,

$$P_n(x) = \sum_{j=1}^{n} (x|e_j)\, e_j.$$

Three questions naturally arise: what is the image of

$$\mathcal{F}(x) := \{(x|e_j)\}_j, \qquad x \in H?$$

Does the Fourier series of x converge? Does it converge to x? The rest of this section will answer these questions.

10.17 Proposition (Bessel's inequality). *Let $\{e_n\}$ be an orthonormal set in the pre-Hilbert space. Then*

$$\sum_{k=1}^{\infty} |(x|e_k)|^2 \leq ||x||^2 \qquad \forall x \in H. \tag{10.2}$$

Proof. Since for all n the orthogonal projection of x on the finite-dimensional subspace $V_n := \text{Span}\,\{e_1, e_2, \dots, e_n\}$ is $P_n(x) = \sum_{k=0}^{n}(x|e_k)e_k$, the Pythagorean theorem yields

$$\sum_{k=0}^{n} |(x|e_k)|^2 = ||P_n(x)||^2 = ||x||^2 - ||x - P_n(x)||^2 \leq ||x||^2.$$

When $n \to \infty$, we get the Bessel inequality (10.2). $\qquad\square$

10.18 Proposition. *Let $\{e_n\}$ be an at most denumerable set in a pre-Hilbert space H. The following claims are equivalent.*

(i) *$\{e_n\}$ is complete.*
(ii) *$\forall x \in H$ we have $x = \sum_{k=0}^{\infty}(x|e_k)e_k$ in H, equivalently $||x - P_n(x)|| \to 0$ as $n \to \infty$.*
(iii) *(PARSEVAL'S FORMULA), $||x||^2 = \sum_{k=0}^{\infty} |(x|e_k)|^2 \; \forall x \in H$ holds.*
(iv) *$\forall x, y \in H$ we have*

$$(x|y) = \sum_{j=1}^{\infty}(x|e_j)\,\overline{(y|e_j)}.$$

In this case $x = 0$ if $(x|e_k) = 0 \; \forall k$.

Proof. (i) $\Leftrightarrow$ (ii). Suppose the set $\{e_n\}$ is complete. For every $x \in H$ and $n \in \mathbb{N}$, we find finite combinations of $e_1, e_2, \dots, e_n$ that converge to x,

$$s_n := \sum_{k=0}^{n} \alpha_k^n e_k, \qquad ||x - s_n|| \to 0.$$

If $P_n(x) = \sum_{k=1}^{n}(x|e_k)e_k$ is the orthogonal projection of x in $V_n = \text{Span}\,\{e_1, \dots, e_n\}$, we have, as $s_n \in V_n$,

$$||x - P_n(x)|| \leq ||x - s_n|| \to 0,$$

therefore $x = \sum_{k=0}^{\infty}(x|e_k)e_k$ in H. The converse (ii) $\Rightarrow$ (i) is trivial.
(ii) $\Leftrightarrow$ (iii) follows from

$$\sum_{k=0}^{n} |(x|e_k)|^2 = ||P_n(x)||^2 = ||x||^2 - ||x - P_n(x)||^2$$

when $n \to \infty$.
(ii) implies (iv) since the inner product is continuous,

$$(x|y) = \Big(\sum_{i=1}^{\infty}(x|e_i)e_i \,\Big|\, \sum_{j=1}^{\infty}(x|e_j)e_j \Big)$$

$$= \sum_{i,j=1}^{\infty}(x|e_i)\,\overline{(y|e_j)}\,(e_i|e_j) = \sum_{j=1}^{\infty}(x|e_j)\,\overline{(y|e_j)}.$$

and (iv) trivially implies (iii).
Finally (iii) implies that $x = 0$ if $(x|e_k) = 0 \; \forall k$. $\qquad\square$

10.19 Proposition. *Let H be a Hilbert space and let $\{e_n\}$ be an orthonormal set of H. Given any sequence $\{c_k\}$ such that $\sum_{j=0}^{\infty} |c_k|^2 < \infty$, then the series $\sum_{j=0}^{\infty} c_j e_j$ converges to H. If moreover $\{e_n\}$ is complete, then*

$$x = \sum_{j=1}^{\infty} (x|e_j)\, e_j \qquad \forall x \in H.$$

Proof. Define $x_n := \sum_{j=1}^{n} c_j e_j$. As

$$\|x_{n+p} - x_n\|^2 = \sum_{j=n+1}^{n+p} |c_j|^2,$$

$\{x_n\}$ is a Cauchy sequence in H, hence it converges to $y := \sum_{j=0}^{\infty} c_j e_j \in H$. On account of the continuity of the scalar product

$$(y|e_j) = (\sum_{i=1}^{\infty} c_i\, e_i | e_j) = \sum_{i=1}^{\infty} c_i\, (e_i|e_j) = c_j$$

for all j. If $x \in H$ and $c_j := (x|e_j)\ \forall j$, then $(x - y|e_j) = 0\ \forall j$, and, since $\{e_n\}$ is complete, Proposition 10.18 yields $x = y$. □

Let H be a pre-Hilbert space. Let us explicitly interpret the previous results as information on the linear map defined by

$$\mathcal{F}(x) := \{(x|e_j)\}_j, \qquad x \in H,$$

that maps $x \in H$ into the sequence of its Fourier coefficients.

○ Bessel's inequality says that $\mathcal{F}(x) \in \ell_2\ \forall x \in H$ and that $\mathcal{F} : H \to \ell_2$ is continuous, actually

$$\|\mathcal{F}(x)\|_{\ell_2}^2 := \sum_{j=1}^{\infty} |(x|e_j)|^2 \le \|x\|^2,$$

○ if $\{e_n\}$ is a complete orthonormal set in H, then Parseval's formula says that $\mathcal{F} : H \to \ell_2$ is an isometry between H and its image $\mathcal{F}(H) \subset \ell_2$, in particular $\mathcal{F} : H \to \ell_2$ is injective,
○ if H is complete and $\{e_n\}$ is a complete orthonormal set, then, according to Proposition 10.19,
   – the series $\sum_{j=1}^{\infty} c_j e_j$ converges in H for every choice of the sequence $\{c_j\} \subset \ell_2$, that is, $\mathcal{F}$ is surjective onto ℓ_2,
   – the inverse map of $\mathcal{F}$, $\mathcal{F}^{-1} : \ell_2 \to H$, is given by

$$\mathcal{F}^{-1}\Big(\{c_j\}\Big) = \sum_{j=1}^{\infty} c_j e_j.$$

Therefore, we can state the following.

10.20 Theorem. *Every separable Hilbert space H over $\mathbb{R}$ (respectively over $\mathbb{C}$) is isometric to $\ell_2(\mathbb{R})$ (respectively to $\ell_2(\mathbb{C})$). More precisely, given an orthonormal basis $\{e_n\} \subset H$, the coordinate map $\mathcal{E} : \ell_2(\mathbb{K}) \to H$ ($\mathbb{K} = \mathbb{R}$ if H is real, resp. $\mathbb{K} = \mathbb{C}$ if H is complex), given by*

$$\mathcal{E}\Big(\{c_k\}\Big) := \sum_{k=0}^{\infty} c_k e_k,$$

is a surjective isometry of Hilbert spaces and its inverse maps any $x \in H$ into the sequence of the corresponding Fourier coefficients $\{(x|e_j)\}_j$.

Finally, we conclude with the following.

10.21 Theorem (Riesz–Fisher). *Let H be a Hilbert space and let $\{e_n\}$ be an at most denumerable orthonormal set of H. Then the following statements are equivalent.*

- (i) *$\{e_n\}$ is a basis of H.*
- (ii) *$\forall x \in H$ we have $x = \sum_{j=1}^{\infty}(x|e_j)e_j$.*
- (iii) *$||x - P_n(x)|| \to 0$, where P_n is the orthogonal projection onto $V_n :=$ Span $\{e_1, e_2, \ldots, e_n\}$.*
- (iv) *(PARSEVAL'S FORMULA or ENERGY EQUALITY) $||x|| = \sum_{j=1}^{\infty} |(x|e_j)|^2$ holds $\forall x \in H$.*
- (v) *$(x|y) = \sum_{j=1}^{\infty}(x|e_j)\,(y|e_j)$.*
- (vi) *if $(x|e_j) = 0$ $\forall j$ then $x = 0$.*

Proof. The equivalences of (i), (ii), (iii), (iv), (v) and of (i) $\Rightarrow$ (vi) were proved in Proposition 10.18. It remains to show that (vi) implies (i). Suppose that $\{e_n\}$ is not complete. Then there is $y \in H$ with $||y||^2 > \sum_{j=1}^{\infty} |(y|e_j)|^2$, while, on the other hand, Bessel's inequality and Proposition 10.19 show that there is $z \in H$ such that $z := \sum_{j=0}^{\infty}(y|e_j)e_j$, and by Parseval's formula, $||z||^2 = \sum_{j=1}^{\infty} |(y|e_j)|^2$. Consequently $||z||^2 < ||y||^2$. But, on account of the continuity of the scalar product

$$(z|e_k) = \Big(\sum_{j=1}^{\infty}(y|e_j)e_j \,\Big|\, e_k\Big) = \sum_{j=1}^{\infty}(y|e_j)(e_j|e_k) = (y|e_k)$$

i.e., $(y - z|e_k) = 0$ $\forall k$. Then by (vi) $y = z$, a contradiction. □

d. Some orthonormal polynomials in L^2

Let I be an interval on $\mathbb{R}$ and let $p : I \to \mathbb{R}$ be a continuous function that is positive in the interior of I and such that for all $n \geq 0$

$$\int_I |t|^n p(t)\, dt < +\infty.$$

The function p is often called a *weight* in I. The subspace V_p of $C^0(I, \mathbb{C})$ of functions $x(t)$ such that

$$\int_I |x(t)|^2 p(t)\, dt < \infty$$

is a linear space and

$$(x|y) := \int_I x(t)\overline{y(t)}p(t)\,dt$$

defines a Hermitian product on it. This way V_p is a pre-Hilbert space. Also, one easily sees that the monomials $\{t^n\}$ $n \geq 0$, are linearly independent; Gram–Schmidt's orthonormalization process then produces orthonormal polynomials $\{P_n(t)\}$ of degree n with respect to the weight p. Classical examples are

○ JACOBI POLYNOMIALS J_n. They correspond to the choice

$$I := [-1, 1], \qquad p(t) := (1-t)^\alpha(1+t)^\beta, \qquad \alpha, \beta > -1.$$

○ LEGENDRE POLYNOMIALS P_n. They correspond to the choice $\alpha = \beta = 0$ in Jacobi polynomials J_n, i.e.,

$$I = [-1, 1], \qquad p(t) := 1.$$

○ CHEBYCHEV POLYNOMIALS T_n. They correspond to the choice $\alpha = \beta = -1/2$ in Jacobi polynomials J_n, i.e.

$$I = [-1, 1], \qquad p(t) := \frac{1}{\sqrt{1-t^2}}.$$

○ LAGUERRE POLYNOMIALS L_n. They correspond to the choice

$$I = [0, +\infty], \qquad p(t) := e^{-t}.$$

○ HERMITE POLYNOMIALS H_n. They correspond to the choice

$$I := [-\infty, +\infty], \qquad p(t) := e^{-t^2}.$$

One can show that the polynomials $\{J_n\}, \{P_n\}, \{T_n\}, \{L_n\}, \{H_n\}$ form respectively, a basis in V_p.

Denoting by $\{R_n\}$ the system of orthonormal polynomials with respect to $p(t)$ obtained by applying the Gram–Schmidt procedure to $\{t^n\}$, $n \geq 0$, the R_n's have interesting properties. First, we explicitly notice the following properties

○ (A1) for all n, R_n is orthogonal to any polynomial of degree less than n,
○ (A2) for all n the polynomial $R_n(t) - t\,R_{n-1}(t)$ has degree less than n, hence

$$(tR_{n-1}|R_n) = (R_n|R_n),$$

○ (A3) for all $x, y, z \in V_p$ we have $(xy|z) = (xy\bar{z}|1) = (x|\bar{y}z)$.

10.22 Proposition (Zeros of R_n). *Every R_n has n real distinct roots in the interior of I.*

Proof. Since $\int_I R_n(t)p(t)\,dt = 0$, it follows that R_n changes sign at least once in I. Let $t_1 < \cdots < t_r$ be the points in int (I) in which R_n changes sign. Let us show that $r = n$. Suppose $r < n$ and let

$$Q(t) := (t - t_1)(t - t_2)\ldots(t - t_r),$$

then $R_n Q$ has constant sign, hence, $(R_n|Q) \neq 0$, that contradicts property (A1). □

10.23 Proposition (Recurrence relations). *There exist two sequences* $\{\lambda_n\}$, $\{\mu_n\}$ *of real numbers such that for* $n \geq 2$

$$R_n(t) = (t + \lambda_n)R_{n-1}(t) - \mu_n R_{n-2}(t).$$

Proof. Since $\deg(R_n - tR_{n-1}) \leq n - 1$, then

$$R_n(t) - tR_{n-1}(t) = \sum_{i=0}^{n-1} c_i R_i(t),$$

and for $i \leq n - 1$, we have $-(tR_{n-1}|R_i) = c_i(R_i|R_i)$. By (A3), we have $(tR_{n-1}|R_i) = (R_i|R_i)$, hence, if $i+1 < n-1$, then $(tR_{n-1}|R_i) = 0$ from which $c_i = 0$ for $i = n-2, n-1$. For $i = n - 2$, property (A2) shows that

$$-(R_{n-1}|R_{n-1}) = c_{n-2}(R_{n-2}|R_{n-2}),$$

hence $c_{n-2} < 0$. □

10.24 ¶. Define

$$Q_n(t) := \frac{1}{2^n n!}\frac{d^n}{dt^n}(t^2 - 1)^n.$$

(i) Integrating by parts show that $\{Q_n\}$ is an orthogonal system in $[-1,1]$ with respect to $p(t) = 1$, and that $(Q_n|Q_n) = 2/(2n + 1)$.
(ii) Show that $Q_n(1) = 1$ and that Q_n is given in terms of Legendre polynomias $\{P_n\}$ by $Q_n(t) = P_n(t)/P_n(1)$. Finally, compute $P_n(1)$.
(iii) Show that the polynomials $\{Q_n\}$ satisfy the recurrence relation

$$nQ_n = (2n - 1)Q_{n-1} - (n - 1)Q_{n-2}$$

and solve the linear ODE

$$\frac{d}{dt}\left((1 - t^2)\frac{dQ_n}{dt}\right) + n(n + 1)Q_n = 0.$$

10.25 ¶. In V_p with $p(t) := e^{-t}$ and $I = [0, +\infty[$, define

$$Q_n(t) := \frac{1}{n!}e^t\frac{d^n}{dt^n}(e^{-t}t^n).$$

(i) Show that $\deg Q_n = n$ and that $\{Q_n\}$ is a system of orthogonal polynomials in V_p. Then compute $(Q_n|Q_n)$.
(ii) Show that $Q_n(0) = 1$ and, in terms of Laguerre polynomials,

$$Q_n(t) \neq \frac{L_n(t)}{L_n(0)}.$$

Compute then $L_n(0)$.
(iii) Show that $e^{\alpha t} \in V_p$ for all $\alpha \geq 0$ and compute its Fourier coefficients $\gamma_{n,\alpha}$ with respect to $\{Q_n\}$.
(iv) Show that $\sum \gamma_{n,\alpha}Q_n(t) = e^{\alpha t}$ in V_p.

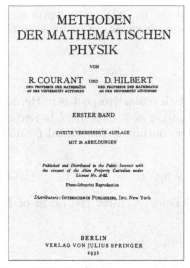
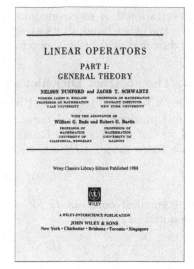

Figure 10.3. The frontispieces of two classical monographs.

(v) Changing variable and using the Stone–Weierstrass theorem, show that $\{e^{-nt}\}$ is a basis in V_p.

10.26 ¶. Define the polynomials $Q_n(t)$ by

$$\frac{d^n}{dt^n}e^{-t^2} = (-1)^n Q_n(t)e^{-t^2}.$$

(i) Show that $\{Q_n\}$ is an orthogonal system in V_p with $I = [0, +\infty[$ and $p(t) = e^{-t^2}$. Show that each $Q_n(t)$ is proportional to the Hermite polynomial H_n.

(ii) Show that $Q_0 = 1$, $Q_1 = 2t$ and that for $n \geq 2$

$$Q_n(t) = 2tQ_{n-1}(t) - 2(n-1)Q_{n-1}(t).$$

(iii) Show that Q_n satisfies $Q_n''(t) - 2tQ_n'(t) + 2nQ_n(t) = 0$ and that $Q_n'(t) = 2nQ_{n-1}(t)$.

10.2 The Abstract Dirichlet's Principle and Orthogonality

The aim of this section is to illustrate some aspects of the linear geometry of Hilbert spaces mainly in connection with the abstract formulation of the *Dirichlet principle*. In its concrete formulation, this principle has played a fundamental role in the geometric theory of functions by Riemann, in the theory of partial differential equations, for instance, when dealing with gravitational or electromagnetic fields and in the calculus of variations. On the other hand, in its abstract formulation it turns out to be a simple orthogonal projection theorem.

a. The abstract Dirichlet's principle

Let H be a real (complex) Hilbert space with scalar (respectively Hermitian) product $(\ |\ )$ and norm $||u|| := \sqrt{(u|u)}$. Let $\mathbb{K} = \mathbb{R}$ if H is real or $\mathbb{K} := \mathbb{C}$ if H is complex. Recall that a linear continuous functional L on H is a linear map $L : H \to \mathbb{R}$ such that

$$|L(u)| \leq K\,||u|| \qquad \forall\, u \in H; \tag{10.3}$$

the smallest constant K for which (10.3) holds is called the *norm* of L, denoted $||L||$ so that

$$|L(u)| \leq ||L||\,||u||, \qquad \forall u \in H,$$

and, see Section 9.4,

$$||L|| = \sup_{||u||=1} |L(u)| = \sup_{u \neq 0} \frac{|L(u)|}{||u||}.$$

We denote by $H^* = \mathcal{L}(H, \mathbb{K})$ the space of linear continuous functionals on H, called the *dual* of H.

10.27 Theorem (Abstract Dirichlet's principle). *Let H be a real or complex Hilbert space and let $L \in H^*$. The functional $\mathcal{F} : H \to \mathbb{R}$ defined by*

$$\mathcal{F}(u) := \frac{1}{2}||u||^2 - \Re(L(u)) \tag{10.4}$$

achieves a unique minimum point $\overline{u}$ in H, and every minimizing sequence, i.e., every sequence $\{u_k\} \subset H$ such that

$$\mathcal{F}(u_k) \to \inf_{v \in H} \mathcal{F}(v),$$

converges to $\overline{u}$ in H. Moreover $\overline{u}$ is characterized as the unique solution of the linear equation

$$(\varphi|\overline{u}) = L(\varphi) \qquad \forall \varphi \in H. \tag{10.5}$$

In particular $||\overline{u}|| = ||L||$.

Proof. Let us prove that $\mathcal{F}$ has a minimum point. First we notice that $\mathcal{F}$ is bounded from below, since, recalling the inequality $2ab \leq a^2 + b^2$, we have for all $v \in H$

$$\mathcal{F}(v) \geq \frac{1}{2}||v||^2 - ||L||\,||v|| \geq \frac{1}{2}||v||^2 - \frac{1}{2}||v||^2 - \frac{1}{2}||L||^2 = -\frac{1}{2}||L||^2 \in \mathbb{R},$$

hence $\lambda := \inf_{v \in H} \mathcal{F}(v) \in \mathbb{R}$. Then we observe that, by the parallelogram law,

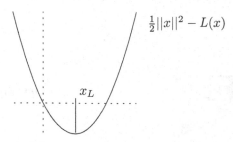

Figure 10.4. The Dirichlet's principle.

$$\frac{1}{4}||u - v||^2 = \frac{1}{2}||u||^2 + \frac{1}{2}||v||^2 - \left\|\frac{u+v}{2}\right\|^2$$
$$- \Re(L(u) - \Re(L(v)) + 2\Re\left(L\left(\frac{u+v}{2}\right)\right) \qquad (10.6)$$
$$= \mathcal{F}(u) + \mathcal{F}(v) - 2\mathcal{F}\left(\frac{u+v}{2}\right).$$

Thus, if $\{u_k\}$ is a minimizing sequence, by (10.6)

$$\frac{1}{4}||u_k - u_h||^2 = \mathcal{F}(u_k) + \mathcal{F}(u_h) - 2\mathcal{F}\left(\frac{u_k + u_h}{2}\right)$$
$$\leq \mathcal{F}(u_k) + \mathcal{F}(u_h) - 2\lambda \to 0$$

as $h, k \to \infty$. Therefore $\{u_k\}$ is a Cauchy sequence in H and converges to some $\overline{u} \in H$; by continuity $\mathcal{F}(u_k) \to \mathcal{F}(\overline{u})$ hence $\mathcal{F}(\overline{u}) = \lambda$. This proves existence of the minimizer $\overline{u}$.

If $\{v_k\}$ is another minimizing sequence for $\mathcal{F}$, (10.6) yields $||u_k - v_k|| \to 0$, and this proves that the minimizer is unique and that every minimizing sequence converges to $\overline{u}$ in H.

Let us show that $\overline{u}$ solves (10.5). Fix $\varphi \in H$, and consider the real function $\epsilon \to \mathcal{F}(\overline{u} + \epsilon\varphi)$, that is the second order polynomial in ϵ

$$\mathcal{F}(\overline{u} + \epsilon\varphi) = \frac{1}{2}||\varphi||^2\epsilon^2 + \Re[(\varphi|\overline{u}) - L(\varphi)]\epsilon + \mathcal{F}(\overline{u})$$

with minimum point at $\epsilon = 0$. We deduce $\Re((\varphi|\overline{u}) - L(\varphi)) = 0 \, \forall \varphi \in H$ hence, as $z = 0$ if $\Re(\lambda z) = 0 \, \forall \lambda \in \mathbb{C}$,

$$(\varphi|\overline{u}) - L(\varphi) = 0 \qquad \forall \varphi \in H. \qquad (10.7)$$

Conversely, if v solves (10.5), then for every $\varphi \in H$

$$\mathcal{F}(v + \varphi) = \frac{1}{2}||v||^2 + \Re(v|\varphi) + \frac{1}{2}||\varphi||^2 - \Re(L(v)) - \Re(L(\varphi))$$
$$= \mathcal{F}(v) + \Re((\varphi|v) - L(\varphi)) + \frac{1}{2}||\varphi||^2 = \mathcal{F}(v) + \frac{1}{2}||\varphi||^2,$$

hence $\mathcal{F}(v + \varphi) \geq \mathcal{F}(v)$, $\forall \varphi \in H$, i.e., v is a minimum point for $\mathcal{F}$ in H. This proves that (10.5) has a unique solution, the minimum point $\overline{u}$ of $\mathcal{F} : H \to \mathbb{R}$. Finally, we infer from (10.5)

$$||L|| = \sup_{\varphi \neq 0} \frac{|L(\varphi)|}{||\varphi||} = \sup_{\varphi \neq 0} \frac{|(\varphi|\overline{u})|}{||\varphi||} = ||\overline{u}||.$$

$\square$

Figure 10.5. Frigyes Riesz (1880–1956) and the frontispiece of a classical monograph.

b. Riesz's theorem

In particular, we have proved the following.

10.28 Theorem (Riesz). *For every linear continuous functional $L \in H^*$ there exists a unique $u_L \in H$ such that*

$$L(\varphi) = (\varphi|u_L) \qquad \forall \varphi \in H. \tag{10.8}$$

Moreover $||u_L|| = ||L||$.

Actually, we have also proved that Riesz's theorem and the abstract Dirichlet's principle are equivalent.

10.29 Continuous dependence and Riesz's operator. If u solves the minimum problem (10.4), or equation (10.8), we have $||u|| = ||L||$. This implies that the solution of (10.4) or (10.8) depends continuously on L. In fact, if $L_n, L \in H^*$ and $||L_n - L|| \to 0$, and if $u_n, u \in H$ solve $(\varphi|u_n) = L(\varphi)$ and $(\varphi|u) = L(\varphi) \ \forall \varphi \in H$, then

$$(\varphi|u_n - u) = (L_n - L)(\varphi) \qquad \forall \varphi \in H,$$

hence

$$||u_n - u|| = ||L_n - L|| \to 0.$$

10.30 Riesz's operator. The map $\Gamma : H^* \to H$ that associates to each $L \in H^*$ the solution u_L of (10.8) is called *Riesz's map*. It is easily seen that $\Gamma : H^* \to H$ is linear and by Riesz's theorem we have $||\Gamma(L)|| = ||u_L|| = ||L||$, i.e., not only is Γ continuous, but

10.31 Theorem. *Riesz's map* $\Gamma : H^* \to H$ *is an isometry between* H^* *and* H.

c. The orthogonal projection theorem

Let us now extend the orthogonal projection theorem onto finite-dimensional subspaces, see Chapter 3, to *closed* subspaces of a Hilbert space.

Let H be a Hilbert space and V a subspace of H. If $f \in H$, then the map $L : V \to \mathbb{K}$, $v \to L(v) = (f|v)$ is a linear continuous operator on V with $||L|| \le ||f||$ since $|(f|v)| \le ||f|| \, ||v|| \; \forall v \in V$ by the Cauchy–Schwarz inequality. Since a *closed* linear subspace V of a Hilbert space H is again a Hilbert space with the induced inner product, a simple consequence of Theorem 10.27 is the following.

10.32 Theorem (Projection theorem). *Let* V *be a closed linear subspace of a Hilbert space* H. *Then for every* $f \in H$ *there is a unique point* $u \in V$ *of minimum distance from* f, *that is*

$$||f - u|| = \mathrm{dist}\,(f, V) := \inf\Big\{|f - \varphi|\,\Big|\,\varphi \in V\Big\}.$$

Moreover, u *is characterized as the unique point such that* $f - u$ *is orthogonal to* V, *i.e.,*

$$(f - u|\varphi) = 0 \qquad \forall \varphi \in V.$$

Proof. We have for all $v \in V$ $||v - f||^2 = ||v||^2 - 2\Re(v|f) + ||f||^2$. Theorem 10.27, when applied to $\mathcal{F}(v) := ||v||^2 - 2\Re(f|v)$, $v \in V$, yields existence of a unique minimizer $u \in V$ of $||v - f||^2$, hence of $v \to ||v - f||$. The characterization of u given by Riesz's theorem states, in our case, that u is also the unique solution of

$$2(\varphi|u) = 2(\varphi|f) \qquad \forall \varphi \in V.$$

$\square$

Let V be a subspace of a Hilbert space H. We denote by $V^\perp$ the class of vectors of H orthogonal to V

$$V^\perp := \Big\{x \in H \,\Big|\, (x|v) = 0 \; \forall v \in V\Big\}.$$

Clearly $V^\perp$ is a closed subspace of H.

10.33 Corollary. *If* V *is a linear closed subspace of a Hilbert space* H, *then* $H = V \oplus V^\perp$, *i.e., every* $u \in H$ *uniquely decomposes as* $u = v + w$, *where* $v \in V$ *and* $w \in V^\perp$.

10.34 ¶. Show that, if V is a linear subspace of a Hilbert space H, then $V^\perp$ is closed and that $(V^\perp)^\perp$ is the closure of V.

10.35 ¶. Show that the orthogonal projection theorem is in fact equivalent to Riesz's theorem and, consequently, to the abstract Dirichlet's principle. [*Hint:* We give the scheme of the proof leaving it to the reader to add details. Uniqueness and $||u_L|| = ||L||$ follow from (10.8). Let us prove the existence of a solution of (10.8). Suppose L is not identically zero, then $\ker L = L^{-1}(\{0\})$ is a linear closed proper subspace of H and there exists $u_0 \in \ker L^\perp$ such that $u_0 \neq 0$ and $L(u_0) = 1$. Since $u - L(u)u_0 \in \ker L$ $\forall u \in H$, we have

$$u = w + L(u)u_0 \qquad \text{with } w \in \ker L \text{ and } u_0 \in \ker L^\perp.$$

Multiplying by u_0, we then find $L(u) = \left(u \mid \frac{u_0}{|u_0|^2} \right)$.

d. Projection operators

Let V be a linear closed subspace of a Hilbert space H. The projection theorem defines a linear continuous operator $P_V : H \to H$ that maps $f \in H$ into its orthogonal projection $P_V f \in V$; of course $||P_V|| \leq 1$ and $\text{Im}(P_V) = V$. Also

$$P_V^2 = P_V \circ P_V = P_V$$

and the formula $H = V \oplus V^\perp$ can be written as

$$\text{Id} = P_V + P_{V^\perp}, \qquad P_V P_{V^\perp} = P_{V^\perp} P_V = 0.$$

For the reader's convenience, we only prove that $P_V(f + g) = P_V(f) + P_V(g)$. Trivially, $f + g - P_V(f + g) \perp V$ and $f + g - P_V f - P_V g \perp V$; since there is a unique $u \in V$ such that $f + g - h \perp V$, we conclude $P_V(f + g) = P_V(f) + P_V(g)$.

10.36 ¶. Let $P : H \to H$ be a linear operator such that $P^2 = P$. Then P is continuous if and only if $\ker P$ and $\text{Im} P$ are closed.

10.37 ¶. If V is a closed subspace of a Hilbert space H and $\{e_n\}$ is a denumerable orthonormal basis of V, then the orthogonal projection of $x \in H$ is given by

$$Px := \sum_{j=1}^{\infty} (Px|e_j)\, e_j = \sum_{j=1}^{\infty} (x|e_j)\, e_j.$$

10.3 Bilinear Forms

From now on we shall only consider real vector spaces, though one could develop similar results for sesquilinear forms on complex vector spaces.

10.3.1 Linear operators and bilinear forms

a. Linear operators

Let H be a real Hilbert space. As we know, the space $\mathcal{L}(H, H)$ of linear continuous operators, also called *bounded operators* from H into H, is a Banach space with the norm

$$||T|| = \sup_{x \neq 0} \frac{||Tx||}{||x||}.$$

If $T \in \mathcal{L}(H, H)$, we denote by $N(T)$ and $R(T)$ respectively the kernel and the image or range of T. Since T is continuous, $N(T) = T^{-1}(\{0\})$ is closed in H, while in general $R(T)$ is not closed. The restriction of T to $N(T)^\perp$, $T : N(T)^\perp \to R(T)$ is of course a linear bijection, therefore, from Banach's open mapping theorem, cf. Section 9.4, we infer the following.

10.38 Proposition. *Let $T \in \mathcal{L}(H, H)$. Then T has a closed range in H if and only if there exists $C > 0$ such that $||x|| \leq C\, ||Tx||\ \forall x \in N(T)^\perp$, that is, if and only if $T^{-1} : R(T) \to N(T)^\perp$ is a bounded operator, or, equivalently, if and only if $T : N(T)^\perp \to R(T)$ is an isomorphism.*

b. Adjoint operator

Let X, Y be two real Hilbert spaces endowed with their inner products $(\ |\ )_X$ and $(\ |\ )_Y$, and let $T \in \mathcal{L}(X, Y)$. For any $y \in Y$ the map $x \to (Tx|y)_Y$ is a linear continuous form on X, hence Riesz's theorem yields a unique element $T^*y \in X$ such that

$$(x|T^*y)_X = (Tx|y)_Y \qquad \forall x \in X,\ \forall y \in Y. \tag{10.9}$$

It is easily seen that the map $T^* : Y \to X$ just defined is a linear operator called the *adjoint* of T. Moreover, from (10.9) T^* is a bounded operator with $||T^*|| = ||T||$. Obviously, if $S, T \in \mathcal{L}(H, H)$

$$(TS)^* = S^*T^*, \qquad (T^*)^* = T.$$

10.39 ¶. Suppose that $P : H \to H$ is a linear continuous operator such that $P^2 = P$ and $P^* = P$. Show that $V := P(H)$ is a closed subspace of H and that P is the orthogonal projection onto V.

An operator $L : H \to H$ on a Hilbert space H is called *self-adjoint* if $T^* = T$, i.e.,

$$(x|Ty) = (Tx|y) \qquad \forall x, y \in H.$$

It follows from (10.9) that $R(T)^\perp = N(T^*)$. Consequently $\overline{R(T)} = N(T^*)^\perp$ and using the open mapping theorem, we conclude the following.

10.40 Corollary. *Let $T \in \mathcal{L}(H, H)$ be a bounded operator with closed range. Then we have*

(i) *The equation $Tx = y$ is solvable if and only if $y \perp N(T^*)$.*
(ii) *T is an isomorphism between $N(T)^{\perp}$ and $R(T) = N(T^*)^{\perp}$. In particular the Moore–Penrose inverse $T^{\dagger} : H \to H$, defined by composing the orthogonal projection onto $R(T)$ with the inverse of $T_{|N(T)^{\perp}}$, is a bounded operator.*
(iii) *We have $H = R(T) \oplus N(T^*)$.*

Proof. (i) For $x, y \in H$ we have $(y|Tx) = (T^*y|x)$. Hence, $R(T)^{\perp} = N(T^*)$. Therefore, by considering the orthogonals,

$$\overline{R(T)} = (R(T)^{\perp})^{\perp} = N(T^*)^{\perp}.$$

(ii) follows from the open mapping theorem and (iii) follows from (i) by considering the projections onto $N(T^*)^{\perp}$ and $N(T^*)$. □

10.41 ¶. Let $A \in \mathcal{L}(X, Y)$ be a bounded operator between Hilbert spaces. Show that
   (i) $N(A) = N(A^*A)$,
   (ii) $R(A^*) \supset R(A^*A)$,
   (iii) $R(A^*) = R(A^*A)$ if and only if $R(A) = N(A^*)^{\perp}$, i.e., if and only if $R(A)$ is closed,
   (iv) if $R(A^*A)$ is closed, then $R(A^*)$ and $R(A)$ are closed.

10.42 ¶. Let H be a Hilbert space and let T be a self-adjoint operator. Show that T is continuous. [*Hint:* Show that T has a closed graph.]

c. Bilinear forms

Let H be a real vector space. A map $\mathcal{B} : H \times H \to \mathbb{R}$ which is linear on each factor is called a bilinear form. A bilinear form $\mathcal{B} : H \times H \to \mathbb{R}$ is called *continuous* or *bounded* if, for some constant Λ, we have

$$|\mathcal{B}(u, v)| \leq \Lambda \, ||u|| \, ||v|| \qquad \forall u, v \in H,$$

and it is called *coercive* if there is $\lambda > 0$ such that

$$\mathcal{B}(u, u) \geq \lambda \, ||u||^2 \qquad \forall u \in H.$$

Finally, $\mathcal{B}(u, v)$ is said to be *symmetric* if

$$\mathcal{B}(u, v) = \mathcal{B}(v, u) \qquad \forall u, v \in H.$$

Any linear operator $T : H \to H$ defines a bilinear form by

$$\mathcal{B}(v, u) := (v|Tu),$$

and $\mathcal{B}$ is bounded if T is bounded since $|\mathcal{B}(v, u)| \leq ||T|| \, ||v|| \, ||u||$. Conversely, given a continuous bilinear form $\mathcal{B} : H \times H \to \mathbb{R}$ on a real Hilbert space H,

$$|\mathcal{B}(u, v)| \leq L \, ||u|| \, ||v|| \qquad \forall u, v \in H, \tag{10.10}$$

for any given $u \in H$, the map $v \to \mathcal{B}(v, u)$ is a linear continuous operator on H, hence by Riesz's theorem, there exists $Tu \in H$ such that

$$\mathcal{B}(v, u) = (v|Tu) \qquad \forall v \in H. \tag{10.11}$$

It is easy to see that T is linear and, from (10.10) that $||T|| \leq \Lambda$ since

$$||Tu||^2 = \mathcal{B}(Tu, u) \leq \Lambda \, ||Tu|| \, ||u||.$$

Consequently, by (10.11), there is a complete equivalence between bilinear continuous forms on a real Hilbert space H and bounded linear operators from H into H.

Also, by (10.11), coercive bilinear continuous forms correspond to bounded operators called *coercive*, i.e., such that for some $\lambda > 0$ $(u|Tu) \geq \lambda ||u||^2$ $\forall u \in H$. Moreover, self-adjoint operators correspond to bilinear symmetric forms, in fact

$$\mathcal{B}(v, u) - \mathcal{B}(u, v) = (v|Tu) - (u|Tv) = (v|(T - T^*)u) \qquad \forall u, v \in H.$$

10.3.2 Coercive symmetric bilinear forms

a. Inner products

Clearly, every symmetric continuous coercive bilinear form on H defines in H a new scalar product, which in turn induces a norm that is equivalent to the original, since

$$\lambda \, ||u||^2 \leq \mathcal{B}(u, u) \leq \Lambda \, ||u||^2 \qquad \forall u \in H.$$

Replacing $(u|v)$ with $\mathcal{B}(u, v)$, Dirichlet's principle and Riesz's theorem read as follows.

10.43 Theorem. *Let H be a real Hilbert space with inner product $(\ |\ )$ and norm $||u|| := \sqrt{(u \mid u)}$ and let $\mathcal{B} : H \times H \to \mathbb{R}$ be a symmetric, continuous and coercive bilinear form on H, i.e., $\mathcal{B}(u, v) = \mathcal{B}(v, u)$ and for some $\Lambda \geq \lambda > 0$*

$$|\mathcal{B}(u, v)| \leq \Lambda \, ||u|| \, ||v||, \quad \mathcal{B}(u, u) \geq \lambda \, ||u||^2, \qquad \forall u, v \in H;$$

finally, let L be a continuous linear form on H. Then the following equivalent claims hold:

(i) (ABSTRACT DIRICHLET'S PRINCIPLE). *The functional*

$$\mathcal{F}(u) := \frac{1}{2}\mathcal{B}(u, u) - L(u)$$

  has a unique minimizer $\overline{u} \in H$, every minimizing sequence converges to $\overline{u}$, $\overline{u}$ in H, $\overline{u}$ solves

$$\mathcal{B}(\varphi, \overline{u}) = L(\varphi) \qquad \forall \varphi \in H.$$

(ii) (RIESZ'S THEOREM) *The equation*

$$\mathcal{B}(\varphi, u) = L(\varphi) \qquad \forall \varphi \in H. \tag{10.12}$$

  has a unique solution $u_L \in H$.

Moreover $\bar{u} = u_L$ and $||u_L|| \leq \frac{1}{\lambda}||L||$.

The continuity estimate for u_L follows from

$$\sqrt{\lambda}||u_L|| \leq \sqrt{\mathcal{B}(u_L, u_L)} = \sup_{u \neq 0} \frac{|L(u)|}{\sqrt{\mathcal{B}(u, u)}} \leq \frac{1}{\sqrt{\lambda}}||L||.$$

In terms of operators Theorem 10.43 may be rephrased as follows.

10.44 Theorem. *Let T be a continuous, coercive, self-adjoint operator on H, i.e.,*

$$||Tu|| \leq ||T||\,||u||, \qquad (u|Tu) \geq \lambda||u||^2, \quad \lambda > 0, \ \forall u \in H, \qquad T = T^*.$$

Then T is invertible with continuous inverse, and $||T^{-1}|| \leq 1/\lambda$.

Proof. From coercivity we infer that

$$\lambda||u||^2 \leq |(u|Tu)| \leq ||u||\,||Tu||, \tag{10.13}$$

hence $N(T) = \{0\}$ and $T^{-1} : R(T) \to H$ is continuous. T is therefore an isomorphism between H and $R(T)$, and therefore $R(T)$ is closed, $R(T) = \overline{R(T)} = N(T^*)^\perp = N(T)^\perp = H$ and (10.13) rewrites as $||T^{-1}u|| \leq \frac{1}{\lambda}||u|| \ \forall u \in H$. $\qquad\square$

A variational proof of Theorem 10.44. For any $y \in H$, consider the bounded operator $L : H \to \mathbb{R}$, $L(\varphi) := (\varphi|y)$ and the bilinear form $\mathcal{B} : H \times H \to \mathbb{R}$

$$\mathcal{B}(\varphi, u) := (\varphi|Tu).$$

$\mathcal{B}$ is bounded and symmetric, T being bounded and self-adjoint. Moreover, the coercivity implies that $\mathcal{B}(\varphi, u)$ is an inner product on H which induces a norm $\sqrt{\mathcal{B}(u, u)}$ equivalent to the original norm since

$$\lambda||u||^2 \leq \mathcal{B}(u, u) \leq ||T||\,||u||\,||u||.$$

By the abstract Dirichlet principle for the functional

$$\mathcal{F}(u) := \frac{1}{2}\mathcal{B}(u, u) - L(u), \qquad u \in H$$

or Riesz's theorem, Theorem 10.43, we find $x \in H$ such that

$$(\varphi|Tx) = \mathcal{B}(\varphi, x) = (\varphi|y) \qquad \forall \varphi \in H,$$

that is, $Tx = y$. Finally, from the coercivity assumption we infer

$$\lambda||x||^2 \leq |(x|Tx)| \leq ||x||\,||Tx||,$$

that is, $||T||^{-1} \leq \frac{1}{\lambda}$. $\qquad\square$

b. Green's operator

Given a bilinear form in a real Hilbert space as above, the *Green operator associated to $\mathcal{B}$* is the operator $\Gamma_{\mathcal{B}} : H^* \to H$ that maps $L \in H^*$ into the unique solution $u_{L,\mathcal{B}} \in H$ of $\mathcal{B}(\varphi, u_{L,\mathcal{B}}) = L(\varphi) \ \forall \varphi \in H$. It is easily seen that $\Gamma_{\mathcal{B}}$ is linear and the estimate $||u_{L,\mathcal{B}}|| \leq \frac{1}{\lambda}||L||$ says that $\Gamma_{\mathcal{B}}$ is continuous. Of course, if Γ is the Riesz operator and $T : H \to H$ is an isomorphism such that $\mathcal{B}(v, u) = (v|Tu)$, then $\Gamma_{\mathcal{B}} = T^{-1} \circ \Gamma$.

10.45 ¶. Under the assumptions of Theorem 10.43, let $K \subset H$ be a closed convex set of a real Hilbert space. Show that
  (i) the functional $\mathcal{F}(u)$ has a unique minimizer $\bar{u} \in K$,
  (ii) $\bar{u}$ is the unique solution $u \in K$ of the *variational inequality*

$$u \in K, \qquad \mathcal{B}(u, u - v) \leq L(v) \qquad \forall\, v \in K.$$

c. Ritz's method

The Dirichlet principle answers the question of the existence and uniqueness of the minimizer of

$$\mathcal{F}(u) := \frac{1}{2}\mathcal{B}(u, u) - L(u)$$

and characterizes such a minimizer as the unique solution of $\mathcal{B}(v, u_L) = L(v) \ \forall v \in H$. But, how can one compute u_L? If H is a *separable* Hilbert space, there is an easy answer. In fact, since $\mathcal{B}(u, v)$ is an inner product, we can find a complete system in H which is orthonormal with respect to $\mathcal{B}$,

$$\mathcal{B}(e_i, e_j) = \delta_{ij},$$

such that every $u \in H$ uniquely writes as $u = \sum_{j=1}^{\infty} \mathcal{B}(u, e_j)e_j$, compare Theorem 10.20. If $\mathcal{B}(\varphi, u) = L(\varphi)$, $\forall \varphi \in H$, then $\mathcal{B}(e_j, u) = L(e_j)$, thus we have the following.

10.46 Theorem (Ritz's method). *Let H be a separable real Hilbert space, $\mathcal{B}$ a symmetric coercive bilinear form, $L \in H^*$ and $\{e_n\}$ a complete orthonormal system with respect to $\mathcal{B}$. Then $L(v) = \mathcal{B}(v, u) \ \forall v \in H$ has the unique solution*

$$u_L = \sum_{j=1}^{\infty} L(e_j)e_j.$$

This, of course, allows us to settle a procedure that, starting from a denumerable dense set of vectors $\{x_n\}$, computes a system of orthonormal vectors with respect to $\mathcal{B}(\ ,\ )$ by the Gram–Schmidt method, and yields the approximations $\sum_{j=1}^{N} L(e_j)\,e_j$ of u_L.

10.47 ¶. With the notation of Theorem 10.46, show that for every integer $N \geq 1$, $u_N := \sum_{j=1}^{N} L(e_j)e_j$ is the solution in Span $\{e_1, e_2, \ldots, e_N\}$ of the system of N-linear equations

$$\mathcal{B}(v, u_N) = L(v), \qquad \forall v \in \text{Span}\left\{e_1, e_2, \ldots, e_N\right\}$$

and the unique minimizer of

$$\frac{1}{2}\mathcal{B}(v, v) - \Re(L(v)), \qquad v \in \text{Span}\left\{e_1, e_2, \ldots, e_N\right\}.$$

10.48 ¶. Show that the following *error estimate for Ritz's method* holds:

$$\frac{\lambda}{2}|u - u_N|^2 \leq \mathcal{F}(u_N) - \inf_H \mathcal{F},$$

where $\mathcal{F}(u) := \frac{1}{2}\mathcal{B}(u, u) - L(u)$. [*Hint*: Compute $\mathcal{F}(\overline{u} + v) - \mathcal{F}(\overline{u})$.]

d. Linear regression

Let H, Y be Hilbert spaces and let $A \in \mathcal{L}(H, Y)$. Given $y \in Y$, we may find the minimum points $u \in H$ of the functional

$$\mathcal{F}(u) := \|Au - y\|_Y^2 \qquad u \in H. \tag{10.14}$$

From the orthogonal projection theorem, we immediately infer the following.

10.49 Proposition. *Let $A \in \mathcal{L}(H, Y)$ be a bounded operator with closed range. Then the functional (10.14) has a minimum point $\overline{u} \in H$ and, if $u \in H$ is another minimizer of (10.14), then $u - \overline{u} \in N(A)$. Moreover, all minimum points are characterized as the points $u \in H$ such that $Au - y \perp R(A)$ i.e., as the solutions of*

$$A^*(Au - y) = 0. \tag{10.15}$$

If $N(A) = \{0\}$, as $N(A) = N(A^*A)$, (10.15) has a unique solution, $\overline{u} = (A^*A)^{-1}A^*y$. If $N(A) \neq \{0\}$, the minimizer is not unique, so it is worth computing the minimizer of least norm, equivalently the only minimizer that belongs to $N(A)^\perp$, or the solution of

$$\begin{cases} A^*(A\overline{u} - y) = 0, \\ \overline{u} \in N(A)^\perp. \end{cases} \tag{10.16}$$

Recall that, being $R(A)$ closed, the map $A_{|N(A)^\perp} \to R(A)$ is an isomorphism by the open mapping theorem. Consequently

$$A^\dagger y := \left(A_{|N(A)^\perp} \right)^{-1} Qy, \qquad y \in Y,$$

where Q is the orthogonal projection onto $R(A)$, defines a bounded linear operator $A^\dagger : Y \to H$ called the *Moore–Penrose inverse* of A. It is trivial to check that the solution $\overline{u}$ of (10.16) is $\overline{u} = A^\dagger y$.

In the simplest case, $N(A^*) = \{0\}$, we have $R(A) = Y$ and (10.15) is equivalent to solving $Au = y$. Since we want to find a solution in $N(A)^\perp$, it is worth solving $AA^*z = y$ so that $\overline{u} = A^\dagger y = A^*(AA^*)^{-1}y$.

In general, however, both AA^* and A^*A are singular and, in order to compute $A^\dagger y$, we resort to an approximation argument. Consider the *penalized functional*

$$\mathcal{F}_\lambda(u) := \|Au - y\|_Y^2 + \lambda \|u\|_H^2 \qquad u \in H, \tag{10.17}$$

where $\lambda > 0$, that we may also write as

$$\mathcal{F}_\lambda(u) = \|y\|^2 - 2(Au|y)_Y + \|Au\|_Y^2 + \lambda \|u\|_H^2.$$

Observing that $L(u) := (Au|y)_Y = (A^*y|u)_H$ belongs to $\mathcal{L}(H, \mathbb{R})$ and that

$$\mathcal{B}(v, u) := (Av|Au)_Y + \lambda(v|u)_H = \lambda(v|u)_H + (v|A^*Au)_H$$

is a symmetric, bounded, coercive, bilinear form on H, it follows from the abstract Dirichlet principle, Theorem 10.43, that $\mathcal{F}_\lambda$ has a unique minimizer $u_\lambda \in H$ given by the unique solution of

$$(\varphi | A^* A u_\lambda)_H + \lambda(\varphi | u_\lambda)_H = (\varphi | A^* y)_H \qquad \forall \varphi \in H,$$

i.e.,

$$(\lambda \operatorname{Id} + A^* A) u_\lambda = A^* y. \tag{10.18}$$

We also get, multiplying both sides of (10.18) by u_λ,

$$\lambda \|u_\lambda\|_H^2 + \|A u_\lambda\|_Y^2 = (y | A u_\lambda)_Y$$

from which we infer the estimate independent on λ

$$\|A u_\lambda\|_Y \leq \|y\|_Y. \tag{10.19}$$

10.50 Proposition. *Let $A \in \mathcal{L}(H, Y)$ be a bounded operator with closed range and for $\lambda > 0$, let*

$$u_\lambda := (\lambda \operatorname{Id} + A^* A)^{-1} A^* y \in H,$$

be the unique minimizer of (10.17). Then $\{u_\lambda\}$ converges to $A^\dagger y$ in H and

$$\left\| \left(\lambda \operatorname{Id} + A^* A \right)^{-1} - A^\dagger \right\| \to 0 \qquad \text{as } \lambda \to 0^+.$$

Proof. Since $R(A)$ is closed by hypothesis, there exists $C > 0$, such that

$$\|v\|_H \leq C \|Av\|_Y \qquad \forall v \in N(A)^\perp. \tag{10.20}$$

Since $\lambda u_\lambda = A^*(y - A u_\lambda) \in R(A^*) \subset N(A)^\perp$, we get in particular from (10.19) and (10.20)

$$\|u_\lambda\|_H \leq C \|A u_\lambda\|_Y \leq C \|y\|_Y,$$

i.e., $\{u_\lambda\}$ *is uniformly bounded in* H.

Let $\lambda, \mu > 0$. From (10.18) we have

$$-(\lambda u_\lambda - \mu u_\mu) = A^* A(u_\lambda - u_\mu)$$

from which we infer

$$\|A(u_\lambda - u_\mu)\|_Y^2 = (u_\lambda - u_\mu | \lambda u_\lambda - \mu u_\mu)_Y \leq \|u_\lambda - u_\mu\|_H \|\lambda u_\lambda - \mu u_\mu\|_H$$

$$\leq \|u_\lambda - u_\mu\|_H \left(|\lambda| \|u_\lambda - u_\mu\|_H + |\lambda - \mu| \|u_\mu\|_H \right)$$

$$\leq \lambda \|u_\lambda - u_\mu\|_H^2 + |\lambda - \mu| \|u_\mu\|_H \|u_\lambda - u_\mu\|_H.$$

Taking into account (10.20) and the boundedness of the u_μ's we then infer

$$\|u_\lambda - u_\mu\|_H^2 \leq C^2 \|A(u_\lambda - u_\mu)\|_Y^2$$

$$\leq C^2 \lambda \|u_\lambda - u_\mu\|_H^2 + C^3 |\lambda - \mu| \|y\|_Y \|u_\lambda - u_\mu\|_H$$

that is,

$$\|u_\lambda - u_\mu\|_H \leq 2 C^3 \|y\|_Y |\lambda - \mu| \tag{10.21}$$

provided $2C^2 \lambda < 1$.

For any $\{\lambda_k\}$, $\lambda_k \to 0^+$, we then infer from (10.21) that $\{u_{\lambda_k}\}$ is a Cauchy sequence in $N(A)^\perp$, hence converges to $\overline{u} \in N(A)^\perp$. Passing to the limit in (10.18), we also get $A^*(A\overline{u} - y) = 0$, since $\{u_\lambda\}$ is bounded, i.e., $\overline{u} := A^\dagger y$, as required. $\qquad \square$

10.3.3 Coercive nonsymmetric bilinear forms

Riesz's theorem extends to nonsymmetric bilinear forms.

a. The Lax–Milgram theorem

As for finite systems of linear equations in a finite number of unknowns, in order to solve $Tx = y$, it is often worth first solving $TT^*x = y$ or $T^*Tx = T^*y$, since TT^* and TT^* are self-adjoint. We proceed in this way to prove the following.

10.51 Theorem (Lax–Milgram). *Let $\mathcal{B}(u, v)$ be a continuous and coercive bilinear form on a Hilbert space H, i.e., there exists $\Lambda \geq \lambda > 0$ such that*

$$|\mathcal{B}(u, v)| \leq \Lambda \, ||u|| \, ||v||, \qquad \mathcal{B}(u, u) \geq \lambda \, ||u||^2 \qquad \forall u, v \in H.$$

Then for all $L \in H^*$ there exists a unique $u_L \in H$ such that

$$\mathcal{B}(v, u_L) = L(v) \qquad \forall v \in H; \tag{10.22}$$

moreover $||u_L|| \leq 1/\lambda ||L||$, i.e., Green's operator associated to $\mathcal{B}$, $\Gamma_{\mathcal{B}}$: $H^* \to H$, $\Gamma_{\mathcal{B}}(L) := u_L$, is continuous.

Proof. Let $T : H \to H$ be the continuous linear operator associated to $\mathcal{B}$ by

$$\mathcal{B}(v, u) = (v|Tu) \qquad \forall u, v \in H.$$

The bilinear form

$$\widetilde{\mathcal{B}}(u, v) := (TT^*u \mid v) = (T^*u \mid T^*v)$$

is trivially continuous and symmetric; it is also coercive, in fact,

$$\lambda^2 ||u||^4 \leq |\mathcal{B}(u, u)|^2 = |(u \mid T^*u)|^2 \leq ||u||^2 ||T^*u||^2 = ||u||^2 \widetilde{\mathcal{B}}(u, u).$$

Riesz's theorem, Theorem 10.43, then yields a unique $\widetilde{u}_L \in H$ such that

$$\widetilde{\mathcal{B}}(v, \widetilde{u}_L) = L(v) \qquad \forall v \in H.$$

Thus $u_L := T^*\widetilde{u}_L$ is a solution for (10.22). Uniqueness follows from the coercivity of $\mathcal{B}$. □

Equivalently we can state the following.

10.52 Theorem. *Let $T : H \to H$ be a continuous and coercive linear operator,*

$$||Tu|| \leq ||T|| \, ||u||, \quad (u|Tu) \geq \lambda ||u||^2 \qquad \forall u \in H$$

where $\Lambda \geq \lambda > 0$. Then T is injective and surjective; moreover its inverse T^{-1} is a linear continuous and coercive operator with $||T^{-1}|| \leq \lambda^{-1}$.

10.53 ¶. Show the equivalence of Theorems 10.51 and 10.52.

10.54 ¶. Read Theorem 10.52 when $H = \mathbb{R}^n$; in particular, interpret coercivity in terms of eigenvalues of the symmetric part of the matrix associated to T.

b. Faedo–Galerkin method

If H is a *separable* Hilbert space, the solution u_L of the linear equation (10.22) can be approximated by a procedure similar to the one of Ritz.

Let H be a separable Hilbert space and let $\{e_n\}$ be a complete orthonormal system in H. For every integer N, we define $V_N := \operatorname{Span}\{e_1, \ldots, e_N\}$ and let $P_N : H \to H$ be the orthogonal projection on V_N and u_N to be the solution of the equation

$$\mathcal{B}(\varphi, u_N) = L(\varphi) \qquad \forall \varphi \in V_N, \tag{10.23}$$

i.e., in coordinates, $u_N := \sum_{i=1}^{N} x^i e_i$ where

$$\sum_{j=1}^{N} \mathcal{B}(e_i, e_j)x^j = L(e_i), \qquad \forall i = 1, \ldots, N.$$

Notice that the system has a unique solution since the matrix $\mathbf{B}$, $\mathbf{B}_{ij} = \mathcal{B}(e_i, e_j)$ has N linearly independent columns as $\mathcal{B}$ is coercive.

10.55 Theorem (Faedo–Galerkin). *The sequence $\{u_N\}$ converges to u_L in H.*

Proof. We have

$$\begin{aligned}
\lambda \|u_N - u_L\|^2 &\leq \mathcal{B}(u_N - u_L, u_N - u_L) \\
&= \mathcal{B}(u_N, u_N) + \mathcal{B}(u_L, u_L) - \mathcal{B}(u_N, u_L) - \mathcal{B}(u_L, u_N) \\
&= \mathcal{B}(u_L, u_L - u_N),
\end{aligned}$$

since $\mathcal{B}(u_N, u_L) = L(u_N) = \mathcal{B}(u_N, u_N)$. It suffices to show that for every $\varphi \in H$

$$\mathcal{B}(\varphi, u_N - u_L) \to 0 \qquad \text{as} \qquad N \to \infty. \tag{10.24}$$

We first observe that the sequence $\{u_N\}$ is bounded in H by $\|L\|/\lambda$ since $\lambda \|u_N\|^2 \leq \mathcal{B}(u_N, u_N) = L(u_N) \leq \|L\| \|u_N\|$. On the other hand, we infer from (10.22) that

$$\mathcal{B}(P_N\varphi, u_N - u_L) = 0 \qquad \forall \varphi \in H, \tag{10.25}$$

hence

$$\begin{aligned}
\mathcal{B}(\varphi, u_N - u_L) &= \mathcal{B}(\varphi - P_N\varphi, u_N - u_L) + \mathcal{B}(P_N\varphi, u_N - u_L) \\
&= \mathcal{B}(\varphi - P_N\varphi, u_N - u_L),
\end{aligned}$$

and

$$|\mathcal{B}(\varphi, u_N - u_L)| \leq \Lambda \|u_N - u_L\| \|\varphi - P_N\varphi\| \leq 2\frac{\Lambda}{\lambda}\|L\| \|\varphi - P_N\varphi\|.$$

Then (10.24) follows since $\|\varphi - P_N\varphi\| \to 0$ as $N \to \infty$. $\qquad \square$

10.4 Linear Compact Operators

In Chapter 4 we presented a rather complete study of linear operators in finite-dimensional spaces. The study of linear operators in infinite-dimensional spaces is more complicated. As we have seen, several important linear operators are not continuous, and moreover, linear continuous operators may have a nonclosed range: we may have $\{x_k\} \subset H$, $\{y_k\} \subset Y$ such that $Tx_k = y_k$, $y_k \to y \in Y$, but the equation $Tx = y$ has no solution.

Here we shall confine ourselves to discussing *compact perturbations of the identity* for which we prove that the range or image is closed. We notice however, that for some applications this is not sufficient, and a *spectral theory* for both bounded and unbounded self-adjoint operators has been developed. But we shall not deal with these topics.

10.4.1 Fredholm–Riesz–Schauder theory

a. Linear compact operators

Let H be a real or complex Hilbert space. Recall, cf. Chapter 9,

10.56 Definition. *A linear operator $K : H \to H$ is said to be compact if and only if K is continuous and maps bounded sets into sets with compact closure. The set of compact operators in H is denoted by $\mathcal{K}(H, H)$.*

Therefore $K : H \to H$ is compact if and only if K is continuous and every bounded sequence $\{u_n\} \subset H$ has a subsequence $\{u_{h_n}\}$ such that $K(u_{h_n})$ converges in H. Also $\mathcal{K}(H, H) \subset \mathcal{L}(H, H)$. Moreover, every linear continuous operator with finite range is a compact operator, in particular every linear operator on H is compact if H has finite dimension. On the other hand, since the identity map on H is not compact if $\dim H = +\infty$, we conclude that $\mathcal{K}(H, H)$ is a proper subset of $\mathcal{L}(H, H)$ if $\dim H = +\infty$.

Exercise 10.89 shows that compact operators need not have finite-dimensional range. However, cf. Theorem 9.140,

10.57 Theorem. $\mathcal{K}(H, H)$ *is the closure of the space of the linear continuous operators of finite-dimensional range.*

Proof. Suppose that the sequence of linear continuous operators with finite-dimensional range $\{A_n\}$ converges to $A \in \mathcal{L}(H, H)$, $||A_n - A|| \to 0$. Then by (i) Theorem 9.140 A is compact. Conversely, suppose that A is compact, and let B be the unit ball of H. Then $A(B)$ has compact closure, hence for all n there is a $1/n$-net covering $A(B)$, i.e., there are points $y_1, y_2, \ldots, y_N \in \overline{A(B)}$, $N = N(n)$, such that $A(B) \subset \cup_{j=1}^{N} B(y_j, 1/n)$. Define

$$V_n := \mathrm{Span}\,\{y_1, y_2, \ldots, y_N\},$$

let $P_n : H \to V_n$ be the orthogonal projection onto V_n and $A_n := P_n \circ A$. Clearly each A_n has finite-dimensional range, thus it suffices to prove that $||A_n - A|| \to 0$.

For all $x \in B$ we find $i \in \{1, 2, \ldots, N\}$ such that $||Ax - y_i|| \leq 1/n$, hence, since $P_n y_i = y_i$ and $||P_n z|| \leq ||z||$,

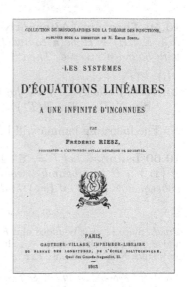

Figure 10.6. Marcel Riesz (1886–1969) and the frontispiece of a volume by Frigyes Riesz (1880–1956).

$$||P_n Ax - Ax|| \leq ||P_n Ax - P_n y_i|| + ||P_n y_i - Ax|| \leq 2||Ax - y_i|| \leq 2/n$$

for all $x \in B$. $\qquad\qquad\qquad\qquad\qquad\qquad\qquad\qquad\qquad\qquad$ $\square$

10.58 Proposition. *Let* $K \in \mathcal{K}(H,H)$. *Then the adjoint* K^* *of* K *is compact and* AK *and* KA *are compact provided* $A \in \mathcal{L}(H,H)$.

Proof. The second part of the claim is trivial. We shall prove the first part. Let $\{u_n\} \subset H$ be a bounded sequence, $||u_n|| \leq M$. Then $\{K^*u_n\}$ is also bounded, hence $\{KK^*u_n\}$ has a bounded subsequence, still denoted by $\{KK^*u_n\}$, that converges. This implies that $\{K^*u_n\}$ is a Cauchy sequence since

$$||K^*u_i - K^*u_j||^2 = (K^*(u_i - u_j)|K^*(u_i - u_j)) = (u_i - u_j|KK^*(u_i - u_j))$$
$$\leq 2M\,||KK^*(u_i - u_j)||.$$

$\qquad\qquad\qquad\qquad\qquad\qquad\qquad\qquad\qquad\qquad\qquad\qquad\qquad\qquad$ $\square$

b. The alternative theorem

Let $A \in \mathcal{L}(H,H)$ be a bounded operator with bounded inverse. A linear operator $T \in \mathcal{L}(H,H)$ of the form $T = A + K$, where $K \in \mathcal{K}(H,H)$, is called a *compact perturbation of* A. Typical examples are the compact perturbations of the identity, $T = \mathrm{Id} + K$, $K \in \mathcal{K}(H,H)$, to which we can always reduce $T = A + K = A(\mathrm{Id} + A^{-1}K)$.

The following theorem, that we already know in finite dimension, holds for compact perturbations of the identity. It is due to Frigyes Riesz (1880–1956) and extends previous results of Ivar Fredholm (1866–1927).

10.59 Theorem (Alternative). *Let H be a Hilbert space and let $T = A + K : H \to H$ be a compact perturbation of an operator $A \in \mathcal{L}(H, H)$ with bounded inverse. Then*

(i) *$R(T)$ is closed,*
(ii) *$N(T)$ and $N(T^*)$ are finite-dimensional linear subspaces; moreover, $\dim N(T) - \dim N(T^*) = 0$.*

The following lemma will be needed in the proof of the theorem.

10.60 Lemma. *Let $T = \mathrm{Id} + K$ be a compact perturbation of the identity. If $\{x_n\} \subset H$ is a bounded sequence such that $Tx_n \to y$, then there exist a subsequence $\{x_{k_n}\}$ of $\{x_n\}$ and $x \in H$ such that*

$$x_{k_n} \to x \qquad \text{and} \qquad Tx = y.$$

Proof. Since $\{x_n\}$ is bounded and K is compact, we find a subsequence $\{x_{k_n}\}$ of $\{x_n\}$ and $z \in H$ such that $Kx_{k_n} \to z$. It follows that $x_{k_n} = Tx_{k_n} - Kx_{k_n} \to y - z =: x$, and, since K is continuous, $Tx_{k_n} \to Tx = x + Kx = x - z = y$.   □

Proof of Theorem 10.59. Since $T = A + K = A(\mathrm{Id} + A^{-1}K)$, A has a bounded inverse, $A^{-1}K$ is compact, we shall assume without loss of generality that $A = \mathrm{Id}$.

Step 1. First we show that there is a constant $C > 0$ such that

$$||x|| \le C \, ||Tx|| \qquad \forall x \in N(T)^\perp. \tag{10.26}$$

Suppose this is not true. Then there exists a sequence $\{x_n\} \subset N(T)^\perp$ such that $||x_n|| = 1$ and $||T(x_n)|| \to 0$. $T(x_n) \to 0$ and Lemma 10.60 yield a subsequence $\{x_{k_n}\}$ of $\{x_n\}$ and $x \in H$ such that $x_{k_n} \to x$ and $Tx = 0$. The first condition yields $x \in N(T)^\perp$ and $||x|| = 1$, while the second $x \in N(T)$. A contradiction.

It follows from (10.26) that T is an isomorphism between the Hilbert space $N(T)^\perp$ and $R(T)$, hence $R(T)$ is complete, thus closed. This proves (i).

Step 2. By Lemma 10.60 every bounded sequence in $N(T)$ has a convergent subsequence. Riesz's theorem, Theorem 9.21, then yields that $\dim N(T) < +\infty$. Similarly, one shows that $\dim N(T^*) < \infty$. The rest of the claim is trivial if K is self-adjoint. Otherwise, we may proceed as follows, also compare 10.62 below.

We use the fact that every compact operator is the limit of operators with finite-dimensional range, Theorem 10.57. First we assume $T = \mathrm{Id} + K$, K of finite-dimensional range. In this case $K : N(K)^\perp \to R(K)$ is an isomorphism, in particular $\dim R(K^*) = \dim N(K)^\perp = \dim R(K)$. Since $T = \mathrm{Id} + K$, we have $N(T) \subset R(K)$ and $N(T^*) \subset R(K)$, hence $N(T)$ and $N(T^*)$ are finite-dimensional, and (ii) is proved. Let $V := R(K) + R(K^*)$. Trivially $N(T), N(T^*) \subset V$ and T and T^* map V into itself. The rank theorem then yields

$$\dim N(T) = \dim N(T^*).$$

This proves the claim (ii) if K has finite-dimensional range.

Step 3. Returning to the case K compact, by the approximation theorem, Theorem 10.57, there is a linear continuous operator K_1 with finite-dimensional range such that $||K - K_1|| < 1$. If $Q := K_1 - K$, the series $\sum_{j=1}^\infty Q^j$ converges in $\mathcal{L}(H, H)$ and

$$(\mathrm{Id} - Q) \sum_{j=1}^\infty Q^j = \mathrm{Id}.$$

In particular, $\mathrm{Id} - Q$ is invertible with bounded inverse $\sum_{j=1}^\infty Q^j$. Therefore we can write

$$T = \mathrm{Id} + K = \mathrm{Id} - Q + K_1 = (\mathrm{Id} - Q)(\mathrm{Id} + (\mathrm{Id} - Q)^{-1}K_1) =: A(\mathrm{Id} + B)$$

where B has finite-dimensional range; the claim (ii) then follows from Step 2.   □

c. Some facts related to the alternative theorem

We collect here a few different proofs of some of the claims of the alternative theorem, since they are of interest by themselves.

10.61 $R(\,\mathrm{Id} + K)$ **is closed.** As we know, this is equivalent to $R(T) = N(T^*)^\perp$, i.e., to show that for every $f \in N(T^*)^\perp$ the equation $Tu := u + K(u) = f$ is solvable. To show this, we can use Riesz's theorem.

Given $f \in N(T^*)^\perp$, we try to solve $TT^*v = f$, i.e.,

$$b(\varphi, v) = (\varphi \mid v) \qquad \forall \varphi \in H, \tag{10.27}$$

where

$$b(\varphi, v) := (TT^*v|\varphi) = (T^*v|T^*\varphi).$$

If $v \in H$ solves (10.27), then $u := T^*v$ solves $Tu = f$.

$$(Tu|\varphi) = (TT^*v|\varphi) = b(v, \varphi) = (f|\varphi), \qquad \forall \varphi \in H.$$

We notice that $N(TT^*) = N(T^*)$, therefore the bilinear bounded form $b(\varphi, v)$ is symmetric if H is real (sesquilinear if H is complex) and well defined on the Hilbert space $N(TT^*)^\perp$. We claim that $b(\varphi, v)$ is coercive on $N(TT^*)^\perp$,

$$b(\varphi, \varphi) \geq c||\varphi||^2 \qquad \forall \varphi \in N(TT^*)^\perp. \tag{10.28}$$

Otherwise, there exists a sequence $\{e_n\} \subset N(T^*)^\perp$ with $||e_n|| = 1$ and

$$b(e_n, e_n) = ||e_n + K^*e_n||^2 \to 0.$$

By Lemma 10.60, there exists $e \in H$ and a subsequence $\{e_{k_n}\}$ of $\{e_n\}$ such that

$$e_{k_n} \to e, \qquad Te = e + Ke = 0;$$

in particular $||e|| = 1$, $e \in N(T^*)$ and $e \in N(T)$, a contradiction. We then conclude that $b(\varphi, u)$ is an inner product on H (a Hermitian product if H is complex), equivalent to the original one.

Applying Riesz's theorem, we then find $v \in N(T^*)^\perp$ such that

$$b(\varphi, v) = (\varphi \mid f) \quad \forall \varphi \in N(T^*)^\perp, \qquad ||v|| \leq \frac{1}{c}||f||. \tag{10.29}$$

It remains to show that v solves (10.27). If P is the orthogonal projection of H into $N(T^*)$, then (10.29) is equivalent to

$$b(P\varphi, v) = (P\varphi|f) \quad \forall \varphi \in H.$$

On the other hand,

$$(\varphi - P\varphi|f) = 0, \qquad b(\varphi - P\varphi, v) = (\varphi - P\varphi|TT^*v) = 0,$$

since f and v are in $N(T^*)^\perp$, hence $b(\varphi, v) = b(P\varphi, v) = (P\varphi|f) = (\varphi|f)$.

10.62 Another proof of $\dim N(T) = \dim N(T^*)$. *Step 1.* Let us prove the equality if T or T^* is injective. Let $H_1 := R(T)$ and, by induction $H_{j+1} := T(H_j)$. H_j is a nonincreasing sequence of closed subspaces of H. We claim that there exists $\overline{n}$ such that $H_{\overline{n}} = H_n \,\forall n \geq \overline{n}$. If not, we can find $\{e_n\} \subset R(H)$ with $||e_n|| = 1$ and $e_n \in H_n \cap H_{n+1}^\perp$. Since for $n > m$, $T(e_n), T(e_m), e_n \in H_{m+1}$, $e_m \in H_{m+1}^\perp$, and

$$Ke_n - Ke_m = (e_n + Ke_n) - (e_m + Ke_m) - e_n + e_m := z + e_m,$$

we may infer

$$||Ke_n - Ke_m||^2 = ||z||^2 + ||e_m||^2 \geq 1:$$

a contradiction, since $\{K(e_n)\}$ has a convergent subsequence.

If $N(T) = \{0\}$ and $H_1 = R(T) \neq H$, then necessarily $H_{j+1} \neq H_j \ \forall j$ since T is injective, and this is not possible, as we have seen. Hence $H = R(T)$ and $N(T^*) = R(T)^\perp = \{0\}$.

If $N(T^*) = \{0\}$, then repeating the above consideration for $\mathrm{Id} + K^*$ we get $N(T) = \{0\}$.

Step 2. Let us prove that $\dim N(T) \geq \dim R(T)^\perp$. Assume that $\dim N(T) < \dim R(T)^\perp$. Then there exists a linear continuous operator L that maps the finite-dimensional space $N(T)$ into the finite-dimensional space $R(T)^\perp$ with L injective but not surjective. Let us extend L as a linear operator from H to $R(T)^\perp$ by setting $Lx = 0$ $\forall x \in N(T)^\perp$. Then L has a finite-dimensional range, thus it is compact. Now we claim that $N(\mathrm{Id} + L + K) = \{0\}$. In fact $u + Ku + Lu = 0$ implies $Tu = u + Ku = -Lu$ and, since $Tu \in R(T)$ and $Lu \in R(T)^\perp$, we infer $Tu = Lu = 0$, i.e., $u \in N(T)$ and $u \in N(T)^\perp$, since L is injective when restricted to $N(T)$; in conclusion $u = 0$. Step 1 then says that $\mathrm{Id} + K + L$ is surjective. This is a contradiction, since the equation $u + Ku + Lu = v$ has no solution when $u \in R(T)^\perp$, $v \notin R(L)$.

Step 3. Replacing K by K^* in the above proves that

$$\dim R(T)^\perp = \dim N(T^*) \geq \dim R(T^*)^\perp = \dim N(T),$$

which completes the proof.

10.63 Yet another proof of $\dim N(T) = \dim N(T^*)$. Let H be a separable Hilbert space, $T = \mathrm{Id} + K$ be a compact perturbation of the identity, and let $\{e_i\}$ be a complete orthonormal system for H, ordered in such a way that $N(T) + N(T^*)$ is generated by the first elements $e_1, e_2, \ldots, e_k$.

Proposition. *Let $V_n = \mathrm{Span}\,\{e_1, e_2, \ldots, e_n\}$, P_n be the orthogonal projection over V_n. Then there exists a constant $\gamma > 0$ and an integer n_0 such that $\forall n \geq n_0$*

$$||P_n T(\varphi)|| \geq \gamma ||\varphi|| \qquad \forall \varphi \in V_n \cap N(T)^\perp.$$

Proof. Suppose the conclusion is not true; then for a sequence $n_i \to \infty$ of vectors $\varphi_i \in V_{n_i} \cap N(T)^\perp$ we have

$$||P_{n_i} T \varphi_i|| \to 0, \qquad ||\varphi_i|| = 1. \tag{10.30}$$

By Lemma 10.60 for a subsequence $\{\varphi_{k_i}\}$ and $\varphi \in H$ we then have $K\varphi_{k_i} \to -\varphi$. Since $P_n x \to x$ as $n \to \infty$, we infer

$$||P_{n_{k_i}} K \varphi_{k_i} + \varphi|| \leq ||P_{n_{k_i}} K \varphi_{k_i}|| + ||K \varphi_{k_i} + \varphi|| \to 0$$

hence $\varphi_{k_i} \to \varphi$ in H, since $\varphi_i = P_{n_i} T \varphi_i - P_{n_i} K(\varphi_i)$, and finally $\varphi + K\varphi = 0$. In particular $||\varphi|| = 1$ and $\varphi \in N(T) \cap N(T)^\perp$, a contradiction. $\qquad \square$

From the previous proposition, if $\{\varphi_1, \varphi_2, \ldots, \varphi_s\}$ is a family of linearly independent vectors, then $P_n T(\varphi_1), \ldots, P_n T(\varphi_s)$ are also linearly independent, at least for n large enough; on the other hand, since $R(T) = N(T^*)^\perp$, the vectors $P_n T(\varphi_1), \ldots, P_n T(\varphi_s)$ belong to $P_n R(T) = V_n \cap N(T^*)^\perp$. Hence we have

$$\dim V_n \cap N(T^*)^\perp \geq \dim V_n \cap N(T)^\perp$$

for n large enough. Similarly one proves $\dim V_n \cap N(T)^\perp \geq \dim V_n \cap N(T^*)^\perp$, hence

$$\dim V_n \cap N(T)^\perp = \dim V_n \cap N(T^*)^\perp$$

for n large enough. The claim then follows by considering the orthogonal complements.

d. The alternative theorem in Banach spaces

The alternative theorem generalizes to the so-called *Fredholm operators* between Banach spaces X and Y of which compact perturbations of the identity are special cases.

Let X be a real Banach space on $\mathbb{K} = \mathbb{R}$ or $\mathbb{K} = \mathbb{C}$ and $X^* := \mathcal{L}(X, \mathbb{K})$ its dual space, which is a Banach space with the *dual norm*

$$||\varphi|| = \sup_{||x||=1} |\varphi(x)|, \qquad \forall \varphi \in X^*.$$

If $\varphi \in X^*$ and $x \in X$, we often write $< \varphi, x >$ for $\varphi(x)$. Clearly, the bilinear map $< \, , \, >: X^* \times X \to \mathbb{K}$, defined by $< \varphi, x >= \varphi(x)$, is continuous,

$$| < \varphi, x > | \le ||\varphi|| \, ||x|| \qquad \forall \varphi \in X^*, \, \forall x \in X.$$

In general, X^* *is not* isomorphic to X, contrary to the case of Hilbert spaces. If X and Y are Banach spaces and if $T : X \to Y$ is a linear bounded operator, the *dual* or *adjoint operator* $T^* : Y^* \to X^*$ is defined by

$$< T^*(\varphi), x > := < \varphi, Tx > . \tag{10.31}$$

T^* is continuous and $||T^*|| = ||T||$.

10.64 ¶. Let $T \in \mathcal{L}(H, H)$, where H is a Hilbert space. We then have two notions of adjoint operators: as the operator $T^* : H \to H$ in (10.9) Chapter 10 and as the operator $\tilde{T} : H^* \to H^*$ defined in (10.31). Show that, if $G : H^* \to H$ is Riesz's operator, then $\tilde{T} = G^{-1} \circ T^* \circ G$.

For a subset $V \subset X$ of a Banach space X, we define

$$V^\perp := \left\{ \varphi \in X^* \, \middle| \, < \varphi, x >= 0 \; \forall x \in V \right\}$$

called the *annihilator* of V. Notice that $V^\perp$ is closed in X^*. We have

10.65 Lemma. *Let $T : X \to X$ be a bounded linear operator. Then $N(T^*) = R(T)^\perp$, $\overline{R(T^*)} = N(T)^\perp$.*

The class of linear compact operators on a Banach space, denoted by $\mathcal{K}(X, X)$ is a closed subset of $\mathcal{L}(X, X)$. But in general *these operators are not limits of linear operators with finite-dimensional range*, contrary to the case $X = H$, where H is a Hilbert space as shown by a famous example due to Lindemann and Strauss. Recall that we can always approximate $K \in \mathcal{K}(X, X)$ by *nonlinear* operators with range contained in a finite-dimensional subspace, see Theorem 9.140. We can now state, but we omit the proof, the following result.

10.66 Theorem (Alternative). *Let X be a Banach space and let $T = A + K : X \to X$ be a compact perturbation of an isomorphism $A : X \to X$. Then*

(i) *$R(T)$ is closed,*

(ii) *$N(T)$ and $N(T^*)$ have finite dimension, and $\dim N(T) = \dim N(T^*)$.*

Consequently, we have the following.

10.67 Corollary (Alternative). *Let $A, K \in \mathcal{L}(X, X)$ where A is a linear isomorphism of X and K is compact. Then the equation $Ax + Kx = y$ is solvable if and only if $y \in N(T^*)^\perp$.*

e. The spectrum of compact operators

10.68 Definition. *Let H be a Hilbert space on $\mathbb{K}$, $\mathbb{K} = \mathbb{R}$ or $\mathbb{K} = \mathbb{C}$, and let $L \in \mathcal{L}(H, H)$ be a bounded linear operator on H. The* resolvent $\rho(L)$ *of the operator L is defined as the set*

$$\rho(L) = \left\{ \lambda \in \mathbb{K} \,\middle|\, (\lambda\,\mathrm{Id} - L)^{-1} \text{ is a bounded operator} \right\}$$

and its complement $\sigma(L) = \mathbb{K} \setminus \rho(L)$ is called the spectrum *of L.*

By the open mapping theorem

$$\sigma(L) := \left\{ \lambda \in \mathbb{K} \,\middle|\, \lambda\,\mathrm{Id} - L \text{ is not injective or surjective} \right\}. \qquad (10.32)$$

10.69 Definition. *Let $L \in \mathcal{L}(H, H)$. Then the* pointwise spectrum *of L is defined as*

$$\sigma_p(L) := \left\{ \lambda \in \mathbb{K} \,\middle|\, \lambda\,\mathrm{Id} - L \text{ is not injective} \right\}.$$

The points in $\sigma_p(L)$ are called eigenvalues *of L, and the elements of $N(\lambda\,\mathrm{Id} - L)$ are called the* eigenvectors *of L corresponding to λ.*

Of course, $\sigma_p(L) \subset \sigma(L)$ and, if $\dim H < +\infty$, $\sigma_p(L) = \sigma(L)$ as, in this case, a linear operator is injective if and only if it is surjective. If $\dim H = +\infty$, there exist, as we know, linear bounded operators which are injective but not surjective, hence, in general $\sigma_p(L) \neq \sigma(L)$.

10.70 Remark. In the sequel we shall deal with compact operators L. For these operators the equality $\sigma_p(L) = \sigma(L)$ also follows from the alternative theorem of the previous section.

As in the finite-dimensional case, see Proposition 4.5, *eigenvectors corresponding to distinct eigenvalues are linearly independent.* Moreover

$$\sigma(L) \subset \left\{ \lambda \in \mathbb{K} \,\middle|\, |\lambda| \leq \|L\| \right\}, \qquad (10.33)$$

because, if $|\lambda| > \|L\|$, then $\left\| \frac{1}{\lambda} L \right\| < 1$, therefore, see Proposition 9.106, $\mathrm{Id} + \frac{1}{\lambda} L$, equivalently $\lambda \mathrm{Id} + L$, is invertible and

$$(\lambda \mathrm{Id} + L)^{-1} = \sum_{j=0}^{\infty} (-1)^j \lambda^{n-j} L^j,$$

hence $\lambda \in \rho(L)$.

The following theorem gives a complete description of the spectrum of a *linear compact operator*.

10.71 Theorem. *Let H be a Hilbert space with $\dim H = +\infty$ and let $K \in \mathcal{K}(H, H)$ be a compact operator. Then*

(i) $0 \in \sigma(K)$,

(ii) *K has either a finite number of eigenvalues or an infinite sequence of eigenvalues that converges to 0.*

(iii) *the eigenspaces corresponding to nonzero eigenvalues have finite dimension,*

(iv) *if $\lambda \neq 0$ and λ is not an eigenvalue for K, then $\lambda Id - K$ is an isomorphism of H and $(\lambda Id - K)^{-1}$ is continuous,*

(v) $\sigma(K) \setminus \{0\} = \sigma_p(K) \setminus \{0\}$.

Proof. (i) In fact $R(K) \neq H$, since K is compact.

(ii) From (10.33) the set of eigenvalues Λ is bounded, thus either Λ is finite or Λ has an accumulation point. Let us prove that in the latter case, Λ has only 0 as an accumulation point; we then conclude that Λ is denumerable, actually a sequence converging to zero.

   Suppose $\{\lambda_n\}$ is a sequence of nonzero eigenvalues with corresponding eigenvectors $\{u_n\}$ such that $\lambda_n \to \lambda \neq 0$. Set $\mu_n := 1/\lambda_n$ and $V_n := \text{Span}\{u_1, u_2, \ldots, u_n\}$, and notice that, if $w := \sum_{j=1}^{n} c_j u_j \in V_n$, then $\lambda_n w - Kw = \sum_{j=1}^{n} c_j(\lambda_n - \lambda_j) u_j \in V_{n-1}$. We now construct a new sequence $\{v_n\}$ with $||v_n|| = 1$ by choosing $v_1 \in V_1$ and, for $n \geq 2$, $v_n \in V_n \cap V_{n-1}^{\perp}$. Clearly v_n is an eigenvector corresponding to λ_n and, according to the previous remark, $v_n - \mu_n K v_n \in V_{n-1}$. For $n > m$ we then find $v_n - \mu_n K v_n$, $\mu_m K v_m \in V_{n-1}$, $v_n \in V_{n-1}^{\perp}$ and

$$K(\mu_n v_n - \mu_m v_m) = v_n - (v_n - \mu_n K v_n + \mu_m K v_m) =: v_n - z,$$

with $v_n \in V_{n-1}^{\perp}$ and $z \in V_{n-1}$. Thus we conclude

$$||K(\mu_n v_n) - K(\mu_m v_m)||^2 = ||v_n||^2 + ||z||^2 \geq 1,$$

a contradiction, since $\{\mu_n u_n\}$ is bounded and K is compact. In conclusion $\lambda = 0$.

(iii), (iv) are part of the claims of the alternative theorem, and (v) follows from (iv). □

10.72 Remark. Actually, Theorem 10.71 holds under the more general assumption that H is a Banach space. In this case it is known as the Riesz–Schauder theorem.

10.4.2 Compact self-adjoint operators

Let us discuss more specifically the spectral properties of linear self-adjoint operators.

a. Self-adjoint operators

10.73 Proposition. *Let H be a real Hilbert space and $L : H \to H$ be a bounded self-adjoint linear operator. Set*

$$m := \inf_{|u|=1} (Lu|u), \qquad M := \sup_{|u|=1} (Lu|u).$$

Then

(i) *eigenvectors corresponding to distinct eigenvalues are orthogonal,*
(ii) $m, M \in \sigma(L)$,
(iii) $||L|| = \sup_{||u||=1} |(u|Lu)| = \max(|m|, |M|)$.

Also, if L is a bounded self-adjoint operator in a complex Hilbert space, then $(u|Lu) \in \mathbb{R} \; \forall u \in H$, consequently all eigenvalues are real, moreover (i), (ii) and (iii) hold.

Proof. (i) In fact $Lu = \lambda u$ and $Lv = \mu v$, $\lambda, \mu \in \mathbb{R}$, $\lambda \neq \mu$ yield

$$(\lambda - \mu)(u \mid v) = (Lu \mid v) - (u \mid Lv) = 0.$$

We now prove that for all $u \in H$

$$||Mu - Lu|| \leq |(Mu - Lu|u)|^{1/2}, \qquad ||mu - Lu|| \leq |(mu - Lu|u)|^{1/2}. \qquad (10.34)$$

The bilinear form $b(u, v) := (Mu - Lu|v)$ is symmetric and nonnegative, $b(u, u) \geq 0$; the Cauchy–Schwarz inequality then yields

$$|(Mu - Lu|v)| \leq |(Mu - Lu|u)|^{1/2}|(Mv - Lv|v)|^{1/2} \leq C|(Mu - Lu|u)|^{1/2}||v||.$$

By choosing $v = Mu - Lu$, the first of (10.34) follows. A similar argument yields the second of (10.34).

(ii) Let us prove that $M \in \sigma(L)$; similarly one proves that $m \in \sigma(L)$. Let $\{u_k\}$ be a sequence such that $||u_k|| = 1$ and $(Lu_k|u_k) \to M$. Because of (10.34) $Mu_k - Lu_k \to 0$ in H. If M is in the resolvent, then $Mu - Lu$ is one-to-one and onto with continuous inverse because of the open mapping theorem. Thus

$$u_k := (M \operatorname{Id} - L)^{-1}(Mu_k - Lu_k) \to 0,$$

that contradicts $||u_k|| = 1$.

(iii) Set $\alpha := \sup_{||u||=1} |(Lu|u)|$; of course $\max(|M|, |m|) = \alpha$ and $\alpha \leq ||L||$. Let us show that $||L|| \leq \alpha$. Since L is self-adjoint

$$4\Re(Lu|v) = (L(u + v)|u + v) - (L(u - v)|u - v),$$

hence, according to the parallelogram law,

$$4|(Lu|v)| \leq \alpha(||u + v||^2 + ||u - v||^2) = 2\alpha(||u||^2 + ||v||^2).$$

Replacing u and v with ϵu, v/ϵ respectively, $\epsilon > 0$, we find

$$4|(Lu|v)| \leq 2\alpha \min_\epsilon (\epsilon^2||u||^2 + \frac{||v||^2}{\epsilon^2}) = 4\alpha||u|| \, ||v||.$$

Hence, if $v := Lu$, we have

$$||Lu||^2 \leq \alpha||u|| \, ||Lu||, \qquad \text{i.e.,} \qquad ||L|| \leq \alpha.$$

In the complex case we have

$$(Lu|u) = (u|L^*u) = (u|Lu) = \overline{(Lu|u)}$$

hence $(Lu|u) \in \mathbb{R}$. We leave to the reader the completion of the proof. $\qquad \square$

We notice that the proof of (iii) Proposition 10.73 uses the continuity of $(M \operatorname{Id} - L)^{-1}$ when $M \in \rho(L)$. If L is compact, this is a consequence of the alternative theorem and the open mapping theorem is not actually needed.

10.74 Corollary. *Let $L : H \to H$ be a linear compact self-adjoint operator. Then there exists an eigenvalue λ of L such that $||L|| = |\lambda|$.*

Proof. If $L = 0$, then $\lambda = 0$ is an eigenvalue. If $L \neq 0$, then $||L|| = \max(|m|, |M|) \neq 0$ and $M, m \in \sigma(L)$. Assuming $||L|| = |M|$, then $M \neq 0$ and, according to Theorem 10.71, $M \in \sigma_p(L)$, i.e., M is an eigenvalue of L.

Alternatively, we can proceed more directly as follows. Let $\{u_n\}$ be a sequence with $||u_n|| = 1$ such that $(Lu_n|u_n) \to M$; then $(Mu_n - Lu_n|u_n) \to 0$, and by (10.34) $Mu_n - Lu_n \to 0$ in H. Since L is compact, there is $\overline{u} \in H$ and a subsequence u_{k_n} of $\{u_n\}$ such that $u_{k_n} \to \overline{u}$ hence

$$M\overline{u} - L\overline{u} = 0, \qquad ||\overline{u}|| = 1,$$

i.e., M is an eigenvalue for L.   □

b. Spectral theorem

10.75 Theorem (Spectral theorem). *Let H be a real or complex Hilbert space and K a linear self-adjoint compact operator. Denote by W the family of finite linear combinations of eigenvectors of K corresponding to nonzero eigenvalues. Then W is dense in $N(K)^\perp$. In particular, $N(K)^\perp$ has an at most denumerable orthonormal basis of eigenvectors of K. If P_j is the orthogonal projection on the eigenspace corresponding to the nonzero eigenvalue λ_j, then*

$$K = \sum_{j=1}^{\infty} \lambda_j P_j \qquad in \ \mathcal{L}(H, H).$$

Proof. We order the nonzero eigenvalues as

$$\lambda_i \neq \lambda_j \quad for \ i \neq j, \qquad |\lambda_1| \geq |\lambda_2| \geq |\lambda_3| \geq \cdots$$

and set $N_j := N(\lambda_j \operatorname{Id} - K)$ for the finite-dimensional eigenspace corresponding to λ_j. According to Proposition 10.73

$$N_j \perp N_k \quad for \ j \neq k \quad and \quad N(K) \perp N_j \ \forall j,$$

hence $N(K) \subset W^\perp$. To prove that W is dense in $N(K)^\perp$, it suffices to show that $\overline{W} = N(K)^\perp$ or $W^\perp = N(K)$.

Define

$$W_n = \begin{cases} \{0\} & \text{if } K \text{ has no nonzero eigenvalues,} \\ \cup_{j=1}^n N_j & \text{if } K \text{ has at least } n \text{ nonzero eigenvalues,} \\ W_p & \text{if } K \text{ has only } p < n \text{ nonzero eigenvalues} \end{cases}$$

and $V_n := W_n^\perp$. Trivially $W^\perp = \cap_n V_n$. Notice that, since K is self-adjoint $K(W_n^\perp) \subset W_n^\perp$ if $K(W_n) \subset W_n$ and the linear operator $K_{|V_n} \in \mathcal{L}(V_n, V_n)$ is again compact and self-adjoint. Moreover, the spectrum of $K_{|V_n}$ is made by the eigenvalues of K different from $\{\lambda_1, \lambda_2, \ldots, \lambda_n\}$. Therefore by Corollary 10.74

$$\left\| K_{|V_n} \right\| = \begin{cases} |\lambda_{n+1}| & \text{if } K \text{ has at least } n+1 \text{ eigenvalues,} \\ 0 & \text{otherwise.} \end{cases} \tag{10.35}$$

If K has a finite number of eigenvalues, then $V = V_{\overline{n}}$ and (10.35) yields $K(V_{\overline{n}}) = \{0\}$, i.e., $V = V_{\overline{n}} \subset N(T)$.

If K has a denumerable set $\{\lambda_n\}$ of eigenvalues, then $|\lambda_n| \to 0$ by Theorem 10.71, hence

$$\left\| K_{|V} \right\| \leq \left\| K_{|V_n} \right\| = |\lambda_{n+1}| \to 0$$

and $K(V) = \{0\}$, i.e., $V \subset N(K)$. Choosing an orthonormal set of eigenvectors in each eigenspace N_j, we can produce an orthonormal system $\{e_n\}$ of eigenvectors of K corresponding to nonzero eigenvalues, i.e., such that

$$(e_i|e_j) = \delta_{ij}, \qquad Ke_j = \lambda_j e_j,$$

that is complete in the closure $\overline{W} = N(K)^\perp$ of W.

Let us prove the last part of the claim. Let P_j and Q_n be the orthogonal projections respectively, on N_j and W_n. Since the eigenspaces are orthogonal, we have $K(Q_n(x)) = \sum_{j=1}^n \lambda_j P_j(x) \; \forall x \in H$, hence

$$Kx - \sum_{j=1}^n \lambda_j P_j(x) = K(x - Q_n(x)),$$

and therefore

$$\left\| Kx - \sum_{j=1}^n \lambda_j P_j(x) \right\| \leq \left\| K_{|V_n} \right\| \|x - Q_n(x)\| \leq |\lambda_{n+1}| \, \|x\|,$$

i.e.,

$$\left\| K - \sum_{j=1}^k \lambda_j P_j \right\| \leq |\lambda_{n+1}|;$$

the conclusion then follows since $|\lambda_n| \to 0$ as $n \to \infty$. $\qquad\qquad\square$

c. Compact normal operators

A linear bounded operator $T \in \mathcal{L}(H, H)$ in a complex Hilbert space H is called *normal* if $T^*T = TT^*$. It is easy to show that if T is normal, then

(i) $N(T) = N(T^*T) = N(TT^*) = N(T^*)$,
(ii) $N(T - \lambda \operatorname{Id}) = N(T^* - \overline{\lambda} \operatorname{Id})$,

that is, T and T^* have the same eigenspaces and conjugate eigenvalues.

If T is normal, the operators

$$A := \frac{T + T^*}{2}, \qquad B := \frac{T - T^*}{2i}$$

are self-adjoint and commute, $AB = BA$.

Two linear compact self-adjoint operators that commute have the same *spectral resolution*, see Theorem 4.29 for the finite-dimensional case.

10.76 Theorem. *Let H be a complex Hilbert space and A, B two linear compact self-adjoint operators in H such that $AB = BA$. Then there exists a denumerable orthonormal system $\{e_n\}$ which is complete in $(N(A) \cap N(B))^\perp$ and made by common eigenvectors of A and B. If λ_j and μ_j are respectively, the eigenvalue of A and the eigenvalue of B relative to e_j, and $P_j : H \to H$ is the orthogonal projection onto $\operatorname{Span}\{e_j\}$, $P_j x := (x|e_j)e_j$, then*

$$A = \sum_{j=1}^\infty \lambda_j P_j, \qquad B = \sum_{j=1}^\infty \mu_j P_j \qquad \text{in } \mathcal{L}(H, H).$$

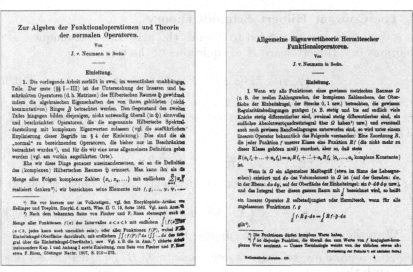

Figure 10.7. Two pages from two papers by John von Neumann (1903–1957) in *Mathematische Annalen.*

Proof. Let W be as in Theorem 10.75. As in the finite-dimensional case, see Proposition 4.27, for every eigenvalue λ of A we find a basis of the corresponding eigenspace $N(\lambda\,\mathrm{Id} - A)$ made by eigenvectors of B. By induction we then find a denumerable orthonormal system which is complete in $\overline{W}$ and made of common eigenvectors $\{e_n\}$ of A and B. By Theorem 10.75 then $\overline{W} = N(A)^\perp$ and $\{e_n\}$ is a basis of $N(A)^\perp$ of common eigenvectors of A and B. Now $AB = BA$ implies that $B(N(A)) \subset N(A)$. Therefore, applying the spectral theorem to $B_{|N(A)}$, we find further eigenvectors $\{u_n\}$ of B corresponding to nonzero eigenvalues that form a basis of $N(A) \cap N(B)^\perp$. The family $\{e_n\} \cup \{u_n\}$ is now a denumerable orthonormal set of eigenvectors common to A and B that is complete in $(N(A) \cap N(B))^\perp$.

The second part of the claim easily follows by applying Theorem 10.75 to A and B. □

10.77 Corollary. *Let H be a complex Hilbert space and let $T : H \to H$ be a compact normal operator. Then there exists a denumerable basis $\{e_n\}$ in H of common eigenvectors of T and T^*. If P_j denotes the orthogonal projection on* $\mathrm{Span}\,\{e_j\}$ *and λ_j is the corresponding eigenvalue, then*

$$T = \sum_{j=1}^{\infty} \lambda_j P_j, \qquad T^* = \sum_{j=1}^{\infty} \overline{\lambda_j} P_j, \qquad in \; \mathcal{L}(H,H).$$

Proof. Set $A := (T + T^*)/2$ and $B := (T - T^*)/(2i)$. We can apply Theorem 10.76 and find a basis $\{e_n\}$ in $(N(A) \cap N(B))^\perp$, i.e., a basis in $\ker(T)^\perp = \ker(T^*)^\perp$ made by common eigenvectors to $T = A + iB$ and $T^* = A - iB$. □

d. The Courant–Hilbert–Schmidt theory

In several instances one is led to discuss the existence and uniqueness of solutions in a Hilbert space H of equations of the type

$$a(\varphi, u) - \lambda k(\varphi, u) = F(\varphi) \qquad \forall \varphi \in H \qquad (10.36)$$

where $F \in H^*$, and $a(\varphi, u)$, $k(\varphi, u)$ are bounded bilinear forms in H. As we have seen, by Riesz's theorem, there exist bounded operators $A, K \in \mathcal{L}(H, H)$ and $f \in H$ such that

$$a(\varphi, u) := (\varphi | Au), \qquad k(\varphi, u) := (\varphi | Ku), \qquad F(\varphi) = (\varphi | f)$$

for all $u, \varphi \in H$. Then (10.36) reads equivalently as the linear equation in H

$$(A - \lambda K)u = f. \qquad (10.37)$$

With the previous notation suppose that

— A is continuous, self-adjoint and coercive on H, i.e., there exists $\nu > 0$ such that
$$a(u, u) \geq \nu \, ||u||^2 \qquad \forall u \in H, \qquad (10.38)$$

— K is compact, self-adjoint and positive, i.e.,

$$k(u, u) = (u | Ku) > 0 \qquad \forall u \neq 0, \ u \in H. \qquad (10.39)$$

With these assumptions, the corresponding bilinear forms are continuous and symmetric; moreover $a(v, u)$ defines an inner product in H equivalent to the original one $(v | u)$ since

$$\nu \, ||u||^2 \leq a(u, u) \leq ||A|| \, ||u||^2.$$

Finally, see Theorem 10.44, A has a continuous inverse. The operator $A - \lambda K$ is therefore a compact perturbation of an isomorphism, and, since A and K are self-adjoint, the alternative theorem yields the following.

10.78 Theorem. *The equation $Au + \lambda Ku = f$ has a solution if and only if f is orthogonal to the solutions of $Au - \lambda Ku = 0$.*

Now we want to study the equation

$$Au - \lambda Ku = 0$$

equivalently,

$$a(\varphi, u) - \lambda k(\varphi, u) = 0 \qquad \forall \varphi \in H,$$

which can be rewritten as

$$\frac{1}{\lambda} u - A^{-1} K u = 0.$$

With the assumptions we have made

○ $A^{-1} K$ is a linear compact operator,

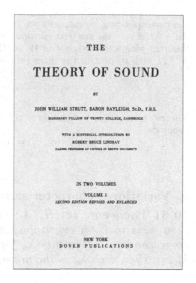

Figure 10.8. Lord William Strutt Rayleigh (1842–1919) and the frontispiece of his *Theory of Sound.*

∘ $A^{-1}K$ is positive, since $a(u, A^{-1}Ku) = (u|AA^{-1}Ku) = (u|Ku) > 0$ for $u \neq 0$,

∘ $A^{-1}K$ is self-adjoint with respect to the inner product $a(v, u)$, since

$$a(v, A^{-1}Ku) = (v|Ku) = (u|Kv) = a(u, A^{-1}Kv).$$

10.79 Definition. *We shall say that $\lambda \neq 0$ is an eigenvalue of (A, K) and that u is a eigenvector of (A, K) corresponding to λ if $1/\lambda$ is an eigenvalue of $A^{-1}K$ and u is a corresponding eigenvector, i.e., a solution of $Au - \lambda Ku = 0$.*

The theory previously developed, when applied to the self-adjoint compact operator $A^{-1}K$ in the Hilbert space H with the inner product $a(v, u)$, yields the following.

10.80 Theorem. *Let H be an infinite-dimensional Hilbert space and let A and $K \in \mathcal{L}(H, H)$ be self-adjoint, for A coercive and K compact. The equation $Au - \lambda Ku = 0$ has zero as its unique solution except for a sequence $\{\lambda_n\}$ of positive real numbers such that $\lambda_n \to +\infty$. For any such λ_n, the vector space of solutions of $Au - \lambda_n Ku = 0$ is finite dimensional. Moreover, if W is the family of finite linear combinations of eigenvectors of (A, K), then W is dense in H. In particular, there exists a complete orthonormal system in H of eigenvectors of (A, K) such that*

$$a(e_i, e_j) = \lambda_j \delta_{ij}, \qquad k(e_i, e_j) = \delta_{ij} \qquad \forall i, j.$$

Proof. The eigenvalues of $A^{-1}K$ are positive since $A^{-1}K$ is positive. Since $A^{-1}K$ is compact, $A^{-1}K$ has a denumerable sequence of eigenvalues $\{\mu_n\}$ and $\mu_n \to 0^+$ and the corresponding eigenspaces are finite dimensional by Theorem 10.71. Consequently $Au - \lambda K u = 0$ has nonzero solutions for the sequence $\lambda_n = 1/\mu_n \to +\infty$. The spectral theorem yields the density of W in H and the existence of an orthonormal basis of $A^{-1}K$ with respect to the inner product $a(v, u)$,

$$a(u_i, u_j) = \delta_{ij}, \qquad \frac{1}{\lambda_j} u_j - A^{-1} K u_j = 0.$$

Therefore, $a(u_i, u_j) = \delta_{ij}$, $\lambda_j k(u_i, u_j) = a(u_i, u_j) = \delta_{ij}$ and, if we set $e_j := \sqrt{\lambda_j} u_j$, we conclude

$$a(e_i, e_j) = \lambda_j \delta_{ij}, \qquad k(e_i, e_j) = \delta_{ij}.$$

$\square$

e. Variational characterization of eigenvalues

10.81 Theorem. *Let H, A and K be as in Theorem 10.80. Let $\{e_n\}$ be a basis in H of eigenvalues of (A, K) ordered in such a way that the corresponding eigenvalues λ_n form a nondecreasing sequence $\{\lambda_n\}$, $\lambda_n \leq \lambda_{n+1}$. Then each λ_n is the minimum of the* Rayleigh's *quotients*

$$\lambda_1 := \min_{u \neq 0} \left\{ \frac{a(u, u)}{k(u, u)} \right\},$$

$$\lambda_n := \min_{u \neq 0} \left\{ \frac{a(u, u)}{k(u, u)} \,\Big|\, k(u, e_i) = 0 \; \forall i = 1, \dots, n-1 \right\}, \quad \text{if } n > 1.$$

Proof. For $u \in H$ write $u = \sum_{j=1}^{\infty} c_j e_j$ so that $k(u, e_j) = c_j$ and $k(u, u) = \sum_{j=1}^{\infty} |c_j|^2$. If $u \in V_n := \{u \,|\, k(u, e_i) = 0, \forall i = 1, \dots, n-1\}$, then $c_i = 0$ for $i = 1, \dots, n-1$, hence

$$k(u, u) = \sum_{j=n}^{\infty} |c_j|^2$$

while

$$a(u, u) = \sum_{j=n}^{\infty} \lambda_j |c_j|^2 \leq \lambda_n \sum_{j=n}^{\infty} |c_j|^2 = \lambda_n \, k(u, u).$$

Therefore $\frac{a(u,u)}{k(u,u)} \leq \lambda_n$ on V_n. On the other hand, $e_n \in V_n$ and $a(e_n, e_n) = \lambda_n \, k(e_n, e_n)$.
$\square$

Moreover with the previous notation and assumptions, we have the following.

10.82 Theorem (Min-max characterization). *Denote by S a generic subspace of dimension $n - 1$. Then we have*

$$\lambda_n = \max_{S} \min \left\{ \frac{a(u, u)}{k(u, u)} \,\Big|\, u \neq 0, \; k(u, z) = 0 \; \forall z \in S \right\}.$$

Proof. The inequality $\leq$ follows from Theorem 10.81. Let S be a linear subspace of H of dimension $n - 1$ and $V_n := \mathrm{Span}\{e_1, e_2, \dots, e_n\}$. Choose a nonzero vector $u_0 := \sum_{i=1}^{n} \alpha_i e_i$ so that

$$k(u_0, z) = 0 \qquad \forall z \in S;$$

this is possible since $\dim S = n - 1$. Then

$$k(u_0, u_0) := \sum_{i=1}^{n} \alpha_i^2,$$

$$a(u_0, u_0) = \sum_{i=1}^{n} \lambda_i a_i^2 \leq \lambda_n \sum_{i=1}^{n} \alpha_i^2 = \lambda_n \, k(u_0, u_0),$$

hence

$$\min\left\{ \frac{a(u, u)}{k(u, u)} \mid k(u, z) = 0 \; \forall z \in S \right\} \leq \frac{a(u_0, u_0)}{k(u_0, u_0)} \leq \lambda_n.$$

□

10.5 Exercises

10.83 ¶. Let $P_V : H \to H$ be the orthogonal projection onto the closed subspace V of a Hilbert space H. Show that
- $P_V + P_W$ is a projection in a closed subspace if and only if $P_V P_W = 0$, and in this case $P_V + P_W = P_{V \oplus W}$,
- $P_V P_W$ is a projection on a closed subspace if and only if $P_V P_W = P_W P_V$, and in this case, $P_V P_W = P_{V \cap W}$.

10.84 ¶. Let $S, T \in \mathcal{L}(H, H)$. Show that $(S+T)^* = S^* + T^*$, $(ST)^* = T^* S^*$, $(\lambda T)^* = \overline{\lambda} T^*$, $\mathrm{Id}^* = \mathrm{Id}$, $T^{**} = T$.

10.85 ¶. Let $T \in \mathcal{L}(H, H)$. Show that $||T||^2 = ||T^*||^2 = ||TT^*|| = ||T^*T||$.

10.86 ¶. Show that *Hilbert's cube* $\{x \in \ell^2 \mid |x_n| \leq 1/n\}$ is compact, while $\{x \in \ell_2 \mid |x_n| \leq 1\}$ is not compact. Show also that Hilbert's cube has no interior points, i.e., its complement is dense.

10.87 ¶. Show the following.

Proposition. *Let $L \in \mathcal{L}(H, H)$ be a bounded self-adjoint operator on a real or complex Hilbert space and*

$$m := \inf_{|u|=1} (Lu|u), \qquad M := \sup_{|u|=1} (Lu|u).$$

Then
(i) *$\sigma_p(L) \subset [m, M]$,*
(ii) *we have $(Lu|u) = M\,||u||^2$ (resp. $(Lu|u) = m\,||u||^2$) if and only if u is an eigenvector of L corresponding to M (respectively m),*

[*Hint:* (i) Proceed by contradiction using Riesz's theorem; in the complex case, first show that $\sigma_p(L) \subset \mathbb{R}$. (ii) Use (10.34).]

10.88 ¶. Show the following.

Proposition. *Let H be a Hilbert space, $\{\lambda_j\}$ a sequence of nonzero real numbers converging to 0, $\{e_j\}$ an orthonormal set in H and $P_j : H \to H$ the orthogonal projection onto $\mathrm{Span}\{e_j\}$. Show that the series $\sum_{j=1}^{\infty} \lambda_j P_j$ converges in $\mathcal{L}(H, H)$. Moreover, if*

$$K := \sum_{j=1}^{\infty} \lambda_j P_j \qquad in \; \mathcal{L}(H, H), \tag{10.40}$$

then

(i) $\forall x \in H$, $Kx = \sum_{j=1}^{\infty} \lambda_j(x|e_j)e_j$,

(ii) *for all j, λ_j is a nonzero eigenvalue of K and $Ke_j = \lambda_j e_j$ for all j,*

(iii) *the sequence $\{\lambda_j\}$ is the set of all nonzero eigenvalues of K,*

(iv) *K is self-adjoint and compact,*

(v) *$\{e_j\}$ is a basis of $\ker K^{\perp}$, in particular $\ker K^{\perp}$ is separable,*

(vi) *if $\lambda \neq 0$ and $\lambda \neq \lambda_j$ $\forall j$, then $(\lambda \mathrm{Id} - K)^{-1}$ is an isomorphism of H into itself.*

If H is a complex Hilbert space and $\lambda_j \in \mathbb{C}$, then (i), (ii), (iii), (iv), (vi) still hold and K is compact and normal.

[*Hint:* (vi) follows from (v) and the alternative theorem. Moreover, an explicit bound for $(\lambda \, \mathrm{Id} - K)^{-1}$ follows from (i), assuming H separable.

Choose an orthonormal basis $\{z_n\}$ in $\ker K$. Then show that the equation $(\lambda \, \mathrm{Id} - K)x = y$ has a unique solution

$$x = (\lambda \, \mathrm{Id} - K)^{-1} = \sum_{k=1}^{\infty} \frac{(y \mid e_k)}{\lambda - \lambda_k} x_k + \sum_{\alpha} \frac{1}{\lambda}(y \mid z_\alpha)z_\alpha.$$

Then

$$|x|^2 = \sum_{k=1}^{\infty} \frac{|(y \mid e_k)|^2}{|\lambda - \lambda_k|^2} + \sum_{\alpha} \frac{|(y \mid z_\alpha)|^2}{\lambda^2} \leq \left(\frac{1}{d^2} + \frac{1}{\lambda^2}\right)|y|^2,$$

if $d := \min_{k \in \mathbb{N}} |\lambda - \lambda_k|$.]

10.89 ¶. Let H be separable and let $\{e_n\}$ be a basis of H. Consider the linear operator $T(e_j) = \frac{1}{j}e_j$, $j \geq 1$, i.e.,

$$T(u) := \sum_{j=1}^{\infty} \frac{1}{j}(u|e_j)e_j.$$

Show that T is compact with a nonclosed range. [*Hint:* Show that T is the limit in $\mathcal{L}(H, H)$ of a sequence of linear operators with finite-dimensional range. Then show that $v \in R(T)$ if and only if $\sum_{j=1}^{\infty} j|(v|e_j)|^2 < +\infty$. Choose $v_0 := \sum_{j=1}^{\infty} j^{-3/2}e_j$, $v_n := \sum_{j=1}^{\infty} j^{-3/2+1/n}e_j$ and show that $v_0 \notin R(T)$, $v_n \in R(T)$ and $|v_n - v_0| \to 0$.]

10.90 ¶. With the notation of Theorem 10.80 show the so-called *completeness relations*

$$k(u, v) = \sum_{i=1}^{\infty} k(u, e_i)k(v, e_i), \qquad a(u, v) = \sum_{i=1}^{\infty} \lambda_i k(u, e_i)k(v, e_i).$$

11. Some Applications

In this chapter we shall illustrate some of the applications of the abstract principles we stated in the previous chapter to specific concrete problems. Our aim is to show the usefulness of the abstract language in formulating and answering specific questions, identifying their specific characteristics and recognizing common features of problems that a priori are very different. Of course, the abstract approach mostly follows and is motivated by concrete questions, but later we see the approach as the most direct way to understand many questions, and even the most natural.

Clearly, the problems we are going to discuss deserve more careful and detailed study because of their relevance, but this is out of our present scope and, in any case, often not possible because of the limited topics we have so far developed. For instance, in this chapter, we shall only use uniform convergence, since the use of integral norms, besides being more complex, requires the notion of Lebesgue's integration, without which any presentation would sound artificial.

11.1 Two Minimum Problems

11.1.1 Minimal geodesics in metric spaces

Let X be a connected metric space so that every two of its points can be connected by a continuous path. One of the simplest problems of calculus of variations is to find a continuous curve of minimal length connecting two given points. A first question is deciding when such a minimal connection exists. Here, we shall see how the Fréchet–Weierstrass theorem, Theorem 6.24, and the Ascoli–Arzelà theorem, Theorem 9.48, lead to an answer.

a. Semicontinuity of the length

Let X be a metric space. Recall that $f : X \to \overline{\mathbb{R}}$ is called *lower semicontinuous* if the level sets $\Gamma_{f,\lambda} := \{x \mid f(x) \leq \lambda\}$ of f are closed for all $\lambda \in \mathbb{R}$, equivalently if $f^{-1}(]\lambda, +\infty[)$ is open for all $\lambda \in \mathbb{R}$. Observing that,

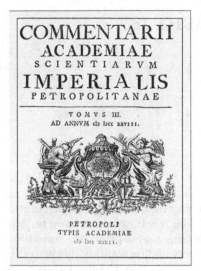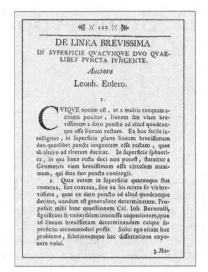

Figure 11.1. Frontispieces of *Commentarii Petropoli* vol. 3 (1732) and of the paper by Leonhard Euler (1707–1783) *De linea brevissima*.

if $f = \sup_i f_i$, then $f^{-1}(]\lambda, +\infty[) = \cup_i f_i^{-1}(]\lambda, +\infty[)$, we conclude the following.

11.1 Proposition. *Let $f_i : X \to \overline{\mathbb{R}}$, $i \in I$, be a family of lower semicontinuous functions on a metric space X. Then $f := \sup_i f_i$ is a lower semicontinuous function.*

Let (X, d) be a metric space. As we have seen, cf. Example 6.25, the *length functional* $L : C^0([a, b], X) \to \overline{\mathbb{R}}$, which for each continuous curve $\varphi : [a, b] \to X$ gives its length, is not a continuous functional with respect to uniform convergence in $C^0([a, b], X)$. But we have the following.

11.2 Theorem (Semicontinuity). *The length functional $L(\varphi)$ is lower semicontinuous in $C^0([a, b], X)$.*

Proof. Recall that we have

$$L(\varphi) = \sup_{S \in \mathcal{S}} V_S(f)$$

where $V_S(f) := \sum_i d_X(f(t_i), f(t_{i+1}))$, $S = \{t_0 = a < t_1 < \cdots < t_N = b\}$. Since the functional $f \to V_S(f)$ is continuous for every fixed subdivision S of $[a, b]$, the result follows.   □

b. Compactness

The intrinsic reparametrization theorem, Theorem 7.44, can be reformulated as: For every family of parametric curves $\{C_i\}$ in X of length strictly less than k, there exists a family of curves $\{C_i'\}$ parametrized in $[0, 1]$, thus belonging in $C^0([0,1], X)$, such that C_i and C'_i are equivalent for all $i \in I$. In particular they have the same length, and the curves C'_i are equi-Lipschitz with Lipschitz constant less than k. Assuming X compact, the Ascoli–Arzelà theorem yields the following.

11.3 Theorem (Compactness). *Let X be a compact metric space and let $\{C_i\}$ be a family of parametrized curves of length strictly less than k. Then the family $\{C_i\}$ is relatively compact with respect to uniform convergence. More precisely, one can reparametrize the curves C_i on $[0,1]$ in such a way that they belong to $\mathrm{Lip}_k([0,1], X)$, and therefore $\{C_i\}$ is a relatively compact subset of $C^0([0,1], X)$.*

c. Existence of minimal geodesics

An immediate consequence of Theorems 11.2 and 11.3, on account of the Frechet–Weierstrass theorem is the following.

11.4 Theorem (Existence). *Let X be an arc-connected compact metric space and P, Q two points of X. There exists a simple rectifiable curve of minimal length joining P to Q, provided there exists at least a rectifiable curve connecting P and Q.*

Proof. Since there exists at least a rectifiable curve connecting P and Q, $\Lambda := \inf\{L(\gamma) \,|\, \gamma \text{ connecting } P \text{ and } Q\} < +\infty$. Let $k > \Lambda$ and let

$$K := \Big\{ \varphi : [0,1] \to X \,\Big|\, \varphi \in \mathrm{Lip}_k([0,1], X),\ \varphi(0) = P,\ \varphi(1) = Q \Big\}.$$

By the Ascoli–Arzelà theorem, K is compact in $C^0([0,1], X)$, hence there is $\varphi_0 \in K$ such that

$$L(\varphi_0) - \inf\Big\{ L(\gamma) \,\Big|\, \gamma \text{ connecting } P \text{ and } Q \Big\} \tag{11.1}$$

by the Fréchet–Weierstrass theorem. The map φ_0 need not be injective a priori. However, the intrinsic reparametrization $\psi : [0, L(\varphi_0)] \to X$, see Theorem 7.44, which is equivalent to φ_0, having Lipschitz constant one, satisfies

$$L(\psi, [x_1, x_2]) = |x_1 - x_2|$$

and is injective. In fact, if $\psi(x_1) = \psi(x_2)$ with $x_1 < x_2$, deleting the loop corresponding to the interval $]x_1, x_2[$, we would still get a curve connecting P and Q, but of length strictly less than $L(\varphi_0)$, contradicting (11.1). □

11.5 ¶. Show that the compactness assumption on X in Theorem 11.4 is necessary. In particular, discuss the cases when X equals the closed unit cube minus an interior open segment and minus a closed interior segment.

11.1.2 A minimum problem in a Hilbert space

In this section we shall show how the theorem ensuring the existence of minimizers for quadratic coercive functionals generalizes to *convex* coercive functionals in a Hilbert space.

a. Weak convergence in Hilbert spaces

Let X be a Banach space. We say that a sequence $\{x_n\} \subset X$ *converges weakly* to $x \in X$, and we write

$$x_n \rightharpoonup x,$$

if $F(x_n) \to F(x) \ \forall F \in X^*$, i.e., for every linear continuous functional $F : X \to \mathbb{R}$ on X.

On account of the Riesz's representation theorem, we have the following.

11.6 Proposition. *A sequence $\{u_n\}$ in a Hilbert space converges weakly to $u \in H$ iff $(u_n|v) \to (u|v) \ \forall v \in H$.*

If H is finite dimensional, weak and strong convergence agree, since weak convergence amounts to the convergence of the components in an orthonormal basis.

On the contrary, if H has infinite dimension, the two notions of convergences differ. In fact, while from the inequality

$$|(u_n - u|v)| \leq ||v|| \, ||u_n - u||$$

we get that strong convergence, $||u_n - u|| \to 0$, implies weak convergence $u_n \rightharpoonup u$; the opposite is not true. Consider, for instance, a denumerable orthonormal set $\{e_n\} \subset H$. Then Bessel inequality yields $(e_n|v) \to 0 \ \forall v \in H$, i.e., $e_n \rightharpoonup 0$, while $\{e_n\}$ does not converge since

$$||e_n - e_m||^2 = ||e_n||^2 - 2(e_n|e_m) + ||e_m||^2 = 2 \qquad \forall n, m.$$

Weak convergence is one of the major tools in modern analysis. Here we only state one of its major useful issues.

11.7 Theorem. *Every bounded sequence in a separable Hilbert space has a subsequence that is weakly convergent.*

Proof. Let $\{x_n\}$, $|x_n| \leq M$, be a bounded sequence in H, and $\{e_i\}$ be a basis of H. $\{x_n\}$ has a subsequence $\{x'_n\}$ such that $(x'_n|e_1) \to \alpha_1$. Similarly $\{x'_n\}$ has a subsequence $\{x''_n\}$ such that $((x''_n|e_2) \to \alpha_2$, and so on. Therefore by a Cantor diagonal procedure, we can find a subsequence $\{x_{k_n}\}$ of $\{x_n\}$ such that $(x_{k_n}|e_i) \to \alpha_i \in \mathbb{R} \ \forall i$, and of course $|a_i| \leq M \ \forall i$. Now consider the map $T : H \to \mathbb{R}$ given by

$$T(y) := \sum_{i=1}^{\infty} (y|e_i)\alpha_i, \qquad y \in H.$$

T is linear and bounded, $||T|| \leq M$ as

$$||T||^2 = \sum_{i=1}^{\infty} |(y|e_i)|^2 |\alpha_i|^2 \leq M^2 ||y||^2,$$

hence the representation theorem of Riesz yields the existence of $x_T \in H$ such that $T(y) = (y|x_T) \ \forall y \in H$ and $||x_T|| = ||T|| \leq M$. In particular $x_T = \sum_{i=1}^{\infty} \alpha_i e_i \in H$. We now prove that $\{x_{k_n}\}$ converges weakly to x_T. For that, set $z_n := x_{k_n} - x_T$ and let y be any vector in H. For any fixed $\epsilon > 0$ choose N sufficiently large so that for $y_N := \sum_{i=1}^{N} (y|e_i)e_i$, we have $||y - y_N|| < \epsilon$. Then

$$|(z_n|y - y_N)| \leq ||z_n|| \, ||y - y_N|| \leq 2M\epsilon;$$

on the other hand $(z_n|e_i) \to 0$, hence $|(z_n|y_N)| < \epsilon$ for n larger than some $\bar{n} = \bar{n}(N, \epsilon)$. Thus

$$|(z_n|y)| \leq (2M + 1)\epsilon \qquad \forall n \geq \bar{n}.$$

$\square$

11.8 Remark. The last part of the proof actually shows that in a separable Hilbert space, weak convergence $(x_n - x|y) \to 0 \; \forall y$ amounts to the convergence of the components $(x_n - x|e_i) \to 0 \; \forall i$ in an orthonormal basis $\{e_i\}$.

11.9 ¶. Show that the compactness theorem, Theorem 10.52, holds in a generic Hilbert space which is not necessarily separable. [*Hint:* Apply Theorem 10.52 to the closure H_0 of the family of finite combinations of $\{x_n\}$, which is a separable Hilbert space. Then find $\bar{x} \in H_0$ and a subsequence $\{x_{k_n}\}$ such that $(x_{k_n} - x|y) \to 0 \; \forall y \in H_0$. Then, use the orthogonal projection theorem onto H_0 to show that actually, $(x_{k_n} - x|y) \to 0 \; \forall y \in H$.]

11.10 Theorem (Banach–Saks). *Every bounded sequence $\{v_n\} \subset H$ weakly convergent to $v \in H$ has a subsequence $\{v_{k_n}\}$ such that*

$$\frac{1}{n} \sum_{i=1}^{n} v_{k_i} \to v \qquad \text{in the norm of } H.$$

Proof. Set $u_n := v_n - v$. Then for a positive M we have $||u_n|| \leq M$ for all n and we extract from $\{u_n\}$ a subsequence $\{u_{k_n}\}$ in such a way that

$$u_{k_1} := u_1,$$
$$(u_{k_2}|u_{k_1}) < 1,$$
$$(u_{k_3}|u_{k_1}), (u_{k_3}|u_{k_2}) < \frac{1}{2},$$
$$\vdots$$
$$(u_{k_{p+1}}|u_{k_i}) < \frac{1}{p} \qquad \forall i = 1, \ldots, p.$$

Therefore

$$\left|\left| \frac{1}{n}\left(\sum_{i=1}^{n} v_{k_i}\right) - v \right|\right|^2 = \frac{1}{n^2} \left|\left| \sum_{i=1}^{n}(v_{k_i} - v) \right|\right|^2 = \frac{1}{n^2} \sum_{i,j=1}^{n}(u_{k_i}|u_{k_j})$$

$$= \frac{2}{n^2} \sum_{j=1}^{n}\sum_{i<j}(u_{k_i}|u_{k_j}) + \frac{1}{n^2}\sum_{j=1}^{n}(u_{k_j}|u_{k_j}) \leq \frac{2}{n} + \frac{M^2}{n}.$$

$\square$

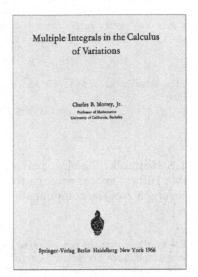

Figure 11.2. Frontispieces of two classical monographs that, in particular, deal with semicontinuity on integral functionals.

b. Existence of minimizers of convex coercive functionals

Let $\mathcal{F} : H \to \mathbb{R}$ be a *convex* functional on a real Hilbert space H. This means that the function

$$\varphi(\lambda) := \mathcal{F}(\lambda u + (1 - \lambda)v)$$

is convex in $[0,1]$ for all $u, v \in H$. A typical example of a convex functional is the quadratic functional

$$\mathcal{F}(u) = \frac{1}{2}||u||^2 - L(u)$$

where L is a bounded linear form on H, that we have encountered in dealing with the abstract Dirichlet principle.

Then we have, compare Proposition 5.61 of [GM1], the following.

11.11 Proposition (Jensen's inequality). *A functional $\mathcal{F} : H \to \mathbb{R}$ is convex if and only if for every finite convex combination*

$$\sum_{i=1}^{m} \alpha_i u_i, \qquad \sum_{i=1}^{m} \alpha_i = 1, \quad \alpha_i \geq 0,$$

of points $u_i \in H$ we have

$$\mathcal{F}\left(\sum_{i=1}^{m} \alpha_i u_i \right) \leq \sum_{i=1}^{m} \alpha_i \mathcal{F}(u_i).$$

Proof. Clearly Jensen's inequality with two points amounts to convexity. So it suffices to prove it assuming $\mathcal{F}$ convex. We give a direct proof by induction on m. Assume the claim holds for $m - 1$ points. Set $\alpha := \alpha_1 + \cdots + \alpha_{m-1}$ and $\alpha_m := 1 - \alpha$. If $\alpha = 0$ or $\alpha = 1$ the claim is proved, otherwise $0 < \alpha < 1$ and, if

$$u := \sum_1^{m-1} \frac{\alpha_i}{\alpha} u_i,$$

then

$$0 \leq \frac{\alpha_i}{\alpha} \leq 1, \qquad \sum_{i=1}^{m-1} \frac{\alpha_i}{\alpha} = 1, \qquad \sum_{i=1}^m \alpha_i u_i = \alpha u + (1 - \alpha) u_m,$$

hence

$$\mathcal{F}\Big(\sum_{i=1}^m \alpha_i u_i\Big) = \mathcal{F}(\alpha u + (1 - \alpha u_m)) \leq \alpha \mathcal{F}(u) + (1 - \alpha)\mathcal{F}(u_m) \leq \sum_{i=1}^m \alpha_i \mathcal{F}(u_i)$$

by the inductive assumption.   □

11.12 Theorem. *Let $\mathcal{F} : H \to \mathbb{R}$ be a continuous, convex, bounded from below and coercive functional, meaning*

$$\inf_{u \in H} \mathcal{F}(u) > -\infty, \qquad \mathcal{F}(u) \to +\infty \qquad as \ |u| \to +\infty.$$

Then $\mathcal{F}$ has a minimizer in H.

Proof. Let $\{u_n\}$ be a minimizing sequence, $\mathcal{F}(u_n) \to \inf_{u \in H} \mathcal{F}(u)$. Since for large n

$$-\infty < \inf_{u \in H} \mathcal{F}(u) \leq \mathcal{F}(u_n) \leq \inf_{u \in H} \mathcal{F}(u) + 1,$$

the sequence $\{u_n\}$ is bounded. Using the Banach–Saks theorem we find $u \in H$, and we can extract a subsequence $\{u_{k_n}\}$ of $\{u_n\}$ such that $u_{k_n} \rightharpoonup u$ and

$$v_n := \frac{1}{n}\Big(\sum_{i=1}^n u_{k_i}\Big) \to u \qquad \text{in the norm of } H.$$

Jensen's inequality yields

$$\mathcal{F}(v_n) = \mathcal{F}\Big(\frac{1}{n}\sum_{i=1}^n u_{k_i}\Big) \leq \frac{1}{n}\sum_{i=1}^n \mathcal{F}(u_{k_i}),$$

thus $\mathcal{F}(v_n) \to \inf_H \mathcal{F}$ since $\mathcal{F}(u_{k_i}) \to \inf_H \mathcal{F}$ as $i \to \infty$, i.e., $\{v_n\}$ is a minimizing sequence, too. Finally

$$\inf_H \mathcal{F} \leq \mathcal{F}(u) = \lim_{n \to \infty} \mathcal{F}(v_n) = \mathcal{F}(u)$$

on account of the continuity of $\mathcal{F}$.   □

11.13 ¶. Show that Theorem 11.12 still holds if $\mathcal{F}$ is convex with values in $\overline{\mathbb{R}}$, bounded from below and lower semicontinuous.

11.2 A Theorem by Gelfand and Kolmogorov

In this section we shall prove that a topological space X is identified by the space of continuous functions on it. If we think of X as a geometric *world* and of a map from X into $\mathbb{R}$ as an *observable* of X, we can say: if we know enough observables, say the continuous observables, then we know our world.

Let us begin by proving the following.

11.14 Proposition. *Every metric space (X, d) can be isometrically embedded in $C^0(X)$.*

Proof. Fix $p \in X$ and consider the map $\varphi : X \to C^0(X, \mathbb{R})$ that maps $a \in X$ into $f_a : X \to \mathbb{R}$ defined by

$$f_a(x) := d(x, a) - d(x, p).$$

Trivially, $f_a \in C^0(X, \mathbb{R})$ and

$$|f_a(x) - f_b(x)| = |d(x, a) - d(x, b)| \leq d(a, b),$$

i.e., $||f_a - f_b||_\infty \leq d(a, b)$; on the other hand for $x = b$ we have $|f_a(b) - f_b(b)| = d(a, b)$, hence φ is an isometry. $\square$

11.15 ¶. Show that every separable metric space (X, d) can be isometrically embedded in l_∞. [*Hint:* Let $\{x_n\}$ be a sequence in X and let $\varphi : X \to l_\infty$ be given by $\varphi(x)_n := d(x, x_n) - d(x_1, x_n)$. Show that φ is an isometry.]

Let X be a topological space, see Chapter 5. The set $C^0(X, \mathbb{R})$ is a linear space and actually a *commutative algebra with identity*, since the product of two continuous functions is continuous. Let R and R' be two commutative algebras. A map $\varphi : R \to R'$ is said to be a *homomorphism* from R into R' if $\varphi(a+b) = \varphi(a)+\varphi(b)$ and $\varphi(ab) = \varphi(a)\varphi(b)$. If, moreover, φ is bijective we say that R and R' are *isomorphic*.

Clearly $C^0(X, \mathbb{R})$ is completely determined by X, in the sense that every topological isomorphism $\varphi : X \to Y$ determines an isomorphism of the commutative algebras $C^0(Y, \mathbb{R})$ and $C^0(X, \mathbb{R})$, the isomorphism being the composition product $f \to f \circ \varphi$. If X is compact, the converse also holds.

11.16 Theorem (Gelfand–Kolmogorov). *Let X be a compact topological space. Then $C^0(X, \mathbb{R})$ determines X.*

We confine ourselves to outlining the proof of Theorem 11.16. An *ideal* $\mathcal{I}$ of the algebra R is a subset of R such that $a, b \in \mathcal{I} \Rightarrow a - b \in \mathcal{I}$ and $a \in \mathcal{I}, r \in R \Rightarrow a \cdot r \in \mathcal{I}$. R is clearly the unique ideal that contains the identity of R. An ideal is called *proper* if $\mathcal{I} \neq R$ and *maximal* if it is not strictly contained in any proper ideal. Finally, we notice that $R/\mathcal{I}$ is a field if and only if $\mathcal{I}$ is maximal.

11.17 Lemma. *Let X be a compact topological space. $\mathcal{I}$ is a proper maximal ideal of $C^0(X)$ if and only if there is $x_0 \in X$ such that $\mathcal{I} = \{f \in C^0(X) \mid f(x_0) = 0\}$.*

Proof. For any $f \in \mathcal{I}$, the set $f^{-1}(0)$ is closed and $f^{-1}(0) \neq 0$. Otherwise $1/f$, hence 1, belongs to $\mathcal{I}$, and $\mathcal{I}$ is not proper. Let $f_1, \ldots, f_n \in \mathcal{I}$. The function $f := \sum_{i=1}^{n} f_i^2$ is in $\mathcal{I}$ and $f^{-1}(0) = \cap f_i^{-1}(0) \neq \emptyset$. Since X is compact, $\cap \{f^{-1}(0) \,|\, f \in \mathcal{I}\} \neq \emptyset$. In particular, there is $x_0 \in X$ such that $f(x_0) = 0 \;\forall f \in \mathcal{I}$. On the other hand, $\{f \,|\, f(x_0) = 0\}$ is an ideal, hence $\mathcal{I} = \{f \,|\, f(x_0) = 0\}$. $\square$

The *spectrum* of a commutative algebra with unity is then defined by

$$\operatorname{spec} R := \Big\{ \mathcal{I} \,\Big|\, \mathcal{I} \text{ maximal ideal of } R \Big\}.$$

Trivially, if R is isomorphic to $C^0(X, \mathbb{R})$, $R \sim C^0(X, \mathbb{R})$, then also the maximal ideals of R and $C^0(X, \mathbb{R})$ are in one-to-one correspondence, hence by Lemma 11.17

$$\operatorname{spec} R \sim \operatorname{spec} C^0(X) \sim X.$$

To conclude the proof of Theorem 11.16, we need to introduce a topology on the space $\operatorname{spec} C^0(X, \mathbb{R})$ in such a way that $\operatorname{spec} C^0(X, \mathbb{R}) \sim X$ becomes a topological isomorphism. For that, we notice that, if $\mathcal{I}$ is a maximal ideal of $C^0(X, \mathbb{R})$, then $C^0(X, \mathbb{R})/\mathcal{I} \simeq \mathbb{R}$, hence the so-called *evaluation maps* $f(\mathcal{I})$, that map $(f, \mathcal{I})$ into $[f] \in C^0(X, \mathbb{R})/\mathcal{I} \simeq \mathbb{R}$, have sign. Now, if we fix the topology on $\operatorname{spec} R \simeq \operatorname{spec} C^0(X, \mathbb{R})$ by choosing as a basis of neighborhoods the finite intersections of

$$U(f) := \Big\{ \mathcal{I} \,\Big|\, \mathcal{I} \text{ proper maximal ideal with } f(\mathcal{I}) > 0 \Big\},$$

it is not difficult to show that the isomorphism $X \to \operatorname{spec} C^0(X, \mathbb{R})$ is continuous. Since X is compact and the points in $\operatorname{spec} C^0(X, \mathbb{R})$ are separated by open neighborhoods, it follows that the isomorphism is actually a topological isomorphism.

Theorem 11.16 has a stronger formulation that we shall not deal with, but that we want to state. A Banach space with a product which makes it an algebra such that $||xy|| \leq ||x|| \, ||y||$ is called a *Banach algebra*. An *involution* on a Banach algebra R is an operation $x \to x^*$ such that $(x + y)^* = x^* + y^*$, $(\lambda x)^* = \overline{\lambda} x^*$, $(xy)^* = (yx)^*$ and $(x^*)^* = x$. A Banach algebra with an involution is called a C^*-algebra. Examples of C^*-algebras are:

(i) the space of complex-valued continuous functions from a topological space with involution $f \to \overline{f}$,

(ii) the space of linear bounded operators on a Hilbert space with the involution given by $A \to A^*$, A^* being the adjoint of A.

Again, the space of proper maximal ideals of a commutative C^*-algebra, endowed with a suitable topology, is called the *spectrum* of the algebra.

Theorem (Gelfand–Naimark). *A C^*-algebra is isometrically isomorphic to the algebra of complex-valued continuous functions on its spectrum.*

11.3 Ordinary Differential Equations

The Banach fixed point theorem in suitable spaces of continuous functions plays a key role in the study of existence, uniqueness and continuous dependence from the data of solutions of ordinary differential equations.

11.3.1 The Cauchy problem

Let D be an open set in $\mathbb{R} \times \mathbb{R}^n$, $n \geq 1$, and $F(t,y) : D \subset \mathbb{R} \times \mathbb{R}^n \to \mathbb{R}^n$ be a continuous function. A *solution* of the *system of ordinary equations*

$$\frac{d}{dt}x(t) = F(t, x(t)) \tag{11.2}$$

is the data of an interval $]\alpha, \beta[\subset \mathbb{R}$ and a function $x \in C^1(]\alpha, \beta[; \mathbb{R}^n)$ such that (11.2) holds for all $t \in]\alpha, \beta[$. In particular, $(t, x(t))$ should belong to D for all $t \in]\alpha, \beta[$. Geometrically, if we interpret $F(t, x)$ as a vector field in D, then $x(t)$ is a solution of (11.2) if and only if its graph curve $t \to (t, x(t))$ is of class C^1, has trajectory in D, and velocity equals to $(1, F(t, x(t)))$ for all t. For this reason, solutions of (11.2) are called *integral curves of the system*.

a. Velocities of class $C^k(D)$

In the sequel, at times we need a fact that comes from the differential calculus for functions of several variables that we are not discussing in this volume. Let $\Omega \subset \mathbb{R}^n$ be an open set. We say that a function $f : \Omega \to \mathbb{R}$ is of class $C^k(\Omega)$, $k \geq 1$, if f possesses continuous partial derivatives up to order k. One can prove that, if $f \in C^k(\Omega)$ and $\gamma : [a, b] \to \Omega$ is a C^k curve in Ω, then $f \circ \gamma : [a, b] \to \mathbb{R}$ is of class $C^k([a, b])$. For $k = 1$ we have the *chain rule*

$$\frac{d}{dt}f(\gamma(t)) = \sum_{i=1}^{n} \frac{\partial f}{\partial x^i}(\gamma(t))\frac{d\gamma^i}{dt}(t) =: \mathbf{D}f(\gamma(t))\gamma'(t).$$

where $\mathbf{D}f(x) := \left(\frac{\partial f}{\partial x^1}(x), \frac{\partial f}{\partial x^2}(x), \ldots, \frac{\partial f}{\partial x^n}(x)\right)$ is the matrix of partial derivatives of f and the product $\mathbf{D}f(\gamma(t))\gamma'(t)$ is the standard matrix product.

A trivial consequence is that integral curves, when they exist, possess one derivative more than the function velocity $F(t, x(t))$. This is true by definition if F is merely continuous. If, moreover, $F(t, x) \in C^k$ and $x(t) \in C^1$, we successively find from the equation $x'(t) = F(t, x(t))$ that $x'(t) \in C^1$, $x'(t) \in C^2, \ldots$, $x'(t) \in C^k$. In particular, if $F(t, x)$ has continuous partial derivatives of any order, then the integral curves are C^∞. It is worth noticing that if $F \in C^1(D)$, then by the chain rule

$$x''(t) = \frac{\partial F}{\partial t}(t, x(t)) + \mathbf{D}F_x(t, x(t))x'(t),$$

where $\mathbf{D}F_x$ is the matrix of partial derivatives with respect to the x's variables and the product $\mathbf{D}F_x(t, x(t))x'(t)$ is understood as the matrix product.

For the sequel, it is convenient to set

11.18 Definition. *We say that a function $F(t, x) : [\alpha, \beta] \times B(x_0, b) \to \mathbb{R}^n$ is Lipschitz in x uniformly with respect to t if there exists $L > 0$ such that*

$$|F(t, x) - F(t, y)| \le L |x - y| \quad \forall (t, x), (t, y) \in [\alpha, \beta] \times B(x_0, b). \quad (11.3)$$

Let D be an open set in $\mathbb{R} \times \mathbb{R}^n$. We say that a function $F(t, x) : D \to \mathbb{R}^n$ is locally Lipschitz in x uniformly with respect to t if for any $\widetilde{D} := [\alpha, \beta] \times B(x_0, b)$ strictly contained in D there exists $L := L(\alpha, \beta, x_0, b)$ such that

$$|F(t, x) - F(t, y)| \le L |x - y| \quad \forall (t, x), (t, y) \in \widetilde{D}.$$

11.19 ¶. Show that the function $f(t, x) = sgn(t)|x|$, $(t, x) \in [-1, 1] \times [-1, 1]$ is Lipschitz in x uniformly with respect to t.

11.20 ¶. Let $D = [a, b] \times [c, d]$ be a closed rectangle in $\mathbb{R} \times \mathbb{R}$. Show that, if for all $t \in [a, b]$, the function $x \to f(t, x)$ has derivative $f_x(t, x)$ on $[c, d]$ and $(t, x) \to f_x(t, x)$ is continuous in D, then f is Lipschitz in x uniformly with respect to t. [*Hint:* Use the mean value theorem.]

11.21 ¶. Show the following. Let D be an open set of $\mathbb{R} \times \mathbb{R}^n$ and let $F(t, x) \in C^1(D)$. Then F is locally Lipschitz in x uniformly with respect to t. [*Hint:* For any $(t_0, x_0) \in D$, choose $a, b \in \mathbb{R}$ such that $\widetilde{D} := \{(t, x) \,|\, |t - t_0| < a, \ |x - x_0| < b\}$ is strictly contained in D. Then, for (t, x_1), $(t, x_2) \in \widehat{D}$, consider the curve $\gamma(s) := (t, (1 - s)x_1 + sx_2)$, $s \in [0, 1]$ whose image is in $\widetilde{D}$ and apply the mean value theorem to $F(\gamma(s))$, $s \in [0, 1]$.]

b. Local existence and uniqueness

Assume $(t_0, x_0) \in D$. We seek a *local solution*

$$x(t) := (x_1(t), \ldots, x_n(t)) \in C^1([t_0 - r, t_0 + r], \mathbb{R}^n)$$

for some $r > 0$ of the *Cauchy problem* relative to the system (11.2), i.e.,

$$\begin{cases} \dfrac{d}{dt}x(t) = F(t, x(t)), \\ x(t_0) = x_0. \end{cases} \quad (11.4)$$

We have the following.

11.22 Proposition. *Let D be an open set in $\mathbb{R} \times \mathbb{R}^n$, $n \ge 1$, and let $F(t, x) : D \to \mathbb{R}^n$ be a continuous function. Then $x(t) \in C^1([t_0 - r, t_0 + r], \mathbb{R}^n)$ solves (11.4) if and only if $x(t)$ belongs to $C^0([t_0 - r, t_0 + r], \mathbb{R}^n)$ and satisfies the integral equation*

$$x(t) = x_0 + \int_{t_0}^{t} F(\tau, x(\tau)) \, d\tau \quad \forall t \in [t_0 - r, t_0 + r]. \quad (11.5)$$

Proof. Set $I := [t_0 - r, t_0 + r]$. If $x \in C^1(I, \mathbb{R}^N)$ solves (11.4), then by integration x satisfies (11.5). Conversely, if $x \in C^0(I, \mathbb{R}^n)$ and satisfies (11.5), then, by the fundamental theorem of calculus, $x(t)$ is differentiable and $x'(t) = F(t, x(t))$ in I, in particular it has a continuous derivative. Moreover, (11.5) for $t = t_0$ yields $x(t_0) = x_0$. $\qquad \square$

Let us start with a local existence and uniqueness result.

11.23 Theorem (Picard–Lindelöf). *Let* $F(t, x) : D \subset \mathbb{R} \times \mathbb{R}^n \to \mathbb{R}^n$ *be a continuous function with domain* $D := \{(t, x) \in \mathbb{R} \times \mathbb{R}^n \,|\, |t - t_0| < a, \; |x - x_0| < b\}$. *Suppose*

(i) $F(t, x)$ *is bounded in* D, $|F(t, x)| \leq M$,
(ii) $F(t, x)$ *is Lipschitz in* x *uniformly with respect to* t,

$$|F(t, x) - F(t, y)| \leq k||x - y|| \qquad \forall (t, x), \; (\theta, y) \in D.$$

Then the Cauchy problem (11.4) has a unique solution in $[t_0 - r, t_0 + r]$ *where*

$$r < \min\left(a, \frac{b}{M}, \frac{1}{k}\right).$$

Proof. Let r be as in the claim and $I_r := [t_0 - r, t_0 + r]$. According to Proposition 11.22, we have to prove that the equation

$$x(t) = x_0 + \int_{t_0}^t F(\tau, x(\tau)) \, d\tau.$$

has a unique solution $x(t) \in C^0(I_r, \mathbb{R}^N)$.

Let $y_1, y_2 \in C^0(I_r, \mathbb{R}^n)$ be two solutions of (11.5). Then for all $t \in I_r$

$$|y_1(t) - y_2(t)| \leq \int_{t_0}^t |F(s, y_1(s)) - F(s, y_2(s))| \, ds \leq k|t - t_0|||y_1 - y_2||_{\infty, I_r}$$

hence

$$||y_1 - y_1||_{\infty, I_r} \leq kr||y_1 - y_1||_{\infty, I_r}.$$

Since $kr < 1$, then $y_1 = y_2$ in I_r.

To show existence, we show that the map $x \to Tx$ given by

$$T[x](t) := x_0 + \int_{t_0}^t F(\tau, x(\tau)) \, d\tau$$

is a contraction on

$$X := \left\{ x \in C^0(I_r, \mathbb{R}^N) \,\Big|\, x(t_0) = x_0, \; |x(t) - x_0| \leq b \; \forall t \in I_r \right\}$$

that is closed in $C^0(I_r, \mathbb{R}^N)$, hence a complete metric space. Clearly $t \to T[x](t)$ is a continuous function in I_r, $T[x](t_0) = x_0$ and

$$|T[x](t) - x_0| \leq \int_{t_0}^t |F(\tau, x(\tau))| \, d\tau \leq M\,|t| \leq Mr \leq b,$$

therefore T maps X into itself. Moreover, it is a contraction; in fact

$$|T[x](t) - T[y](t)| \leq \left| \int_0^t |F(\tau, x(\tau)) - F(\tau, y(\tau))| \, d\tau \right|$$

$$\leq k \left| \int_0^t |x(\tau) - y(\tau)| \, d\tau \right| \leq k|t|||x - y||_\infty \leq kr \, ||x - y||_{\infty, I_r}.$$

The fixed point theorem of Banach, Theorem 9.128, yields a (actually, a unique) fixed point $T[x] = x$ in X. In other words, the equation (11.5) has a unique solution. $\qquad \square$

Taking into account the proof of the fixed point theorem we see that the solution $x(t)$ of (11.4) is the uniform limit of *Picard's successive approximations*

$$x_0(t) := x_0, \qquad \text{and, for } n \geq 1, \quad x_n(t) := x_0 + \int_{t_0}^{t} F(\tau, x_{n-1}(\tau)) \, d\tau.$$

The Picard–Lindelöf theorem allows us to discuss the uniqueness for the initial value problem (11.4).

11.24 Theorem (Uniqueness). *Let $D \subset \mathbb{R} \times \mathbb{R}^n$ be a bounded domain, let $F(t,x) : D \to \mathbb{R}^n$ be a continuous function that is also locally Lipspchitz in x uniformly in t, and let $(t_0, x_0) \in D$. Then any two solutions $x_1 : I \to \mathbb{R}^n$, $x_2 : J \to \mathbb{R}^n$ defined respectively, on open intervals I and J containing t_0 of the inital value problem*

$$\begin{cases} x'(t) = F(t, x(t)), \\ x(t_0) = x_0, \end{cases}$$

are equal on $I \cap J$.

Proof. It is enough to assume $I \subset J$. Define

$$E := \Big\{ t \in I \,\Big|\, x_1(t) = x_2(t) \Big\}.$$

Obviously $t_0 \in E$ and E is closed relatively to I, as x_1, x_2 are continuous. We now prove that E is open in I, concluding $E = I$ since I is an interval, compare Chapter 5.

Let $t^* \in E$, define $x^* := x_1(t^*) = x_2(t^*)$. Let $a, b \in \mathbb{R}_+$ be such that $\tilde{D} := \{(t, x) \in D \,|\, |t - t^*| < a, \ |x - x^*| < b\}$ is strictly contained in D. F being bounded and locally Lipschitz in x uniformly with respect to t in $\tilde{D}$, the Picard–Lindelöf theorem applies on $\tilde{D}$. Since $x_1(t)$ and $x_2(t)$ both solve the initial value problem starting at (t^*, x^*), we conclude that $x_1(t) = x_2(t)$ on a small interval around x^*. Thus E is open. $\qquad\square$

c. Continuation of solutions

We have seen that the initial value problem has a solution that exists on a possibly small interval. Does a solution in a larger interval exist?

As we have seen, given two solutions $x_1 : I \to \mathbb{R}^n$, $x_2 : J \to \mathbb{R}^n$ of the same initial value problem, one can glue them together to form a new function $x : I \cup J \to \mathbb{R}^n$, that is again a solution of the same initial value problem but defined on a possibly larger interval. We say that x is a *continuation* of both x_1 and x_2. Therefore, Theorem 11.24 allows us to define the *maximal solution*, or simply the *solution* as *the* solution defined on the largest possible interval.

11.25 Lemma. *Suppose that $F : D \subset \mathbb{R} \times \mathbb{R}^n \to \mathbb{R}^n$ is continuous in D, and let $x(t)$ be a solution of the initial value problem*

$$\begin{cases} x'(t) = F(t, x(t)), \\ x(t_0) = x_0 \end{cases}$$

in the bounded interval $\gamma < t < \delta$; in particular $(t, x(t)) \in D \ \forall t \in]\gamma, \delta[$. If F is bounded near $(\delta, x(\delta))$, then $x(t)$ can be continuously extended on δ. Moreover, if $(\delta, x(\delta)) \in D$, then the extension is C^1 up to δ. A similar result holds also at $(\gamma, x(\gamma))$.

Proof. Suppose that $|F(t, x)| \le M \ \forall(t, x)$ and let $x(t)$, $t \in]\gamma, \delta[$, be a solution. For $t_1, t_2 \in]\gamma, \delta[$ we have

$$|x(t_2) - x(t_1)| \le \int_{t_1}^{t_2} |F(t, x(t))| \, dt \le M|t_1 - t_2|,$$

i.e., x is Lipschitz on $]\gamma, \delta[$, therefore it can be continuusly extended to $[\gamma, \delta]$. The second part of the claim follows from (11.5) to get for $t < \delta$

$$\frac{x(t) - x(\delta)}{t - \delta} = \frac{1}{t - \delta} \int_d^t F(s, x(s)) \, ds$$

and letting $t \to \delta^+$.   □

Now if, for instance, $(\delta, x(\delta))$ is not on the boundary of D and we can solve the initial value problem with initial datum $x(\delta)$ at $t_0 = \delta$, we can continue the solution in the C^1 sense because of Proposition 11.22, beyond the time δ, thus concluding the following.

11.26 Theorem (Continuation of solutions). *Let $F(t, x)$ be continuous in an open set $D \subset \mathbb{R} \times \mathbb{R}^n$ and locally Lipschitz in x uniformly with respect to t. Then the unique (maximal) solution of $x'(t) = F(t, x(t))$ with $x(t_0) = x_0$ extends forwards and backwards till the closure of its graph eventually meets the boundary of D. More precisely, any (maximal) solution $x(t)$ is defined on an interval $]\alpha, \beta[$ with the following property: for any given compact set $K \subset \Delta$, there is $\delta = \delta(K) > 0$ such that $(t, x(t)) \notin K$ for $t \notin [\alpha + \delta, \beta - \delta]$.*

Recalling Exercise 11.21, we get the following.

11.27 Corollary. *Let D be an open domain in $\mathbb{R} \times \mathbb{R}^n$, and let $F \in C^1(D)$. Then every (maximal) solution of $x'(t) = F(t, x(t))$ can be extended forwards and backwards till the closure of its graph eventually reaches ∂D.*

11.28 Corollary. *Let $D :=]a, b[\times \mathbb{R}^n$ (a and b may be respectively, $+\infty$ and $-\infty$) and let $F(t, x) : D \to \mathbb{R}^n$ be continuous and locally Lipschitz in x uniformly with respect to t. Then every locally bounded (maximal) solution of $x' = F(t, x)$ is defined on the entire interval $]a, b[$.*

Proof. Let $|x(t)| \le M$. Should the maximal solution be defined on $[\alpha, \beta]$ with, say, $\beta < b$, then the graph of x would be contained in the compact set $[\alpha, \beta] \times \overline{B(0, M)}$ strictly contained in $]a, b[\times \mathbb{R}^n$. This contradicts Theorem 11.26.   □

Of course, if F is bounded in $D :=]a, b[\times \mathbb{R}^n$, all solutions of $x' = F(t, x)$ are automatically locally bounded since their velocities are bounded, so the previous theorem applies. For a weaker condition and stronger result, see Exercise 11.33.

11.29 Example. Consider the initial value problem $x' = x^2$, $x(0) = 1$, in $D_a := \{(t, x) \mid t \in \mathbb{R}, |x| < a\}$. Since $|F| \le a^2$ in D_a, the continuation theorem applies. In fact, the maximal solution $1/(1 - t)$, $t \in] - \infty, 1 - \frac{1}{a}[$ has a graph that extends backwards till $-\infty$ and forward until it touches ∂D_a.

d. Systems of higher order equations

We notice that a differential equation of order n in normal form in the scalar unkown $x(t)$

$$\frac{d^n}{dt^n} x(t) = F\left(t, x(t), \frac{d}{dt} x(t), \ldots, \frac{d^{n-1}}{dt^{n-1}} x(t)\right) \qquad (11.6)$$

can be written, by defining

$$x_1(t) := x(t), \quad x_2(t) := \frac{d}{dt} x_1(t), \quad \ldots, \quad x_n(t) := \frac{d^{n-1}}{dt^{n-1}} x(t),$$

as the first order system

$$\begin{cases} x_1'(t) =: x_2(t), \\ x_2'(t) =: x_3(t), \\ \vdots \\ x_{n-1}'(t) =: x_n(t), \\ x_n'(t) =: F(t, x_1(t), x_2(t), \ldots, x_n(t)) \end{cases}$$

or, compactly as,

$$y'(t) = F(t, y(t))$$

for the vector-valued unknown $y(t) := (x_1(t), x_2(t), \ldots, x_n(t))$ and $F : D \subset \mathbb{R} \times \mathbb{R}^n \to \mathbb{R}^n$ given by

$$F(t, x_1, \ldots, x_n) := (x_2, x_3, \ldots, x_n, f(t, x_1(t), x_2(t), \ldots, x_n(t))).$$

Consequently, the Cauchy problem for (11.6) is

$$\begin{cases} x^{(n)}(t) = F(t, x(t), x'(t), x''(t), \ldots, x^{(n-1)}(t)), \\ x(t_0) = x_0, \\ x'(t_0) = x_1, \\ x''(t_0) = x_2, \\ \vdots \\ x^{(n-1)}(t_0) = x_{n-1}. \end{cases} \qquad (11.7)$$

Along the same line, the initial value problem for a system of higher order equations can be reformulated as a Cauchy problem for a system of first order equations, to which we can apply the theory just developed.

e. Linear systems

For linear systems

$$x'(t) = A(t)x(t) + g(t), \tag{11.8}$$

where $A(t)$ is an $n \times n$ matrix and $g(t) \in \mathbb{R}^n$, we have the following.

11.30 Theorem. *Suppose that $A(t)$ and $g(t)$ are continuous in $[a, b]$ and that $t_0 \in [a, b]$ and $x_0 \in \mathbb{R}^n$. Then the solution of (11.8) with initial value $x(t_0) = x_0$ exists on the entire interval.*

Proof. Assume for simplicity that $t_0 \in]a, b[$. The field $F(t, x) := A(t)x + g(t)$ is continuous in $D :=]a, b[\times \mathbb{R}^n$ and locally Lipschitz in x uniformly with respect to t,

$$|F(t, x) - F(t, y)| \leq \sup_{t \in [\alpha, \beta]} ||A(t)|| \, ||x - y|| \qquad \forall a < \alpha < \beta < b, \forall x, y \in \mathbb{R}^n.$$

Therefore, a solution of (11.8) exists in a small interval of time around t_0, according to Picard–Lindelöf theorem. To show that the solution can be continued on the whole interval $]a, b[$, it suffices to show, according to Corollary 11.28, that $x(t)$ is bounded. In fact, we have

$$x(t) - x(t_0) = \int_{t_0}^{t} A(s)x(s) \, ds + \int_{t_0}^{t} g(s) \, ds.$$

For $t > t_0$ we then conclude that

$$|x(t)| \leq |x_0| + \max_{[a,b]} |g|(b - a) + \sup_{t \in [a,b]} ||A(t)|| \int_{t_0}^{t} |x(s)| \, ds,$$

and the boundedness follows from *Grönwall's inequality* below. □

11.31 Proposition (Grönwall's inequality). *Suppose that k is a nonnegative constant and that f and g are two nonnegative continuous functions in $[\alpha, \beta]$ such that*

$$f(t) \leq k + \int_{\alpha}^{t} f(s)g(s) \, ds, \qquad t \in [\alpha, \beta].$$

Then

$$f(t) \leq k \exp\left(\int_{\alpha}^{t} g(s) \, ds \right).$$

Proof. Set $U(t) := k + \int_{\alpha}^{t} f(s)g(s) \, ds$. Then we have

$$f(t) \leq U(t), \qquad U'(t) = f(t)g(t) \leq g(t)U(t), \qquad U(\alpha) = k,$$

in particular

Figure 11.3. Thomas Grönwall (1877–1932) and a page from one of his papers.

$$\frac{d}{dt}\left[U(t)\exp\left(-\int_\alpha^t g(s)\,ds\right)\right] \leq 0,$$

hence

$$U(t)\exp\left(-\int_\alpha^t g(s)\,ds\right) - U(\alpha) \leq 0.$$

$\square$

11.32 ¶. Let $w : [a,b] \to \mathbb{R}^n$ be of class $C^1([a,b])$. Assume that

$$|w'|(t) \leq a(t)\,|w(t)| + b(t) \qquad \forall t \in [a,b]$$

where $a(t), b(t)$ are nonnegative functions of class $C^0([a,b])$. Show that

$$|w(t)| \leq \left(|w(t_0)| + \int_{t_0}^t b(s)\,ds \right)\exp\left(\int_{t_0}^t a(s)\,ds\right)$$

for every $t, t_0 \in [a,b]$. [*Hint:* Apply Grönwall's lemma to $f(t) := |w(t)|$.]

11.33 ¶. Let $F(t,x) : I \times \mathbb{R}^n \to \mathbb{R}^n$ be continuous and locally Lipschitz in x uniformly with respect to t. Suppose that there exist nonnegative continuous functions $a(t)$ and $b(t)$ such that $|F(t,x)| \leq a(t)|x| + b(t)$. Show that all the solutions of $x' = F(t,x)$ can be extended to the entire interval I.

f. A direct approach to Cauchy problem for linear systems

For the reader's convenience we shall give here a more direct approach to the uniqueness and existence of the solution of the initial value problem

$$\begin{cases} t_0 \in [a,b], \\ X(t_0) = X_0, \\ X'(t) = \mathbf{A}(t)X(t) + F(t) \qquad \forall t \in [a,b] \end{cases} \tag{11.9}$$

where $X_0 \in \mathbb{R}^n$ and the functions $t \to \mathbf{A}(t)$ and $t \to F(t)$ are given continuous functions defined in $[a, b]$ with values respectively, in $M_{n,n}(\mathbb{C})$ and $\mathbb{C}^n$. Recall that $||\mathbf{A}(t)|| := \sup_{|x|=1} |\mathbf{A}(t)x|$ denotes the norm of the matrix $\mathbf{A}(t)$ and set

$$M := \sup_{t \in [a,b]} ||\mathbf{A}(t)||.$$

As we have seen, see Proposition 11.22, $X(t)$, $t \in [a, b]$ solves (11.9) if and only if $t \to X(t)$ is of class $C^0([a, b])$ and solves the integral equation

$$X(t) = X_0 + \int_{t_0}^t (\mathbf{A}(s)X(s) + F(s))\, ds \qquad (11.10)$$

that is, iff $X(t)$ is a fixed point for the map

$$T : X(t) \mapsto T(X)(t) := X_0 + \int_{t_0}^t (\mathbf{A}(s)X(s) + F(s))\, ds. \qquad (11.11)$$

Let $\gamma > 0$. The function on $C^0([a, b], \mathbb{R}^n)$ defined by

$$||X||_\gamma := \sup_{t \in [a,b]} \left(|X(t)| e^{-\gamma|t-t_0|} \right)$$

is trivially a norm on $C^0([a, b])$. Moreover, it is equivalent to the uniform norm on $C^0([a, b], \mathbb{R}^n)$ since

$$e^{-\gamma|b-a|}||X||_{\infty,[a,b]} \le ||X||_\gamma \le ||X||_{\infty,[a,b]}.$$

Hence the space $C^0([a, b], \mathbb{R}^n)$ endowed with the norm $|| \ ||_\gamma$, that we denote by C_γ, is a Banach space.

11.34 Proposition. *Let T be the map in (11.11). Then $T(C_\gamma) \subset C_\gamma$ $\forall \gamma \ge 0$. Moreover, T is a contraction map on C_γ if*

$$\gamma > M := \sup_{t \in [a,b]} ||\mathbf{A}(t)||.$$

Proof. In fact, $\forall X, Y \in C_\gamma$ and $t \in [a, b]$, we have

$$|TX(t) - TY(t)| = \left| \int_{t_0}^t \mathbf{A}(s)(X(s) - Y(s))\, ds \right|$$

$$= \left| \int_{t_0}^t \mathbf{A}(s)(X(s) - Y(s)) e^{-\gamma|s-t_0|} e^{\gamma|s-t_0|}\, ds \right|$$

$$\le \int_{t_0}^t ||\mathbf{A}(s)|| \left(|X(s) - Y(s)| e^{-\gamma|s-t_0|} \right) e^{\gamma|s-t_0|}\, ds$$

$$\le M ||X - Y||_\gamma \int_{t_0}^t e^{\gamma|s-t_0|}\, ds \le \frac{M}{\gamma} ||X - Y||_\gamma e^{\gamma|t-t_0|}.$$

Multiplying the last inequality by $e^{-\gamma|t-t_0|}$ and taking the sup norm gives

$$||TX - TY||_\gamma \le \frac{M}{\gamma} ||X - Y||_\gamma.$$

$\square$

11.35 Theorem. *The initial value problem* (11.9) *has a unique solution* $X(t)$ *of class* $C^1([a, b])$, *and*

$$|X(t)| \le \Big(|X_0| + \int_{t_0}^t |F(s)|\, ds\Big) \exp\Big(\int_{t_0}^t \|A(s)\|\, ds\Big).$$

Moreover, $X(t)$ *is the uniform limit in* $C^0([a, b], \mathbb{R}^n)$ *of the sequence* $\{X_n(t)\}$ *of functions defined inductively by*

$$\begin{cases} X_0(t) := X_0, \\ X_{n+1}(t) := X_0 + \int_{t_0}^t (A(s)X_n(s) + F(s))\, ds. \end{cases} \tag{11.12}$$

Proof. Choose $\gamma > M$. Then $T : C_\gamma \to C_\gamma$ is a contraction map. Therefore, by the Banach fixed point theorem T has a unique fixed point. Going into its proof, we get the approximations. Finally, the estimate on $|X(t)|$ follows from (11.10) and the Grönwall Lemma.   □

11.36 Remark. In the special case $a = -\infty$, $b = +\infty$, $t_0 = 0$, $F(t) = 0$ $\forall t$ and $A(t) = A$ constant, then (11.12) reduces to

$$X_n(t) = \Big(\sum_{k=0}^n \frac{A^k}{k!} t^k\Big) X_0$$

hence *the solution of the initial value problem for the homogeneous linear system with constant coefficients*

$$\begin{cases} X'(t) = AX(t), \\ X(0) = X_0 \end{cases}$$

is

$$X(t) = \Big(\sum_{n=0}^\infty \frac{A^k}{k!} t^k\Big) X_0 = \exp(t\, A) X_0 \qquad \forall t \in \mathbb{R}$$

uniformly on bounded sets of $\mathbb{R}$ *and*

$$|X(t)| \le |X_0| \exp(|t - t_0|\, \|A\|) \qquad \forall t \in \mathbb{R}.$$

g. Continuous dependence on data

We now show that the local solution $x(t; t_0, x_0)$ of the Cauchy problem

$$\begin{cases} x' = F(t, x), \\ x(t_0) = x_0 \end{cases}$$

depends continuously on the initial point (t_0, x_0), and in fact is continuous in (t, t_0, x_0).

11.37 Theorem. *Let $F(t,x)$ and $F_x(t,x)$ be bounded and continuous in a region D. Also suppose that in D we have*

$$|F(t,x)| \le M, \qquad |F_x(t,x)| \le k.$$

Then, for any $\epsilon > 0$ there exists $\delta > 0$ such that

$$|x(t;t_0,x_0) - x(\widehat{t};\widehat{t}_0,\widehat{x}_0)| < \epsilon$$

provided $|t - \widehat{t}| < \delta$ and $|x_0 - \widehat{x}_0| < \delta$ and $t, \widehat{t}$ are in a common interval of existence.

Proof. Set $\phi(t) := x(t;t_0,x_0)$, $\psi(t) := x(t;\widehat{t}_0,\widehat{x}_0)$. From

$$\phi(t) = x_0 + \int_{t_0}^{t} F(s,\phi(s))\,ds, \qquad \psi(t) = \widehat{x}_0 + \int_{\widehat{t}_0}^{t} F(s,\psi(s))\,ds,$$

$$\int_{t_0}^{t} F(s,\phi(s))\,ds = \int_{t_0}^{\widehat{t}_0} F(s,\phi(s))\,ds + \int_{\widehat{t}_0}^{t} F(s,\phi(s))\,ds$$

we infer

$$\phi(t) - \psi(t) = x_0 - \widehat{x}_0 + \int_{t_0}^{t} [F(s,\phi(s)) - F(s,\psi(s))]\,ds + \int_{t_0}^{\widehat{t}_0} F(s,\phi(s))\,ds,$$

hence

$$|\phi(t) - \psi(t)| \le |x - \widehat{x}_0| + k\left| \int_{\widehat{t}_0}^{t} |\phi(s) - \psi(s)|\,ds \right| + M|\widehat{t}_0 - t_0|$$

$$\le \delta + k\left| \int_{\widehat{t}_0}^{t} |\phi(s) - \psi(s)|\,ds \right| + M\delta.$$

Grönwall's inequality then yields

$$|\phi(t) - \psi(t)| \le \delta(1 + M)\exp\left(k|t - \widehat{t}_0|\right) \le \delta(1 + M)\exp\left(k(\beta - \alpha)\right).$$

Since

$$|\psi(t) - \psi(\widehat{t})| \le \left| \int_{\widehat{t}}^{t} |F(s,\psi(s))|\,ds \right| \le M|t - \widehat{t}| \le M\delta$$

we conclude

$$|\phi(t) - \psi(\widehat{t})| \le |\phi(t) - \psi(t)| + |\psi(t) - \psi(\widehat{t})|$$
$$\le \delta(1 + M)\exp\left(k(\beta - \alpha)\right) + \delta M$$

if $|t - \widehat{t}| < \delta$. $\qquad\qquad\square$

11.38 ¶. Let $F(t,x)$ and $G(t,x)$ be as in Theorem 11.37, and let $\phi(t)$ and $\psi(t)$ be respectively, solutions of the Cauchy problems

$$\begin{cases} x' = F(t,x), \\ x(t_0) = x_0 \end{cases} \qquad \text{and} \qquad \begin{cases} x' = G(t,x), \\ x(\widehat{t}_0) = \widehat{x}_0. \end{cases}$$

Show that

$$|\phi(t) - \psi(t)| \le (|x_0 - \widehat{x}_0| + \epsilon(\beta - \alpha))\exp\left(k(t - t_0)\right)$$

if $|F(t,x) - G(t,x)| < \epsilon$.

h. The Peano theorem

We shall now prove existence for the Cauchy problem (11.4) assuming only continuity on the velocity field $F(t, x)$. As we know, in this case we cannot have uniqueness, see Example 6.16 of [GM1].

11.39 Theorem (Peano). *Let $F(t, x)$ be a bounded continuous function in a domain D, and let (t_0, x_0) be a point in D. Then there exists at least one solution of*

$$\begin{cases} x' = F(t, x), \\ x(t_0) = x_0. \end{cases}$$

Proof. Let $|F(t, x)| \leq M$ and $\widetilde{D} := \{(t, x) \in \mathbb{R} \times \mathbb{R}^n \mid |t - t_0| < a, \ |x - x_0| < b\}$ be strictly contained in D. If $r < \min\{a, b/M\}$ we have seen that

$$T[x](t) := \int_{t_0}^{t} F(\tau, x(\tau)) \, d\tau$$

maps the closed and convex set

$$X := \Big\{ x \in C^0([x_0 - r, x_0 + r], \mathbb{R}^n) \,\Big|\, x(t_0) = x_0, \ |x - x_0| \leq b \Big\}$$

in itself, see Theorem 11.23. The operator T is continuous; in fact, since F is uniformly continuous in $\widetilde{D}$, $\forall \epsilon > 0 \ \exists \eta$ such that

$$|F(t, x) - F(t, x')| < \epsilon \quad \forall t \in [a, b] \qquad \text{if } |x - x'| < \eta,$$

hence

$$|F(t, x_n(t)) - F(t, x_\infty(t))| < \epsilon \quad \forall t \in [a, b]$$

for large enough n if $x_n(t) \to x_\infty(t)$ uniformly. Then we have

$$||T[x_n] - T[x_\infty]||_\infty \leq \left| \int_{t_0}^{t} |F(t, x_n(t)) - F(t, x_\infty(t))| \, dt \right| < \epsilon \, (b - a).$$

Moreover

$$|T[x](t') - T[x](t)| = \left| \int_{t'}^{t} F(\tau, x(\tau)) \, d\tau \right| \leq M|t - t'|,$$

and we conclude by the Ascoli–Arzelà theorem that $T : X \to X$ is compact. The Caccioppoli–Schauder theorem yields the existence of at least one fixed point $x(t)$, $x(t) = T[x](t)$; this concludes the proof. $\qquad\square$

Notice that the solutions can be continued, cf. Lemma 11.25, possibly in a nonunique way. Therefore any solution can be continued as a solution forwards and backwards in time till the closure of the graph of the extension eventually meets the boundary of the domain D.

11.40 ¶ Comparison principle. Let $f : [a, b] \times \mathbb{R} \to \mathbb{R}$ be a function that is Lipschitz on each rectangle $[a, b] \times [-A, A]$ and let $\alpha(t), \beta(t)$ be two functions such that

$$\alpha(t) \leq \beta(t), \quad \alpha'(t) \leq f(t, \alpha(t)), \quad \beta'(t) \geq f(t, \beta(t)) \quad \forall t \in [a, b].$$

Show that every solution of

$$\begin{cases} x'(t) = f(t, x(t)), \\ x(0) = x_0, \end{cases} \qquad \alpha(a) \le x_0 \le \beta(a),$$

satisfies $\alpha(t) \le x(t) \le \beta(t)$ $\forall t \in [a, b]$. In particular, there is a solution that is defined on the entire interval.

11.41 ¶ Peano's phenomenon. Consider the Cauchy problem

$$x'(t) = f(t, x(t)), \quad x(t_0) = x_0 \quad \text{in } [a, b], \tag{11.13}$$

where $f(t, x)$ is a continuous function. Show that
  (i) there exist a *minimal* and a *maximal solution*, i.e., $\underline{x}(t)$ and $\overline{x}(t)$ solutions of (11.13) such that for any other solution of (11.13) we have $\underline{x}(t) \le x(t) \le \overline{x}(t)$,
  (ii) if the minimal and the maximal solutions of (11.13) exist in $[t_0, t_0 + \delta]$, show that through every point $(\tilde{t}_0, \tilde{x}_0)$ with $\tilde{t} \in [t_0, t_0 + \delta]$ and $\tilde{x} \in [\underline{x}(\tilde{t}), \overline{x}(\tilde{t})]$ there passes a solution of (11.13).
[*Hint:* To show existence of a maximal solution, show that, if $x_n(t)$ solves $x' = f(t, x) + \frac{1}{n}$, then, possibly passing to a subsequence, $\{x_n\}$ converges to a maximal solution.]

11.42 ¶. Study the following Cauchy problem passing to polar coordinates (ρ, θ)

$$\begin{cases} x_1'(t) = x_1(t) - \dfrac{x_2(t)\sqrt{|x_2(t)|}}{\sqrt{x_1^2(t) + x_2^2(t)}}, \\ x_2'(t) = x_2(t) - \dfrac{x_1(t)\sqrt{|x_1(t)|}}{\sqrt{x_1^2(t) + x_2^2(t)}}, \\ x_1(0) = 1, \qquad x_2(0) = 0. \end{cases}$$

11.3.2 Boundary value problems

For second order equations it is useful to consider, besides the initial value problem, so-called *boundary value problems* in which the values of u or u', or a combination of these values, are prescribed at the boundary of the interval. For instance, suppose we want to find the linear motion of a particle under the external force $F(t, x(t), x'(t))$ starting at time $t = 0$ in x_0 and ending at time $t = 1$ in x_1, i.e., we want to solve the *Dirichlet problem*,

$$\begin{cases} x''(t) = F(t, x(t), x'(t)) \quad \text{in }]0, 1[, \\ x(0) = x_0, \\ x(1) = x_1. \end{cases}$$

11.43 ¶. Check that the problem

$$\begin{cases} x'' + x = 0 \quad \text{in } [0, t_1], \\ x(0) = 0, \\ x(t_1) = x_1 \end{cases}$$

  (i) has a unique solution if $t_1 \ne n\pi$, $n \in \mathbb{Z}$ and $x_1 \in \mathbb{R}$,
  (ii) has infinite many solutions if $t_1 = n\pi$, $n \in \mathbb{Z}$ and $x_1 = 0$,
  (iii) has no solutions if $t_1 = n\pi$, $n \in \mathbb{Z}$ and $x_1 \ne 0$.

Discuss also the same problem for the equation $x'' + \lambda x = 0$.

11.44 Theorem. *Let $F(t, x, y)$ be a continuous function in the domain $D := \{(t, x, y) \,|\, t \in [0, 1],\ |x| \leq a,\ |y| \leq a\}$. Moreover, suppose that $F(t, x, y)$ is Lipschitz in (x, y) uniformly with respect to t, i.e., there exists $\mu > 0$ such that*

$$|F(t, x_1, y_1) - F(t, x_2, y_2)| \leq \mu \left(|x_1 - x_2| + |y_1 - y_2| \right)$$

for every (t, x_1, y_1), $(t, x_2, y_2) \in D$. Then for $|\lambda|$ sufficiently small the problem

$$\begin{cases} x'' = \lambda F(t, x, x'), \\ x(0) = x(1) = 0 \end{cases} \tag{11.14}$$

has a unique solution $x(t) \in C^2([0, 1])$. Moreover $|x(t)| \leq a$ and $|x'(t)| \leq a$ $\forall t \in [0, 1]$.

Proof. If $x(t)$ solves $x'' = \lambda F(t, x(t), x'(t))$, then

$$x'(t) = A + \lambda \int_0^t F(\tau, x(\tau), x'(\tau))\, d\tau = A + \lambda \frac{d}{dt} \int_0^t (t - \tau) F(\tau, x(\tau), x'(\tau))\, d\tau,$$

and

$$x(t) = At + B + \lambda \int_0^t (t - \tau) F(\tau, x(\tau), x'(\tau))\, d\tau;$$

the boundary conditions yield

$$B = 0, \qquad A + \lambda \int_0^1 (1 - \tau) F(\tau, x(\tau), x'(\tau))\, d\tau = 0.$$

Thus, $x(t)$ is of class $C^2([0, 1])$ and solves (11.14) if and only if $x(t)$ is of class $C^1([0, 1])$ and solves

$$x(t) = \lambda \int_0^t (t - \tau) F(\tau, x(\tau), x'(\tau))\, d\tau \tag{11.15}$$

$$- \lambda t \int_0^1 (1 - \tau) F(\tau, x(\tau), x'(\tau))\, d\tau.$$

Now consider the class

$$X := \left\{ x \in C^1([0, 1]) \,\Big|\, x(0) = 0,\ \sup_{[0,1]} |x'(t)| \leq a \right\}$$

endowed with the metric

$$d(x_1, x_2) := \sup_{t \in [0, 1]} |x_1'(t) - x_2'(t)|$$

that is equivalent to the C^1 metric $||x_1 - x_2||_{\infty, [0,1]} + ||x_1' - x_2'||_{\infty, [0,1]}$. It is easily seen that (X, d) is a complete metric space and that the map $x(t) \to T[x](t)$ given by

$$T[x](t) := \lambda \int_0^t (t - \tau) F(\tau, x(\tau), x'(\tau))\, d\tau - \lambda t \int_0^1 (1 - \tau) F(\tau, x(\tau), x'(\tau))\, d\tau,$$

maps X into itself and is a contraction provided $|\lambda|$ is sufficiently small. The Banach fixed point theorem then yields a unique solution $x \in X$. On the other hand, (11.15) implies that any solution belongs to X if $|\lambda|$ is sufficiently small, hence the solution is unique. $\qquad \square$

a. The shooting method

A natural approach to show existence of scalar solutions to the boundary value problem

$$\begin{cases} x'' = F(t, x, x') & \text{in }]0, \bar{t}[, \\ x(0) = 0, \\ x(\bar{t}) = \bar{x} \end{cases} \qquad (11.16)$$

consists in showing first existence of solutions $y(t, \lambda)$ of the initial value problem

$$\begin{cases} y'' = F(t, y, y') & \text{in } [0, \bar{t}] \\ y(0) = 0, \\ y'(0) = \lambda, \end{cases} \qquad (11.17)$$

defined in the interval $[0, \bar{t}]$, and then showing that the scalar equation

$$y(\bar{t}, \lambda) = \bar{x}$$

has at least a solution $\bar{\lambda}$; in this case the function $y(t, \bar{\lambda})$ clearly solves (11.16). Since $y(\bar{t}, \lambda)$ is continuous in λ by Theorem 11.37, to solve the last equation it suffices to show that there are values λ_1 and λ_2 such that $y(\bar{t}, \lambda_1) < \bar{x} < y(\bar{t}, \lambda_2)$. This approach is usually referred to as the *shooting method*, introduced in 1905 by Carlo Severini (1872–1951).

11.45 Theorem. *Let $F(t, x, y)$ be a continuous function in a domain D. The problem (11.16) has at least a solution, provided that $\bar{t}$ and $\bar{x}/\bar{t}$ are sufficiently small.*

Proof. Suppose $|F(t, x, y)| \leq M'$, choose $M > M'$ and a sequence of Lipschitz functions $F_k(t, x, y)$ that converge uniformly to $F(t, x, y)$ with

$$|F_k(t, x, y)| \leq M \qquad \forall k, \; \forall t, x, y.$$

Problem (11.17) for F_k transforms into the Cauchy problem for the first order system

$$\begin{cases} z' = G_k(t, z), \\ z(0) = (0, \lambda) \end{cases} \qquad (11.18)$$

where $z(t) = (x(t), y(t))$ and $G_k(x, z) = (y, F_k(t, x, y))$. Now if $b > 0$ is chosen so that $D := \{(t, z) \,|\, |t| < a \text{ e } |z - (0, \lambda)| < b\}$ is in the domain of $G_k(t, z)$, and we proceed as in the proof of Peano's theorem, we find a solution $z_{k, \lambda}$ of (11.18) defined in $[0, r]$ with

$$r < \min\left\{a, \frac{b}{b + |\lambda| + M}\right\}. \qquad (11.19)$$

Since G_k is a Lipschitz function, z_k is in fact the unique solution of (11.18) and depends continuously on $\Lambda := (0, \lambda)$. If $x_{k, \lambda}(t)$ is the first component of $z_{k, \lambda}$, we have, see Theorem 11.44,

$$x_{k, \lambda}(t) = \lambda t + \int_0^t (t - \tau) F(\tau, x_{k, \lambda}(\tau), x'_{k, \lambda}(\tau)) \, d\tau,$$

hence
$$\lambda r - r^2 M \leq x_{k,\lambda}(r) \leq \lambda r + r^2 M$$
and in particular,
$$x_{k,\lambda}(r) < \overline{x} \quad \text{if } \lambda r + r^2 M < \overline{x},$$
$$x_{k,\lambda}(r) > \overline{x} \quad \text{if } \lambda r - r^2 M > \overline{x}. \tag{11.20}$$

It follows from (11.19) that the assumptions in (11.20) hold for two values of λ if r and $\overline{x}/r$ are small enough, concluding that there is a solution $x_k \in C^2([0,r])$ to the boundary value problem

$$\begin{cases} x_k''(t) = F_k(t, x_k, x_k'), \\ x_k(0) = 0, \\ x_k(r) = \overline{x}. \end{cases} \tag{11.21}$$

As in Theorem 11.44, we see that the family $\{x_k(t)\}$ is equibounded with equicontinuous derivatives, thus, by the Ascoli–Arzelà theorem, a subsequence converges to x in the space $C^1([0,r])$, and passing to the limit in the integral form of (11.21), we see actually that $x \in C^2([0,r])$ and solves (11.16) in $[0,r]$. $\qquad\square$

b. A maximum principle

Let $u \in C^2(]0,1[) \cap C^0([0,1])$, but $[0,1]$ can be replaced by any bounded interval. If u has a local maximum point x_0 in the interior of $[0,1]$, then

$$u'(x_0) = 0 \quad \text{and} \quad u''(x_0) \leq 0. \tag{11.22}$$

If, moreover, u satisfies the differential inequality

$$u'' + b(x)u' > 0, \tag{11.23}$$

then clearly (11.22) does not hold at points of $]0,1[$, thus the maximum of u is at 0 or 1, that is, at the boundary of $[0,1]$. If we allow the nonstrict inequality

$$u'' + b(x)u' \geq 0$$

the constant functions that have maximum at every point, are allowed; but this is the only exception. In fact, we have the following.

11.46 Theorem (Maximum principle). *Let u be a function of class $C^2(]x_1,x_2[) \cap C^0([x_1,x_2])$ that satisfies the differential inequality*

$$u'' + b(x)u' \geq 0 \qquad in \ ]x_1,x_2[$$

where $b(x)$ is a function that is bounded below. Then u is constant, if it has an interior maximum point.

Proof. By contradiction, suppose $x_0 \in]x_1,x_2[$ is an interior maximum point and u is not constant so that there is $\overline{x}$ such that $u(\overline{x}) < u(x_0)$. Assume for instance $\overline{x} \in]x_0,x_2[$ and consider the function

$$z(x) := e^{\alpha(x-x_0)} - 1, \qquad x \in [x_1,x_2],$$

where α is a positive constant to be chosen. Trivially $z(x) < 0$ in $]x_1,x_0[$, $z(x_0) = 0$, $z(x) > 0$ in $]x_0,x_2[$ and

$$z'' + b(x)z' = (\alpha^2 + b(x)\alpha)e^{\alpha(x-x_0)} > 0 \qquad \text{in } [x_1, x_2]$$

if $\alpha > \max(0, -\inf_{x \in [x_1, x_2]} b(x))$. Also consider the function

$$w(x) := u(x) + \epsilon z(x)$$

where $\epsilon > 0$ has to be chosen. We have $w(x_0) = u(x_0)$, $w(x) \leq u(x) \leq u(x_0) = w(x_0)$ for $x < x_0$, and $w(\overline{x}) = u(\overline{x}) + \epsilon z(\overline{x}) < u(x_0)$ if $\epsilon < \frac{u(x_0) - u(\overline{x})}{z(\overline{x})}$. With the previous choices of α and ϵ, the function w has an interior maximum point in $]x_1, x_2[$, but $w'' + b(x)w > 0$: a contradiction. $\qquad \square$

11.47 ¶. In the previous proof, $z(x) := e^{\alpha(x-x_0)} - 1$ is one of the possible choices. Show for instance that $z(x) := (x - x_1)^\alpha - (x_0 - x_1)^\alpha$ does it as well.

11.48 Theorem. *Let $u \in C^2(]x_1, x_2[) \cap C^1([x_1, x_2])$ be a nonconstant solution of the differential inequality*

$$u''(x) + b(x)u'(x) \geq 0 \qquad \text{in }]x_1, x_2[$$

where $b(x)$ is bounded from below. Then, $u'(x_1) < 0$ if u has a maximum value at x_1 and $u'(x_2) > 0$ if u has maximum value at x_2.

Proof. As in Theorem 11.46 we find $w'(x_1) = u'(a) + \epsilon\alpha \leq 0$ if u has maximum value at x_1. $\qquad \square$

Similarly we get the following.

11.49 Theorem (Maximum principle). *Let $b(x)$ and $c(x)$ be two functions with $b(x)$ bounded from below and $c(x) \leq 0$ in $[x_1, x_2]$. Suppose that $u \in C^2(]x_1, x_2[) \cap C^0([x_1, x_2])$ satisfies the differential inequality*

$$u'' + b(x)u'(x) + c(x)u \geq 0 \qquad \text{in }]x_1, x_2[.$$

Then

(i) *either u is constant or u has no nonnegative maximum at an interior point,*

(ii) *if u is not constant and has nonnegative maximum at x_1 (respectively, at x_2), then $u'(x_1) < 0$ (respectively, $u'(x_2) > 0$).*

An immediate consequence is the following *comparison* and *uniqueness* theorem for the Dirichlet boundary value problem for linear second order equations.

11.50 Theorem (Comparison principle). *Let u_1 and u_2 be two functions in $C^2(]x_1, x_2[) \cap C^0([x_1, x_2])$ that solve the differential equation*

$$u''(x) + b(x)u'(x) + c(x)u(x) = f(x)$$

where b, c and f are bounded functions and $c(x) \leq 0$.

(i) *If $u_1 \geq u_2$ at x_1 and x_2, then $u_1 \geq u_2$ in $[x_1, x_2]$,*

(ii) *if $u_1 = u_2$ in x_1 and x_2, then $u_1 = u_2$ in $[x_1, x_2]$.*

11.51 ¶. Add details to the proofs of Theorems 11.49 and 11.50. By considering the equations $u'' + u = 0$ e $u'' - u = 0$ show that Theorem 11.49 is optimal.

c. The method of super- and sub-solutions

Consider the boundary value problem

$$\begin{cases} -u'' + \lambda u = f(x) & \text{in }]0, 1[, \\ u(0) = u(1) = 0. \end{cases} \tag{11.24}$$

The comparison principle, Theorem 11.50, says that it has at most one solution if $\lambda \geq 0$, and, since we know the general integral, (11.24) has a unique solution. Let $\mathcal{G}$ be the *Green operator* that maps $f \in C^0([0,1])$ to the unique $C^2([0,1])$ solution of (11.24). $\mathcal{G}$ is trivially continuous; since $C^2([0,1])$ embeds into $C^0([0,1])$ compactly, $\mathcal{G}$ is compact from $C^0([0,1])$ into $C^0([0,1])$; finally by the maximum principle, $\mathcal{G}$ is monotone: if $f \leq g$, then $\mathcal{G}f \leq \mathcal{G}g$.

Consider now the boundary value problem

$$\begin{cases} -u'' = f(x, u), \\ u(0) = u(1) = 0 \end{cases}$$

where we assume $f : [0,1] \times \mathbb{R} \to \mathbb{R}$ to be continuous, differentiable in u for every fixed x, with $f_u(x, u)$ continuous and bounded, $|f_u(x, u)| \leq k$ $\forall (x, u) \in [0,1] \times \mathbb{R}$. By choosing λ sufficiently large, we see that $f(x, u) + \lambda u$ is increasing in u and we may apply to the problem

$$\begin{cases} -u'' + \lambda u = f(x, u) + \lambda u, \\ u(0) = u(1) = 0 \end{cases} \tag{11.25}$$

the argument in Theorem 11.46, inferring that, if $\underline{u}$ and $\overline{u}$ are respectively, a subsolution and a supersolution for $-u'' = f(x, u)$, i.e.,

$$\begin{cases} -\underline{u}'' \leq f(x, \underline{u}), \\ \underline{u}(0), \ \underline{u}(1) \leq 0, \\ \underline{u} \in C^2([0,1]), \end{cases} \qquad \begin{cases} -\overline{u}'' \geq f(x, \overline{u}), \\ \overline{u}(0), \ \overline{u}(1) \leq 0, \\ \overline{u} \in C^2([0,1]) \end{cases}$$

then setting $Tu := \mathcal{G}(f(x, u(x)) + \lambda u(x))$ and

$$\begin{cases} u_0 := \underline{u}, \\ v_0 := \overline{u}, \end{cases} \qquad \begin{cases} u_{n+1} = Tu_n, \\ v_{n+1} = Tv_n \end{cases} \qquad \text{for } n \geq 1,$$

the sequences $\{u_n\}$ and $\{v_n\}$ converge uniformly to a solution of

$$Tu = u,$$

i.e., to a function of class C^2 that solves (11.25). Hence we conclude

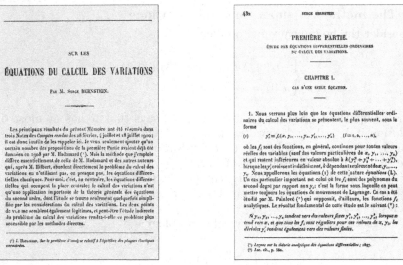

Figure 11.4. Two pages from a paper by Sergei Bernstein (1880–1968).

11.52 Theorem. *Let $f(x,u)$ be a smooth function with $|f_u(x,u)| \leq k$ $\forall(x,u)$. Assume that there exist a subsolution and a supersolution for*

$$\begin{cases} -u'' = f(x,u) & in \ [0,1], \\ u(0) = u(1) = 0. \end{cases}$$

Then there also exists a solution.

We also have the following.

11.53 Theorem. *Let $f(t,p) : [0,+\infty[\times\mathbb{R} \to \mathbb{R}$ be a function of class C^1 that is periodic of period p in t. If the equation $x''(t) = f(t,x(t))$ has a subsolution $\underline{x}(t)$ and a supersolution $\overline{x}(t)$ that are periodic of period p with $\underline{x}(t) \leq \overline{x}(t)$ for all t, then it has also a solution in between, of period p.*

11.54 ¶. Prove Theorem 11.53. [*Hint:* Follow the following scheme.

(i) Choose M so that $f(t,x) - Mx$ is decreasing.

(ii) Inductively define a sequence of p-periodic functions by $x_0(t) := \underline{x}(t)$ and $x_{n+1}(t)$, $n \geq 0$, as solution of

$$x''_{n+1}(t) - Mx_{n+1}(t) = f(t,x_n(t)) - Mx_n(t).$$

(iii) Show that $x_n(t) \leq x_{n+1}(t) \leq \overline{x}(t)$.

(iv) Show that the sequences $\{x''_n\}$ and $\{x'_n\}$ are equibounded, in particular $\{x_n\}$ and $\{x'_n\}$ have subsequences that converge, and actually that $\{x_n\}$, $\{x'_n\}$ and $\{x''_n\}$ converge uniformly to x_∞, x'_∞, x''_∞.

(v) Finally, show that x_∞ is the solution we are looking for.]

d. A theorem by Bernstein

We conclude our excursus in the field of ODEs by the following result.

11.55 Theorem (Bernstein). *Let $F(x, u, p) : [a, b] \times \mathbb{R} \times \mathbb{R} \to \mathbb{R}$ be a continuous function such that*

(i) *there exists $M > 0$ such that $uF(x, u, 0) > 0$ if $|u| > M$,*
(ii) *there exist continuous nonnegative functions $a(x, u)$ and $b(x, u)$ such that*

$$|F(x, u, p)| \le a(x, u)|p|^2 + b(x, u) \qquad \forall (x, u, p) \in [a, b] \times \mathbb{R} \times \mathbb{R}.$$

Then the problem

$$\begin{cases} u'' = F(x, u, u') & in\ ]a, b[, \\ u(a) = u(b) = 0 \end{cases}$$

has a solution.

The original theorem[1] by Bernstein, instead of (i), requires the stronger assumption that F be of class C^1 and for some positive constant k one has $F_u(x, u, p) \ge k > 0$ for all (x, u, p). Its proof uses the shooting method. We shall instead use Schaefer's theorem, Theorem 9.142.

Proof. As we have seen, the operator that maps every $v \in C^2([a, b])$ into the solution of the problem

$$\begin{cases} u'' = F(x, v(x), v'(x)), \\ u(a) = 0, \quad u(b) = 0 \end{cases}$$

is compact. Therefore, according to Schaefer's theorem, it suffices to show that, under the assumptions of Theorem 11.55, there exists $r > 0$ such that, whenever the function $v \in C^2([a, b])$ solves

$$\begin{cases} v'' = \lambda F(x, v, v'), \\ v(a) = v(b) = 0, \end{cases}$$

for some $\lambda \in [0, 1]$, then $||v||_{C^2([a,b])} < r$.

ESTIMATE OF $||v||_\infty$. Let x_0 be a maximum point for $v^2(x)$. We may assume $x_0 \in]a, b[$, otherwise $v \equiv 0$; therefore we have $v'(x_0) = 0$ and

$$0 \ge \frac{d^2}{dx^2}v^2(x)|_{x=x_0} = 2v'^2(x_0) + 2v(x_0)v''(x_0) = \lambda v(x_0)F(x_0, v(x_0), 0);$$

the assumption (i) then implies $|v(x_0)| \le M$, hence $||v||_\infty \le M$.

ESTIMATE OF $||v'||_\infty$. Let μ be a positive constant and let A and B be bounds for $a(x, u)$ and $b(x, u)$ when $x \in [a, b]$ and $|u(x)| \le M$. Multiplying the equation for v by $e^{-\mu v}$ we find

$$\lambda F(x, v, v')e^{-\mu v} = v''e^{-\mu v} = (v'e^{-\mu v})' + \mu v'^2 e^{-\mu v},$$

hence

$$(v'e^{-\mu v})' \le \lambda Av'^2 e^{-\mu v} + \lambda Be^{-\mu v} - \mu v'^2 e^{-\mu v} \le \lambda Be^{-\mu v}$$

if $\mu \ge A$. Similarly, multiplying the equation for v by $e^{\mu v}$, we find

---

[1] S.N. BERNSTEIN, *Sur les équations du calcul des variations*, Ann. Sci. Ec. Norm. Sup. Paris **29** (1912) 481–485.

SUR UNE CLASSE D'ÉQUATIONS FONCTIONNELLES

PAR

IVAR FREDHOLM
À STOCKHOLM.

Dans quelques travaux[1] ABEL s'est occupé avec le problème de déterminer une fonction $\varphi(x)$ de manière qu'elle satisfasse à l'équation fonctionnelle

(a) $\int f(x,y)\varphi(y)dy = \psi(x)$

$f(x,y)$ et $\psi(x)$ étant des fonctions données. ABEL a résolu quelques cas particuliers de cette équation fonctionnelle dont il paraît avoir reconnu le premier l'importance. C'est pour cela que je propose d'appeler l'équation fonctionnelle (a) une *équation fonctionnelle abélienne*.

Dans cette note je ne m'occupe pas en premier lieu de l'équation abélienne mais de l'équation fonctionnelle

(b) $\varphi(x) + \int f(x,y)\varphi(y)dy = \psi(x)$,

qui est étroitement liée à l'équation abélienne.

En effet, si on introduit au lieu de $f(x,y)$ et $\psi(x)$, $\frac{1}{\lambda}f(x,y)$ et $\frac{1}{\lambda}\psi(x)$, l'équation (b) s'écrit

(c) $\lambda\varphi(x) + \int f(x,y)\varphi(y)dy = \psi(x)$,

équation qui se transforme en l'équation (a) en posant $\lambda = 0$. Ainsi la solution de l'équation (a) peut être considérée comme implicitement contenue dans la solution de l'équation (b).

[1] Magazin for Naturvidenskaberne, Kristiania 1823 et Oeuvres complètes.

Figure 11.5. Ivar Fredholm (1866–1927) and a page from one of his papers.

$$(v'e^{\mu v})' \geq -\lambda Be^{\mu v}$$

if $\mu \geq A$. Since v' vanishes at some point in $]a,b[$, integrating we deduce for all $x \in [a,b]$

$$-\lambda Be^{-\mu M}(b-a) \leq v'e^{\mu v} \leq \lambda Be^{\mu M}(b-a),$$

therefore $||v'||_\infty \leq c(A,B,M)$ since $||v||_\infty \leq M$ by step (i).

ESTIMATE OF $||v''||_\infty$. This is now trivial, since from the equation we have

$$|v''(x)| \leq \lambda |F(x,v(x),v'(x))| \leq c(M),$$

F being continuous in $[a,b] \times [0,M] \times [0,c(A,B,M)]$.   □

11.4 Linear Integral Equations

11.4.1 Some motivations

In several instances we have encountered *integral equations*, as convolution operators or, when solving linear equations, as *integral equations* of the type

$$x(t) = y_0 + \int_{t_0}^t f(s,y(s))\,ds;$$

for instance, the linear system $x'(t) = A(t)x(t)$ can be written as

$$x(t) = \int_{t_0}^t A(s)x(s)\,ds. \tag{11.26}$$

(11.26) is an example of *Volterra's equation*, i.e., of equations of the form

$$f(t) = ax(t) + \int_0^t k(t,\tau)x(\tau)\,d\tau. \tag{11.27}$$

a. Integral form of second order equations

The equation $x''(t) - A(t)x(t) = 0$, $t \in [\alpha, \beta]$, can be written as a Volterra equation. In fact, integrating, we get

$$x'(t) = c_1 + \int_{t_0}^t A(s)x(s)\,ds$$

and, integrating again,

$$\begin{aligned}
x(t) &= c_0 + c_1(t - t_0) + \int_{t_0}^t \left(\int_{t_0}^\tau A(s)x(s)\,ds \right) d\tau \\
&= c_0 + c_1(t - t_0) + \int_{t_0}^t (t - s)A(s)x(s)\,ds\,d\tau \tag{11.28} \\
&=: F(t) + \int_{t_0}^t (t - s)A(s)x(s)\,ds,
\end{aligned}$$

with $F(t) := c_0 + c_1(t - t_0)$ and $G : [\alpha, \beta] \times [\alpha, \beta] \to \mathbb{R}$ given by

$$G(t, s) := \begin{cases} (t - s)A(s) & \text{if } s \leq t, \\ 0 & \text{otherwise.} \end{cases}$$

b. Materials with memory

Hooke's law states that the actual *stress* σ is proportional to the actual *strain* ϵ. At the end of 1800, Boltzmann and Volterra observed that the past history of the deformations of the body cannot always be neglected. In these cases the actual stress σ depends not only on the actual strain, but on the whole of the deformations the body was subjected to in the past, hence at every instant t

$$\sigma(t) = a\epsilon(t) + F[\epsilon(\tau)_0^t],$$

where F is a functional depending on *all* values of $\epsilon(\tau)$, $0 < \tau < t$. In the linear context, Volterra proposed the following analytical model for F,

$$F[\epsilon(\tau)_0^t] := \int_0^t k(t,\tau)\epsilon(\tau)\,d\tau.$$

This leads to the study of equations of the type

$$\sigma(t) = a\epsilon(t) + \int_0^t k(t, \tau)\epsilon(\tau)\, d\tau,$$

that are called *Volterra's integral equations* of *first* and *second kind* according to whether $a = 0$ or $a \neq 0$.

c. Boundary value problems

Consider the boundary value problem

$$\begin{cases} x'' - A(t)x = 0, \\ x(0) = a, \\ x(L) = b. \end{cases} \qquad (11.29)$$

From (11.28) we infer

$$x(t) = c_1 + c_2 t + \int_0^t (t - s)A(s)x(s)\, ds$$

and, taking into account the boundary conditions,

$$c_1 = a, \qquad c_2 = \frac{b - a}{L} - \frac{1}{L}\int_0^L (L - s)A(s)x(s)\, ds,$$

we conclude that

$$x(t) = a + \frac{b - a}{L}t - \frac{t}{L}\int_0^L (L - s)A(s)x(s)\, ds + \int_0^t (t - s)A(s)x(s)\, ds$$

$$= a + \frac{b - a}{L}t - \int_0^t \frac{s(L - t)}{L}A(s)x(s)\, ds - \int_t^L \frac{t(L - s)}{L}A(s)x(s)\, ds.$$

In other words, $x(t)$ solves (11.29) if and only if $x(t)$ solves the integral equation, called *Fredholm equation*,

$$x(t) = F(t) + \int_0^L G(t, s)x(s)\, ds$$

where $F(t) := a + \frac{b-a}{L}t$ and $G : [0, L] \times [0, L] \to \mathbb{R}$ is given by

$$G(t, s) := \begin{cases} \dfrac{s(L - t)}{L} & \text{se } s \leq t, \\ \dfrac{t(L - s)}{L} & \text{se } t \leq s. \end{cases}$$

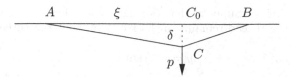

Figure 11.6. An elastic thread.

d. Equilibrium of an elastic thread

Consider an elastic thread of length ℓ which readily changes its shape, but which requires a force $c\,d\ell$ to increase its length by $d\ell$ according to Hooke's law. At rest, the position of the thread is horizontal (the segment AB) under the action of the *tensile force T_0* which is *very large* compared to any other force under consideration.

If we apply a vertical force p at C for which $x = \xi$, the thread will assume the form in Figure 11.6. Assume that $\delta = CC_0$ be very small compared to AC_0 and C_0B (as a consequence of the smallness of p compared with T_0) and, disregarding terms of the order δ^2 (compared with ℓ), the tension of the thread remains equal to T_0. Then the condition of equilibrium of forces is

$$T_0\frac{\delta}{\xi} + T_0\frac{\xi}{l-\xi} = p \qquad \text{i.e.,} \qquad \delta = \frac{p(l-\xi)\xi}{T_0 l}.$$

Denoting by $y(x)$ the vertical deflection at a point of abscissa x, we have

$$y(x) = G(x,\xi)\,p$$

where

$$G(x,\xi) := \begin{cases} \dfrac{x(l-\xi)}{T_0 l} & 0 \le x \le \xi, \\[2mm] \dfrac{(l-\xi)\xi}{T_0 l} & \xi \le x \le l. \end{cases}$$

Now suppose that a continuously distributed force with length density $p(\xi)$ acts on the thread. By the principle of superposition the thread will assume the shape

$$y(x) = \int_0^l G(x,\xi)p(\xi)\,d\xi. \tag{11.30}$$

If we seek the distribution density $p(\xi)$ so that the thread is in the shape $y(x)$, we are led to study Fredholm's integral equation in (11.30).

e. Dynamics of an elastic thread

Suppose now that a force, which varies with the time t and has density at ξ given by

$$p(\xi)\sin\omega t, \qquad \omega > 0,$$

acts on the thread. Suppose that during the motion the abscissa of every point of the thread remains unchanged and that the thread oscillates according to

$$y = y(x) \sin \omega t.$$

Then we find that at time t the piece of thread between ξ and $\xi + \Delta\xi$ is acted upon by the force $p(\xi) \sin(\omega t)\Delta\xi$ plus the force of inertia

$$-p(\xi)\Delta\xi \frac{d^2y}{dt^2} = \rho(\xi)y(x)\omega^2 \sin \omega t \Delta\xi,$$

where ξ is the density of mass of the thread at ξ, and the equation (11.30) takes the form

$$y(x) \sin \omega t = \int_0^l G(x, \xi)[p(\xi) \sin \omega t + \omega^2 \rho(\xi)y(\xi) \sin \omega t]\, d\xi. \qquad (11.31)$$

If we set

$$\int_0^l G(x, \xi)p(\xi)\, d\xi =: f(x), \qquad G(x, \xi)\rho(\xi) =: k(x, \xi), \qquad \omega^2 =: \lambda,$$

(11.31) takes the form of *Fredholm equation*

$$y(x) = \lambda \int_0^l k(x, \xi)y(\xi)\, d\xi + f(x). \qquad (11.32)$$

11.56 ¶. Show that, if in (11.32) we assume $\rho(\xi)$ constant and f smooth, then $y(x)$ solves

$$\begin{cases} y''(x) + \omega^2 c\, y(x) = f''(x), \\ y(0) = 0, \\ y(l) = 0, \end{cases} \qquad (11.33)$$

where $c = \rho/T_0$. Show also that, conversely, if y solves (11.33), then it also solves (11.32).

11.57 ¶. In the case $\rho = \text{const}$, show that the unique solution of (11.33) is

$$y(x) = -\frac{1}{\mu} \frac{\sin \mu x}{\sin \mu l} \int_0^l f''(\xi) \sin \mu(l, \xi)\, d\xi + \frac{1}{\mu} \int_0^x f''(\xi) \sin \mu(x - \xi)\, d\xi$$

if $\sin \mu l \neq 0$, $\mu := \omega\sqrt{c}$. Instead, if $\sin \mu \lambda = 0$, i.e., $\mu = \mu_k$ where

$$\mu_k := \frac{k\pi}{l}, \qquad \omega_k := \frac{k\pi}{l\sqrt{c}}, \qquad \lambda_k := \frac{k^2\pi^2}{l^2 c}, \qquad k \in \mathbb{Z},$$

then (11.33) is solvable if and only if

Figure 11.7. Vito Volterra (1860–1940) and the frontispiece of the first volume of his collected works.

$$\int_0^l f''(\xi) \sin \mu(l - \xi) \, d\xi = 0$$

equivalently, iff

$$\int_0^l f(\xi) \sin \mu\xi \, d\xi = 0.$$

In particular, if $f(x) = 0$ and $\mu = \mu_k$, all solutions are given by

$$y(x) = C \sin \mu_k x \qquad C \in \mathbb{R}$$

and the *natural oscillations* of the thread are given by

$$y = C \sin \mu_k x \sin \omega_k t.$$

Compare the above with the alternative theorem of Fredholm in Chapter 10.

11.4.2 Volterra integral equations

A linear integral equation in the unknown $x(t)$, $t \in [a, b]$ of the type

$$x(t) = f(t) + \int_a^b k(t, \tau) x(\tau) \, d\tau$$

where $f(t)$ and $k(t, x)$ are given functions, is called a *Fredholm equation of second kind*, while a *Fredholm equation of the first kind* has the form

$$\int_a^b k(x,\tau)x(\tau)\,d\tau = f(t).$$

The function $k(t,\tau)$ is called the *kernel* of the integral equation. If the kernel satisfies $k(t,\tau) = 0$ for $t > \tau$, the Fredholm equations of first and second kind are called *Volterra equations*. However it is convenient to treat Volterra equations separately.

11.58 Theorem. *Let $k(t,\tau)$ be a continuous kernel in $[a,b] \times [a,b]$ and let $f \in C^0([a,b])$. Then the Volterra integral equation*

$$x(t) = f(t) + \lambda \int_a^t k(t,\tau)x(\tau)\,d\tau$$

has a unique solution in $C^0([a,b])$ for all values of λ.

Proof. The transformation

$$T[x](t) := f(t) + \lambda \int_a^b k(t,\tau)x(\tau)\,d\tau$$

maps $C^0([a,b])$ into itself. Moreover for all $t \in [a,b]$ we have

$$|T[x_1](t) - T[x_2](t)| \le |\lambda|\,M(t-a)\|x_1 - x_2\|_{\infty,[a,b]}$$

hence

$$|T^2[x_1](t) - T^2[x_2](t)| \le |\lambda|^2\,M^2\frac{(t-a)^2}{2}\|x_1 - x_2\|_{\infty,[a,b]}$$

and by induction, if $T^n := T \circ \cdots \circ T$ n times,

$$|T^n[x_1](t) - T^n[x_2](t)| \le |\lambda|^n\,M^n\frac{(t-a)^n}{n!}\|x_1 - x_2\|_{\infty,[a,b]}.$$

If n is sufficiently large, so that $|\lambda|^n\,M^n\,(b-a)^n/n! < 1$, we conclude that T^n is a contraction, hence it has a unique fixed point $x \in C^0([a,b])$. If $n = 1$ the proof is done, otherwise Tx is also a unique fixed point for T^n, so necessarily we again have $Tx = x$ by uniqueness. $\qquad\square$

11.4.3 Fredholm integral equations in C^0

11.59 Theorem. *Let $k(t,\tau)$ be a continuous kernel in $[a,b] \times [a,b]$ and let $f \in C^0([a,b])$. The Fredholm integral equation*

$$x(t) = f(t) + \lambda \int_a^b k(t,\tau)x(\tau)\,d\tau$$

has a unique solution $x(t)$ in $C^0([a,b])$, provided $|\lambda|$ is sufficiently small.

Proof. Trivially, the transformation

$$T[x](t) := f(t) + \lambda \int_a^b k(t,\tau)x(\tau)\, d\tau$$

maps $C^0([a,b])$ into itself and is contractive for λ close to zero, in fact, if $M :=$ $\max |k(t,\tau)|$,

$$|T[x_1](t) - T[x_2](t)| \le |\lambda| \int_a^b |k(t,\tau)|\, |x_1(\tau) - x_2(\tau)|\, d\tau$$

$$\le |\lambda|\, M(b-a)\, \|x_1(t) - x_2(t)\|_{\infty,[a,b]}$$

$$< \frac{1}{2}\|x_1(t) - x_2(t)\|_{\infty,[a,b]}$$

if $|\lambda|\, M(b-a) < 1/2$. $\qquad\qquad\qquad\qquad\qquad\qquad\qquad\qquad\qquad\qquad\square$

In order to understand what happens for large λ, observe that the transformation

$$T[x](t) := f(t) + \int_a^b k(t,\tau)x(\tau)\, d\tau$$

is linear, continuous and compact, see Example 9.139. The Riesz–Schauder theorem in Remark 10.72 then yields the following.

11.60 Theorem. *Let $k(t,\tau) \subset C^0([a,b] \times [a,b])$ and $f \in C^0([a,b])$. The equation*

$$\lambda x(t) = f(t) + \int_a^b k(t,\tau)x(\tau)\, d\tau \qquad\qquad (11.34)$$

has a set of eigenvalues Λ with the only accumulation point $\lambda = 0$. Each eigenvalue $\lambda \ne 0$ has finite multiplicity and for any λ, $\lambda \ne 0$ and $\lambda \notin \Lambda$, (11.34) has a unique solution.

Further information concerning the eigenvalue case requires the use of a different space norm, the integral norm $\|\ \|_2$, and therefore a description of the completion $L^2((a,b))$ of $C^0((a,b))$ that we have not yet treated.

11.5 Fourier's Series

In 1747 Jean d'Alembert (1717–1783) showed that the general solution of the *wave equation*

$$\frac{\partial^2 u}{\partial t^2} = a^2 \frac{\partial^2 u}{\partial x^2}, \qquad\qquad (11.35)$$

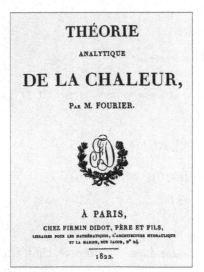

Figure 11.8. Frontispieces of two celebrated works by Joseph Fourier (1768–1830) and G. F. Bernhard Riemann (1826–1866).

that transforms into

$$\frac{\partial^2 u}{\partial r \partial s} = 0$$

by the change of variables $r = x + at$, $s = x - at$, is given by

$$u(t, x) = \varphi(x + at) + \psi(x - at),$$

where φ and ψ are, in principle, generic functions.

Slightly later, in 1753, Daniel Bernoulli (1700–1782) proposed a different approach. Starting with the observation of Brook Taylor (1685–1731) that the functions

$$\sin\left(\frac{n\pi x}{\ell}\right)\cos\left(\frac{n\pi a(t - \beta)}{\ell}\right), \qquad n = 1, 2, \ldots \qquad (11.36)$$

are solutions of the equation (11.35) and satisfy the boundary conditions $u(t, 0) = u(t, \ell) = 0$, Bernoulli came to the conclusion that all solutions of (11.35) could be represented as superpositions of the tones in (11.36). An outcome of this was that every function could be represented as a sum of analytic functions, and, indeed,

$$\frac{4}{\pi}\sum_{n=0}^{\infty}\frac{\sin(2n+1)x}{2n+1} = \begin{cases} 1 & \text{if } 0 < x < \pi, \\ 0 & \text{if } x = \pi, \\ -1 & \text{if } \pi < x < 2\pi. \end{cases}$$

Bernoulli's result caused numerous disputes that lasted well into the nineteenth century that even included the notion of function and, eventually, was clarified with the contributions of Joseph Fourier (1768–1830), Lejeune Dirichlet (1805–1859), G. F. Bernhard Riemann (1826–1866) and many other mathematicians. The methods developed in this context, in particular the idea that a physical system near its equilibrium position can be described as superposition of vibrations and the idea that space analysis can be transformed into a frequency analysis, turned out to be of fundamental relevance both in physics and mathematics.

11.5.1 Definitions and preliminaries

We denote by $L_{2\pi}^1$ the space of complex-valued 2π-periodic functions in $\mathbb{R}$ that are summable on a period, for instance in $[-\pi, \pi]$. For $k \in \mathbb{Z}$, the kth *Fourier coefficient* of $f \in L_{2\pi}^1$ is the complex number

$$c_k = c_k(f) := \frac{1}{2\pi} \int_{-\pi}^{\pi} f(t)e^{-ikt}\, dt$$

often denoted by $f\widehat{\ }(k)$ or $\widehat{f}_k$.

11.61 Definition. *The* Fourier nth partial sum *of $f \in L_{2\pi}^1$ is the trigonometric polynomial of order n given by*

$$S_n f(x) := \sum_{k=-n}^{n} c_k e^{ikx}, \quad x \in \mathbb{R}, \qquad c_k = \frac{1}{2\pi} \int_{-\pi}^{\pi} f(t)e^{-ikt}\, dt.$$

The Fourier series *of f is the sequence of its Fourier partial sums and their limit*

$$Sf(x) = \sum_{k=-\infty}^{\infty} c_k e^{ikx} := \lim_{n \to \infty} S_n f(x) = \lim_{n \to \infty} \sum_{k=-n}^{n} c_k e^{ikx}.$$

If $f \in L_{2\pi}^1$ is real-valued, then

$$c_k = \overline{c_k} \qquad \forall k \in \mathbb{Z}$$

since $f(t) = \overline{f(t)}$ and

$$\int_{-\pi}^{\pi} f(t)e^{ikt}\, dt = \int_{-\pi}^{\pi} \overline{f(t)e^{-ikt}}\, dt = \overline{\int_{-\pi}^{\pi} f(t)e^{-ikt}\, dt}.$$

The partial sums of the Fourier series of a real-valued function have the form

$$S_n f(x) = c_0 + \sum_{k=1}^{n}(c_k e^{ikx} + \overline{c_k} e^{-ikx}) = c_0 + \sum_{k=1}^{n} \Re(2c_k e^{ikx}),$$

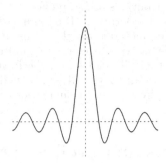

Figure 11.9. The Dirichlet kernel with $n = 5$. Observe that the zeros of $D_n(t)$ are equidistributed $x_n := \frac{2\pi}{2n+1}j$, $j \neq 2k\pi$, $k \in \mathbb{Z}$.

thus, decomposing c_k in its real and imaginary parts, $c_k =: (a_k - ib_k)/2$, that is, setting

$$a_k := \frac{1}{\pi} \int_{-\pi}^{\pi} f(t) \cos(kt)\, dt, \qquad b_k := \frac{1}{\pi} \int_{-\pi}^{\pi} f(t) \sin(kt)\, dt,$$

we find the *trigonometric series*

$$S_n f(x) = \frac{a_0}{2} + \sum_{k=1}^{n} \Re((a_k - ib_k)(\cos(kx) + i\sin(kx))) \qquad (11.37)$$

$$= \frac{a_0}{2} + \sum_{k=1}^{n} (a_k \cos kx + b_k \sin kx).$$

However, the complex notation is handier even for real-valued functions.

11.62 ¶. Show that the operator $\widehat{}$ mapping every function in $L_{2\pi}^1$ into the sequence of its Fourier coefficients, $f \to \{\widehat{f}(k)\}$, has the following properties:

(i) it is linear $(\lambda f + \mu g)\widehat{}(k) = \lambda \widehat{f}(k) + \mu \widehat{g}(k)$ $\forall \lambda, \mu \in \mathbb{C}$, $\forall f, g \in L_{2\pi}^1$,

(ii) $(fg)\widehat{}k = (\widehat{f} * \widehat{g})(k)$, see Proposition 4.46,

(iii) $(f * g)\widehat{}(k) = \widehat{f}(k)\widehat{g}(k)$, see Proposition 4.48,

(iv) if $g(t) = f(-t)$, then $\widehat{g}(k) = \widehat{f}(-k)$,

(v) if $g(t) = f(t - \varphi)$, then $\widehat{g}(k) = e^{-ik\varphi} \widehat{f}(k)$,

(vi) if f is real and even, then its Fourier coefficients are real and its Fourier series is a cosine series,

(vii) if f is real and odd, its Fourier coefficients are imaginary and its Fourier series is a sine series.

(viii) if f has continuous derivative, or more generally f is continuous and f' is piecewise continuous, then $c_k(f') = ik\, c_k(f)$ $\forall k \in \mathbb{Z}$.

a. Dirichlet's kernel

The *Dirichlet kernel* or order n is defined by

$$D_n(x) := 1 + 2\sum_{k=1}^{n} \cos(kx) = \sum_{k=-n}^{n} e^{ikx}, \qquad x \in \mathbb{R}.$$

As we have seen in Section 5.4 of [GM2], $D_n(t)$ is a trigonometric polynomial of order n and 2π-periodic, $D_n(t)$ is even,

$$\frac{1}{2\pi}\int_{-\pi}^{0} D_n(t)\, dt = \frac{1}{2\pi}\int_{0}^{\pi} D_n(t)\, dt = \frac{1}{2},$$

and

$$D_n(t) = \begin{cases} 2n + 1 & \text{if } t = 2k\pi,\ k \in \mathbb{Z}, \\[2mm] \dfrac{\sin((n + 1/2)t)}{\sin(t/2)} & \text{if } t \neq 2k\pi. \end{cases}$$

The Fourier coefficients of $\{D_n(t)\}$ are trivially

$$c_k(D_n) = \begin{cases} 1 & \text{if } |k| \leq n \\ 0 & \text{if } |k| > n. \end{cases}$$

Therefore it is not surprising that we have the following.

11.63 Lemma. *For every* $f \in L^1_{2\pi}(\mathbb{R})$ *we have*

$$S_n f(x) = \frac{1}{2\pi}\int_{0}^{\pi} (f(x + t) + f(x - t)) D_n(t)\, dt \qquad \forall x \in \mathbb{R}.$$

Proof. In fact

$$S_n f(x) = \sum_{k=-n}^{n} c_k e^{ikx} = \frac{1}{2\pi}\int_{-\pi}^{\pi} f(t) c^{ik(x-t)}\, dt = \frac{1}{2\pi}\int_{-\pi}^{\pi} f(t) D_n(x - t)\, dt$$

$$= \frac{1}{2\pi}\int_{-\pi}^{\pi} f(t) D_n(t - x)\, dt = \frac{1}{2\pi}\int_{-\pi-x}^{\pi-x} f(x + t) D_n(t)\, dt$$

$$= \frac{1}{2\pi}\int_{-\pi}^{\pi} f(t + x) D_n(t)\, dt = \frac{1}{2\pi}\int_{0}^{\pi} (f(x + t) + f(x - t)) D_n(t)\, dt,$$

where we used, in the fourth equality, that $D_n(t)$ is even and in the second to last equality that for a 2π-periodic function we have

$$\int_{a}^{a+2\pi} u(t)\, dt = \int_{-\pi}^{\pi} u(t)\, dt \qquad \forall a \in \mathbb{R}.$$

$\square$

Finally we explicitly notice that, though $\int_{-\pi}^{\pi} D_n(t)\, dt = 2\pi$, we have

$$\int_{-\pi}^{\pi} |D_n(t)|\, dt = O(\log n).$$

This prevents us from estimating the modulus of integrals involving $D_n(t)$ by estimating the integral of the modulus.

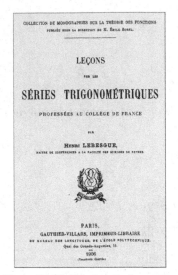

Figure 11.10. The frontispieces of two volumes on trigonometric series by Henri Lebesgue (1875–1941) and Leonida Tonelli (1885–1946).

11.5.2 Pointwise convergence

If P is a trigonometric polynomial, $P \in \mathcal{P}_{n,2\pi}$, then P agrees with its Fourier series, $P(x) = \sum_{k=-n}^{n} c_k e^{ikx} \; \forall x \in \mathbb{R}$, see Section 5.4 of [GM2]. But this does not hold for every $f \in L^1_{2\pi}$. Given $f \in L^2_{2\pi}$, we then ask ourselves under which assumptions on f the Fourier series of f converges and converges to f.

a. The Riemann–Lebesgue theorem

The theorem below states that a rapidly oscillating function with a summable profile has an integral that converges to zero when the frequency of its oscillations tends to infinity, as a result of the compensation of positive and negative contributions due to oscillations, even though the L^1 norms are far from zero.

11.64 Theorem (Riemann–Lebesgue). *Let $f :]a, b[\to \mathbb{R}$ be a Riemann summable function in $]a, b[$. For every interval $]c, d[\subset]a, b[$ we have*

$$\int_c^d f(t) e^{i\lambda t} \, dt \to 0 \qquad as \; |\lambda| \to \infty$$

uniformly with respect to c and d.

Proof. (i) Assume first that f is a step function, and let $\sigma := \{x_0 = a, x_1, \ldots, x_n = b\}$ be a subdivision of $]a, b[$ so that $f(x) = a_k$ on $[x_{k-1}, x_k]$. Then

$$\int_c^d f(t)e^{i\lambda t}\,dt = \sum_{k=1}^n a_k \frac{e^{i\lambda t}}{i\lambda}\Big|_c^d \le \frac{2}{\lambda}\sum_{k=1}^n |a_k|.$$

This proves the theorem in this case.

(ii) Let f be summable in $]a, b[$ and $\epsilon > 0$. By truncating f suitably, we find a bounded Riemann integrable function h_ϵ such that $\int_a^b |f(t) - h_\epsilon(t)|\,dt < \epsilon$, and in turn a step function $g_\epsilon : (a, b) \to \mathbb{R}$ with $\int_a^b |h_\epsilon(t) - g_\epsilon(t)|\,dt < \epsilon$. Consequently $\int_a^b |f(t) - g_\epsilon(t)|\,dt < 2\epsilon$ and from

$$\int_c^d f(t)e^{i\lambda t}\,dt = \int_c^d g_\epsilon(t)e^{i\lambda t}\,dt + \int_c^d (f(t) - g_\epsilon(t))e^{i\lambda t}\,dt$$

we infer

$$\left|\int_c^d f(t)e^{i\lambda t}\,dt\right| \le \left|\int_c^d g_\epsilon(t)e^{i\lambda t}\,dt\right| + \int_a^b |f(t) - g_\epsilon(t)|\,dt$$

$$\le \left|\int_c^d g_\epsilon(t)e^{i\lambda t}\,dt\right| + \epsilon.$$

The conclusion then follows by applying part (i) to g_ϵ. $\square$

11.65 Corollary. *Let f be Riemann summable in $]a, b[$. Then*

$$\int_c^d f(s)\sin\left(\left(n + \frac{1}{2}\right)s\right)\,ds \to 0 \qquad as\ n \to \infty$$

uniformly with respect to the interval $]c, d[\subset]a, b[$.

11.66 ¶. Show the following.

Proposition. *Let $f \in L_{2\pi}^1$. Then we have*

$$\int_{\delta < |s| \le \pi} f(t)D_n(t)\,dt \to 0 \qquad as\ n \to \infty$$

for every $\delta > 0$.

11.67 ¶. Show Theorem 11.64 integrating by parts if f is of class $C^1([a, b])$.

11.68 ¶. Let $f \in L_{2\pi}^1$ and let $\{c_k(f)\}$ be the sequence of its Fourier coefficients. Show that $|c_k(f)| \to 0$ as $k \to \pm\infty$.

b. Regular functions and Dini test

11.69 Definition. *We say in this context that $f \in L_{2\pi}^2$ is regular at $x \in \mathbb{R}$ if there exist real numbers $L^\pm(x)$ and $M^\pm(x)$ such that*

$$\lim_{t\to 0+} f(x+t) = L^+(x), \qquad \lim_{t\to 0-} f(x+t) = L^-(x), \qquad (11.38)$$

$$\lim_{t\to 0+} \frac{f(x+t) - L^+(x)}{t} = M^+(x), \qquad \lim_{t\to 0-} \frac{f(x+t) - L^+(x)}{t} = M^-(x).$$

Of course, if f is differentiable at x, then f is regular at x with $L^{\pm}(x) = f(x)$ and $M^{\pm}(x) = f'(x)$. Discontinuous functions with left and right limits at x and bounded slope near x are evidently regular at x. In particular square waves, sawtooth ramps and C^1 functions are regular at every $x \in \mathbb{R}$.

It is easy to see that if f is regular at x then the function

$$\varphi_x(t) := \frac{f(x+t) + f(x-t) - L^+(x) - L^-(x)}{t} \tag{11.39}$$

is bounded hence Riemann integrable in $]0, \pi]$.

11.70 Definition. *We say that a 2π-periodic piecewise-continuous map $f : \mathbb{R} \to \mathbb{C}$ is Dini-regular at $x \in \mathbb{R}$ if there exist real numbers $L^{\pm}(x)$ such that*

$$\int_0^{\pi} \left| \frac{f(x+t) + f(x-t) - L^+(x) - L^-(x)}{t} \right| dt < +\infty. \tag{11.40}$$

11.71 Theorem (Dini's test). *Let $f \in L^1_{2\pi}(\mathbb{R})$ be Dini-regular at $x \in \mathbb{R}$ and let $L^+(x), L^-(x)$ be as in (11.40). Then $S_n f(x) \to (L^+(x) + L^-(x))/2$.*

Proof. We may assume that $x \in [-\pi, \pi]$. Since $\frac{1}{2\pi} \int_{-\pi}^0 D_n(t)\, dt = \frac{1}{2\pi} \int_0^{\pi} D_n(t)\, dt = 1/2$, we have

$$S_n f(x) - \frac{L^+(x) + L^-(x)}{2} = \frac{1}{2\pi} \int_0^{\pi} (f(x+t) + f(x-t) - L^+ - L^-) D_n(t)\, dt \tag{11.41}$$

$$= \frac{1}{2\pi} \int_0^{\pi} \varphi_x(t)\, t\, D_n(t)\, dt$$

where $\varphi_x(t)$ is as in (11.39). Set $h(t) := \varphi_x(t) \frac{t}{\sin(t/2)}$, so that $|h(t)| \leq \pi |\varphi_x(t)|$ in $[0, \pi]$ and consequently $h(t)$ is summable. Since $\varphi_x(t)\, t\, D_n(t) = h(t) \sin((n + 1/2)t)$, (11.41), the Corollary 11.65 yields

$$S_n f(x) - \frac{L^+(x) + L^-(x)}{2} = \frac{1}{2\pi} \int_0^{\pi} h(t) \sin((n + 1/2)t)\, dt \to 0.$$

$\square$

In particular, if f is continuous, 2π-periodic and satisfies the Dini condition at every x, then $S_n f(x) \to f(x)\ \forall x \in \mathbb{R}$ pointwise.

11.72 Example. Let $0 < \alpha \leq 1$ and $A \subset \mathbb{R}$. Recall that $f : A \to \mathbb{R}$ is said to be α-Hölder-continuous if there exists $K > 0$ such that

$$|f(x) - f(y)| \leq K|x - y|^{\alpha} \qquad \forall x, y \in A.$$

We claim that a 2π-periodic α-Hölder-continuous function on $[a, b]$ satisfies the Dini test at every $x \in]a, b[$. In fact, if $\delta = \delta_x := \min(|x - a|, |x - b|)$, then

$$\int_0^{\pi} \left| \frac{f(x+t) - f(x) + f(x-t) - f(x)}{t} \right| dt \leq \int_0^{\delta} \dots dt + \int_{\delta}^{\pi} \dots dt$$

$$\leq 2K \int_0^{\delta} t^{-1+\alpha}\, dt + \frac{4\|f\|_{\infty, [a,b]}}{\delta} < +\infty.$$

11.73 ¶. Show that the 2π-periodic extension of $\sqrt{|t|}$, $t \in [-\pi, \pi]$ is $1/2$-Hölder-continuous.

11.74 Example. Show that, if f is continuous and satisfies the Dini test at x, then $L^+(x) = L^-(x) = f(x)$.

11.75 ¶. Show that the 2π-periodic extension of $f(t) := 1/\log(1/|t|)$, $t \in [-\pi, \pi]$ does not satisfy the Dini test at 0.

11.5.3 L^2-convergence and the energy equality

a. Fourier's partial sums and orthogonality

Denote by $||f||_2$ the quadratic mean over a period of f

$$||f||_2^2 := \frac{1}{2\pi} \int_{-\pi}^{\pi} |f(t)|^2 \, dt,$$

and with $L_{2\pi}^2$ the space of integrable functions with $||f||_2 < \infty$. The Hermitian bilinear form and the corresponding "norm"

$$(f|g) := \frac{1}{2\pi} \int_{-\pi}^{\pi} f(t)\overline{g(t)} \, dt, \qquad ||f||_2 := \left(\frac{1}{2\pi} \int_{-\pi}^{\pi} |f(t)|^2 \, dt \right)^{1/2},$$

are *not* a Hermitian product and a norm in $L_{2\pi}^2$, since $||f||_2 = 0$ does not imply $f(t) = 0 \; \forall t$, but they do define a Hermitian product and a norm in $L_{2\pi}^2 \cap C^0(\mathbb{R})$, since $||f||_2 = 0$ implies $f = 0$ if f is continous. Alternatively, we may identify functions f and g in $L_{2\pi}^2$ if $||f - g||_2 = 0$, and again $(f|g)$ and $||f||_2$ define a Hermitian product and a norm on the equivalence classes of $L_{2\pi}^2$ if, as it is usual, we still denote by $L_{2\pi}^2$ the space of equivalence classes. It is easily seen that $L_{2\pi}^2$ is a pre-Hilbert space with $(f|g)$. Notice that two nonidentical continuous functions belong to different equivalence classes.

Since e^{ikx}, $k \in \mathbb{Z}$, belong to $L_{2\pi}^2$ and

$$(e^{ihx}|e^{ikx}) = \frac{1}{2\pi} \int_{-\pi}^{\pi} e^{i(h-k)x} \, dx = \delta_{hk},$$

we have the following.

11.76 Proposition. *The trigonometric system $\{e^{ikt} \,|\, k \in \mathbb{Z}\}$ is an orthonormal system in $L_{2\pi}^2$.*

Since

$$(f|e^{ikx}) = \frac{1}{2\pi} \int_{-\pi}^{\pi} f(t)e^{-ikt}\, dt = c_k(f),$$

we have

$$S_n f(x) = \sum_{k=-n}^{n} (f|e^{ikx})e^{ikx}, \qquad x \in \mathbb{R},$$

i.e., the Fourier series of f is the abstract Fourier series with respect to the trigonometric orthonormal system. Therefore the results of Section 10.1.2 apply, in particular the *Bessel inequality* holds

$$\sum_{k=-\infty}^{\infty} |c_k|^2 \le ||f||_2^2$$

as well as Proposition 10.18, in particular

$$||f - S_n f||_2 \le ||f - P||_2 \qquad \forall P \in \mathcal{P}_{n,2\pi}.$$

Recall also that for a trigonometric polynomial $P \in \mathcal{P}_{n,2\pi}$ the Pythagorean theroem holds

$$\frac{1}{2\pi} \int_{-\pi}^{\pi} |P(t)|^2\, dt = \sum_{k=-n}^{n} |c_k(P)|^2.$$

b. A first uniform convergence result
11.77 Theorem. *Let $f \in C^1(\mathbb{R})$. Then $S_n f \to f$ uniformly in $\mathbb{R}$.*

Proof. Since $S_n f(x) \to f(x)\ \forall x$, it suffices to show the uniform convergence of $S_n f$. We notice that $f' \in L_{2\pi}^1$ and that, by integration by parts,

$$c_k(f') := ikc_k(f) \qquad \forall k \in \mathbb{Z},$$

hence , if $k \ne 0$,

$$|c_k(f)| \le \frac{|c_k(f')|}{k} \le |c_k(f')|^2 + \frac{1}{k^2}$$

where we have used the inequality $|ab| \le a^2 + b^2$. Since $\sum_{k=-\infty}^{\infty} |c_k(f')|^2$ converges by Bessel's inequality, we therefore conclude that $\sum_{k=-\infty}^{\infty} |c_k(f)|$ converges, consequently

$$\sum_{k=-\infty}^{\infty} c_k e^{ikx}$$

converges absolutely in $C^0(\mathbb{R})$ since $||e^{ikx}||_{\infty,\mathbb{R}} = 1\ \forall k$. $\square$

11.78 ¶. Let $f \in C^n(\mathbb{R})$ and let $\{c_k\}$ be its Fourier coefficients. Show that $k^n|c_k| \to 0$ as $|k| \to \infty$.

For stronger results about uniform convergence of Fourier series see Section 11.5.4.

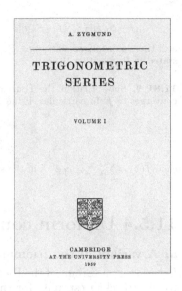

A. ZYGMUND

TRIGONOMETRIC
SERIES

VOLUME I

CAMBRIDGE
AT THE UNIVERSITY PRESS
1959

Figure 11.11. Antoni Zygmund (1900–1992) and the frontispiece of the first edition of volume I of his *Trigonometric Series*.

c. Energy equality

We have, compare Chapter 9, the following.

11.79 Lemma. $C^1(\mathbb{R}) \cap L^2_{2\pi}$ *is dense in* $L^2_{2\pi}$.

Proof. Let $f \in L^2_{2\pi}$ and $\epsilon > 0$. There is a Riemann integrable function h_ϵ with $||f - h_\epsilon||_2 < \epsilon$ and a step function k_ϵ in $[-\pi, \pi]$ such that $||k_\epsilon|| \le M_\epsilon$ and $||h_\epsilon - k_\epsilon||_1 \le (\pi\epsilon^2)/M_\epsilon$ where $M_\epsilon := ||h_\epsilon||_\infty$, consequently

$$||h_\epsilon - k_\epsilon||_2^2 = \frac{1}{2\pi} \int_{-\pi}^{\pi} |h_\epsilon - k_\epsilon|^2 \, dt \le \frac{1}{2\pi} 2M_\epsilon \int_{-\pi}^{\pi} |h_\epsilon - k_\epsilon| \, dt \le \epsilon^2.$$

First, approximating k_ϵ by a Lipschitz function, then smoothing the edges, we find $l_\epsilon \in C^1([-\pi, \pi])$ with $||k_\epsilon - l_\epsilon||_2 < \epsilon$. Finally we modify l_ϵ near π and $-\pi$ to obtain a new function g_ϵ with $g_\epsilon(-\pi) = g_\epsilon(\pi) = g'(-\pi) = g'(\pi) = 0$. Extending g_ϵ to a periodic function in $\mathbb{R}$, we finally get $g_\epsilon \in C^1(\mathbb{R}) \cap L^2_{2\pi}$ and $||f - g_\epsilon||_2 < 4\epsilon$. $\qquad\square$

Now we can state the following.

11.80 Theorem. *For every* $f \in L^2_{2\pi}$ *we have* $||S_n f - f||_2 \to 0$. *Therefore, the trigonometric system* $\{e^{ikx}\}$, $k \in \mathbb{Z}$, *is orthonormal and complete in* $L^2_{2\pi}$; *moreover, for any* $f \in L^2_{2\pi}$ *the energy equality or Parseval's identity holds:*

$$\sum_{k=-\infty}^{\infty} |c_k|^2 = \frac{1}{2\pi} \int_{-\pi}^{\pi} |f(t)|^2 \, dt.$$

Proof. Given $f \in L^2_{2\pi}$ and $\epsilon > 0$, let $g \in C^1(\mathbb{R}) \cap L^2_{2\pi}$ be such that $||f - g||_2 < \epsilon$. Since $S_n g$ is a trigonometric polynomial of order at most n, and $S_n f$ is the point of minimal $L^2_{2\pi}$ distance in $L^2_{2\pi}$ from f we have

$$||f - S_n f||_2 \leq ||f - S_n g||_2 \leq ||f - g||_2 + ||g - S_n g||_2 < \epsilon + ||g - S_n g||_\infty$$

and the claim follows since $||g - S_n g||_\infty \to 0$ as $n \to \infty$. The rest of the claim is now stated in Proposition 10.18. $\square$

11.81 ¶. Show that, if the Fourier series of $f \in L^2_{2\pi}$ converges uniformly, then it converges to f. In particular, if the Fourier coefficients c_k of f satisfy

$$\sum_{k=-\infty}^{+\infty} |c_k| < +\infty,$$

then $f(x) = \sum_{k=-\infty}^{+\infty} c_k e^{ikx}$ in the sense of uniform convergence in $\mathbb{R}$.

11.5.4 Uniform convergence

a. A variant of the Riemann–Lebesgue theorem

Let us state a variant of the Riemann–Lebesgue theorem that is also related to the Dirichlet estimate for the series of products.

11.82 Proposition (Second theorem of mean value). *Let f and g be Riemann integrable functions in $]a, b[$. Suppose moreover that f is not decreasing, and denote by M and m respectively, the maximum and the minimum values of $x \to \int_x^b g(t)\, dt$, $x \in [a, b]$. Then we have*

$$m\, f(b) \leq \int_a^b f(t) g(t)\, dt \leq M f(b).$$

In particular, there exists $c \in]a, b[$ such that

$$\int_a^b f(t) g(t)\, dt = f(b) \int_c^b g(t)\, dt.$$

Proof. Choose a constant d such that $g(t) + d > 0$ in $]a, b[$. If f is differentiable, the claim follows easily integrating by parts $\int_a^b f(t)(g(t)+d)\, dt$. The general case can be treated by approximation (but we have not developed the correct means yet) or using the formula of summation by parts, see Section 6.5 of [GM2]. For the reader's convenience we give the explicit computation. Let $\sigma = \{x_0 = a, x_1, \ldots, x_n = b\}$ be a partition of $[a, b]$. Denote by Δ_k the interval $[x_{k-1}, x_k]$ and set $\sigma_k := \sum_{k=1}^n f(x_k)(x_k - x_{k-1})$. We have

$$\int_a^b f(t)(g(t) + d)\, dt = \sum_{k=1}^n \int_{\Delta_k} f(t)(g(t) + d))\, dt \leq \sum_{k=1}^n f(x_k)(G(x_{k-1}) - G(x_k)) + d\sigma_k$$

$$= f(x_1) G(x_0) + \sum_{k=1}^{n-1} G(x_k)(f(x_{k+1}) - f(x_k)) + d\sigma_k$$

$$\leq M\Big(f(x_1) + \sum_{k=1}^{n-1} (f(x_{k+1}) - f(x_k))\Big) + d\sigma_k$$

$$= M f(b) + d\sigma_k.$$

Figure 11.12. Ulisse Dini (1845–1918) and the frontispiece of his *Serie di Fourier*.

Since $\sigma_k \to \int_a^b g(t)\, dt$ as $k \to \infty$, we infer

$$\int_a^b f(t)g(t)\, dt \le Mf(b).$$

Similarly, we get $\int_a^b f(t)g(t)\, dt \ge mf(b)$. The second part of the claim follows from the intermediate value theorem since $\int_x^b g(t)\, dt$ is continuous. □

From the Riemann–Lebesgue lemma, see Exercise 11.66, for any $f \in L_{2\pi}^1$ and $\delta > 0$ we have for every fixed x

$$\int_\delta^\pi f(x+t)D_n(t)\, dt \to 0 \qquad \text{as } n \to \infty.$$

For future use we prove the following.

11.83 Proposition. *Let $f \in L_{2\pi}^1$ and $\delta > 0$. Then*

$$\int_\delta^\pi f(x+t)D_n(t)\, dt \to 0 \qquad \text{as } n \to \infty$$

uniformly in $x \in \mathbb{R}$.

Proof. Since $1/\sin(t/2)$ is decreasing in $]0, \pi]$, the second theorem of mean value yields $\xi = \xi(x) \in [\delta, \pi]$ such that

$$\int_\delta^\pi f(x+t)D_n(t)\, dt = \frac{1}{\sin(\delta/2)} \int_\delta^\xi f(x+t)\sin((n+1/2)t\, dt.$$

On the other hand,

$$\int_\delta^\xi f(x+t)\sin((n+1/2)t\,dt = \int_{\delta+x}^{\xi+x} f(t)\sin((n+1/2)(t-x))\,dt$$

$$= \cos((n+1/2)x)\int_{\delta+x}^{\xi+x} f(t)\sin((n+1/2)t)\,dt$$

$$- \sin((n+1/2)x)\int_{\delta+x}^{\xi+x} f(t)\cos((n+1/2)t)\,dt$$

and the last two integrals converge uniformly to zero in $[-\pi,\pi]$, see Exercise 11.62. Thus $\int_\delta^\pi f(x+t)D_n(t)\,dt \to 0$ uniformly in $[-\pi,\pi]$, hence in $\mathbb{R}$.   □

b. Uniform convergence for Dini-continuous functions

Let $f \in C^{0,\alpha}(\mathbb{R}) \cap L^1_{2\pi}$ be a 2π-periodic and α-Hölder-continuous function. It is easy to see that f is continuous and Dini-regular at every $x \in \mathbb{R}$. In fact, if $\delta = \delta_x := \min(|x-a|,|x-b|)$, then

$$\int_0^\pi \left| \frac{f(x+t)-f(x)+f(x-t)-f(x)}{t} \right| dt \leq \int_0^\delta \dots dt + \int_\delta^\pi \dots dt$$

$$\leq 2K \int_0^\pi t^{-1+\alpha}\,dt + \frac{4\|f\|_{\infty,[a,b]}}{\delta} < +\infty.$$

Therefore $S_n f(x) \to f(x)\ \forall x \in \mathbb{R}$ by the Dini test theorem, Theorem 11.71. We have the following.

11.84 Theorem. *If f is 2π-periodic and of class $C^{0,\alpha}(\mathbb{R})$, $0 < \alpha \leq 1$, then $S_n f(x) \to f(x)$ uniformly in $\mathbb{R}$.*

Proof. Let $\delta > 0$ to be chosen later. We have

$$S_n f(x) - f(x) = \int_0^\delta (f(x+t)-f(x))D_n(t)\,dt + \int_\delta^\pi (f(x+t)-f(x))D_n(t)\,dt$$

$$=: I_1(\delta,n,x) + I_2(\delta,n,x)$$

Let $\epsilon > 0$. Since f is α-Hölder-continuous there exists $K > 0$ such that

$$|f(x+t)-f(x)| \leq K|t|^\alpha \qquad \forall x \in \mathbb{R},\ \forall t \in [0,2\pi],$$

hence

$$|I_1(\delta,n,x)| \leq K \int_0^\delta t^\alpha \frac{1}{\sin(t/2)}|\sin((n+1/2)t)|\,dt \leq 2K \int_0^\delta t^{-1+\alpha}\,dt = \frac{2K}{\alpha}\delta^\alpha.$$

We can therefore choose δ in such a way that $|I_1(\bar\delta,n,x)| < \epsilon$ uniformly with respect to x and n. On the other hand $|I_2(\bar\delta,n,x)| < \epsilon$ uniformly with respect to x as $n \to +\infty$ by Proposition 11.83 concluding that

$$|S_n f(x) - f(x)| \leq 2\epsilon \qquad \text{uniformly in } x$$

for n sufficiently large.   □

With the same proof we also infer the following.

11.85 Theorem (Dini's test). *Let $f \in C^0(\mathbb{R}) \cap L^1_{2\pi}$ be a 2π-periodic and continuous function with modulus of continuity $\omega(\delta)$, $|f(x)-f(y)| \leq \omega(\delta)$ if $|x-y| \leq \delta$, such that $\omega(\delta)/\delta$ is summable in a neighborhood of $\delta = 0$. Then $S_n f \to f$ uniformly in $\mathbb{R}$.*

c. Riemann's localiziation principles

The convergence of Fourier's partial sums is a *local* property in the following sense

11.86 Proposition. *If $g, h \in L^1_{2\pi}$ and $g = h$ in a neighborhood of a point x, then $S_n g(x) - S_n h(x) \to 0$ as $n \to \infty$.*

Proof. Assume $f := g - h$ vanishes in $[x - \delta, x + \delta]$, $\delta > 0$. Then, for every $t \in [0, \delta]$ we have $f(x + t) = f(x - t) = 0$, hence

$$S_n f(x) - f(x) = \frac{1}{2\pi} \int_\delta^\pi (f(x + t) + f(x - t)) D_n(t) \, dt.$$

Since $(f(x + t) + f(x - t)) / \sin(t/2)$ is summable in (δ, π), the result follows from the Riemann–Lebesgue theorem. □

11.87 Proposition. *If $f \in L^1_{2\pi}$ and $f = 0$ in $]a, b[$, then $S_n f(x) \to 0$ uniformly on every interval $[c, d]$ with $a < c < d < b$.*

Proof. Let us show that $S_n f(x) \to 0$ uniformly in $[a + \delta, b - \delta]$, $0 < \delta < (b - a)/2$. For $x \in [a + \delta, b - \delta]$ and $0 < t < \delta$ we have $f(x + t) = f(x - t) = 0$, hence

$$S_n f(x) = \frac{1}{2\pi} \int_\delta^\pi (f(x + t) + f(x - t)) D_n(t) \, dt.$$

The claim follows from Proposition 11.83. □

The localization principle says that, when studying the pointwise convergence in an open interval $]a, b[$ or the uniform convergence in a closed interval inside $]a, b[$ of the Fourier series of a function f, we can modify f outside of $]a, b[$. With this observation we easily get the following.

11.88 Corollary. *Let $f \in L^1_{2\pi}$ be a function that is of class $C^1([a, b])$. Then $\{S_n f(x)\}$ converges uniformly to $f(x)$ in any interval strictly contained in $]a, b[$.*

11.5.5 A few complementary facts

a. The primitive of the Dirichlet kernel

Denote by $G_n(x)$ the primitive of the Dirichlet kernel,

$$G_n(x) := \int_0^x D_n(t) \, dt.$$

It is easy to realize that $G_n(x)$ is odd and nonnegative in $[0, \pi]$ and takes its maximum value in $[0, \pi]$ at the first zero $x_n := \frac{2\pi}{2n+1}$ of D_n. Thus

$$||G_n||_{\infty,[0,\pi]} = G_n\left(\frac{2\pi}{2n+1}\right) = \int_0^{2\pi/(2n+1)} \frac{\sin((n + 1/2)s)}{\sin(s/2)} \, ds$$

$$= \int_0^\pi \frac{\sin s}{\sin(s/(2n+1))} \, ds \leq \frac{2(n+1)}{2n+1} \pi \leq 2\pi$$

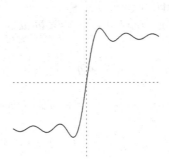

Figure 11.13. The graph of $G_5(x)$ in $[-\pi, \pi]$.

independently of n; in particular,

$$\left| \int_c^d D_n(t)\,dt \right| \le 2\pi \tag{11.42}$$

for all $c, d \in [0, \pi]$. Also, by Exercise 11.66, or directly by an integration by parts, it is easily seen that, given any $\delta > 0$, there is a constant $c(\delta)$ such that

$$|G_n(\pi) - G_n(x)| = \left| \int_x^\pi D_n(t)\,dt \right| \le c(\delta)\frac{1}{n} \tag{11.43}$$

for all $x \in [0, \delta]$.

For future use we now show that

$$\lim_{n \to \infty} G_n(x_n) = 2 \int_0^\pi \frac{\sin s}{s}\,ds. \tag{11.44}$$

In order to do that, we first notice that

$$\frac{2}{\pi}t \le \sin t \le t, \qquad 0 \le t - \sin t \le \frac{1}{6}t^3$$

i.e.,

$$\left| \frac{1}{\sin t} - \frac{1}{t} \right| \le \frac{\pi}{12}t \qquad t \in\,]0, \pi],$$

hence

$$\left| \int_0^{x_n} D_n(t)\,dt - \int_0^{x_n} \frac{\sin((n+1/2)t)}{t/2}\,dt \right| \le \frac{\pi}{12}\left(\frac{2\pi}{2n+1} \right)^2 \to 0 \tag{11.45}$$

as $n \to \infty$. Equality (11.44) then follows as

$$\int_0^{x_n} \frac{\sin((n+1/2)t)}{t/2}\,dt = \frac{2}{2n+1} \int_0^\pi \frac{\sin s}{s/(2n+1)}\,ds = 2 \int_0^\pi \frac{\sin s}{s}\,ds. \tag{11.46}$$

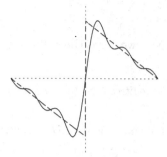

Figure 11.14. The sawthooth $h(x)$ and its Fourier partial sum of order 5 in $[-\pi, \pi]$.

b. Gibbs's phenomenon

Consider the 2π-periodic function h defined by periodically extending the function

$$h(t) := \begin{cases} -\pi - t & \text{if } -\pi \le t < 0, \\ 0 & \text{if } t = 0, \\ \pi - t & \text{if } 0 < t \le \pi. \end{cases} \qquad (11.47)$$

Its Fourier coefficients are easily computed to be

$$c_0 = 0, \qquad c_k := \frac{1}{ik}, \ k \ne 0,$$

hence

$$S_n h(x) = \sum_{\substack{k=-n,n \\ k \ne 0}} \frac{e^{ikx}}{ik} = \int_0^x D_n(t)\, dt - x \qquad (11.48)$$

or

$$S_n h(x) = 2 \sum_{k=1}^{n} \frac{\sin kx}{k}.$$

In particular $S_n h(0) = 0 \ \forall n$, and, by Dini's test, Theorem 11.71,

$$2 \sum_{k=1}^{\infty} \frac{\sin kx}{k} = h(x) \qquad \text{pointwise in } \mathbb{R}.$$

The energy inequality yields

$$\frac{\pi^2}{3} = \frac{1}{2\pi} \int_{-\pi}^{\pi} |h(t)|^2\, dt = \sum_{k=-\infty}^{\infty} |c_k|^2 = 2 \sum_{k=1}^{\infty} \frac{1}{k^2}$$

or

$$\sum_{k=1}^{\infty} \frac{1}{k^2} = \frac{\pi^2}{6}. \qquad (11.49)$$

As we have already seen, we have the following, of which we give a direct proof.

11.89 Theorem. *For any positive $\delta > 0$ the Fourier series of h converges uniformly to h in $[\delta, \pi]$.*

Proof. We know that $S_n h(x)$ converge pointwise to h, therefore it suffices to show that

$$\sum_{\substack{k=-\infty \\ k \neq 0}}^{\infty} \frac{e^{ikx}}{ik} \qquad (11.50)$$

converges uniformly in $[\delta, \pi]$. We apply Dirichlet's theorem for series of products, see Section 6.5 of [GM2], respectively, to the series with positive and negative indices with $a_k = 1/(ik)$ and $b_k := e^{ikx}$, to find that

$$|S_n h(x) - h(x)| = \Big| \sum_{|k| \geq n+1} \frac{e^{ikx}}{ik} \Big| \leq \frac{4}{|1 - e^{ix}|} \frac{1}{n+1}$$

hence

$$\|S_n h - h\|_{\infty, [\delta, \pi]} \leq \frac{4}{|1 - e^{i\delta}|} \frac{1}{n+1} \to 0 \qquad \text{as } n \to \infty.$$

Alternatively, from (11.48) we infer

$$S_n h(x) - h(x) = \int_0^x D_n(t) - \pi = -\int_x^\pi D_n(s) \, ds$$

and, by (11.43),

$$|S_n h(x) - h(x)| \leq c(\delta) \frac{1}{n} \qquad \text{uniformly in } [\delta, \pi].$$

However the Fourier series of h does not converge uniformly in $[0, \pi]$. □

11.90 Proposition. *We have*

$$\|S_n h\|_{\infty, [0, \pi]} \to 2 \int_0^\pi \frac{\sin s}{s} \, ds.$$

Proof. Let y_n be the point where $S_n h(x)$ obtains its maximum value in $[0, \pi]$,

$$M_n := \sup_{[0,\pi]} S_n h(x) = S_n(y_n),$$

and let $x_n := \frac{2\pi}{2n+1}$. Since x_n is the maximum point of $G_n(x)$, we have

$$G_n(y_n) - x_n \leq G_n(x_n) - x_n = S_n f(x_n) \leq S_n f(y_n) = G_n(y_n) - y_n \leq G_n(x_n) - y_n.$$

This implies $0 \leq y_n \leq x_n$ and $-x_n \leq S_n(y_n) - G_n(x_n) \leq -y_n$, hence

$$|M_n - G_n(x_n)| \leq x_n = \frac{2\pi}{2n+1}. \qquad (11.51)$$

The conclusion follows from (11.44). □

We can rewrite the statement in Proposition 11.90 as

$$||S_n h||_{\infty,[0,\pi]} \rightarrow \left(\frac{2}{\pi} \int_0^\pi \frac{\sin s}{s} \, ds \right) ||h||_{\infty,[0,\pi]}.$$

Since

$$\frac{2}{\pi} \int_0^\pi \frac{\sin s}{s} \, ds = 1.089490\ldots$$

we see that, while $S_n h(x) \rightarrow h(x)$ for all $x \in \mathbb{R}$, near 0, $S_n h(x)$ always has a maximum which stays away from the maximum of h that is $||h||_{\infty,[0,\pi]} = \pi$ by a positive quantity: this is the *Gibbs phenomenon*, which is in fact typical of Fourier series at jump points; but we shall not enter into this subject.

11.5.6 The Dirichlet–Jordan theorem

The pointwise convergence of the Fourier series of a continuous or summable function is a subtle question and goes far beyond Dini's test, Theorem 11.71. An important result, proved by Lejeune Dirichlet (1805–1859), shows in fact that a 2π-periodic function which has only a finite number of jumps and maxima and minima, has a Fourier series that converges pointwise to $(L^+ + L^-)/2$ where $L^\pm := \lim_{y \to x^\pm} f(y)$; in particular $S_n f(x) \rightarrow f(x)$ at the points of continuity. The same proof applies to functions with bounded variation, see Theorem 11.91. In 1876 Paul du Bois–Reymond (1831–1889) showed a continuous function whose Fourier series diverges at one point, and, therefore, that the continuity does not solely suffice for the pointwise convergence of the Fourier series. We shall present a different example due to Lipót Fejér (1880–1959). Starting from this example one can show continuous functions whose Fourier series do not converge in a denumerable dense set, for instance, the rationals. In the 1920's Andrey Kolmogorov (1903–1987) showed a continuous function with Fourier series divergent on a set with the power of the continuum, and Hugo Steinhaus (1887–1972) showed a continuous function whose Fourier series converges pointwise everywhere, but does not converge uniformly in any interval. Eventually, the question was clarified in 1962 by Lennart Carleson. Here we collect some complements.

a. The Dirichlet–Jordan test

11.91 Theorem (Dirichlet–Jordan). *Let f be a 2π-periodic function with bounded total variation in $[a, b]$.*

(i) *For every $x \in]a, b[$ we have $S_n f(x) \rightarrow (L^+ + L^-)/2$ where $L^\pm := \lim_{y \to x^\pm} f(y)$.*

(ii) *If f is also continuous in $]\alpha, b[$, then $S_n f(x) \rightarrow f(x)$ uniformly in any closed interval strictly contained in $]a, b[$.*

Figure 11.15. The amplitude of the harmonics of $Q_{n,\mu}(x)$.

Proof. Let $[a, b]$ be an interval with $b - a < 2\pi$. Since every function with bounded variation in $[a, b]$ is the sum of an increasing function and of a decreasing function, we may also assume that f is nondecreasing in $[a, b]$.

(i) Let $x \in]a, b[$, $g_x(t) := f(x + t) - L^+ + f(x - t) - L^-$ where $L^\pm := \lim_{y \to x^\pm} f(y)$. We have

$$
\begin{aligned}
S_n f(x) - \frac{L^+ - L^-}{2} &= \frac{1}{2\pi} \int_0^\pi (f(x + t) - L^+ + f(x - t) - L^-) D_n(t)\, dt \\
&= \frac{1}{2\pi} \int_0^\delta g_x(t) D_n(t)\, ds + \frac{1}{2\pi} \int_\delta^\pi g_x(s) D_n(s)\, ds \\
&= I_1 + I_2.
\end{aligned}
\tag{11.52}
$$

where $\delta > 0$ is to be chosen later. Since $f(x + t) - L^+$ and $-(f(x - t) - L^-)$ are nondecreasing near $t = 0$ and nonnegative, the second theorem of mean value and (11.42) yield

$$
|I_1| \leq 2\pi |f(x + \delta) - L^+| + |f(x - \delta) - L^-|
\tag{11.53}
$$

while (11.43) yields

$$
|I_2| \leq c(\delta) \frac{1}{n}.
\tag{11.54}
$$

Therefore, given $\epsilon > 0$, we can choose $\delta > 0$ in such a way that

$$
|f(x + \delta) - L^+| + |f(x - \delta) - L^-| < \epsilon
$$

to obtain from (11.53) and (11.54) that

$$
\left| S_n f(x) - \frac{L^+ - L^-}{2} \right| \leq 2\pi\epsilon + c(\delta) \frac{1}{n}.
$$

That proves the pointwise convergence at x.

(ii) In this case for every $x \in [a, b]$, we have $L^+ = L^- = f(x)$ and it suffices to estimate uniformly in $[a + \sigma, b - \sigma]$, $0 < \sigma < (b - a)/2$, I_1 and I_2 in (11.52). Since f is uniformly continuous in $[a, b]$, given $\epsilon > 0$, we can choose δ, $0 < \delta < \sigma$, in such a way that $|f(x + \delta) - f(x)| + |f(x - \delta) - f(x)| < \epsilon$, uniformly with respect to x in $[a + \sigma, b - \sigma]$, hence from (11.53) $|I_1| < 2\pi\epsilon$ uniformly in $[a + \sigma, b - \sigma]$. The uniform estimate of $|I_2|$ is instead the claim of Proposition 11.83.

Finally, if $b - a > 2\pi$, it suffices to write $[a, b]$ as a finite union of intervals of length less than 2π and apply the above to them. $\quad\square$

11.92 Remark. Notice that the Dirichlet–Jordan theorem is in fact a claim on monotone functions. Monotone functions are continuous except on a denumerable set of jump points, that is not necessarily discrete.

b. Féjer example

Let $\mu \in \mathbb{N}$ be a natural number to be chosen later. For every $n \in \mathbb{N}$ consider the trigonometric polynomial of degree n

$$Q_{n,\mu}(x) := \sum_{k=1}^{n} \frac{\cos(n+\mu-k)x - \cos(n+\mu+k)x}{k}$$

$$= 2\sin((n+\mu)x) \sum_{k=1}^{n} \frac{\sin kx}{x},$$

see Figure 11.15. It is a cosine polynomial with harmonics of order $\mu, \mu+1, n+\mu-1, n+\mu+1, \ldots, n+2\mu$. Now choose

- a sequence $\{a_k\}$ of positive numbers in such a way that $\sum_{k=1}^{\infty} a_k < +\infty$,
- a sequence $\{n_k\}$ of nonnegative integers such that $a_k \log n_k$ does not converge to zero,
- a sequence $\{\mu_k\}$ of nonnegative integers such that $\mu_{k+1} > \mu_k + 2n_k$,

and set

$$Q_k(x) := Q_{n_k,\mu_k}(x).$$

Since the sums $\sum_{k=1}^{n} \frac{\sin kx}{k}$ are equibounded, see (11.42) and (11.48), the polynomials $Q_{n,\mu}(x)$ are equibounded independently of $n, \mu \in \mathbb{N}$ and $x \in \mathbb{R}$. Consequently $\sum_{k=1}^{\infty} a_k Q_k(x)$ converges absolutely in $C^0(\mathbb{R})$ to a continuous function $f(x)$, $x \in \mathbb{R}$,

$$f(x) = \sum_{k=1}^{\infty} a_k Q_{n_k,\mu_k}(x),$$

which is 2π-periodic and even, for f a sum of cosines. The Fourier series of f is then a cosine series

$$Sf(x) = \frac{a_0}{2} + \sum_{k=1}^{\infty} a_k \cos(kx).$$

We now show that $S_n f(0)$ has no limit as $n \to \infty$.

Since f is a uniform limit, we can integrate term by term to get Fourier coefficients

$$c_j := \frac{1}{\pi} \int_{-\pi}^{\pi} f(t) \cos(jt)\, dt = \sum_{k=1}^{\infty} \frac{a_k}{\pi} \int_{-\pi}^{\pi} Q_k(t) \cos(jt)\, dt$$

because of the choice of the μ_k, the harmonics of Q_k and Q_h, $h \neq k$ are distinct, in particular

$$\sum_{\mu_k}^{\mu_k+n_k-1} c_j = a_k \sum_{j=1}^{n_k} \frac{1}{j} \geq a_k \int_1^{n_k} \frac{dt}{t} = a_k \log n_k.$$

Consequently, we deduce for the Fourier partial sums of f at 0

$$S_{n_k+\mu_k} f(0) - S_{\mu_k-1} f(0) = \sum_{\mu_k}^{m_k+n_k-1} c_j \geq a_k \log n_k.$$

Therefore $S_n f(0)$ does not converge, because of our choice of $\{n_k\}$.

A possible choice of the previous constants is

$$a_k := \frac{1}{k^2}, \qquad n_k = 2^{k^2}, \qquad \mu_k = 2^{k^2},$$

which yields $a_k \log(n_k) = \log 2$.

Figure 11.16. Paul du Bois–Reymond (1831–1889) and Lipót Fejér (1880–1959).

11.5.7 Féjer's sums

Let f be a continuous and 2π-periodic function. The Fourier partial sums of f need not provide a good approximation of f, neither uniformly nor pointwise; on the other hand f can be approximated uniformly by trigonometric polynomials, see Theorem 9.58. A specific interesting approximation was pointed out by Lipót Fejér (1880–1959).

Let $f \in L^1_{2\pi}$ and $S_n f(x) = \sum_{k=-n}^{n} c_k e^{ikx}$. Féjer's sums of f are defined by

$$F_n f(x) := \frac{1}{n+1} \sum_{k=0}^{n} S_n f(x).$$

Trivially $F_n f(x)$ are trigonometric polynomials of order n that can be written as

$$F_n(x) = \frac{1}{n+1} \sum_{k=0}^{n} \sum_{j=-k}^{k} c_j e^{ijx} = \frac{1}{n+1} \sum_{j=-n}^{n} (n - |j|) c_j e^{ijx}.$$

We have

11.93 Theorem (Féjer). *Let $f \in L^1_{2\pi} \cap C^0(\mathbb{R})$. The Féjer sums $F_n f(x)$ converge to f uniformly in $\mathbb{R}$.*

Before proving Féjer's theorem, let us state a few properties of the *Féjer kernel* defined by

$$F_n(x) := \frac{1}{n+1} \sum_{k=0}^{n} D_k(x)$$

where D_k denotes the Dirichlet's kernel of order k.

11.94 Proposition. *We have*

$$F_n(x) = \begin{cases} n+1 & \text{if } x = 2k\pi, \ k \in \mathbb{Z}, \\ \dfrac{1}{n+1} \left(\dfrac{\sin((n+1)x/2)}{\sin(x/2)} \right)^2 & \text{otherwise.} \end{cases}$$

Proof. Trivially

$$F_n(0) = \frac{1}{n+1} \sum_{k=0}^{n} D_k(0) = \frac{1}{n+1} \sum_{k=0}^{n} (2k+1) = \frac{(n+1)^2}{n+1} = n+1.$$

Observing that in

$$F_n(x) = \left(\frac{\sin(x/2) + \cdots + \sin((n+1/2)x)}{\sin(x/2)} \right) \Big/ (n+1)$$

the expression in parentheses is the imaginary part of

$$\frac{e^{ix/2} + e^{i3x/2} + \cdots + e^{i(2n+1)x/2}}{\sin(x/2)} = \frac{e^{ix/2}(e^{i(n+1)x} - 1)}{\sin(x/2)(e^{ix} - 1)}$$

$$= \frac{e^{i(n+1)x/2}}{\sin(x/2)} \frac{2i \sin((n+1)x/2)}{2i \sin(x/2)} = e^{i((n+1)x/2)} \frac{\sin((n+1)x/2)}{\sin^2(x/2)},$$

we see that

$$F_n(x) = \frac{1}{n+1} \left(\frac{\sin((n+1)x/2)}{\sin(x/2)} \right)^2.$$

$\square$

11.95 Proposition. *Féjer's kernel has the following properties*

(i) $F_n(x) \geq 0$,
(ii) $F_n(x)$ *is even,*
(iii) $\frac{1}{2\pi} \int_{-\pi}^{\pi} F_n(t) \, dt = 1$,
(iv) $F_n(x)$ *attains its maximum value at* $2k\pi$, $k \in \mathbb{Z}$,
(v) *for all* $\delta > 0$, $F_n(x) \to 0$ *uniformly in* $[\delta, \pi]$ *as* $n \to \infty$,
(vi) *there exists a constant* $A > 0$ *such that* $F_n(x) \leq \frac{A}{(n+1)x^2}$ *for all* $n \in \mathbb{N}$
    and $x \neq 0$ *in* $[-\pi, \pi]$,
(vii) $\{F_n\}$ *is an approximation of the Dirac mass* δ.

Proof. (i),(ii),(iii), (iv), (v) are trivial; (vi) follows from the estimate $\sin t \geq 2t/\pi$ in $]0, \pi/2]$. Finally (vii) follows from (iii) and (v). $\square$

Proof of Féjer's theorem, Theorem 11.93. First we observe that

$$F_n f(x) - f(x) = \frac{1}{2\pi} \int_0^{\pi} (f(x+t) + f(x-t) - 2f(x)) F_n(t) \, dt.$$

Thus, if we set $g(t) := f(x+t) + f(x-t) - 2f(x)$,

$$F_n f(x) - f(x) = \frac{1}{2\pi} \int_0^{\delta} g(t) F_n(t) \, dt + \frac{1}{2\pi} \int_{\delta}^{\pi} g(t) F_n(t) \, dt =: I_1 + I_2.$$

Now, given $\epsilon > 0$, we can choose δ so that $|f(x+t) + f(x-t) - 2f(x)| < 2\epsilon$ for all $t \in [0, \delta]$ uniformly in x, since f is uniformly continuous. Hence

$$|I_1| \leq 2\epsilon \int_0^{\delta} F_n(t) \leq 2\epsilon \int_0^{\pi} F_n(t) \, dt = 2\pi\epsilon.$$

On the other hand $|I_2| \leq 4\|f\|_\infty A/((n+1)\delta^2)$, hence

$$|F_n f(x) - f(x)| \leq \epsilon + 4\|f\|_\infty \frac{A}{(n+1)\delta^2}.$$

$\square$

A. Mathematicians and Other Scientists

Maria Agnesi (1718–1799)
Pavel Alexandroff (1896–1982)
James Alexander (1888–1971)
Archimedes of Syracuse (287BC–212BC)
Cesare Arzelà (1847–1912)
Giulio Ascoli (1843–1896)
René-Louis Baire (1874–1932)
Stefan Banach (1892–1945)
Isaac Barrow (1630–1677)
Giusto Bellavitis (1803–1880)
Daniel Bernoulli (1700–1782)
Jacob Bernoulli (1654–1705)
Johann Bernoulli (1667–1748)
Sergei Bernstein (1880–1968)
Wilhelm Bessel (1784–1846)
Jacques Binet (1786–1856)
George Birkhoff (1884–1944)
Bernhard Bolzano (1781–1848)
Emile Borel (1871–1956)
Karol Borsuk (1905–1982)
L. E. Brouwer (1881–1966)
Renato Caccioppoli (1904–1959)
Georg Cantor (1845–1918)
Alfredo Capelli (1855-1910)
Lennart Carleson (1928–)
Lazare Carnot (1753–1823)
Élie Cartan (1869–1951)
Giovanni Cassini (1625–1712)
Augustin-Louis Cauchy (1789–1857)
Arthur Cayley (1821–1895)
Eduard Čech (1893–1960)
Pafnuty Chebyshev (1821–1894)
Richard Courant (1888–1972)
Gabriel Cramer (1704–1752)
Jean d'Alembert (1717–1783)
Georges de Rham (1903–1990)
Richard Dedekind (1831–1916)
René Descartes (1596–1650)
Ulisse Dini (1845–1918)
Diocles (240BC–180BC)
Paul Dirac (1902–1984)
Lejeune Dirichlet (1805–1859)
Paul du Bois–Reymond (1831–1889)
James Dugundji (1919–1985)

Albrecht Dürer (1471–1528)
Euclid of Alexandria (325BC–265BC)
Leonhard Euler (1707–1783)
Alessandro Faedo (1914–2001)
Herbert Federer (1920–)
Lipót Fejér (1880–1959)
Pierre de Fermat (1601–1665)
Sir Ronald Fisher (1890–1962)
Joseph Fourier (1768–1830)
Maurice Fréchet (1878–1973)
Ivar Fredholm (1866–1927)
Georg Frobenius (1849–1917)
Boris Galerkin (1871–1945)
Galileo Galilei (1564–1642)
Carl Friedrich Gauss (1777–1855)
Israel Moiseevitch Gelfand (1913–)
Camille-Christophe Gerono (1799-1891)
J. Willard Gibbs (1839–1903)
Jorgen Gram (1850–1916)
Hermann Grassmann (1808–1877)
George Green (1793–1841)
Thomas Grönwall (1877–1932)
Jacques Hadamard (1865–1963)
Hans Hahn (1879–1934)
Georg Hamel (1877–1954)
William R. Hamilton (1805–1865)
Felix Hausdorff (1869–1942)
Oliver Heaviside (1850–1925)
Eduard Heine (1821–1881)
Charles Hermite (1822–1901)
David Hilbert (1862–1943)
Otto Hölder (1859–1937)
Robert Hooke (1635–1703)
Heinz Hopf (1894–1971)
Guillaume de l'Hôpital (1661–1704)
Christiaan Huygens (1629–1695)
Carl Jacobi (1804–1851)
Johan Jensen (1859–1925)
Camille Jordan (1838–1922)
Oliver Kellogg (1878–1957)
Felix Klein (1849–1925)
Helge von Koch (1870–1924)
Andrey Kolmogorov (1903–1987)
Leopold Kronecker (1823–1891)

Kazimierz Kuratowski (1896–1980)
Joseph-Louis Lagrange (1736–1813)
Edmond Laguerre (1834–1886)
Pierre-Simon Laplace (1749–1827)
Gaspar Lax (1487–1560)
Henri Lebesgue (1875–1941)
Solomon Lefschetz (1884–1972)
Adrien-Marie Legendre (1752–1833)
Gottfried von Leibniz (1646–1716)
Jean Leray (1906–1998)
Sophus Lie (1842–1899)
Ernst Lindelöf (1870–1946)
Rudolf Lipschitz (1832–1903)
Jules Lissajous (1822–1880)
L. Agranovich Lyusternik (1899–1981)
James Clerk Maxwell (1831–1879)
Edward McShane (1904–1989)
Arthur Milgram (1912–1961)
Hermann Minkowski (1864–1909)
Carlo Miranda (1912–1982)
August Möbius (1790–1868)
Harald Marston Morse (1892–1977)
Mark Naimark (1909–1978)
Nicomedes (280BC–210BC)
des Chênes M.– A. Parseval (1755–1836)
Blaise Pascal (1623–1662)
Etienne Pascal (1588–1640)
Giuseppe Peano (1858–1932)
Oskar Perron (1880–1975)
Émile Picard (1856–1941)
J. Henri Poincaré (1854–1912)
Diadochus Proclus (411–485)
Pythagoras of Samos (580BC–520BC)
Hans Rademacher (1892–1969)
Tibor Radó (1895–1965)
Lord William Strutt Rayleigh (1842–1919)

Kurt Reidemeister (1893–1971)
G. F. Bernhard Riemann (1826–1866)
Frigyes Riesz (1880–1956)
Marcel Riesz (1886–1969)
Eugène Rouché (1832–1910)
Adhémar de Saint Venant (1797–1886)
Stanislaw Saks (1897–1942)
Helmut Schaefer (1925–)
Juliusz Schauder (1899–1943)
Erhard Schmidt (1876–1959)
Lev G. Schnirelmann (1905–1938)
Hermann Schwarz (1843–1921)
Karl Seifert (1907–1996)
Takakazu Seki (1642–1708)
Carlo Severini (1872–1951)
Hugo Steinhaus (1887–1972)
Thomas Jan Stieltjes (1856–1894)
Marshall Stone (1903–1989)
James Joseph Sylvester (1814–1897)
Brook Taylor (1685–1731)
Heinrich Tietze (1880–1964)
Leonida Tonelli (1885–1946)
Stanislaw Ulam (1909–1984)
Pavel Urysohn (1898–1924)
Charles de la Vallée-Poussin (1866–1962)
Egbert van Kampen (1908–1942)
Alexandre Vandermonde (1735–1796)
Giuseppe Vitali (1875–1932)
Vito Volterra (1860–1940)
John von Neumann (1903–1957)
Karl Weierstrass (1815–1897)
Norbert Wiener (1894–1964)
Kôsaku Yosida (1909–1990)
William Young (1863–1942)
Nikolay Zhukovsky (1847–1921)
Max Zorn (1906–1993)
Antoni Zygmund (1900–1992)

There exist many web sites dedicated to the history of mathematics, we mention, e.g.,
http://www-history.mcs.st-and.ac.uk/~history.

B. Bibliographical Notes

We collect here a few suggestions for the readers interested in delving deeper into some of the topics treated in this volume.

Concerning *linear algebra* the reader may consult
- P. D. Lax, *Linear Algebra*, Wiley & Sons, New York, 1997,
- S. Lang, *Linear Algebra*, Addison-Wesley, Reading, 1966,
- A. Quarteroni, R. Sacco, F. Saleri, *Numerical Mathematics*, Springer-Verlag, New-York, 2000,
- G. Strang, *Introduction to Applied Mathematics*, Wellesley–Cambridge Press, 1961.

Of couse, *curves* and *surfaces* are discussed in many textbooks. We mention
- M. do Carmo, *Differential Geometry of Curves and Surfaces*, Prentice Hall Inc., New Jersey, 1976,
- A. Gray, *Modern Differential Geometry of Curves and Surfaces*, CRC Press, Boca Raton, 1993.

Concerning *general topology* and *topology* the reader may consult among the many volumes that are available
- J. Dugundji, *Topology*, Alyn and Bacon, Inc., Boston, 1966,
- K. Jänich, *Topology*, Springer-Verlag, Berlin, 1994,
- I. M. Singer, J. A. Thorpe, *Lecture Notes on Elementary Topology and Geometry*, Springer-Verlag, New York, 1967,
- J. W. Vick, *Homology Theory. An Introduction to Algebraic Topology*, Springer-Verlag, New York, 1994.

With special reference to *degree theory* and *existence of fixed points* we mention
- A. Granas, J. Dugundji, *Fixed Point Theory*, Springer-Verlag, New York, 2003.
- L. Nirenberg, *Topics in Nonlinear Functional Analysis*, Courant Institute of Mathematical Sciences, New York University, 1974.

The literature on Banach and Hilbert spaces, linear operators, spectral theory and linear and nonlinear functional analysis is incredibly wide. Here we mention only a few titles
- N. J. Akhiezer, I. M. Glazman, *Theory of Linear Operators in Hilbert Spaces*, Dover, New York, 1983,
- H. Brezis, *Analyse Fonctionelle*, Masson, Paris, 1983,
- A. Friedman, *Foundations of Modern Analysis*, Dover, New York, 1970,

and also
- N. Dundford, J. Schwartz, *Linear Operators*, John Wiley, New York, 1988,
- K. Yosida, *Functional Analysis*, Springer-Verlag, Berlin, 1974,

as well as the celebrated
- R. Courant, D. Hilbert, *Methods of Mathematical Physics*, Interscience Publishers, 1953,
- F. Riesz, B. Sz. Nagy, *Leçons d'Analyse Fonctionelle*, Gauthier–Villars, Paris, 1965.

C. Index